Lecture Notes in Computer Sci

Commenced Publication in 1973
Founding and Former Series Editors:
Gerhard Goos, Juris Hartmanis, and Jan van Leeuwen

Andreas Brandstädt   Dieter Kratsch
Haiko Müller (Eds.)

Graph-Theoretic Concepts in Computer Science

33rd International Workshop, WG 2007
Dornburg, Germany, June 21-23, 2007
Revised Papers

 Springer

Volume Editors

Andreas Brandstädt
Universität Rostock
Institut für Informatik
18051 Rostock, Germany
E-mail: ab@informatik.uni-rostock.de

Dieter Kratsch
Université Paul Verlaine - Metz
LITA, UFR MIM
Ile du Saulcy, 57045 Metz Cedex 01, France
E-mail: kratsch@univ-metz.fr

Haiko Müller
University of Leeds
School of Computing, Leeds, LS2 9JT, UK
E-mail: hm@comp.leeds.ac.uk

Library of Congress Control Number: 2007941251

CR Subject Classification (1998): F.2, G.2, G.1.6, G.1.2, E.1, I.3.5

LNCS Sublibrary: SL 1 – Theoretical Computer Science and General Issues

ISSN   0302-9743
ISBN-10   3-540-74838-5 Springer Berlin Heidelberg New York
ISBN-13   978-3-540-74838-0 Springer Berlin Heidelberg New York

Springer is a part of Springer Science+Business Media

springer.com

© Springer-Verlag Berlin Heidelberg 2007
Printed in Germany

Typesetting: Camera-ready by author, data conversion by Scientific Publishing Services, Chennai, India
Printed on acid-free paper   SPIN: 12120109   06/3180   5 4 3 2 1 0

Preface

The 33rd International Conference "Workshop on Graph-Theoretic Concepts in Computer Science" (WG 2007) took place in the Conference Center in old castle in Dornburg near Jena, Germany, June 21–23, 2007. The approximately 80 participants came from various countries all over the world, among them Brazil, Canada, the Czech Republic, France, UK, Greece, Hungary, Italy, Japan, The Netherlands, Norway, Sweden, Taiwan, and the USA.

WG 2007 continued the series of 32 previous WG conferences. Since 1975, the WG conference has taken place 20 times in Germany, four times in The Netherlands, twice in Austria as well as once in Italy, Slovakia, Switzerland, the Czech Republic, France and in Norway.

The WG conference traditionally aims at uniting theory and practice by demonstrating how graph-theoretic concepts can be applied to various areas in computer science, or by extracting new problems from applications. The goal is to present recent research results and to identify and explore directions of future research.

The continuing interest in the WG conferences was reflected in the high number of submissions; 99 papers were submitted and in an evaluation process with four reports per submission, 30 papers were accepted by the Program Committee for the conference. Due to the high number of submissions and the limited schedule of 3 days, various good papers could not be accepted.

There were invited talks by Ming-Yang Kao (Evanston, Illinois) on algorithmic DNA assembly, and by Klaus Jansen (Kiel, Germany) on approximation algorithms for geometric intersection graphs.

We are grateful to all those who contributed to WG 2007: First of all, the authors submitting so many good papers, the numerous referees, the speakers, the Program Committee, the Organizing Committee (special thanks go to Katrin Erdmann and Roswitha Fengler, Rostock, to Mathieu Liedloff, Metz, as well as to Nadja Betzler, Falk Hüffner and Marita Venth, Jena, and the whole group of Rolf Niedermeier for hosting WG 2007 in the wonderful Conference Center in the old castle in Dornburg) and last but not least, to our sponsors the German Research Council (DFG), Land Thüringen, Universität Jena, and Stiftung für Innovation, Technologie und Forschung Thüringen (STIFT) as well as to the Universities of Leeds, Metz and Rostock.

October 2007

Andreas Brandstädt
Dieter Kratsch
Haiko Müller

Organization

The Tradition of WG

1975   U. Pape – Berlin, Germany
1976   H. Noltemeier – Göttingen, Germany
1977   J. Mühlbacher – Linz, Austria
1978   M. Nagl, H.J. Schneider – Castle Feuerstein, Germany
1979   U. Pape – Berlin, Germany
1980   H. Noltemeier – Bad Honnef, Germany
1981   J. Mühlbacher – Linz, Austria
1982   H.J. Schneider, H. Göttler – Neuenkirchen, Germany
1983   M. Nagl, J. Perl – Haus Ohrbeck near Osnabrück, Germany
1984   U. Pape – Berlin, Germany
1985   H. Noltemeier – Castle Schwanenberg near Würzburg, Germany
1986   G. Tinhofer, G. Schmidt – Bernried near Munich, Germany
1987   H. Göttler, H.J. Schneider – Kloster Banz near Bamberg, Germany
1988   J. van Leeuwen – Amsterdam, The Netherlands
1989   M. Nagl – Castle Rolduc, The Netherlands
1990   R.H. Möhring – Berlin, Germany
1991   G. Schmidt, R. Berghammer – Fischbachau near Munich, Germany
1992   E.W. Mayr – Wiesbaden-Naurod, Germany
1993   J. van Leeuwen – Utrecht, The Netherlands
1994   G. Tinhofer, E.W. Mayr, G. Schmidt – Herrsching near Munich, Germany
1995   M. Nagl – Aachen, Germany
1996   G. Ausiello, A. Marchetti-Spaccamela – Como, Italy
1997   R.H. Möhring – Berlin, Germany
1998   J. Hromkovič, O. Sýkora – Smolenice-Castle, Slovak Republic
1999   P. Widmayer – Ascona, Switzerland
2000   D. Wagner – Konstanz, Germany
2001   A. Brandstädt – Boltenhagen near Rostock, Germany
2002   L. Kučera – Cesky Krumlov, Czech Republic
2003   H.L. Bodlaender – Elspeet, The Netherlands
2004   J. Hromkovič, M. Nagl Bad Honnef, Germany
2005   D. Kratsch – Metz, France
2006   F.V. Fomin – Bergen, Norway
2007   A. Brandstädt, D. Kratsch, H. Müller – Dornburg near Jena, Germany

Program Committee

Hans L. Bodlaender	Universiteit Utrecht, The Netherlands
Andreas Brandstädt	Universität Rostock, Germany (Co-chair)
Hajo Broersma	University of Durham, UK
Victor Chepoi	Université de la Mediterranée, Marseille, France
Thomas Eiter	Technische Universität Wien, Austria
Thomas Erlebach	University of Leicester, UK
Michel Habib	Université Paris 7 - Denis Diderot, Paris, France
Magnus M. Halldorsson	University of Iceland, Reykjavik, Iceland
Kazuo Iwama	Kyoto University, Japan
Michael Kaufmann	Universität Tübingen, Germany
Dieter Kratsch	Université Paul Verlaine - Metz, France (Co-chair)
Ernst W. Mayr	Technische Universität München, Germany
Ross McConnell	Colorado State University, Fort Collins, Colorado, USA
Kurt Mehlhorn	Max Planck Institut, Saarbrücken, Germany
Haiko Müller	University of Leeds, United Kingdom (Co-chair)
Rolf Niedermeier	Friedrich-Schiller-Universität Jena, Germany
Andrzej Proskurowski	University of Oregon, Eugene, Oregon, USA
Bernhard Westfechtel	Universität Bayreuth, Germany

Additional Reviewers

Tatsuya Akutsu	Pierre Fraigniaud	Matthew Johnson
Tobias Berg	Hubert de Fraysseix	Iyad Kanj
Nadja Betzler	Markus Geyer	Ragnar Karlsson
Paul Bonsma	Leslie Ann Goldberg	Ekkehard Köhler
Timothy M. Chan	Catherine Greenhill	Guy Kortsarz
Victor Chepoi	Jiong Guo	Lefteris M. Kirousis
Derek Corneil	Anupam Gupta	Stephan Kreutzer
Bruno Courcelle	Frank Gurski	Michael Krüger
Nadia Creignou	Torben Hagerup	Van Bang Le
Mary Cryan	Tero Harju	Katharina Lehmann
Elias Dahlhaus	Carsten Henneges	Mathieu Liedloff
Michael Dom	Pinar Heggernes	Vincent Limouzy
Feodor F. Dragan	Jan van den Heuvel	Marina Lipshteyn
Martin Dyer	Chinh T. Hoang	Elena Losievskaja
Bertrand Estellon	Falk Hüffner	David Manlove
Dmitry Feichtner-Kozlov	Christian Hundt	Mark Minas
Henning Fernau	Hiro Ito	Hannes Moser
Fedor Fomin	Mark Jerrum	Stefan Naeher

Shin-ichi Nakano
Karim Nouioua
Sang-il Oum
Maurizio Patrignani
Christophe Paul
Eelko Penninkx
Artem Pyatkin
Bert Randerath
Michaël Rao
Adrian Riskin
Andy Schürr
Morgan Seston

Natalia Shakhlevich
Martin Siebenhaller
Iain Stewart
M. M. Syslo
Roberto Tamassia
Seiichi Tani
Gerard Tel
Jan Arne Telle
Dimitrios M. Thilikos
Ioan Todinca
Akinori Uejima
Walter Unger

Ruth Urner
Jean-Marie Vanherpe
Yann Vaxes
Marinus Veldhorst
Yngve Villanger
Egon Wanke
Peter Wagner
Pawel Winter
Atsuko Yamaguchi
Martin Zachariasen
Pawel Zielinski

Table of Contents

Computational Complexity of Generalized Domination:
A Complete Dichotomy for Chordal Graphs

Petr Golovach[1,*] and Jan Kratochvíl[2,**]

[1] Department of Informatics, University of Bergen, 5020 Bergen, Norway
`petrg@ii.uib.no`
[2] Department of Applied Mathematics and Institute for Theoretical
Computer Science, Charles University, Prague, Czech Republic
`honza@kam.mff.cuni.cz`

Abstract. The so called (σ, ρ)-domination, introduced by J.A. Telle, is a concept which provides a unifying generalization for many variants of domination in graphs. (A set S of vertices of a graph G is called (σ, ρ)-*dominating* if for every vertex $v \in S$, $|S \cap N(v)| \in \sigma$, and for every $v \notin S$, $|S \cap N(v)| \in \rho$, where σ and ρ are sets of nonnegative integers and $N(v)$ denotes the open neighborhood of the vertex v in G.) It was known that for any two nonempty finite sets σ and ρ (such that $0 \notin \rho$), the decision problem whether an input graph contains a (σ, ρ)-dominating set is NP-complete, but that when restricted to chordal graphs, some polynomial time solvable instances occur. We show that for chordal graphs, the problem performs a complete dichotomy: it is polynomial time solvable if σ, ρ are such that every chordal graph contains at most one (σ, ρ)-dominating set, and NP-complete otherwise. The proof involves certain flavor of existentionality - we are not able to characterize such pairs (σ, ρ) by a structural description, but at least we can provide a recursive algorithm for their recognition. If ρ contains the 0 element, every graph contains a (σ, ρ)-dominating set (the empty one), and so the nontrivial question here is to ask for a maximum such set. We show that MAX-(σ, ρ)-domination problem is NP-complete for chordal graphs whenever ρ contains, besides 0, at least one more integer.

Keywords: Computational complexity, graph algorithms.

1 Introduction and Overview of Results

We consider finite undirected graphs without loops or multiple edges. The vertex set of a graph G is denoted by $V(G)$ and its edge set by $E(G)$. The open neighborhood of a vertex is denoted by $N(u) = \{v : uv \in E(G)\}$. A graph is chordal if it does not contain an induced cycle of length greater than three.

---

* On leave from Department of Applied Mathematics, Syktyvkar State University, Syktyvkar, Russia. Most of the results were obtained during the research stay of the first author at DIMATIA Prague in 2006.
** Supported by the Czech Ministry of Education as Research Project No. 1M0545.

A. Brandstädt, D. Kratsch, and H. Müller (Eds.): WG 2007, LNCS 4769, pp. 1–11, 2007.

1.1 (σ, ρ)-Domination

Let σ, ρ be a pair of nonempty sets of nonnegative integers. A set of vertices of G is called (σ, ρ)-*dominating* if for every vertex $v \in S$, $|S \cap N(v)| \in \sigma$, and for every $v \notin S$, $|S \cap N(v)| \in \rho$. The concept of (σ, ρ)-domination was introduced by J.A. Telle [14,15] (and further elaborated on in [12,9]) as a unifying generalization of many previously studied variants of the notion of dominating sets (see [8] for an extensive bibliography on domination in graphs). In particular, $(\mathbb{N}_0,\mathbb{N})$-dominating sets are ordinary dominating sets, $(\{0\},\mathbb{N}_0)$-dominating sets are independent sets, $(\mathbb{N}_0,\{1\})$-dominating sets are efficient dominating sets, $(\{0\},\{1\})$-dominating sets are 1-perfect codes (or independent efficient dominating sets), $(\{0\},\{0,1\})$-dominating sets are strong stable sets, $(\{0\},\mathbb{N})$-dominating sets are independent dominating set, $(\{1\},\{1\})$-dominating sets are total perfect dominating set, or $(\{r\},\mathbb{N}_0)$-dominating sets are induced r-regular subgraphs ($\mathbb{N}$ and $\mathbb{N}_0$ denote the sets of positive and nonnegative integers, respectively).

We are interested in the complexity of the problem of existence of a (σ, ρ)-dominating set in an input graph, and we denote this problem by $\exists(\sigma, \rho)$-DOMINATION. It can be easily seen that if $0 \in \rho$, then the $\exists(\sigma, \rho)$-DOMINATION problem has a trivial solution $S = \emptyset$. So throughout the main part of the paper (and unless not explicitly stated otherwise) we suppose that $0 \notin \rho$.

1.2   Our Results

In view of the above given examples, it is not surprising that for any nontrivial combination of finite sets σ and ρ (considered as fixed parameters of the problem), $\exists(\sigma, \rho)$-DOMINATION is NP-complete [14]. It is then natural to pay attention to restricted graph classes for inputs of the problem. It was observed in [11] that for any pair of finite sets σ and ρ, the problem is solvable in polynomial time for interval graphs, but that it becomes NP-complete when restricted to chordal graphs (for some parameter sets σ and ρ). In particular, it was shown that for one-element sets $\sigma = \{p\}, \rho = \{q\}$, $\exists(\sigma, \rho)$-DOMINATION is polynomial time solvable if $q > 2p + 1$ and NP-complete if $q \le p + 1$. We close this gap by showing that all the remaining cases are also polynomial time solvable. Moreover, we extend this polytime/NP-completeness dichotomy to any pair of finite sets σ, ρ by showing the following characterization:

Theorem A. *For finite sets σ, ρ, $\exists(\sigma, \rho)$-DOMINATION is polynomial time solvable for chordal graphs if every chordal graph has at most one (σ, ρ)-dominating set, and it is NP-complete otherwise.*

This theorem provides a full characterization and dichotomy, with both the polynomial time solvable and NP-complete cases including nontrivial and interesting samples (as we show by discussing some examples in Section 4). Dichotomy results are valued and intensively looked for (e.g., the classification of Boolean satisfiability by Schaefer [13], further dichotomy results for larger classes of the Constraint Satisfaction Problem by Bulatov et al. [2] paving the way to the utmost CSP dichotomy conjecture of Feder and Vardi [4], or several results for graph homomorphisms [10,3,6,5].) The characterization is nonconstructive in the sense that we are not able to provide a structural description of ambivalent (or non-ambivalent) pairs σ, ρ (we call a pair σ, ρ *ambivalent* if there exists a chordal graph containing two different (σ, ρ)-dominating sets), and there is indication that such a description will not be simple. Indeed, for any pair of σ and ρ, there are infinitely many chordal graphs to be checked if any of them, by chance, contains two different (σ, ρ)-dominating sets. Perhaps somewhat surprisingly we show that this fact can be overcome at least from the computational point of view:

Theorem B. *It can be decided in finite time (i.e., by a recursive algorithm) whether for a given pair of finite sets σ, ρ, there exists a chordal graph containing two different (σ, ρ)-dominating sets.*

The NP-hardness part of Theorem A is proved in Section 2 by a reduction from a variant of the EXACT COVER problem. Its polynomial part is proved in Section 3 by providing an explicit dynamic programming algorithm. Theorem B is proved by providing an explicit upper bound on the minimum size of an ambivalent graph in Section 4. In Section 5, we discuss the case when $0 \in \rho$. As we have already mentioned, the $\exists(\sigma, \rho)$-DOMINATION problem is then trivial (the empty set is always (σ, ρ)-dominating), and the natural question here is the optimization variant. However, we show this is always a hard problem:

Theorem C. *Given a chordal graph graph G and a number k, it is NP-complete to decide if G contains a (σ, ρ)-dominating set of size at least k, provided σ, ρ are finite sets of nonnegative integers and $\rho \neq \{0\}$.*

Throughout the paper $n = |V(G)|$, $p_{min} = \min \sigma$, $p_{max} = \max \sigma$, $q_{min} = \min \rho$ and $q_{max} = \max \rho$, where G is the graph and σ, ρ the sets under consideration. In case of single-element sets σ or ρ, we write simply $p = p_{min} = p_{max}$ and $q = q_{min} = q_{max}$.

2   NP-Complete Cases

This section is devoted to the proof of the following theorem.

Theorem 1. *Let σ, ρ be finite sets of nonnegative integers, $0 \notin \rho$. If there is a chordal graph with at least two different (σ, ρ)-dominating sets, then the $\exists(\sigma, \rho)$-DOMINATION problem is NP-complete for chordal graphs.*

2.1   An Auxiliary Complexity Lemma

We are going to reduce from a special variant of the COVER BY TRIPLES problem (or EXACT COVER)(see [7]).

Let r be a positive integer. An instance of the COVER BY NO MORE THAN r TRIPLES is a pair (X, M), where X is a nonempty finite set and M is a set of triples of elements of X. We ask about the existence of a set $M' \subset M$ such that every element of X belongs to at least one and to at most r triples of M'. Such a set we call a *cover of X by no more than r triples*. For space limitations the proof of the following auxiliary lemma is omitted.

Lemma 1. *For every fixed $r \geq 1$, the* COVER BY NO MORE THAN r TRIPLES *problem is NP-complete.*

2.2   The Forcing Gadget

Our next step of the proof is the construction of a gadget which "enforces" on a given vertex the property of "not belonging to any (σ, ρ)-dominating set".

It is known (cf. [11]) that if $q_{\min} \geq 2p_{\max}+2$, then every chordal graph contains at most one (σ, ρ)-dominating set. Hence we assume that $q_{\min} \leq 2p_{\max} + 1$. We construct a rooted graph F as follows.

Suppose first that $q_{\min} \leq p_{\max} + 1$. We start with a complete graph $K_{p_{\max}+1}$ with vertices $u_1, u_2, \ldots, u_{p_{\max}+1}$. Let $\{S_1, S_2, \ldots, S_t\}$ be a set of $q_{\min}$-tuples which covers the set $\{u_1, u_2, \ldots, u_{p_{\max}+1}\}$ (i.e., each u_j belongs to at least one S_i). For every $i = 1, 2, \ldots, t$, we add $q_{\max} + 1$ new vertices $v_1^{(i)}, v_2^{(i)}, \ldots, v_{q_{\max}+1}^{(i)}$ and connect them to all vertices of S_i by edges.

If $q_{\min} > p_{\max} + 1$, the construction is slightly different. We again start with a complete graph $K_{p_{\max}+1}$ with vertices $u_1, u_2, \ldots, u_{p_{\max}+1}$. We add $q_{\max} + 1$ new vertices $v_1, v_2, \ldots, v_{q_{\max}+1}$ and $q_{\max}+1$ copies of $K_{p_{\max}+1}$, say $Q_1, Q_2, \ldots, Q_{q_{\max}+1}$, and connect every v_j by edges to all vertices $u_1, u_2, \ldots, u_{p_{\max}+1}$ and to $q_{\min}-p_{\max}+1$ vertices of the corresponding Q_j.

In both cases the vertex u_1 is the root of F.

Lemma 2. *The graph F has at least one (σ, ρ)-dominating set, and for every (σ, ρ)-dominating set S in F, $u_1, u_2, \ldots, u_{p_{\max}+1} \in S$. Moreover, if F is an induced subgraph of a graph F' such that u_1 is the only vertex of F adjacent to vertices of $F' \setminus F$, then the vertices of $F' \setminus F$ that are adjacent to u_1 do not belong to any (σ, ρ)-dominating set in F'.*

Proof. Suppose that $q_{\min} \leq p_{\max}+1$. Obviously $\{u_1, u_2, \ldots, u_{p_{\max}+1}\}$ is a (σ, ρ)-dominating set in F. For the second statement, assume that S is a (σ, ρ)-dominating set in F and $u_i \notin S$ for some i. Let S_j be a $q_{\min}$-tuple which contains u_i. It is readily seen that $v_1^{(j)}, v_2^{(j)}, \ldots, v_{q_{\max}+1}^{(j)} \in S$. But then u_i is adjacent to at least $q_{\max} + 1$ vertices of S, a contradiction.

If $q_{\min} > p_{\max} + 1$, the proof of the second statement is similar. For the first part, note that the vertices $u_1, u_2, \ldots, u_{p_{\max}+1}$ and all vertices of the added cliques Q_j form a (σ, ρ)-dominating set.

For the last statement, note that we have proved that in both cases u_1 is in S and has $p_{\max}$ neighbors in S, for any (σ, ρ)-dominating set S in F, but the argument survives for any (σ, ρ)-dominating set S in F' as well.   □

2.3   The Reduction

Let H be a graph which has at least two different (σ, ρ)-dominating sets $S, \widetilde{S}$. We choose a vertex $u \in S \div \widetilde{S}$, where $\div$ denotes the symmetric difference of sets, and pronounce u the root of H. Let $k = \max\{i \in \mathbb{N}_0 : i \notin \rho, i+1 \in \rho\}$. Since $0 \notin \rho$, k is correctly defined.

Let a set $X = \{x_1, x_2, \ldots, x_n\}$ and a set $M = \{t_1, t_2, \ldots, t_m\}$ of triples on X be given as an instance of COVER BY NO MORE THAN r TRIPLES for $r = q_{\max} - k > 0$.

We start the construction of a graph G with a complete graph K_n with vertices $x_1, x_2, \ldots, x_n$. For every triple $t_i = \{x_a, x_b, x_c\}$, a copy H_i of the graph H with root u_i is added, and u_i is connected by edges to x_a, x_b, x_c. If $k = 0$, we further add $q_{\max}$ copies of the graph F with roots $v_1, v_2, \ldots, v_{q_{\max}}$, add a new extra vertex y, and join y with $x_1, x_2, \ldots, x_n$ and $v_1, v_2, \ldots, v_{q_{\max}}$ by edges. If $k > 0$, then k copies of F with roots $v_1, v_2, \ldots, v_k$ are added, and vertices $v_1, v_2, \ldots, v_k$ are connected with $x_1, x_2, \ldots, x_n$ by edges.

We claim that the graph G constructed in this way has a (σ, ρ)-dominating set if and only if (X, M) allows a cover by no more than r triples. Since the graphs H and F depend only on σ and ρ, G has $O(n + m)$ vertices, our reduction is polynomial and the proof will be concluded.

Suppose first that G has a (σ, ρ)-dominating set S. Let $M' = \{t_i \in M : u_i \in S\}$. If $k = 0$, then $y \notin S$ and $v_1, v_2, \ldots, v_{q_{\max}} \in S$ by Lemma 2. Hence $x_1, x_2, \ldots, x_n \notin S$. Since $0 \notin \rho$, for every $i = 1, 2, \ldots, n$, the vertex x_i has at least one S-neighbor in the set $\{u_1, u_2, \ldots, u_m\}$, but no more than $r = q_{\max}$ such neighbors. So M' is a cover of X by no more than r triples.

If $k > 0$, then $v_1, v_2, \ldots, v_k \in S$ and $x_1, x_2, \ldots, x_n \notin S$ by Lemma 2 again. Since $k \notin \rho$, for every $i = 1, 2, \ldots, n$, the vertex x_i has at least one S-neighbor in the set $\{u_1, u_2, \ldots, u_m\}$, but no more than $r = q_{\max} - k$ such neighbors. Hence again, M' is a cover of X by no more than r triples.

Suppose now that $M' \subseteq M$ is a cover of X by no more than r triples. For every $i = 1, 2, \ldots, m$, we choose a (σ, ρ)-dominating set S_i in H_i such that $u_i \in S_i$ if and only if $t_i \in M'$. Let $S_1', S_2', \ldots$ be (σ, ρ)-dominating sets in the copies of F. Since $\{k+1, k+2, \ldots, q_{\max}\} \subseteq \rho$, $S = S_1 \cup S_2 \cup \cdots \cup S_n \cup S_1' \cup S_2' \cup \ldots$ is a (σ, ρ)-dominating set in G.

3   The Polynomial Cases

In this section we prove the complementary part of Theorem A by presenting a polynomial time algorithm that decides the existence of a (σ, ρ)-dominating set in a chordal graph, provided the parameters σ and ρ are such that every chordal graph contains at most one (σ, ρ)-dominating set. It is perhaps of some interest that our algorithm can be formulated in a general way so that it is based only on the promise of a unique solution. On the contrary, in many situations the assumption of uniqueness of the solution does not help.

In fact we present two algorithms in this section. In the first subsection we give the general algorithm, and in the latter one we deal with a special case of one-element set σ. The running time of the second algorithm is much better and moreover, this algorithm explicitly closes the gap between polynomial and NP-complete cases for single-element parameter sets left open in [11].

3.1   The General Algorithm

In this subsection it is assumed that σ and ρ are such that every chordal graph contains no more than one (σ, ρ)-dominating set. The algorithm uses dynamic programming and is based on the clique-decomposition of the input graph.

Let $\mathcal{K}$ be the set of all maximal cliques of an input chordal graph G, and let T be a clique tree of G, i.e., $V(T) = \mathcal{K}$ and for every $u \in V(G)$, the subgraph of T induced by $\{K \in \mathcal{K} : u \in K\}$ is connected. It is well known (see, for example, [1]) that a clique tree of a chordal graph graph is not unique, but can be constructed in linear time. We choose a clique $R_0 \in \mathcal{K}$ and consider the clique tree T rooted in R_0. This induces a parent-child relation in the tree, in which all vertices are descendants of the root. For any clique $R \in \mathcal{K}$, we denote by T_R the subtree of T rooted in R and containing all descendants of R, and we denote by G_R the subgraph of G induced by the vertices contained in the cliques of $V(T_R)$.

The key idea of the algorithm is the fact that every clique $R \in \mathcal{K}$ contains at most $p_{\max} + 1$ vertices of any (σ, ρ)-dominating set, and hence for every clique we can list all possible intersections with a solution set S in polynomial time. We need to keep track of how many S-neighbors these vertices have. Towards this end we build the following array. Let $R \in \mathcal{K}$ and let $X = \{x_1, x_2, \ldots, x_r\}$ be an ordered subset of R, $0 \leq r \leq p_{\max} + 1$ (X can also be empty). Further let $P = (p_1, p_2, \ldots, p_r)$ be a sequence of nonnegative integers, $p_i \leq p_{\max}$ for $i = 1, 2, \ldots, r$. For each such triple R, X, P, our algorithm constructs a set $S(R, X, P) \subseteq V(G_R)$ which satisfies

- $S(R, X, P) \cap R = X$,
- $|N(x_i) \cap S(R, X, P)| = p_i$ for $i = 1, 2, \ldots, r$,
- for every $v \in V(G_R) \setminus R$, $|N(v) \cap S(R, X, P)| \in \begin{cases} \sigma & \text{if } v \in S(R, X, P), \\ \rho & \text{if } v \notin S(R, X, P); \end{cases}$

(i.e., $S(R, X, P)$ is a candidate for $S \cap V(G_R)$) or $S(R, X, P) =$ NIL if we can deduce that no such set can be extended to a solution S for the entire G. The details of the algorithm will appear in the full version of the paper. The recursive step is technical but straightforward. The crucial fact is that for each triple R, X, P, we store at most one candidate set, which follows from the following lemma.

Lemma 3. *If S_1 and S_2 are distinct subsets of $V(G_R)$ satisfying the candidate conditions for the same triple R, X, P, then none of them can be extended to a (σ, ρ)-dominating set S in the entire graph G.*

Proof. Suppose S_1 can be extended to a (σ, ρ)-dominating set S. Then S and $(S \setminus S_1) \cup S_2$ are two distinct (σ, ρ)-dominating sets in $G[V(G) \setminus (R \setminus X)]$, which is a contradiction to the assumption that every chordal graph contains at most one (σ, ρ)-dominating set.   □

The algorithm can be implemented to run in time $O(n^{p_{\max}^2 + 2p_{\max} + 3})$. Hence we have proved the polynomial part of Theorem A:

Theorem 2. *If σ and ρ are finite sets such that every chordal graph contains at most one (σ, ρ)-dominating set, then the $\exists(\sigma, \rho)$-DOMINATION problem is solvable in polynomial time for chordal graphs.*

3.2   Single-Element Sigma

In the case when the set σ contains only one element, we are able to design a more efficient greedy algorithm. This algorithm uses the simple structure of (σ, ρ)-dominating sets in such a case.

Lemma 4. *Let $\sigma = \{p\}$, let ρ be arbitrary and suppose that S is a (σ, ρ)-dominating set in a chordal graph G. Then S is the union of disjoint cliques of size $p + 1$, and vertices of different cliques are nonadjacent.*

Proof. Let $G[S]$ be the subgraph of G induced by S. This graph is chordal, so it has a simplicial vertex. The closed neighborhood of this vertex is a clique of size $p + 1$, and this clique is the vertex set of one component of $G[S]$. By repeating these arguments we prove that all components of $G[S]$ are induced by cliques of size $p + 1$. □

Though we use the following observation for the case of single-element σ, we state it in a more general form:

Lemma 5. *Suppose that $p_{\max} + 2 \leq q_{\min}$. Let S be a (σ, ρ)-dominating set in G. Then all simplicial vertices of G belong to S.*

Proof. Let v be a simplicial vertex of G. If $v \notin S$, then $|N(v) \cap S| \in \rho$, and since $p_{\max} + 2 \leq q_{\min}$, $|N(v) \cap S| \geq p_{\max} + 2$. So S contains a clique of size $p_{\max} + 2$, a contradiction. □

The core observation for our algorithm is the following lemma, which is a straightforward corollary of Lemmas 4 and 5.

Lemma 6. *Let $\sigma = \{p\}$ and let $p + 2 \leq q_{\min}$. Let S be a (σ, ρ)-dominating set in a chordal graph G. Further let T be a clique tree of G, let X be a leaf of T, and Y the neighbor of X in T. Then*

- *$|X \setminus Y| \leq p + 1$,*
- *if $|X \setminus Y| = p + 1$, then $X \setminus Y \subseteq S$ and $(Y \cap X) \cap S = \emptyset$,*
- *if $|X \setminus Y| < p + 1$, then $(Y \setminus X) \cap S = \emptyset$.*

Given a chordal graph, our algorithm first builds a clique tree and then consecutively reduces it by deleting vertices which must or must not belong to any (σ, ρ)-dominating set. In the final step it is necessary to check whether the only candidate (if any) for S is really a (σ, ρ)-dominating set. The technical details will again appear in the full version of the paper. We only note that with some extra care the algorithm can be designed so that in each reduction step, a clique tree of the reduced graph can be easily derived from the clique tree of the previous one. Thus we can claim:

Theorem 3. *If $\sigma = \{p\}$ and $p + 2 \leq q_{\min}$, then the (σ, ρ)-DOMINATION problem can be solved in time $O(n^2)$.*

4   Uniqueness of (σ, ρ)-Dominating Sets

It would be most desirable to have a full classification of the pairs of parameter sets σ, ρ for which there exist chordal graphs with two different (σ, ρ)-dominating sets. Such a classification is not currently known and perhaps not easy to obtain. In the first subsection of this section we summarize the known results in this direction. A positive result is proven in the second subsection. We show a bound on the size of a minimal chordal graph containing two different (σ, ρ)-dominating sets, thus showing that the existence of such a graph can be decided by a finite algorithm.

Recall that we call a pair (σ, ρ) *ambivalent* if there exists a graph containing at least two different (σ, ρ)-dominating sets. Such a graph will be called (σ, ρ)-*ambivalent*.

4.1   On the Way to Classification

First observation about the uniqueness of a (σ, ρ)-dominating set in a chordal graph was made in [11]. More cases are covered by the following theorem, but the picture is far from being complete. Fully characterized are the cases of $\sigma = \{p\}$ and $\sigma = \{0, p_{\max}\}$.

Theorem 4. *The following table presents examples of ambivalent and non-ambivalent pairs of σ and ρ:*

ambivalent	non-ambivalent
$q_{\min} \leq p_{\max} + 1$	$q_{\min} \geq 2p_{\max} + 2$
$\exists i : \{i, i+1\} \subseteq \sigma, \ q_{\min} \leq 2p_{\max} + 1$	$\sigma = \{p\}, \ q_{\min} \geq p + 2$
$\sigma = \{0, p_{\max}\}, \ q_{\min} \leq p_{\max} + 2$	$\sigma = \{0, p_{\max}\}, \ q_{\min} \geq p_{\max} + 3$

Proof. Will appear in the full version.   □

4.2   Deciding the Ambivalence

The main goal of this subsection is to prove Theorem B. We do so by proving an upper bound on the number of vertices of any minimum chordal graph containing two different (σ, ρ)-dominating sets, in terms of $p_{\max}$ and $q_{\max}$.

Theorem 5. *Let σ, ρ be finite sets of nonnegative integers, $0 \notin \rho$. Suppose that G is a minimum chordal (σ, ρ)-ambivalent graph. Then*

- *for every maximal clique K of G, $|K| \leq 2p_{\max} + 2$,*
- *for every vertex $v \in V(G)$, $\deg v \leq \max\{2p_{\max}, p_{\max} + q_{\max}\}$,*
- *the diameter of G is $O(p_{\max}^{2p_{\max} + 2q_{\max} + 7})$.*

Proof. Will appear in the full version.   □

Since every graph of maximum degree Δ and diameter d has at most Δ^{d+1} vertices, we have proven the following corollary and hence also Theorem B.

Corollary 1. *Let σ, ρ be finite sets of nonnegative integers, $0 \notin \rho$. Then the size of every minimum (σ, ρ)-ambivalent chordal graph is bounded by a function of p_{max} and q_{max} and the existence of such a graph can be tested algorithmically by a finite procedure.*

5   MAX-(σ, ρ)-Domination

So far we have been considering the question of existence of (σ, ρ)-dominating sets. One could also pose the optimization questions, i.e., asking for the sizes of minimum or maximum (σ, ρ)-dominating sets. Since optimization problems are at least as difficult as the existence ones, and since the polynomial part of our Theorem A is based on uniqueness of the solution, our results translate directly to the optimization variants. Namely, if $0 \notin \rho$, then both MIN-(σ, ρ)-DOMINATION and MAX-(σ, ρ)-DOMINATION problems are NP-hard when restricted to chordal graphs for ambivalent (σ, ρ) and polynomial time solvable for the non-ambivalent pairs.

If $\rho = \{0\}$, the only possible (σ, ρ)-dominating sets in a connected graph G are $S = \emptyset$ and $S = V(G)$. The latter is the maximum (σ, ρ)-dominating set if $\deg v \in \sigma$ for every $v \in V(G)$, otherwise $S = \emptyset$ is the only (and hence also the maximum) (σ, ρ)-dominating set in G. This is, however, the only polynomially solvable case, as Theorem C claims. The rest of this section is devoted to its proof.

5.1   Proof of Theorem C

We begin with an auxiliary construction. Let F consist of q_{max} copies of the complete graph $K_{p_{max}+1}$, say $Q_1, Q_2, \ldots, Q_{q_{max}}$, and one extra vertex r, the root of F, which is adjacent to exactly one vertex from each Q_i. The following technical lemma is straightforward.

Lemma 7. *The set $S = V(Q_1) \cup V(Q_2) \cup \cdots \cup V(Q_{q_{max}})$ is a maximum (σ, ρ)-dominating set in F, and it has cardinality $q_{max}(p_{max} + 1)$. Moreover, suppose that a graph F' is created by uniting F with some graph (with different vertices) and joining the root of F to some new vertices $u_1, u_2, \ldots, u_s$. If S' is a (σ, ρ)-dominating set in F', $r \notin S'$, and $u_i \in S'$ for some i, then $|V(F) \cap S'| < q_{max}(p_{max} + 1)$.*

Now we prove Theorem C by a reduction from the EXACT h-COVER problem, whose is input is a pair (X, M), where $X = \{x_1, x_2, \ldots, x_n\}$ is a finite set and $M = \{t_1, t_2, \ldots, t_m\}$ is a set of triples on X, and the question is if M contains a subsystem $M' \subset M$ such that every element of X belongs to exactly h triples of M'. This problem is NP-complete for every fixed $h > 0$ (cf. e.g., [11]). For our reduction, we use $h = q_{max}$. For a given instance (X, M), we may assume without loss of generality that $nq_{max} = 3l$ and $l \leq m$.

We start the construction of a graph G with a complete graph K_n with vertices $x_1, x_2, \ldots, x_n$. For every triple $t_i = \{x_a, x_b, x_c\}$, a copy H_i of the complete graph $K_{p_{\min}+1}$ is added, and one vertex of this graph is connected by edges to x_a, x_b, x_c. We further add $s = (m - l)(p_{\min} + 1) + 2p_{\max} + 2$ copies $F_1, \ldots, F_s$ of the graph F, with their roots $r_1, r_2, \ldots, r_s$ being adjacent to all $x_1, x_2, \ldots, x_n$. This graph G has $m(p_{\min} + 1) + n + s(q_{\max}(p_{\max} + 1) + 1)$ vertices and it is constructed from (X, M) in polynomial time. We claim that G contains a (σ, ρ)-dominating set of size $\geq k = l(p_{\min} + 1) + sq_{\max}(p_{\max} + 1)$ if and only if (X, M) contains an exact $q_{\max}$-cover.

Let first M' be a $q_{\max}$-cover of X. Clearly, $|M'| = l$, and it is straightforward to check that $S = \bigcup_{i: t_i \in M'} V(H_i) \cup \bigcup_{i=1}^{s}(V(F_i) \setminus \{r_i\})$ is a (σ, ρ)-dominating set in G of cardinality k.

Assume now that S is a (σ, ρ)-dominating set in G, and $|S| \geq k$. Suppose that some vertex x_j is in S. Then S can contain no more than $m(p_{\min} + 1)$ vertices of the graphs H_i, and no more than $p_{\max} + 1$ vertices from the set $\{x_1, x_2, \ldots, x_n\}$. Also at least $s - p_{\max}$ vertices from $\{r_1, r_2, \ldots, r_s\}$ do not belong to S. So, according to the preceding lemma, $|S| \leq m(p_{\min}+1)+p_{\max}+1+p_{\max}q_{\max}(p_{\max}+1)+(s-p_{\max})(q_{\max}(p_{\max}+1)-1) = m(p_{\min}+1)+2p_{\max}+1+sq_{\max}(p_{\max}+1)-(m-l)(p_{\min}+1)-2p_{\max}-2 = l(p_{\min}+1)+sq_{\max}(p_{\max}+1)-1 < k$. So, none of the vertices $x_1, x_2, \ldots, x_n$ belongs to S. Note that in this case $V(H_i) \subset S$ or $V(H_i) \cap S = \emptyset$ for all $i = 1, 2, \ldots, m$, and vertices of no more than l graphs H_i belong to S. Since S can contain no more than $q_{\max}(p_{\max}+1)$ vertices from every graph F_i, and $|S| \geq k$, vertices of exactly l graphs H_i are included to S. Every vertex x_j can have no more than $q_{\max}$ adjacent vertices from S. Hence each x_j is adjacent to exactly $q_{\max}$ vertices from H_i's and the set $M' = \{t_i : V(H_i) \subset S\}$ is a $q_{\max}$-cover of X.

6 Concluding Remarks and Open Problems

The complete classification of ambivalent pairs (σ, ρ) remains the first and main open problem. We believe that it is an interesting combinatorial problem by itself, and that it deserves attention. Perhaps it is impossible to formulate simple necessary and sufficient conditions for the general problem, but it would be interesting to obtain a complete solution at least for some special cases. For example for two-element sets $\sigma = \{p_1, p_2\}$ (it seems that cardinality of σ is more important).

A related complexity question is if the ambivalence of (σ, ρ) can be tested in polynomial time.

Another interesting question is a fixed parameter tractability of the $\exists(\sigma, \rho)$-DOMINATION. If the maximal value of σ is supposed to be the parameter, then the Theorem 3 shows that this problem is in FPT for $|\sigma| = 1$ and $p + 2 \leq q_{\min}$ (in fact our algorithm is polynomial in p and n). On the other hand, our general algorithm from Subsection 3.1 has the parameter $p_{\max}$ in the exponent of the running time, and hence is not FPT-algorithm. Fixed parameter tractability (or intractability) of the general case remains an open problem. Also it would be interesting to consider the problem parametrized by the size of the (σ, ρ)-dominating set.

References

1. Blair, J.R.S., Peyton, B.W.: An introduction to chordal graphs and clique tree. In: George, J.A., Gilbert, J.R., Liu, J.W.H. (eds.) Sparse Matrix Computation: Graph Theory Issues and Algorithms. IMA Volumes in Mathematics and its Applications, vol. 56, pp. 1–30. Springer, Heidelberg (1993)
2. Bulatov, A.A., Jeavons, P., Krokhin, A.A.: Classifying the complexity of constraints using finite algebras. SIAM J. Comput. 34(3), 720–742 (2005)
3. Feder, T., Hell, P., Huang, J.: Bi-arc graphs and the complexity of list homomorphisms, J. Graph Theory 42, 61–80 (2003)
4. Feder, T., Vardi, M.Y.: The computational structure of momotone monadic SNP and constraint satisfaction: A sudy through datalog and group theory. SIAM Journal of Computing 1, 57–104 (1998)
5. Fiala, J., Kratochvil, J.: Locally injective graph homomorphism: Lists guarantee dichotomy. In: Fomin, F.V. (ed.) WG 2006. LNCS, vol. 4271, pp. 15–26. Springer, Heidelberg (2006)
6. Fiala, J., Paulusma, D.: A complete complexity classification of the role assignment problem. Theoretical Computer Science 1 349, 67–81 (2005)
7. Garey, M., Johnson, D.: Computers and intractability. W.H.Freeman, New York (1979)
8. Haynes, T., Hedetniemi, S., Slater, P.: Domination in Graphs: The Theory. Marcel Dekker, New York (1997)
9. Heggernes, P., Telle, J.A.: Partitioning graphs into generalized dominating sets. Nordic J. Comput. 5, 173–195 (1998)
10. Hell, P., Nešetřil, J.: On the complexity of H-colouring. Journal of Combinatorial Theory B 48, 92–110 (1990)
11. Kratochvíl, J., Manuel, P., Miller, M.: Generalized domination in chordal graphs. Nordic Journal of Computing 2, 41–50 (1995)
12. Proskurowski, A., Telle, J.A.: Algorithms for vertex partitioning problems on partial k-trees. SIAM J. Discrete Math. 10, 529–550 (1997)
13. Schaefer, T.J.: The complexity of the satisfability problem. In: Proceedings of the 10th Annual ACM Symposium on Theory of Computing, pp. 216–226. ACM Press, New York (1978)
14. Telle, J.A.: Complexity of domination-type problems in graphs. Nordic Journal of Computing 1, 157–171 (1994)
15. Telle, J.A.: Vertex partitioning problems: characterization, complexity and algorithms on partial k-trees, PhD thesis, Department of Computer Science, Universiy of Oregon, Eugene (1994)

Recognizing Bipartite Tolerance Graphs in Linear Time

Arthur H. Busch[1,*] and Garth Isaak[2]

[1] Unviersity of Dayton, Dayton OH 45419-2316
art.busch@notes.udayton.edu
[2] Lehigh University, Bethlehem PA 18015
gi02@lehigh.edu

Abstract. A graph $G = (V, E)$ is a tolerance graph if each vertex $v \in V$ can be associated with an interval of the real line I_v and a positive real number t_v in such a way that $(uv) \in E$ if and only if $|I_v \cap I_u| \geq \min(t_v, t_u)$. No algorithm for recognizing tolerance graphs in general is known. In this paper we present an $O(n + m)$ algorithm for recognizing tolerance graphs that are also bipartite, where n and m are the number vertices and edges of the graph, respectively. We also give a new structural characterization of these graphs based on the algorithm.

1 Introduction and Notation

A graph $G = (V, E)$ consists of a set V, called vertices and a collection E of edges, which are unordered pairs of elements of V. We assume throughout this paper that our graphs are simple and finite. In other words, $|V|$ is always finite, and E is a set which contains no edge of the form (vv). The *order* of G is $|V|$ and we will denote this throughout the paper as n. Similarly, the *size* of G is $|E|$ which we will denote by m. A graph is a *tree* if is connected and contains no cycles, and a tree in which there is at most one vertex incident with multiple edges will be called a *star*. A graph is *bipartite* when the vertex set can be partitioned into two sets so that no edge connects two vertices from the same set. When G is a bipartite graph, we will represent a bipartition of V as V_x and V_y, with $n_x = |V_x|$ and $n_y = |V_y|$. For the bipartite graph $G = (V_x, V_y, E)$ with $V_x = \{x_1, \ldots x_{n_x}\}$ and $V_y = \{y_1, \ldots, y_{n_y}\}$, we will use $A(G)$ to denote the *reduced adjacency matrix* of G. This is the $n_x \times n_y$ matrix with $a_{ij} = 1$ if $(x_i y_j) \in E$ and $a_{ij} = 0$ otherwise.

We will call a collection of sets $\mathcal{U}$ *consecutively orderable* if the sets can be indexed $U_1, U_2, \ldots U_k$ so that whenever $x \in U_i \cap U_k$ then $x \in U_j$ for every $i \leq j \leq k$. A collection of sets together with such an ordering will be referred to as *consecutively ordered*. In this paper, the collections of sets will generally be subsets of the vertex set of a graph and in order to conserve notation we will say that a set of subgraphs $G_1, \ldots G_k$ is consecutively ordered when $V(G_1), \ldots, V(G_k)$ is consecutively ordered.

---

* Research for this article was performed while the author was a Visiting Assistant
Professor at Lehigh University.

A. Brandstädt, D. Kratsch, and H. Müller (Eds.): WG 2007, LNCS 4769, pp. 12–20, 2007.

1.1 Tolerance Graphs

Tolerance graphs were introduced in 1982 by Golumbic and Monma [8] to model certain scheduling problems. A graph $G = (V, E)$ is a *tolerance graph* if each vertex $v \in V$ can be associated with an interval of the real line I_v and a positive real number t_v in such a way that $(uv) \in E$ if and only if $|I_v \cap I_u| \geq \min(t_v, t_u)$. The collection $\langle \mathcal{I}, t \rangle$ of intervals and tolerances is called a *tolerance representation* of the graph G. A tolerance representation is called *bounded* when $|I_v| \leq t_v$ for every $v \in V$, and when G has such a bounded tolerance representation, we will say that G is a *bounded tolerance graph*.

Note that some authors (see [3], [7] and [18]) have studied a class of graphs that they call "bipartite tolerance graphs" but which is properly contained in the intersection of the classes of tolerance graphs and bipartite graphs (the graph T_2 in Fig. 1 is a separating example, as it is both bipartite and a tolerance graph, but is not a "bipartite tolerance graph" as defined in [7]). This smaller class of graphs was shown to be equivalent to bipartite permutation graphs in [3] and [18], and it follows from a theorem of Langley [12] that the class of bipartite permutation graphs is equivalent to bipartite bounded tolerance graphs. As a result, we will follow the convention used in [10], and we will use the phrase *bipartite tolerance graph* for the intersection of tolerance graphs and bipartite graphs, and the phrase *bipartite bounded tolerance graph* for the smaller class that is equivalent to bipartite permutation graphs.

Additional background and results on tolerance graphs can be found in the recent book by Golumbic and Trenk [10]. Although tolerance graphs and related topics have been studied extensively, the problem of characterizing tolerance graphs remains open, as does tolerance graph recognition [10]. It was shown in [11] that every tolerance graph has a polynomial sized tolerance representation, and hence Tolerance Graphs recognition is in the class NP. However, this result gives no information on how to construct an algorithm that recognizes when a graph has a tolerance representation.

The class of cycle free tolerance graphs was characterized in [9].

Theorem 1 (Golumbic, Monma and Trotter, [9]). *Let T be a tree. Then T is a tolerance graph if and only if T contains no induced subgraph isomorphic to T_3, in Fig. 1.*

For bipartite graphs which contain cycles, Busch [5] gave the following characterization.

Theorem 2 (Busch [5]). *A bipartite graph G is a tolerance graph if and only if there exists a set of consecutively ordered stars $S_1, S_2, \ldots, S_t$ which partition the edges of G.*

Fig. 1. The trees T_2 and T_3

1.2 Asteroidal Triples and Consecutive Orderings

A $(0,1)$-matrix M has the *consecutive 1's property for rows* if the columns of M can be permuted in such a way that the 1's in every row occur consecutively. Analogously, a matrix M has the *consecutive 1's property for columns* if the rows of M can be permuted in such a way that the 1's in every column occur consecutively. Tucker [20] investigated when the reduced adjacency matrix $A(G)$ of a bipartite graph $G = (V_x, V_y, E)$ has the consecutive 1's property for rows or columns. A consecutive ordering of the columns or rows of $A(G)$ represents an ordering of either V_x or V_y such that the collection of neighborhoods $\mathcal{N}_x = N(x_1), N(x_2), \ldots, N(x_{n_x})$ or $\mathcal{N}_y = N(y_1), N(y_2), \ldots, N(y_{n_y})$ is consecutively ordered. Tucker calls bipartite graphs where $\mathcal{N}_x$ is consecutive orderable *X-consecutive*, and analogously, a bipartite graph is *Y-consecutive* when $\mathcal{N}_y$ is consecutive orderable. Bipartite graphs which are either X-consecutive or Y-consecutive are known as *convex*, while bipartite graphs that are both X-consecutive and Y-consecutive are *biconvex*.

An *asteroidal triple* in a graph $G = (V, E)$ is a triple of distinct vertices v_0, v_1, v_2 with the property that for each $i = 0, 1, 2$, there is a path from v_{i+1} to v_{i+2} in G that contains no vertex adjacent to v_i (subscript addition is performed modulo 3). Tucker showed the following connection between consecutive orderings and asteroidal triples.

Theorem 3 (Tucker [20]). *A bipartite graph $G = (V_x, V_y, E)$ is X-consecutive if and only if G has no asteroidal triple contained in V_x. Similarly G is Y-consecutive if and only if G has no asteroidal triple contained in V_y.*

Algorithms which determine if a matrix has the consecutive 1's property for rows form the basis for the first linear recognition algorithm for interval graphs, due to Booth and Lueker [1], and closely related algorithms also recognize convex graphs in linear time (although such algorithms generally avoid using adjacency matrices to preserve linear running times even for sparse graphs). Algorithms to identify the consecutive 1's property of a matrix can also easily be used to determine when a collection of subgraphs $\mathcal{G} = G_1 \ldots G_t$ of a given graph G is consecutively orderable. We simply construct the "vertex-graph incidence matrix" $M_{n \times t} = [m_{ij}]$ which has $m_{ij} = 1$ if the vertex v_i is contained in $V(G_j)$ and $m_{ij} = 0$ otherwise. Then $\mathcal{G}$ is consecutively orderable if and only if M has the consecutive 1's property for rows. Thus, when $\mathcal{G}$ is part of the input, and $\mathcal{G}$ is a collection of stars which partition the edges of G, a consecutive 1's algorithm can be used to show that G is a tolerance graph using Thm. 2. However, a tolerance graph G generally has many star partitions (the set E, for example), not all of which can be consecutively ordered. As a result, the above procedure cannot be used to decide if an arbitrary bipartite graph is a tolerance graph. In the following section, we investigate the structure of bipartite graphs whose edges can be partitioned into sets which induce stars which are consecutively orderable. We call such a partition a *consecutive star partition* (CSP), and in the process obtain both a conceptually simple linear time algorithm ($O(n+m)$) that recognizes the class of bipartite tolerance graphs, and a new characterization of this class.

2   Bipartite Tolerance Graphs

We begin with some basic observations about consecutive star partitions and tolerance graphs. Throughout this section we will denote a consecutive star partition (CSP) of a graph G as $\mathcal{S} = S_1, S_2, \ldots, S_t$, where each S_i is a star, and we will call t the *length* of $\mathcal{S}$. We will denote the vertex and edge set of the star S_j as $V(S_j)$ and $E(S_j)$, respectively. If S_i is a single edge, we will arbitrarily designate one endpoint of this edge as c_i. Otherwise, let c_i be the unique central vertex of S_i.

Observation 4. *Let $G = (V_x, V_y, E)$ be a connected bipartite tolerance graph with CSP $\mathcal{S} = S_1, S_2, \ldots, S_t$. Then $V(S_i) \cap V(S_{i+1})$ is a cut-set of G for each $1 \le i < t$.*

Proof. Let $L = V \setminus \bigcup_{j=1}^{i} V(S_j)$ and $R = V \setminus \bigcup_{j=i+1}^{t} V(S_j)$. Since $\mathcal{S}$ partitions $E(G)$ and G is bipartite, both L and R are non-empty. Each edge incident with a vertex of L is contained in a star S_j with $j \le i$, and any edge incident with a vertex of R is contained in a star S_j with $j \ge i+1$. Thus, no edge connects L and R and $V - (L \cup R) = V(S_i) \cap V(S_{i+1})$ is a cut-set. ∎

Observation 5. *Let G be a 2-connected bipartite graph. Then G is a tolerance graph if and only if G is convex.*

Proof. Assume $G = (V_x, V_y, E)$ is a tolerance graph. Then G has a CSP $\mathcal{S} = S_1, S_2, \ldots, S_t$, and since G is 2-connected, we must have $|V(S_i) \cap V(S_{i+1})| \ge 2$ for $i = 1, \ldots, t-1$. Without loss of generality, assume $c_1 \in V_x$. If $c_i \in V_x$ for each $1 \le i \le t$, then G is clearly X-consecutive. Otherwise, let i be the minimal index such that $c_i \in V_y$. It follows immediately that $V(S_i) \cap V(S_{i-1}) = \{c_i, c_{i-1}\}$ and thus the edge $(c_i c_{i-1})$ is in both S_i and S_{i-1}, a contradiction.

For the converse, assume that $\mathcal{N}_x$ is consecutively ordered, and let $V_x = \{x_1, v_2, \ldots x_{n_x}\}$ correspond to this ordering. Let S_i be the subgraph of G induced on $N[x_i] = N(x_i) \cup \{x_i\}$ for $1 \le i \le t$. Since G is bipartite, each induced subgraph is a star, and because each x_i is in a unique $V(S_i)$, the set of stars are consecutively ordered and clearly partition the edges of G. Thus, G is a tolerance graph by Thm. 2. ∎

Observation 5, together with the hereditary property of bipartite tolerance graphs, shows that every 2-connected subgraph of a bipartite tolerance graph must be convex. Recall that a *block* of a graph G is a maximal subgraph of G with no cut-vertex. It is easy to see that every block of a connected graph is either 2-connected, a cut-edge or an isolated vertex (in the trivial case where $G = K_1$). We will define the *boundary* of a block B, denoted $\mathcal{B}(B)$ as the set of vertices in B with $N(v) \nsubseteq V(B)$. In other words, the boundary of a block

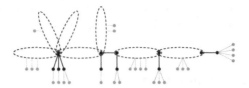

Fig. 2. The block structure of $G - P_G$ for a bipartite tolerance graph G. The dashed ovals represent 2-connected blocks and the gray vertices are pendant in G.

B is the set of cut-vertices of G that are also vertices of B. If P_G is the set of pendant vertices of G, we will partition the boundary of B into two sets $\mathcal{B}^1(B) = \{v \in \mathcal{B}(B) \mid (N(v) \setminus B) \subseteq P_G\}$ and $\mathcal{B}^2(B) = \mathcal{B}(B) \setminus \mathcal{B}^1(B)$.

Let B be a block of a graph G. We will define B' as the graph induced by the vertices of B, together with the vertices in P_G adjacent to $\mathcal{B}^1(B)$. We then define a graph H_B from B' by adding two new vertices v' and v'' to B' for each vertex $v \in \mathcal{B}^2(B)$, along with the edges (vv') and $(v'v'')$. Note that H_B is an induced subgraph of G, and that if $\mathcal{B}^2(B) = \emptyset$ then $H_B = B' = G$.

Our algorithm is based on the following results, the first of which is a slight extension of Obs. 5.

Lemma 6. *If G is a bipartite tolerance graph, then for every block B of G, H_B is convex.*

Lemma 7. *If G is a bipartite tolerance graph and B is a 2-connected block of G with $|\mathcal{B}^2(B)| \geq 2$, then $\mathcal{B}^2(B) = \{u, v\}$ and B' has a CSP with that begins with a star containing u and ends at a star containing v.*

Corollary 8. *If G is a bipartite graph and B is a 2-connected block of G with $|\mathcal{B}^2(B)| > 2$, then H_B is not convex, and G is not a tolerance graph.*

Lemma 9. *If G is a bipartite tolerance graph, and B is a block of G with $\mathcal{B}^2(B) = \{v\}$ and v is at distance two or less from every other vertex of B', then B' has a CSP such that v is contained in every star.*

In broad terms, for a bipartite tolerance graph G, the above results impose a structure on the blocks of $G - P_G$, as well as a structure on the arrangement of those blocks. We illustrate this informally in Figure 2. After removing the pendant vertices, Cor. 8 and Lem. 9 imply that the cut-vertices of $G - P_G$ can be arranged linearly.

3   Recognizing Bipartite Tolerance Graphs

We now present an algorithm based on the results from the previous section that recognizes bipartite tolerance graphs.

Algorithm 1. Determine if G is a bipartite tolerance graph

Require: G is a connected, bipartite graph
 1: Find P_G, the blocks of $G - P_G$, and the block-cutpoint graph T of $G - P_G$.
 2: **for all** blocks B of $G - P_G$ **do**
 3:     Construct B' and H_B
 4:     **if** H_B is not convex **then**
 5:         **return false**
 6:     **else**
 7:         **if** $d_T(B) = 1$ and $\epsilon_{B'}(c)$? 2 for the unique cut-vertex c ? B **then**
 8:             Delete B from T
 9:         **end if**
10:     **end if**
11: **end for**
12: **if** T is a path **then**
13:     **return true**
14: **else**
15:     **return false**
16: **end if**

Theorem 10. *Algorithm 1 is correct, and runs in $O(n + m)$ steps.*

Proof. First, we show that the algorithm is correct. If $G - P_G$ contains a block B such H_B is not convex, then G is not a tolerance graph by Lem. 6. If H_B is convex for every block B, but T is not a path after all applicable blocks have been deleted, then T contains a vertex of degree three or more. This vertex does not represent a block of $G - P_G$, since for such a block B, $|\mathcal{B}^2(B)| \geq 3$ which implies that H_B is not convex by Lem. 8. So this vertex in T represents a cut-vertex v, and because none of the blocks adjacent to v were deleted, v is at the end of at least three paths of length three or more. Furthermore, the edges of these paths incident with v are each in different blocks of G, and as a result G contains an induced subgraph isomorphic to T_3, and hence is not a tolerance graph by Thm. 1. In all other cases, the algorithm returns true. In such cases, either G is a star and hence obviously a tolerance graph, or we can construct a CSP for G as follows:

Since each block of T has a corresponding H_B that is convex, each H_B is clearly also a tolerance graph. Thus, for each block that is adjacent to two cut-vertices u and v on this path, the associated graph B' has a CSP that begins at a star containing u and ends at a star containing v by Lem. 7 as applied to H_B. For any blocks B on the ends of this path adjacent to only one cut vertex v, the same argument used in the proof of Lem. 7 guarantees that B' will have a CSP that begins or ends at a star containing v. Thus, we can easily combine each of these CSPs into a CSP that contains every edge in an undeleted block. Finally, each pendant edge of G that is not in any B' and each block B that was deleted from T is adjacent to a single cut-vertex v and has a CSP with v contained in every star (trivially, in the case of a pendant edge, or by Lem. 9

as applied to H_B). Thus we can insert this CSP into our combined CSP at the beginning (or end) if the first (or last) star already contains v, or between any two stars that both contain v. Two such stars must exist if the first and last star do not already contain v, since in this case v must be contained in two blocks that were not deleted from T. Because every edge of G is in at least one graph B', this covers the edges of G. Further, any edge in more than one B' must be pendant, and by deleting these duplicates we obtain a CSP for G. Therefore, G is a tolerance graph by Thm. 2.

It now remains to show that the algorithm requires $O(n + m)$ steps. Finding the set P_G requires $O(n)$ time, and finding the blocks and cut-vertices of $G - P_G$ and building the block-cutpoint graph T can be done in $O(n + m)$ time [19]. Using these blocks, we can also construct each graph B' and the associated graph H_B in $O(n_b)$ time, and verifying that each H_B is convex requires $O(n_b + m_b)$ time, where n_b and m_b are the order and size of B', respectively. Determining if $d_T(B) = 1$, and if so, identifying the cut vertex c adjacent to B in T can be done in constant time, and because B is bipartite we can determine if B' contains an induced path of length three or more that begins at v in $O(n_b)$ time. Thus, the total running time for all of these tests is $\sum_{B'} O(n_b + m_b)$.

After all these tests are complete and the blocks satisfying the condition in line 8 have been deleted from T, testing that the graph remaining is a path requires $O(|V(T)|) = O(n + m)$ steps.

Each edge is in at most one B', and an easy induction proof shows that $\sum_{V(G)} b_v \leq 2n$, where b_v is the number of blocks which contain the vertex v. Hence the total running time of the algorithm is

$$\sum_{B'} O(n_b + m_b) = O(\sum_{B'} n_b + m_b) = O(m + \sum_{V(G)} b_v) = O(m + n)$$

as desired.

Note that this algorithm also gives a new structural characterization of bipartite tolerance graphs, which we give in the following theorem:

Theorem 11. *Let G be a connected bipartite graph. Then G is a tolerance graph if and only if, for every block B of G, H_B is convex, and G contains no induced subgraph isomorphic to the graph T_3 in Figure 1.*

As indicated in the proof of Theorem 10, Algorithm 1 can easily be modified to provide a CSP of G when G is a bipartite tolerance graph. This CSP can then be combined with the algorithm in [5], to give a tolerance representation of the graph G. Although this representation is not guaranteed to be polynomial in the size of G, such a polynomial sized representation is guaranteed by the result in [11]. Since the algorithm in [5] is clearly not optimal, it seems likely that there is an efficient algorithm that will convert the CSP into a polynomial sized tolerance representation of G, which could then be used to certify the correctness of the algorithm.

When Algorithm 1 returns false, we can also certify this result, either by identifying an induced subgraph of G isomorphic to T_3, or by giving an induced subgraph H_B of G that is not convex. Although we do not have complete list of such obstructions, we can identify an asteroidal triple of H_B that is contained in V_x and an asteroidal triple of H_B contained in V_y. This certifies that H_B is not convex, and hence that G is not a tolerance graph by the contrapositive of Lem. 6.

4   Class Hierarchies

In conclusion, we consider how various sub-classes of chordal bipartite graphs relate to the class of bipartite tolerance graphs and the potential extensions of our algorithm to related classes. We begin by noting some basic inclusions from [3]. Recall that the class of bipartite permutation graphs is equivalent to the class which we denote as bipartite bounded tolerance graphs.

$$permutation \cap bipartite \subset biconvex \subset convex \subset chordal\ bipartite$$

Next, we note a refinement of the above hierarchy due to a combination of the results of Brown [4], Busch [5], Müller [14], and Sheng [16].

$$convex \subset probe\ int. \cap bip. \subset tolerance \cap bip. \subset int.\ bigraph \subset chordal\ bip.$$

Both convex and biconvex graphs can be recognized in linear time [2], and chordal bipartite graphs may be recognized in $O(\min\{m \log(n), n^2\})$ time [13,15,17]. The best known algorithms for the class of bipartite probe-interval graphs and for the class of interval bigraphs are polynomial [6,14]. In the case of 2-connected bipartite graphs, Obs. 5 shows that $convex = tolerance \cap bipartite$, and it is this equivalence that forms the basis of the algorithm in the previous section. Furthermore, it is easy to show that within the subclass of 2-edge-connected bipartite graphs, $convex = probe\text{-}interval \cap bipartite$. This equivalence suggests a similar approach to the one used above may provide a linear time algorithm for the class of bipartite probe-interval graphs. It is less certain that our approach can be extended to give a linear time recognition algorithm for the class of interval bigraphs or the class of chordal bipartite graphs.

Acknowledgment

The authors would like to thank R. Sritharan for his suggestions and comments.

References

1. Booth, K., Lueker, G.: Linear Algorithms to Recognize Interval Graphs and Test for the Consecutive Ones Property. In: Proceedings of the Seventh Annual ACM Symposium on Theory of Computing, pp. 255–265 (1975)
2. Brandstädt, A., Le, V.B., Spinrad, J.P.: Graph Classes, A Survey. Soc. for Industrial and Applied Math., Philadelphia, PA (1999)

3. Brandstädt, A., Spinrad, J.P., Stewart, L.: Bipartite permutation graphs are bipartite tolerance graphs. Congress. Numer. 58, 165–174 (1987)
4. Brown, D.E.: Variations on Interval Graphs, Ph.D. thesis, University of Colorado at Denver (2004)
5. Busch, A.H.: A characterization of bipartite tolerance graphs. Discrete Applied Math. 154, 471–477 (2006)
6. Chang, G., Kloks, A., Liu, J., Peng, S.: The PIGS Full Monty - A Floor Show of Minimal Separators. In: Diekert, V., Durand, B. (eds.) STACS 2005. LNCS, vol. 3404, pp. 521–532. Springer, Heidelberg (2005)
7. Derigs, U., Goecke, O., Schrader, R.: Bisimplicial edges, gaussian elimination and matchings in bipartite graphs. In: Inter. Workshop on Graph-Theoretic Concepts in Comp. Sci., pp. 79–87 (1984)
8. Golumbic, M.C., Monma, C.L.: A generalization of interval graphs with tolerances. Congress. Numer. 35, 321–331 (1982)
9. Golumbic, M.C., Monma, C.L., Trotter, W.T.: Tolerance Graphs. Discrete Applied Math. 9, 157–170 (1984)
10. Golumbic, M.C., Trenk, A.N.: Tolerance Graphs. Cambridge University Press, Cambridge, UK (2004)
11. Hayward, R.B., Shamir, R.: A note on tolerance graph recognition. Discrete Appl. Math. 143, 307–311 (2004)
12. Langley, L.: Interval tolerance orders and dimension, Ph.D. thesis, Dartmouth College (1993)
13. Lubiw, A.: Doubly lexical orderings of matrices. Siam J. Comput. 16, 854–879 (1987)
14. Müller, H.: Recognizing Interval digraphs and interval bigraphs in polynomial time. Discrete Appl. Math. 78, 189–205 (1997)
15. Paige, R., Tarjan, R.E.: Three partition refinements of algorithms. SIAM J. Comput. 16, 973–989 (1987)
16. Sheng, L.: Cycle free probe interval graphs. Congr. Numer. 140, 33–42 (1999)
17. Spinrad, J.P.: Doubly lexical ordering of dense 0-1 matrices. Inf. Proc. Lett. 45, 229–235 (1993)
18. Spinrad, J.P., Brandstädt, A., Stewart, L.: Bipartite permutation graphs. Discrete Appl. Math. 18, 279–292 (1987)
19. Tarjan, R.E.: Depth-First Search and Linear Graph Algorithms. SIAM J. Comput. 1, 146–160 (1972)
20. Tucker, A.C.: A Structure Theorem for the Consecutive 1's Property. J. Combinatorial Theory Ser. B 12, 153–162 (1972)

Graph Searching in a Crime Wave*

David Richerby** and Dimitrios M. Thilikos

Department of Mathematics, University of Athens, Panepistimioupolis, GR-15784
Athens, Greece
{davidr,sedthilk}@math.uoa.gr

Abstract. We define helicopter cop and robber games with multiple robbers, extending previous research, which only considered the pursuit of a single robber. Our model is defined for robbers that are visible (the cops know their position) and active (able to move at every turn) but is easily adapted to other common variants of the game. The game with many robbers is non-monotone: more cops are needed if their moves are restricted so as to monotonically decrease the space available to the robbers. Because the cops may decide their moves based on the robbers' current position, strategies in the game are interactive but the game becomes, in a sense, less interactive as the initial number of robbers increases. We prove that the main parameter emerging from the game captures a hierarchy of parameters between proper pathwidth and proper treewidth. We give a complete characterization of the parameter for trees and an upper bound for general graphs.

1 Introduction

During recent decades, the problem of searching a graph has attracted much attention, because of its purely graph-theoretic interest and its numerous applications in modelling problems in communication networks. In general, graph searching problems are modelled as a game played between a team of cops and a robber, whom the cops attempt to capture by moving systematically through the graph. We wish to know the minimum number of cops required to catch the robber, under various constraints on the players' behaviour. Several versions have been examined varying, for example, in whether the cops know the robber's position, whether the robber can move at will or only when disturbed by a cop, and how the cops move through the graph.

One of the main models of graph searching, the helicopter cops and robber game, was introduced by Seymour and Thomas [8]. In this model, the robber occupies a vertex of a graph and is *active*, in the sense that he may move at any round of the game, along any path whose vertices are not guarded by cops. The cops do not have to stay within the graph and can be placed on or removed

---

* This research is supported by the Project "CAPODISTRIAS" (AΠ 736/24.3.2006) of the National and Capodistrian University of Athens (project code: 70/4/8757).
** Funded by the European Social Fund and Greek National Resources — (EΠEAEK II) PYTHAGORAS II.

A. Brandstädt, D. Kratsch, and H. Müller (Eds.): WG 2007, LNCS 4769, pp. 21–32, 2007.
© Springer-Verlag Berlin Heidelberg 2007

from its vertices, as if flying by helicopter. The cops win when a cop lands on the vertex occupied by the robber and he cannot escape.

Critically, the robber is *visible*: the cops know his current position at all times and can use this information to decide their moves, so the strategy they use is *interactive*. The least number of cops guaranteed to be able to catch the robber in a graph, regardless of how the robber attempts to escape, is one greater than the graph's treewidth [8]. The proof of this is based on a proof of the game's monotonicity; i.e., the fact that the cops do not become weaker when their moves are restricted to those that monotonically decrease the space available to the robber.

In an important variant of the game, the robber is a *lazy* fugitive who may move only when a cop moves to his vertex. To compensate, the robber is *invisible*: the cops do not know his position. Strategies are no longer interactive but may be given in advance as a predetermined sequence of moves. This version of the game is equivalent to the Seymour–Thomas game in the sense that, for any graph, the two games require the same number of cops [3]. It follows easily from [3] (see also [1]) that the number of cops required to ensure the capture of an active, invisible robber is one greater than the pathwidth of the graph, another graph parameter of equal importance to treewidth.

We examine and unify the above models under the natural extension where the graph contains many robbers rather than just one. This is the first time that multiple robbers have been considered and we believe that our results will motivate a corresponding study for other searching models as well.

We describe our graph searching model using the most general setting of *mixed search*, proposed by Bienstock and Seymour [1] (see also [9,10,11,12]). In this model, as well as being placed on or removed from the graph, the cops can slide along its edges. This may reduce by one the number of cops required to search a graph but the version without sliding can be reduced to mixed search by replacing each edge in the graph with two parallel edges or a triangle including a new vertex [1]. As well as being more general, using mixed search makes the presentation of our results cleaner.

It is not obvious how to generalize the concept of monotonicity for the setting with many robbers. Each robber has his own individual free space, so it is not clear whether monotonicity should be defined individually or collectively. We give three natural definitions (individual and collective, and a definition based on the cops' behaviour) and show them to be equivalent.

Monotonicity is crucial in the multiple-robber case. If we do not require monotonicity, we can catch r visible, active robbers one at a time by repeating the strategy to catch a single robber, without requiring a greater number of cops. However, when we restrict to monotone strategies, the number of cops required, which we denote $\mathbf{mvams}(G, r)$ (for monotone, visible, active mixed search number against r robbers), can be greater and depends on the number of robbers. In particular, $\mathbf{mvams}(G, 1)$, the mixed search number for a single visible, active robber, is equal to the parameter of *proper treewidth* [2,9]. On the other hand, if n is the order of G, then $\mathbf{mvams}(G, n)$ is equal to the mixed search number for a single invisible, active robber, which is equal to the parameter of *proper pathwidth* [11]. We show that $\mathbf{mvams}(G, r)$ can, for appropriate values of r, take all values between proper treewidth and proper pathwidth. As our main result, we give the exact value of our parameter on trees T and an upper bound for general graphs G:

$$\mathbf{mvams}(T, r) = \min\left\{\mathbf{ppw}(T), \lfloor\log r\rfloor + 1\right\}$$
$$\mathbf{mvams}(G, r) \leq \min\left\{\mathbf{ppw}(G), \mathbf{ptw}(G) \cdot (\lfloor\log r\rfloor + 1)\right\}.$$

Our result for trees extends the analogous characterizations for pathwidth and proper pathwidth given in [10] and [4], respectively. Our results can be seen as showing that the number of robbers tunes the amount of interactivity in search strategies, spanning all intermediate levels from pathwidth (fully predetermined) to treewidth (fully interactive). Another way of defining this tuning was given by Fomin, Fraigniaud and Nisse, who considered a single active robber but restricted the number of rounds at which the cops can ask for the robber's position [5].

The remainder of the present paper is organized as follows. Our searching model is defined in Section 2. In Section 3, we show the equivalence of three reasonable definitions of monotonicity and explore the rôle of monotonicity in the game. To relate our hierarchy of parameters to the proper pathwidth and proper treewidth, we briefly consider games with a single invisible robber in Section 4. Our complete characterization of $\mathbf{mvams}(T, r)$ for trees T follows in Section 5, where we also give an upper bound for general graphs. Several consequences of our results and open problems are presented in Section 6.

2   The Searching Model

All graphs in this paper are finite, simple and undirected.

In a helicopter search game with many visible robbers, the opponents are a group of cops and a group of robbers, who each occupy vertices of the graph. The cops and robbers have full information about each other's location and may use this information to decide their moves. The goal of the cops is to capture the robbers. Initially, there are no cops in the graph but, at all times, any robber who has not been captured is on some vertex. A play of the game consists of a sequence of rounds, with each round being composed of three parts, as follows.

Announcement. The cops announce their intended move to the robbers. One cop moves in each round, by one of the following operations.
- Placing a cop on a vertex v, not currently occupied by a cop, denoted by $\mathbf{place}(v)$.
- Removing a cop from an occupied vertex v, denoted by $\mathbf{remove}(v)$.
- Sliding a cop who is on one endpoint u of an edge $\{u, v\}$ to the other, which is initially unoccupied, denoted by $\mathbf{slide}(u \rightarrow v)$.

Avoidance. Each of the robbers can move to any vertex reachable from his current position by a path not blocked by cops, as long as this vertex will not be occupied by a cop once the cops' current move has been realized. If the announced move is a placement to or removal from some vertex, that vertex is not considered blocked for the purposes of the robbers' movement in the round; if it is **slide**$(u \rightarrow v)$, the edge $\{u, v\}$ is considered to be blocked for this round but the vertices u and v are not.

Realization. The cops carry out the announced action.

A robber is captured if the cops announce that they will move (by placement or sliding) to his vertex and he cannot move away.

To formalize the game, we use a string $\mathbf{R} \in (V(G) \cup \{*\})^r$ to denote the positions of the r robbers in the graph. The ith character of $\mathbf{R}$ is either the vertex occupied by the ith robber or is "$*$", when the ith robber has been captured. We write $V(\mathbf{R})$ for the set of characters of $\mathbf{R}$ except $*$. Since, at any time, there is at most one cop on any vertex, we may represent the position of the cops as a set $S \in V(G)^{[\leq k]}$ (i.e., $S \subseteq V(G)$ and $|S| \leq k$).

A *play* of the game on a graph is an infinite sequence of positions $\mathcal{P} = S_0, \mathbf{R}_0, S_1, \mathbf{R}_1, \ldots$, where, for each i, the transition from having the cops at S_i and robbers at $\mathbf{R}_i$ to the cops at S_{i+1} and robbers at $\mathbf{R}_{i+1}$ is a valid move of the game. Specifically,

- $S_0 = \emptyset$;
- $S_1 = \{v\}$ for some vertex v — the first move is **place**(v); and
- if S_i, S_{i+1} are consecutive sets, then one of the following holds:
  - $S_{i+1} - S_i = \{v\}$ — the move is **place**(v),
  - $S_i - S_{i+1} = \{v\}$ — the move is **remove**(v),
  - $S_{i+1} \triangle S_i = \{u, v\} \in E(G)$ — the move is **slide**$(u \rightarrow v)$ where $S_i - S_{i+1} = \{u\}$ and $S_{i+1} - S_i = \{v\}$.

We call such a sequence of cop positions *consistent*. Given two consecutive sets S_i and S_{i+1} of a consistent sequence, say that a path P of G is (S_i, S_{i+1})-*avoiding* if its internal vertices avoid $S_i \cap S_{i+1}$ and its edges avoid the edge $e = S_{i+1} \triangle S_i$, in the case that $|e| = 2$. If the location of the robbers at the i-th step is $\mathbf{R}_i = [a_1 \ldots a_r]$, the set of *free* locations for the jth robber after step i is $F_{i+1}^j = \emptyset$ if $a_j = *$ and, otherwise,

$$F_{i+1}^j = \left\{ y \in V(G) - S_{i+1} \mid G \text{ has an } (S_i, S_{i+1})\text{-avoiding } (a_j, y)\text{-path} \right\}.$$

As a response to the ith move of the cops, the robbers can choose their new location to be any string $\mathbf{R}_{i+1} = [a'_1 \ldots a'_r]$ such that for $j \in \{1, \ldots, r\}$, $a'_j = *$ if $F_{i+1}^j = \emptyset$ and $a'_j \in F_{i+1}^j$, otherwise. (Note, in particular, that, if $a_j = *$, then $F_{i+1}^j = \emptyset$, so $a'_j = *$, also. Thus, a robber who has been captured can never return to the game.)

We set $F_0 = V(G)$, and for $i \geq 1$, we define $F_i = \bigcup_{j \in \{1, \ldots, r\}} F_i^j$. We say that the sequence $F_0, F_1, \ldots$ is the sequence of *free positions* for the robbers. If, for every $i \geq 0$, $F_{i+1} \subseteq F_i$ we say that $\mathcal{P}$ is a *monotone* play.

A play $\mathcal{P} = S_0, \mathbf{R}_0, S_1, \mathbf{R}_1, \ldots$ is *winning* (for the cops) if $V(\mathbf{R}_i) = \emptyset$ for some $i \geq 0$: that is, all the robbers are eventually captured. The *essential part* of a winning play is the subsequence $S_0, \mathbf{R}_0, \ldots, S_\ell, \mathbf{R}_\ell$, where ℓ is minimal such that $V(\mathbf{R}_\ell) = \emptyset$.

According to our description of the game, each move of the cops may depend on the current position of the cops and robbers in the graph. A (k, r)-*strategy* is a function[1]

$$\mu : V(G)^{[\leq k]} \times (V(G) \cup \{*\})^r \to V(G)^{[\leq k]},$$

whose inputs are the position S of the cops and the positions $\mathbf{R}$ of the robbers and whose output is S', the new position of the cops, with the requirement that, for all S and $\mathbf{R}$, the sets S and $S' = \mu(S, \mathbf{R})$ obey the restrictions given in the definition of consistency for sequences. That is, there is a single move which transforms the cop position S to S'.

Given a strategy μ, a μ-*play*, is any play $S_0, \mathbf{R}_0, S_1, \mathbf{R}_1, \ldots$ where $S_{i+1} = \mu(S_i, \mathbf{R}_i)$ for all $i \geq 0$. A strategy μ is monotone (respectively, winning) if all μ-plays are.

We define the *non-monotone* and *monotone, visible, active mixed search number*, respectively, of a graph G as follows:

$$\mathbf{vams}(G, r) = \min \left\{ k \mid G \text{ has a winning } (k, r)\text{-strategy} \right\}$$

$$\mathbf{mvams}(G, r) = \min \left\{ k \mid G \text{ has a monotone winning } (k, r)\text{-strategy} \right\}.$$

It is equivalent, and usually easier, to describe a search strategy as a *search program* Π, that makes move decisions depending only on the current position of the cops and robbers. The program receives the positions of the robbers by calling a routine *robbers_positions()*.

As an example, Program 1, defines a monotone winning search program for one cop against one robber in a tree T. At each step, the robber must choose his position in the component T' where he resides, excluding the vertex w that is the target of the cop's move. Thus, the robber's set of free positions becomes strictly smaller, so the strategy is both winning and monotone.

In this program, the cops only care about the component that contains the robber and not about which vertex he occupies within that component. With an eye to the situation with more than one robber, we can say that the move of the cops from position S depends only on the number of robbers in each component of $T - S$. The cops do not lose any power if their information is restricted in this way. Say that a (k, r)-strategy for a graph G is *smooth* if, for all $S \in V(G)^{[\leq k]}$ and $\mathbf{R}, \mathbf{R}' \in (V(G) \cup \{*\})^r$, such that each component of $G - S$ contains the same number of robbers in $\mathbf{R}_1$ as in $\mathbf{R}_2$, we have $\mu(S, \mathbf{R}_1) = \mu(S, \mathbf{R}_2)$.

---

[1] When we define strategies, we will not define the action of the cops in positions that can never occur when the strategy is executed. Thus, we give only a partial function. Formally, the strategy is any total extension of this partial function, assigning arbitrary moves to the cops in situations that can never happen. Further, it can be shown that more general notions of strategy, allowing the cops to decide their move depending on previous positions in the game, as well as the current position, do not increase the cops' power.

Search Program 1. $\Pi(T,1)$ to capture one robber in a tree T.

---

place(v) where v is any vertex of T.
Let $\mathbf{R} \leftarrow$ *robbers_positions()*.
Let $T' \leftarrow T$.
While $V(\mathbf{R}) \neq \emptyset$,
    Let T' be the connected component of $T - v$ containing $V(\mathbf{R})$
        and let w be the (unique) vertex of T' adjacent to v.
    slide$(v \rightarrow w)$.
    Let $v \leftarrow w$.
    Let $\mathbf{R} \leftarrow$ *robbers_positions()*.
remove(v).

Lemma 1. *There is a winning (k,r)-strategy in G if, and only if, there is a smooth winning (k,r)-strategy in G.*

Using smoothness, the proof of the following important is straightforward. We write $G \preccurlyeq H$ to indicate that G is a minor of H (i.e., that G can be obtained from a subgraph of H by a sequence of edge contractions).

Proposition 2. *If $G \preccurlyeq H$ and $r \geq 1$, then* $\mathbf{mvams}(G,r) \leq \mathbf{mvams}(H,r)$.

3   Variants of Monotonicity

We have defined monotonicity for plays and strategies. These definitions are natural extensions of the single-robber case but not the only ones. We now consider two other natural definitions, which turn out to be equivalent to the first, and begin an investigation of the cost of monotonicity.

    Let $\mathcal{P} = S_0, \mathbf{R}_0, S_1, \mathbf{R}_1, \ldots$ be a μ-play. $\mathcal{P}$ is *pointwise monotone* if, for each $j \in \{1, \ldots, r\}$ and each $i \geq 0$, $F_{i+1}^j \subseteq F_i^j$; i.e., no robber's set of free positions ever increases. $\mathcal{P}$ is *cop monotone* if, for each $v \in V(G)$, the set $\{i \mid v \in S_i$ and $V(\mathbf{R}_i) \neq \emptyset\}$ is an interval of $\mathbb{N}$: once the cops have left a vertex, they never return to it while there are robbers in the graph. Any cop-monotone μ-play must be winning because the cops must revisit a vertex if a robber lives forever. A strategy is monotone according one of the above definitions if all its plays are.

Lemma 3. *For any graph G and $k,r \geq 1$, the following are equivalent:*

1. *there is a monotone winning (k,r)-strategy in G;*
2. *there is a pointwise-monotone winning (k,r)-strategy in G;*
3. *there is a cop-monotone (k,r)-strategy in G.*

Proof (sketch). (2)$\Rightarrow$(1) trivially. The main idea for (1)$\Rightarrow$(3) is that any vertex that is revisited cannot be in the free space of the robbers so, with some work and assuming smoothness (using Lemma 1), the strategy can be modified to omit the revisit. For (3)⇒(2), if some robber's free space increases, a cop must have left some vertex on its boundary and the robbers can force the cops to revisit that vertex.   □

The natural definitions of monotonicity are equivalent but is monotonicity important? If there are r robbers in a tree T, we can modify Program 1 to let T' be the component containing the ith robber, where i is minimal among those robbers who have not yet been caught. This program catches the first robber (and any other robbers foolish enough to follow him), then the second, and so on. It is not monotone for more than one robber but wins against any number of robbers, still with only one cop.

The same technique can be applied to transform any search program for one robber (on an arbitrary collection of graphs), into a non-monotone program for any number of robbers. On the other hand, it is clear that monotonically searching for $r > 1$ robbers requires at least as many cops as does monotonically searching for a single robber. We summarize these observations in the following lemma.

Lemma 4. *For any graph G and any $r \geq 1$,*

$$\textbf{vams}(G, r) = \textbf{vams}(G, 1)$$
$$\textbf{mvams}(G, r) \geq \textbf{mvams}(G, 1) \,.$$

Thus, allowing non-monotone strategies may make it easier to search for many robbers. This raises the question of what is the cost of requiring monotonicity when facing a crime wave. Given a graph G and $r \geq 1$, what is the ratio $\textbf{mvams}(G, r)/\textbf{mvams}(G, 1)$? In Section 5, we give a full answer for trees and an upper bound for general graphs. We postpone this until we have established some necessary results in the next section.

4 Invisible Robbers

In this section, we give brief descriptions of two game variants where the single robber is invisible and the cops must, therefore, determine their moves without reference to the robbers' position. In one of the variants we consider, the robber is active (that is, he can move at every round); in the other, he is lazy (he can move only when a cop moves onto his vertex).

In both cases, as the robber is now invisible, the game is no longer interactive and the cops' moves may be given in advance as a *predetermined* strategy. Thus, we define a k-*strategy* for k cops to be any consistent sequence $\mathcal{S} = S_0, S_1, \dots$ of sets in $V(G)^{[\leq k]}$.

Given such a strategy, the free space for an invisible, active robber is the sequence given by $F_0 = \emptyset$ and

$$F_{i+1} = \{y \in V(G) - S_{i+1} \mid \text{there is an } (S_i, S_{i+1})\text{-avoiding}$$
$$x\text{-}y \text{ path for some } x \in F_i\} \,.$$

A strategy $\mathcal{S}$ is *monotone* if $F_{i+1} \subseteq F_i$ for all $i \geq 0$ and is *winning* if $F_i = \emptyset$ for some $i \geq 1$. Because the game is not interactive, plays do not feature in our analysis so we do not define them.

The *non-monotone* and *monotone, invisible, active mixed search numbers* are, respectively,

$$\mathbf{iams}(G) = \min\{k \mid \text{there is a winning } k\text{-strategy for } G\}$$
$$\mathbf{miams}(G) = \min\{k \mid \text{there is monotone winning } k\text{-strategy for } G\}.$$

It is known that $\mathbf{iams}(G) = \mathbf{miams}(G)$ [1]. It is not hard to see that searching for an invisible, active robber in a n-vertex graph G is equivalent to searching for n visible, active robbers. Informally, an invisible robber could be on any vertex in the free space but with n robbers, we may assume that there really is at least one robber on each vertex of the free space.

Lemma 5. *For any graph G of order n,* $\mathbf{miams}(G) = \mathbf{mvams}(G, n)$.

The case of an invisible, lazy robber is similar to the active case but with the difference that now, the robber moves only when a cop lands on his vertex. Define free space, k-strategies, monotonicity and winning against a lazy, invisible robber identically to the active case and write $\mathbf{milms}(G)$ and $\mathbf{ilms}(G)$ for the corresponding *monotone* and *non-monotone, invisible, lazy mixed search number*, respectively.

Lemma 6. *For any graph G,* $\mathbf{milms}(G) = \mathbf{mvams}(G, 1)$.

Proof (sketch). Given a monotone, winning $(k, 1)$-strategy μ for G, any concatenation of the essential parts of all possible μ-plays is a monotone winning k-strategy against an invisible, lazy robber.

For the converse, given a k-strategy $\mathcal{S} = S_0, S_1, \ldots, S_n$ against an invisible, lazy robber in G, the monotone, winning search program first makes the placement to the vertex in S_1. At each subsequent move, it removes a cop that is not on the boundary of the robber's free space, if any such cop exists; if not, it plays the first move of $\mathcal{S}$ that moves a cop (by placement or sliding) into the robber's free space. It can be shown that this is always possible and that the resulting strategy is both monotone and winning.   □

Define the *proper pathwidth* $\mathbf{ppw}(G)$ of a graph G to be the least k for which $G \preccurlyeq K_k \times P$ for some path P, where $K_k \times P$ is the graph formed from P by replacing the vertices with disjoint copies of K_k and adding a matching between each pair of cliques corresponding to adjacent vertices. Similarly, define the *proper treewidth* $\mathbf{ptw}(G)$ of a graph G to be the least k for which $G \preccurlyeq K_k \times T$ for some tree T. It is known that $\mathbf{miams}(G) = \mathbf{ppw}(G)$ and $\mathbf{milms}(G) = \mathbf{ptw}(G)$ (see, e.g., [6,9]). The following is straightforward.

Corollary 7. *For any graph G and $r \geq 1$,* $\mathbf{ptw}(G) \leq \mathbf{mvams}(G, r) \leq \mathbf{ppw}(G)$.

We will not formally consider multiple invisible robbers. For lazy robbers, we can consider either that all robbers may move whenever any robber is disturbed

---

Search Program 2. $\Pi(T, v, r)$ to monotonically capture r robbers in a tree T.

---

place(v).
Let $\mathbf{R} \leftarrow$ *robbers_positions()*.
While $V(\mathbf{R}) \neq \emptyset$,
    Let $T_1, \ldots, T_\ell$ be the connected components of $T - v$
        containing at least one and at most $\lfloor \frac{r}{2} \rfloor$ robbers.
    For $i \in \{1, \ldots, \ell\}$,
        Choose any vertex $v_i \in V(T_i)$.
        Let r_i be the number of robbers in T_i.
        Call $\Pi(T_i, v_i, r_i)$.
    Let $\mathbf{R} \leftarrow$ *robbers_positions()*.
    if $V(\mathbf{R}) \cap V(T) \neq \emptyset$ (i.e., robbers remain in T), then
        Let T' be the unique connected component of $T - v$ where $V(\mathbf{R}) \subseteq V(T')$
            and let w be the vertex of T' adjacent to v in T.
        slide($v \rightarrow w$).
        Let $v \leftarrow w$ and let $T \leftarrow T'$.
remove(w).

---

or that just the disturbed robbers may move. In both of these cases and the case for multiple active, invisible robbers, the search number for any graph G is either $\mathbf{ppw}(G)$ or $\mathbf{ptw}(G)$, depending only on the number of robbers. Thus, the classical equivalence of one lazy, invisible robber with one active, visible robber fails to hold when we move to multiple robbers. This is not the first case where the equivalence does not hold: it also fails for directed graphs, even with just one robber [7].

5   Catching Multiple Robbers

We now present our exact characterization of the number of cops required to monotonically catch r robbers in a tree T. Recall that a single cop suffices for one robber.

Lemma 8. *For any tree T and any $r \geq 1$,*

$$\mathbf{mvams}(T, r) \leq \min \{\mathbf{ppw}(G), \lfloor \log r \rfloor + 1\}.$$

Proof (sketch). The fact that $\mathbf{mvams}(T, r) \leq \mathbf{ppw}(G)$ follows from Corollary 7. To prove that $\mathbf{mvams}(T, r) \leq \lfloor \log r \rfloor + 1$, let $\Pi(T, r)$ be the search program that calls Program 2 with v assigned to be any vertex of T. It can be shown that $\Pi(T, r)$ is winning, monotone and uses at most $\lfloor \log r \rfloor + 1$ cops. $\square$

A *3-star composition* of disjoint, connected graphs G_1, G_2 and G_3 is any graph $\mathsf{Y}(G_1, G_2, G_3)$ formed by adding a new vertex v to $G_1 \cup G_2 \cup G_3$ and adding one edge from v to each of the three component graphs. The following lemma is the key technical result in this section and allows us to quickly deduce lower bounds for the search numbers of trees.

Lemma 9. *Let* $G = \mathsf{Y}(G_1, G_2, G_3)$, *where* $\mathbf{mvams}(G_i, \lfloor \frac{r}{2} \rfloor) \geq k$ *for each* $i \in \{1, 2, 3\}$. *Then,* $\mathbf{mvams}(G, r) > k$.

Proof (sketch). Suppose, towards a contradiction, that μ is a smooth, monotone, winning (k, r)-strategy for G. By smoothness, there is some vertex v such that the cops' first move is $\mathbf{place}(v)$, for any initial position of the robbers. We may assume, without loss of generality, that $v \notin V(G_1) \cup V(G_2)$ and that $\lceil \frac{r}{2} \rceil$ robbers go to G_1 and $\lfloor \frac{r}{2} \rfloor$ to G_2.

μ can be shown to induce a "nondeterministic strategy" (in which the cops may have more than one move in some positions) on G_1. By the inductive hypothesis, this requires at least k cops. Thus, on G_1, the robbers can always force k cops to be in the graph. We can also define a nondeterministic strategy μ_2 for G_2 such that every μ-play $\mathcal{P}$ corresponds to an interleaving of a μ_1-play $\mathcal{P}_1$, a μ_2-play $\mathcal{P}_2$ and moves outside G_1 and G_2.

For $i \in \{1, 2\}$, $\mathcal{P}_i$ may be assumed to have k cops in G_i at some point. Assuming, without loss of generality, that the robbers are cleared from G_1 before G_2, there is a point when all k cops are in G_1 so, in particular, there is no longer a cop on v. But this contradicts monotonicity because there are still robbers in G_2 and these robbers can now reach v, which they could not reach before.   □

Corollary 10. *For any tree* T, *and any* $r \geq 1$,

$$\mathbf{mvams}(T, r) \geq \min \{\mathbf{ppw}(T), \lfloor \log r \rfloor + 1\}.$$

Proof. Induction on $q = \lfloor \log r \rfloor + 1$. For the base case, $q = 1$, Program 1 shows that $\mathbf{mvams}(T, r) = 1 = q$. For the inductive step, suppose the result holds for all values smaller than q and let T be a tree. If $\mathbf{ppw}(T) = 1$, then T is a path and $\mathbf{mvams}(T, r) = 1$ for any r, as required. Otherwise, by [10], we may write $T = \mathsf{Y}(T_1, T_2, T_3)$ where, for each i, $\mathbf{ppw}(T_i) = \mathbf{ppw}(T) - 1$. By the inductive hypothesis, for each i, $\mathbf{mvams}(T_i, \lfloor \frac{r}{2} \rfloor) \geq \min \{\mathbf{ppw}(T) - 1, q - 1\}$. By Lemma 9, $\mathbf{mvams}(T, r) \geq \min \{\mathbf{ppw}(T) - 1, q - 1\} + 1 = \min \{\mathbf{ppw}(T), q\}$, as required.   □

This completes our characterization of the number of cops required to catch r robbers in a tree.

Theorem 11. *For any tree* T *and any* $r \geq 1$,

$$\mathbf{mvams}(T, r) = \min \{\mathbf{ppw}(T), \lfloor \log r \rfloor + 1\}.$$

For general graphs, we are able to give the following upper bound.

Theorem 12. *For any graph* G *and any* $r \geq 1$,

$$\mathbf{mvams}(G, r) \leq \min \{\mathbf{ppw}(G), \mathbf{ptw}(G) \cdot (\lfloor \log r \rfloor + 1)\}.$$

Proof (sketch). $\mathbf{mvams}(G, r) \leq \mathbf{ppw}(G)$ by Corollary 7. For the other case, let $q = \mathbf{ptw}(G)$. We have $G \preccurlyeq K_q \times T$ for some tree T and we can find r robbers in $K_q \times T$ by following the strategy for T but using q cops to cover the clique in $K_q \times T$ corresponding to a single vertex in T. The result follows by Lemma 8 and Proposition 2.   □

6   Conclusions and Open Problems

We have presented our results in the setting of mixed search (searching with placement, removal and sliding of cops). For node search (searching with only placement and removal), we can similarly define parameters $\mathbf{vans}(G, r)$ and $\mathbf{mvans}(G, r)$ for the general and monotone node search number for r visible, active robbers. Similarly, we can adapt all our other mixed-search parameters to node search. The difference between mixed search and node search is small: node search can be reduced to mixed search and requires at most one more cop.

We could, in principle, rewrite the present paper in terms of node search. Writing $\mathbf{pw}(G)$ and $\mathbf{tw}(G)$ for the well-known parameters of pathwidth and treewidth, it can be shown that, for a graph of order n, $\mathbf{vians}(G, 1) = \mathbf{tw}(G) + 1$ and $\mathbf{vians}(G, n) = \mathbf{pw}(G) + 1$ (using the results in [3, 8]). Our core results, in this setting, are that, for any tree T and graph G,

$$\mathbf{mvans}(T, r) = \min\left\{\mathbf{pw}(T), \lfloor \log r \rfloor + 1\right\} + 1$$
$$\mathbf{mvans}(G, r) \leq \min\left\{\mathbf{pw}(G) + 1, (\mathbf{tw}(G) + 1) \cdot (\lfloor \log r \rfloor + 2)\right\}.$$

The framework of this paper can also be applied to edge search. As this version can also be reduced to mixed search [1, 9] we make no further comments to this direction.

The problem settled in this paper can be stated in the following way: what is the maximum number of visible, active robbers that can be captured monotonically by k cops in some graph G? This number is unbounded if $k \geq \mathbf{ppw}(G)$ and, otherwise, 2^{k-1} robbers can be caught in a tree and at least $2^{k/\mathbf{ptw}(G)-1}$ in a general graph. This interpretation of our results may be useful for estimating how many sweeps of a graph a small number of cops need to catch a large number of robbers.

We identify three main open problems in graph searching for many robbers. The first is to find good lower bounds for $\mathbf{mvams}(G, r)$ in terms of G and r, for general graphs, corresponding to the bounds for trees found in this paper. We believe that this is a hard task as it appears to require the identification of obstructions for $\mathbf{mvams}(G, r)$ for all r.

Another open problem is to find graph decompositions corresponding to the game, tuning between (proper) tree decompositions, the case of one robber, and (proper) path decompositions, the case of one robber per vertex. It is unclear what form such a family of decompositions would take.

Finally, it would be interesting to know whether there is any relation between our results and the search game defined by Fomin et al.[5]. That game has only one robber but tunes between pathwidth and treewidth by limiting the number of rounds at which the cops may ask for the robber's position. This provides an alternative way of tuning interactivity: the game is fully interactive if the cops may ask for the robber's position at every move and fully predetermined if they may never ask for his position. Correspondingly, our game is fully interactive with a single robber and fully predetermined with a robber for each vertex.

References

1. Bienstock, D., Seymour, P.D.: Monotonicity in graph searching. J. Algorithms 12(2), 239–245 (1991)
2. Colin de Verdière, Y.: Multiplicities of eigenvalues and tree-width of graphs. J. Combin. Theory Ser. B 74(2), 121–146 (1998)
3. Dendris, N.D., Kirousis, L.M., Thilikos, D.M.: Fugitive-search games on graphs and related parameters. Theoret. Comput. Sci. 172(1–2), 233–254 (1997)
4. Ellis, J.A., Sudborough, I.H., Turner, J.S.: The vertex separation and search number of a graph. Information and Computation 113(1), 50–79 (1994)
5. Fomin, F.V., Fraigniaud, P., Nisse, N.: Nondeterministic graph searching: from pathwidth to treewidth. In: Jedrzejowicz, J., Szepietowski, A. (eds.) MFCS 2005. LNCS, vol. 3618, pp. 364–375. Springer, Heidelberg (2005)
6. Fomin, F.V., Thilikos, D.M.: Multiple edges matter when searching a graph. Unpublished manuscript
7. Hunter, P., Kreutzer, S.: Digraph measures: Kelly decompositions, games, and orderings. In: SODA. 18th ACM-SIAM Symp. on Disc. Algorithms, pp. 637–644 (2007)
8. Seymour, P.D., Thomas, R.: Graph searching and a min-max theorem for tree-width. J. Combin. Theory Ser. B 58(1), 22–33 (1993)
9. Stamatiou, Y.C., Thilikos, D.M.: Monotonicity and inert fugitive search games. Electronic Notes in Discrete Mathematics 3 (1999)
10. Takahashi, A., Ueno, S., Kajitani, Y.: Minimal acyclic forbidden minors for the family of graphs with bounded path-width. Disc. Math. 127(1–3), 293–304 (1994)
11. Takahashi, A., Ueno, S., Kajitani, Y.: Mixed searching and proper-path-width. Theoret. Comput. Sci. 137(2), 253–268 (1995)
12. Thilikos, D.M.: Algorithms and obstructions for linear-width and related search parameters. Discrete Applied Math. 105, 239–271 (2000)

Monotonicity of Non-deterministic Graph Searching[*]

Frédéric Mazoit[1] and Nicolas Nisse[2]

[1] LABRI, University of Bordeaux, 33405 Talence, France
[2] LRI, University of Paris Sud, 91405 Orsay, France

Abstract. In graph searching, a team of searchers is aiming at capturing a fugitive moving in a graph. In the initial variant, called *invisible graph searching*, the searchers do not know the position of the fugitive until they catch it. In another variant, the searchers know the position of the fugitive, i.e. the fugitive is visible. This latter variant is called *visible graph searching*. A search strategy that catches any fugitive in such a way that, the part of the graph reachable by the fugitive never grows is called *monotone*. A priori, monotone strategies may require more searchers than general strategies to catch any fugitive. This is however not the case for visible and invisible graph searching. Two important consequences of the monotonicity of visible and invisible graph searching are: (1) the decision problem corresponding to the computation of the smallest number of searchers required to clear a graph is in NP, and (2) computing optimal search strategies is simplified by taking into account that there exist some that never backtrack.

Fomin *et al.* (2005) introduced an important graph searching variant, called *non-deterministic graph searching*, that unifies visible and invisible graph searching. In this variant, the fugitive is invisible, and the searchers can query an oracle that knows the current position of the fugitive. The question of the monotonicity of non-deterministic graph searching is however left open.

In this paper, we prove that non-deterministic graph searching is monotone. In particular, this result is a unified proof of monotonicity for visible and invisible graph searching. As a consequence, the decision problem corresponding to non-determinisitic graph searching belongs to NP. Moreover, the exact algorithms designed by Fomin *et al.* do compute optimal non-deterministic search strategies.

Keywords: Graph searching, Treewidth, Monotonicity.

1 Introduction

Introduced in [14], graph searching is a game in which a team of searchers aims at catching a fugitive moving in a graph. At each step of the game, a searcher

---

[*] The second author received additional supports from the project "Alpage" of the ACI Masses de Données, from the project "Fragile" of the ACI Sécurité Informatique, and from the project "Grand Large" of INRIA.

A. Brandstädt, D. Kratsch, and H. Müller (Eds.): WG 2007, LNCS 4769, pp. 33–44, 2007.

can either be placed at or removed from a vertex of the graph [10]. The fugitive is invisible, arbitrary fast and aware of the positions of the searchers. It can move along paths of the graph as long as it does not cross any vertex occupied by a searcher. The fugitive is caught when a searcher is placed at the vertex it occupies, and it cannot flee because all the neighbors are occupied by searchers. A *search strategy* for a graph G is a sequence of basic operations, i.e., place or remove a searcher, that results in catching any invisible fugitive in G. The *node search number* of a graph G, denoted by $\mathbf{s}(G)$, is the smallest integer k such that it exists a search strategy for G using at most k searchers. Given a graph G, the *graph searching problem* consists in computing an *optimal* search strategy for G, i.e., a strategy that clears G using at most $\mathbf{s}(G)$ searchers.

During a search strategy, the vertices that are accessible by the fugitive are said *contaminated*. A non-contaminated vertex is said *clear*. A strategy is *monotone* if it does not allow *recontamination*, i.e., after having been cleared, a vertex remains clear until the end of the strategy. LaPaugh [11] proved that "recontamination does not help" to catch an invisible fugitive. That is, for any graph G, there exists a monotone search strategy of G using at most $\mathbf{s}(G)$ searchers. We say that invisible graph searching is monotone. LaPaugh's proof was later simplified by Bienstock and Seymour [4] using the concept of *crusades*. Both these proofs are constructive. Indeed, they transform any strategy into a monotone one without increasing the number of searchers.

In [17], Seymour and Thomas introduce the *visible graph searching*. In this variant [5,17], the searchers are aware of the position of the fugitive. Hence, they can adapt their strategy according to its position. The *visible search number* of a graph G, denoted by $\mathbf{vs}(G)$, is the smallest integer k such that k searchers are sufficient to catch any visible fugitive in G. Seymour and Thomas [17] proved that visible graph searching is monotone. However, Seymour and Thomas' proof is not constructive. They show that, if no monotone strategies using k searchers exist for a graph G, then there exists an escape strategy for the fugitive which actually is a general escape strategy, and thus, no non-monotone strategies using at most k searchers allow to catch any visible fugitive in G.

Monotonicity plays a crucial role in graph searching. First, a monotone strategy conludes in a polynomial number of steps and thus, gives a certificate of polynomial size for the decision problem corresponding to the graph searching problem. Since the decision problems corresponding to the visible and invisible graph searching problems are known to be NP-hard [12,17], they are NP-complete. Second, it appears algorithmically difficult to design strategies that are not monotone. Last but not least, monotone strategies for catching an invisible (resp., visible) fugitive in a graph G correspond exactly to path-decompositions (resp., tree-decompositions) [15] of G.

Indeed, the importance of visible graph searching and invisible graph searching comes from their close relationship with crucial notions of graph theory: *treewidth* and *pathwidth* [15]. Roughly speaking, the treewidth $\mathbf{tw}(G)$ (resp., the pathwidth $\mathbf{pw}(G)$) of a graph G measures how close this graph is from a tree (resp., a path). The correspondence between search numbers and width parameters provides different interpretations of these parameters, and thus, different ways of handling them. More precisely, $\mathbf{s}(G) = \mathbf{pw}(G) + 1$ (see [6,10]), and $\mathbf{vs}(G) = \mathbf{tw}(G) + 1$ (see [5,17]).

In [7], Fomin *et al.* provide a unique approach to the pathwidth and the treewidth of a graph. For any graph G and any $q \geq 0$, they define the notions of *q-branched tree decomposition* and *q-branched treewidth*, denoted by $\mathbf{tw}_q(G)$. Roughly speaking, a q-branched tree decomposition of a graph is a parametrized tree decomposition such that the number of branching nodes of the tree is limited. In particular, path-decompositions are exactly 0-branched tree decompositions, and tree-decompositions are exactly ∞-branched tree decompositions.

Fomin *et al.* also provide an interpretation of q-branched tree decompositions in terms of graph searching. More precisely, they provide a unique approach to both visible and invisible search problems, called *non-deterministic* graph searching. In this variant, the fugitive is invisible. However, the searchers can query an oracle that knows the position of the fugitive. Given the set $S \subseteq V(G)$ of clear vertices, a query returns a connected component C of $G \setminus S$, and all vertices in $G \setminus C$ are cleared. The choice of C is nondeterministic. Intuitively, the oracle gives the position of the fugitive to the searchers. More formally, the searchers can perform one of the following three basic operations called *search steps*: (1) place a searcher at a vertex of the graph, (2) remove a searcher from a vertex of the graph, and (3) perform a query to the oracle.

The number of query steps that the searchers can perform is however limited. For $q \geq 0$, the monotone q-limited search number $\mathbf{ms}_q(G)$ of a graph G is the smallest number of searchers required to catch any fugitive in G in a monotone way performing at most q queries. The main result of Fomin *et al.* [7] is the following generalization of the above two equations:

$$\text{For any graph } G \text{ and any } q \geq 0, \ \mathbf{tw}_q(G) + 1 = \mathbf{ms}_q(G). \tag{1}$$

Moreover, Fomin *et al.* [7] prove the NP-completness of the problem of computing $\mathbf{ms}_q(G)$, for any $q \geq 0$. Using the correspondence between monotone q-limited graph searching and q-branched treedecomposition, they also design an exact exponential algorithm that computes $\mathbf{tw}_q(G)$ and the corresponding decomposition, for any graph G and any $q \geq 0$.

However, Fomin *et al.* only consider monotone non-deterministic search strategies. They left open the problem whether recontamination helps for q-limited graph searching, for any $q \geq 0$. This paper answers this question.

1.1 Our Results

Let G be a graph and $q \geq 0$. Let $\mathbf{s}_q(G)$ denotes the smallest number of searchers required to catch any fugitive in G performing at most q queries. We prove that, for any graph G and any $q \geq 0$, recontamination does not help to catch a fugitive in G performing at most q queries. In other words, we prove that for any graph G and any $q \geq 0$, there exists a monotone search strategy of G using at most $\mathbf{s}_q(G)$ searchers, i.e. $\mathbf{s}_q(G) = \mathbf{ms}_q(G)$. In particular, this implies that the decision problem related to non-deterministic graph searching is in NP. This also implies that the exponential exact algorithm designed in [7] actually computes $s_q(G)$ for any graph G and any $q \geq 0$. More interestingly, our result unifies the proof of the monotonicity of invisible graph searching [4] and the proof of the monotonicity of visible graph searching [17]. The original proof of the monotonicity of visible graph searching is not constructive, while our proof is constructive and turns any general strategy into a monotone one.

1.2   Related Works

The monotonicity property of several graph searching variants has been studied before. In [2], Barrière *et al.* have defined the connected graph searching. A search strategy is *connected* if, at any step of the strategy, the subgraph induced by the clear vertices is connected. Barrière *et al.* [2] proved that connected graph searching is monotone as long as the input graph is restricted to be a tree. However, this does not remain true in case of arbitrary graphs. Yang *et al.* [18] proved that there exist graphs for which "recontamination does help" to catch an invisible fugitive in a connected way. In [8], Fraigniaud and Nisse proved that recontamination does help as well to catch a visible fugitive in a connected way.

In [9], Johnson *et al.* introduced *directed* graph searching. In this variant of the game, a visible fugitive is moving in a digraph. However, it is only permitted to move to vertices where there exists a directed searcher-free path from its intended destination back to its current position. The authors exhibit a graph for which recontamination does help. Obdržálek [13] and Berwanger *et al.* [3] independamently defined a new visible graph searching game in a digraph by relaxing the latter constraint. The question of the monotonicity of this latter variant is however left open. In [1], Barát studies the monotonicity property of a search strategy for catching invisible fugitive moving in a digraph. He proves that mixed-graph searching is monotone in directed graph.

2   Formal Definitions

In this paper, $G = (V, E)$ will denote a connected graph with vertex-set V and edge-set E. For $A \subseteq E$, we denote by $V[A]$ the set of vertices incident to at least one edge in A. The border of two disjoint edge sets A and B is the set $\delta(A, B) = V[A] \cap V[B]$ of the vertices incident both to an edge in A and to an edge in B. We extend this definition to any family of pair-wise disjoint edge sets $\{X_1, \ldots, X_p\}$ by setting $\delta(X_1, \cdots, X_n) = \bigcup_{1 \leq i < j \leq n} \delta(X_i, X_j)$. The *border* $\delta(X)$ of $X \subseteq E$ denotes the set $\delta(X, E \setminus X)$.

2.1   Non-deterministic Graph Searching

Now, we formally define the notion of non-deterministic search strategy. Intuitively, given a graph G, a non-deterministic search strategy (or simply a non-deterministic strategy) for G is a sequence of pairs, such that each pair consists of a subset of V, the positions of searchers, and a subset of E, the clear part of G. More precisely, a *non-deterministic strategy* is a sequence of ordered pairs $(Z_i, A_i)_{i \in [0,l]}$ such that

- for any $0 \le i \le l$, $Z_i \subseteq V$ and $A_i \subseteq E$;
- $Z_0 = \emptyset$ and $A_0 = \emptyset$;
- for any $0 \le i < l$ one of the following holds
  - (placing searchers) there is $X_{i+1} \subseteq V$, such that $Z_{i+1} = Z_i \cup X_{i+1}$ and $A_{i+1} = A_i \cup B_{i+1}$ with B_{i+1} the set of edges with both ends in Z_{i+1}, or
  - (removing searchers) there is $X_{i+1} \subseteq V$, such that $Z_{i+1} = Z_i \setminus X_{i+1}$ and A_{i+1} is obtained from A_i by recursively removing the edges $\{x, y\} \in A_i$ with $y \notin Z_{i+1}$ and such that there is $z \in V$ with $\{y, z\} \notin A_{i+1}$, or
  - (performing a query) $Z_{i+1} = Z_i$ and A_{i+1} is the set of edges not incident to a vertex of C, one of the connected component of $G \setminus Z_i$ not incident to a vertex of A_i. The choice of C is non-deterministic.

For any $0 \le i \le l$, (Z_i, A_i) is the *configuration* reached by the strategy at the ith step. A strategy $(Z_i, A_i)_{i \in [0,l]}$ uses at most $k \ge 1$ searchers if, for any $0 \le i \le l$, $|Z_i| \le k$. A *non-deterministic search program* is a non-deterministic program that takes as input a graph G and an integer $k \ge 1$, and returns a non-deterministic strategy for G using at most k searchers. A non-deterministic search program *wins* if for every possible fugitive moves, at least one of the strategies that the program computes catches the fugitive. That is, for any non-deterministic choice of the component C during the "performing a query" steps, the computed strategy insures that $A_l = E$. A non-deterministic search program is *monotone* if the strategies that it computes are monotone. The number of searchers required by a non-deterministic strategy is the maximum number of searchers required by the strategies that it computes.

A *q-limited non-deterministic search program* (or simply, a *q-program*) is a non-deterministic search program that computes strategies using at most q query steps. The *q-limited search number* (or simply the *q-search number*) of a graph G, denoted by $\mathbf{s}_q(G)$, is the smallest number of searchers required by a q-program to win against any fugitive in G. Similarily, we define the *monotone q-limited search number* of a graph G, denoted by $\mathbf{ms}_q(G)$, as the smallest number of searchers required by a monotone q-program to win against any fugitive in G.

If $q = 0$, no non-deterministic steps are allowed, and the previous definition is similar to the usual definition of an invisible search strategy [10]. Note that, in this case, the deterministic strategy $(Z_i, A_i)_{i \in [0,l]}$ wins, if and only if, there is $0 < i \le l$ such that, for any $j \ge i$, $A_j = E$.

2.2   Branched Treewidth

Fomin *et al.* [7] defined a parametrized version of the tree-decomposition of a graph. Their main result is the interpretation of this decomposition in terms of graph searching.

A *tree decomposition* [15] of graph G is a pair $(T, \mathcal{X})$ where T is a tree of node set I, and $\mathcal{X} = \{X_i, i \in I\}$ is a collection of subsets of $V(G)$ satisfying the following three conditions:

i. $V(G) = \cup_{i \in I} X_i$;
ii. for any edge e of G, there is a set $X_i \in \mathcal{X}$ containing both end-points of e;
iii. for any $i_1, i_2, i_3 \in I$ with i_2 is on the path from i_1 to i_3 in T, $X_{i_1} \cap X_{i_3} \subseteq X_{i_2}$.

The *width* $\mathbf{w}(T, \mathcal{X})$ of a tree decomposition is $\max_{i \in I}\{|X_i| - 1\}$ and the *treewidth* of a graph is the minimum width over all its tree-decompositions.

A *rooted* tree decomposition of a graph G, denoted by $(T, \mathcal{X}, r)$, is a tree decomposition $(T, \mathcal{X})$ of G such that T is a rooted tree and r is its root. A *branching node* of a rooted tree decomposition is a node with at least two children. For any $q \geq 0$, a *q-branched tree decomposition* [7] (or simply, a *q-tree decomposition*) of a graph G is a rooted tree decomposition $(T, \mathcal{X}, r)$ of G such that every path in T from the root r to a leaf contains at most q branching nodes. Thus a path decomposition rooted at one of its extremities is a 0-branched tree decomposition, and a usual tree decomposition is a ∞-branched tree decomposition. For any graph G, the *q-branched treewidth* (or simply, the *q-treewidth*) of G, denoted by $\mathbf{tw}_q(G)$, is the minimum width of any q-tree decomposition of G.

Theorem 1. *[7] Let G be a graph, $q \geq 0$ and $k \geq 1$. There is a winning monotone q-program using at most k searchers in G if and only if $\mathbf{tw}_q(G) < k$.*

2.3   Search-Tree

To prove the monotonicity of non-deterministic graph searching, we define an auxiliary structure called *search-tree* which is inspired by the tree-labelling defined by Robertson and Seymour [16].

A *search-tree* of a graph G is a triple (T, α, β) with T a tree, α a mapping from the incidence (between vertices and edges) of T into the subsets of E and β a mapping from the vertices of T into the subsets of E such that:

- for any edge $e = \{u, v\}$ of T, $\alpha(u, e) \cap \alpha(v, e) = \emptyset$;
- for no leaf v of T incident to an edge e of T is such that $\alpha(v, e) = E$;
- for any node v of T incident to $e_1, \ldots, e_p$, $\{\beta(v)\} \cup \mu(v)$ is a (possibly degenerated) partition of E with $\mu(v) = \{\alpha(v, e_1), \ldots, \alpha(v, e_p)\}$.

We extend the function β to any sub-tree T' of T by setting $\beta(T') = \cup_{v \in T'} \beta(v)$. The *width* of a search-tree is defined as $\mathbf{w}(T, \alpha, \beta) = \max_{v \in V(T)}\{|\chi_v|\}$ where $\chi_v = V[\beta(v)] \cup \delta(\mu(v))$ and $|\chi_v|$ denotes the *weight* of the node $v \in V(T)$. As for tree decompositions, we consider *rooted* search-trees, denoted by (T, α, β, r), that are search-trees over rooted trees. A *branching node* of a rooted search-tree is a node with at least two children. For any $q \geq 0$, a *q-branched* search-tree is a rooted search-tree (T, α, β, r) of G such that every path in T from the root r to a leaf contains at most q branching nodes. An edge $e = \{u, v\}$ of a search-tree is *monotone* if $\alpha(u, e) = E \setminus \alpha(v, e)$, and a search-tree is *monotone* if all its edges are monotone. Edges that are not monotone are said *dirty*.

3   Monotonicity of Non-deterministic Graph Searching

The remaining part of the paper is devoted to prove the monotonicity of non-deterministic graph searching. For this purpose, we prove that, from any winning

q-program using at most k searchers in a graph G, we can build a q-branched search-tree of width at most k for G (Lemma 1). Then, by performing local optimisations, we transform any q-search-tree into a monotone one (Lemma 3) without increasing its width. To conclude, any monotone q-branched search-tree, of width k, of a graph G can be transformed into a q-branched tree-decomposition, of width at most $k - 1$, of G (Lemma 5). Then, the proof of the monotonicity property of non-deterministic graph searching follows from Theorem 1. More formally, we prove the following theorem:

Theorem 2. *Let G be a connected graph, $q \geq 0$ and $k \geq 2$. The following are equivalent:*

 i. There is a winning q-program for G using at most k searchers;
 ii. There is a q-search-tree of width at most k for G;
 iii. There is a monotone q-search-tree of width at most k for G;
 iv. There is a q-tree decomposition of width at most $k - 1$ for G;
 v. There is a winning monotone q-program for G using at most k searchers.

Proof. We prove that $(i) \Rightarrow (ii)$ (Lemma 1), $(ii) \Rightarrow (iii)$ (Lemma 3), $(iii) \Rightarrow (iv)$ (Lemma 5). Proposition $(iv) \Rightarrow (v)$ follows from Theorem 1 and $(v) \Rightarrow (i)$ is obvious.

3.1   From Strategies to Search-Trees

Lemma 1. *Let G be a connected graph, $q \geq 0$ and $k \geq 2$, $(i) \Rightarrow (ii)$.*

 i. There is a winning q-program using at most k searchers for G;
 ii. There is a q-search-tree of width at most k for G.

Proof. In this proof, we consider extended search programs whose the *starting configuration* is not necessarily the $(\emptyset, \emptyset)$ configuration. That is, we consider search programs whose strategies start from a configuration (Z_0, A_0) that satisfies $\delta(A_0) \subseteq Z_0$. The *length* of a search program is the maximum number of steps of the strategies it computes. Let us define the *partial width* of a rooted search tree as the maximum weight of its nodes, the maximum being taken over all the nodes of the search-tree but its root. We prove the following claim by induction on the length of the search program.

Claim. For every winning q-program using at most k searchers with (Z_0, A_0) as starting configuration, there is a rooted q-search-tree (T, α, β, r) of partial width at most k, and such that, r is incident to a unique edge $e \in E(T)$, and $\alpha(r, e) = E \setminus A_0$.

Let $q \geq 0$ and let $\mathcal{S}$ be a q-program on G with k searchers and with (Z_0, A_0) as starting configuration.

• Suppose that $\mathcal{S}$ has length 1.
  The only search step has to be a "placing searchers" step. Thus, $\mathcal{S}$ computes only the following 0-strategy: $(Z_0, A_0), (Z_1, A_1)$ in which $Z_1 = Z_0 \cup X_1$ and $A_1 = A_0 \cup B_1 = E$.

Define the tree T with only one edge $\{r, v\}$, $\beta(v) = \alpha(r, \{r, v\}) = E \setminus A_0$ and $\beta(r) = \alpha(v, \{r, v\}) = A_0$. Since $V[\beta(v)] \cup \delta(\mu_v) = V[E \setminus A_0]$ which is a subset of Z_1, (T, α, β, r) is a rooted 0-search tree of partial width at most k.

- Suppose that S has length $l > 1$. Let us assume that for any winning q-program S' using at most k searchers with (Z, A) as starting configuration (with $\delta(A) \subseteq Z$) and such that S' has length $l' < l$, there is a rooted q-search-tree (T, α, β, r) of partial width at most k, and such that, r is incident to a unique edge $e \in E(T)$, and $\alpha(r, e) = E \setminus A$. Consider S' obtained by removing the first configuration of the sequences of S. Note that, S' is strictly shorter than S. We consider three cases according to the type of the first step of S.

  a. if the first step of S is a "removing searchers" step, S' is a q-program with (Z_1, A_1) as a starting configuration, $Z_1 \subseteq Z_0$ and $A_1 \subseteq A_0$. According to the induction hypothesis, there is a rooted q-search-tree $(T', \alpha', \beta', r')$ of partial width at most k and such that there is an edge e' incident to r' with $\alpha'(r', e') = E \setminus A_1$.

     Define a new q-search-tree (T, α, β, r) from $(T', \alpha', \beta', r')$ as follows:
     - add a new leaf r linked to r' in T', and set r as the new root,
     - put $\alpha(r, \{r, r'\}) = E \setminus A_0$, $\alpha(r', \{r, r'\}) = A_1$ and $\alpha = \alpha'$ otherwise;
     - put $\beta(r) = A_0$, $\beta(r') = \emptyset$ and $\beta = \beta'$ otherwise.

     Since $A_1 \subseteq A_0$, $\alpha(r, \{r, r'\}) \cap \alpha(r', \{r, r'\}) = \emptyset$ and (T, α, r) is a q-search-tree. Moreover, $V[\beta(r')] \cup \delta(\mu(r')) \subseteq Z_1$ and (T, α, β, r) satisfies the required conditions.

  b. if the first step of S is a "placing searchers" step, S' is a q-program with (Z_1, A_1) as a starting configuration, $Z_1 = Z_0 \cup X_1$ and $A_1 = A_0 \cup B_1$. According to the induction hypothesis, there is a rooted q-search-tree $(T', \alpha', \beta', r')$ of partial width at most k and such that there is an edge e' incident to r' with $\alpha'(r', e') = E \setminus A_1$.

     Define a new q-search-tree (T, α, β, r) from $(T', \alpha', \beta', r')$ as follows:
     - add a new leaf r linked to r' in T', and set r as the new root,
     - put $\alpha(r, \{r, r'\}) = E \setminus A_0$, $\alpha(r', \{r, r'\}) = A_0$ and $\alpha = \alpha'$ otherwise;
     - set $\beta(r) = A_0$, $\beta(r') = B_1$ and $\beta = \beta'$ otherwise.

     By construction, (T, α, β, r) is a q-search-tree that satisfies the required conditions.

  c. if the first step of S is a "performing a query" step, there are $p \geq 1$ distinct $(q-1)$-programs $S_1, \ldots, S_p$ for G such that: $\{A_0, E \setminus Y_1, \ldots, E \setminus Y_p\}$ is a partition of E, and, for any $1 \leq i \leq p$, S_i is a winning $(q-1)$-program for G, starting from the configuration (Z_i, Y_i) and using at most k searchers. For any $1 \leq i \leq p$, since the $(q-1)$-programs S_i are shorter than S, there exists a rooted $(q-1)$-search-tree $(T_i, \alpha_i, \beta_i, r_i)$ of partial width at most k, and such that there is an edge e_i incident to r_i with $\alpha_i(r_i, e_i) = E \setminus Y_i$.

     Define a new q-search-tree (T, α, β, r) from these search-trees as follow:
     - identify the roots r_i with a node r', add a new leaf r linked to r' in T', and set r as the new root,.
     - put $\alpha(r, \{r, r'\}) = E \setminus A_0$, $\alpha(r', \{r, r'\}) = A_0$ and $\alpha(u, e) = \alpha_i(u, e)$ for every edge e of T_i;

– put $\beta(r) = A_0$, $\beta(r') = \emptyset$, and, for any $1 \leq i \leq p$ and any node u of T_i, $\beta(u) = \beta_i(u)$.

The rooted search-tree (T, α, β, r) has one more branching node than any search-tree $(T_i, \alpha_i, \beta_i, r_i)$ and, since each of them has at most $q - 1$ branching nodes, (T, α, β, r) satisfies the required conditions.

Therefore, for any winning q-program S using at most k searchers with (Z_0, A_0) as starting configuration (with $\delta(A_0) \subseteq Z_0$) and such that S has length l, there is a rooted q-search-tree (T, α, β, r) of partial width at most k, and such that, r is incident to a unique edge $e \in E(T)$, and $\alpha(r, e) = E \setminus A_0$. This concludes the induction and the proof of the claim.

To conclude the proof, is it sufficient to note that, if $A_0 = \emptyset$, the weight of the root of the search-tree equals 0. Thus, its partial width equals its width.

3.2   From Search-Trees to Monotone Search-Trees

To prove the second step of the proof, we need the following technical lemma.

Lemma 2. *Let $G = (V, E)$ be a connected graph, $\mu = \{E_1, \ldots, E_p\}$ be a (possibly degenerated) partition of E and $F \subseteq E \setminus E_1$. Set $E_1' = E \setminus F$, $E_i' = E_i \cap F$ for $2 \leq i \leq p$ and $\mu' = \{E_1', \ldots, E_p'\}$.*

$$If \ |\delta(F)| \leq |\delta(E_1)| \ then \ |\delta(\mu')| \leq |\delta(\mu)|$$
$$If \ |\delta(F)| < |\delta(E_1)| \ then \ |\delta(\mu')| < |\delta(\mu)|$$

Proof. Since $\delta(E_1) \subseteq \delta(\mu)$ and $\delta(F) \subseteq \delta(\mu')$, we get that $|\delta(\mu)| = |\delta(\mu) \setminus \delta(E_1)| + |\delta(E_1)|$ and $|\delta(\mu')| = |\delta(\mu') \setminus \delta(F)| + |\delta(F)|$. This implies that

$$|\delta(\mu)| - |\delta(\mu')| = \Big(|\delta(\mu) \setminus \delta(E_1)| + |\delta(E_1)|\Big) - \Big(|\delta(\mu') \setminus \delta(F)| + |\delta(F)|\Big)$$
$$= |\delta(E_1)| - |\delta(F)| + \Big(|\delta(\mu) \setminus \delta(E_1)| - |\delta(\mu') \setminus \delta(F)|\Big)$$

To complete the proof, it is sufficient to show that $\delta(\mu') \setminus \delta(F) \subseteq \delta(\mu) \setminus \delta(E_1)$.

To prove this latter assertion, first note that any vertex $w \in \delta(E_1) \cap \delta(\mu')$ belongs to $\delta(F)$. Indeed, $w \in \delta(\mu')$ implies, by definition of μ', the existence of $e_1 \in F$ incident to w. Beside, $w \in \delta(E_1)$ implies the existence of $e_2 \in E_1$ incident to w. Since $E_1 \subseteq E \setminus F$, we have $e_1 \in F$ and $e_2 \notin F$. Therefore, $w \in \delta(F)$. Hence, we obtain that $\big(\delta(\mu') \setminus \delta(F)\big) \cap \delta(E_1) = \emptyset$. Finally, since $\delta(\mu') \setminus \delta(F) \subseteq \delta(\mu)$, it implies that $\delta(\mu') \setminus \delta(F) \subseteq \delta(\mu) \setminus \delta(E_1)$. That concludes the proof.

Lemma 3. *Let $n > 0$. Let G be a n-node connected graph, $q \geq 0$ and $k \geq 2$, $(ii) \Rightarrow (iii)$.*

ii. *There is a q-search-tree of width k on G;*
iii. *There is a monotone q-search-tree of width k on G.*

Proof. Let $\mathcal{T} = (T, \alpha, \beta, r)$ be a rooted q-search-tree of G of width k.

For every edge e of T, denote by **dist**(e) the *distance* of e to the root r. The *weight* **wg**$(\mathcal{T})$ of $\mathcal{T}$ is $\sum_{v \in V(T)} |\delta(\mu_\alpha(v))|$ and the *badness* **bn**$(\mathcal{T})$ of $\mathcal{T}$ is

$\sum n^{-\mathbf{dist}(e)}$ the sum being taken over the dirty edges of $\mathcal{T}$. Let $\mathcal{T}_1$ and $\mathcal{T}_2$ be two rooted q-search-trees. $\mathcal{T}_1$ is *tighter* than $\mathcal{T}_2$ if either $\mathbf{wg}(\mathcal{T}_1) < \mathbf{wg}(\mathcal{T}_2)$, or $\mathbf{wg}(\mathcal{T}_1) = \mathbf{wg}(\mathcal{T}_2)$ and $\mathbf{bn}(\mathcal{T}_1) < \mathbf{bn}(\mathcal{T}_2)$.

The remaining part of this lemma is devoted to prove that the tightest search-tree among any q-search-tree of width k of G is monotone. For this purpose, we make local optimisations that are compatible with the above relation. Let $\{u, v\}$ be a dirty edge of $\mathcal{T}$.

a. Let us assume that $\left| \delta\big(\alpha(u, \{u, v\})\big) \right| < \left| \delta\big(\alpha(v, \{u, v\})\big) \right|$.

   - Let us assume that v is a leaf. If $\alpha(u, \{u, v\}) = \emptyset$, just remove the leaf (by setting u as the new root, if $r = v$). Otherwise, set $\alpha(v, \{u, v\}) = E \setminus \alpha(u, \{u, v\})$ and $\beta(v) = \alpha(u, \{u, v\})$. The resulting search-tree is tighter than $\mathcal{T}$.

   - Now, let us assume that v is an internal node of $\mathcal{T}$. Set $u = u_1$, let $u_2, \ldots, u_p$ be the other neighbours of v, set $\alpha(v, \{v, u_i\}) = E_i$, $\mu_v = \{E_1, \ldots, E_p\}$ and $F = \alpha(u, \{u, v\})$ so that the condition on $\{u, v\}$ can be rephrased as $\left| \delta(E_1) \right| > \left| \delta(F) \right|$. Let us modify $\mathcal{T}$ by setting $\beta(v) = \beta(v) \cap F$, $\alpha(v, vu_i) = E_i'$ for $1 \le i \le p$. Since, $E_i' \subseteq E_i$ for $2 \le i \le p$ and $E_1' = E \setminus F$, we obtain a new q-search-tree $\mathcal{T}'$. It remains to prove that $\mathcal{T}'$ is tighteer than $\mathcal{T}$. Let η_v be the partition $\{E_1, \ldots, E_p, \beta(v)\}$ of E. Consider $\eta_v' = \{E_1', \ldots, E_p', \beta(v) \cap F\}$ with $E_1' = E \setminus F$, and $E_i' = E_i \cap F$ for $2 \le i \le p$. By lemma 2, $\left| \delta(\eta_v') \right| < \left| \delta(\eta_v) \right|$.

     Beside, $\left| \chi_{\mathcal{T}}(v) \right| = \left| \delta(\mu_v) \cup V[\beta(v)] \right| = \left| \delta(\eta_v) \cup (V[\beta(v)] \setminus \delta(\beta(v))) \right| = \left| \delta(\eta_v) \right| + \left| V[\beta(v)] \setminus \delta(\beta(v)) \right| > \left| \delta(\eta_v') \right| + \left| V[\beta(v) \cap F] \setminus (\delta(\beta(v) \cap F)) \right| = \left| \delta(E_1', \ldots, E_p') \cup V[\beta(v) \cap F] \right| = \left| \chi_{\mathcal{T}}'(v) \right|$. Thus, $\mathcal{T}'$ has strictly smaller weight than $\mathcal{T}$. Therefore, $\mathcal{T}'$ is tighter than $\mathcal{T}$.

b. Let us assume that $\left| \delta\big(\alpha(u, \{u, v\})\big) \right| = \left| \delta\big(\alpha(v, \{u, v\})\big) \right|$. We can suppose without loss of generality that u is closer to the root r than v.

   - Let us assume that v is a leaf. We set $\alpha(v, \{u, v\}) = E \setminus \alpha(u, \{u, v\})$ and $\beta(v) = \alpha(u, \{u, v\})$. The resulting search-tree $\mathcal{T}'$ is such that $\mathbf{wg}(\mathcal{T}') \le \mathbf{wg}(\mathcal{T})$, and has smaller badness. Thus, $\mathcal{T}'$ is tighter than $\mathcal{T}$.

   - Now, let us assume that v is an internal node of $\mathcal{T}$. We consider exactly the same new q-search-tree $\mathcal{T}'$ as in the second item of case a. The only difference is that using lemma 2, we only get $\left| \delta(\eta_v') \right| \le \left| \delta(\eta_v) \right|$ and thus $\mathbf{wg}(\mathcal{T}') \le \mathbf{wg}(\mathcal{T})$. However, in $\mathcal{T}'$, the edge $\{u, v\}$ is monotone. Moreover, the only edges that were monotone in $\mathcal{T}$, and that could have become dirty are the edges $\{v, u_i\}$ for $2 \le i \le p$. Since $p \le n + 1$ and $\mathbf{dist}(\{v, u_i\}) = \mathbf{dist}(\{v, u\}) + 1$ for $2 \le i \le p$, we have

$$\mathbf{bn}(\mathcal{T}) - \mathbf{bn}(\mathcal{T}') \ge n^{-\mathbf{dist}(\{u,v\})} - \sum_{i=2}^{p} n^{-\mathbf{dist}(\{v,u_i\})}$$

$$\ge n^{-\mathbf{dist}(\{u,v\})} - (n-1)n^{-\mathbf{dist}(\{v,u\})-1} > 0$$

     The q-search-tree $\mathcal{T}'$ is tighter than $\mathcal{T}$.

If a q-search-tree of width k has a dirty edge, we can algorithmically turn it into a new q-search-tree of width at most k which is tighter. Since there are no infinitely decreasing sequences for this relation, there exists a monotone q-search-tree of width at most k.

3.3   From Monotone Search-Trees to Tree Decompositions

The two following lemmas conclude the third step of the proof of Theorem 2.

Lemma 4. *Let G be a connected graph and $\mathcal{T} = (T, \alpha, \beta, r)$ be a monotone search-tree on G. For any edge $\{u, v\}$ of T, $\alpha(u, \{u, v\}) = \beta(T_v)$ with T_v the connected component of $T \setminus \{u, v\}$ that contains v.*

Proof. We prove this by induction of $|V(T_v)|$.

- if $|V(T_v)| = 1$, then $\beta(v) = E \setminus \alpha(v, \{u, v\})$ and since $\alpha(u, \{u, v\}) = E \setminus \alpha(v, \{u, v\})$ ($\mathcal{T}$ is monotone), we have $\alpha(u, \{u, v\}) = \beta(v) = \beta(T_v)$.
- otherwise, let $w_1, \ldots, w_p$ be the neighbours of v in T_v and for $1 \leq i \leq p$, let T_{w_i} be the connected components of $T_v \setminus \{v, w_i\}$ that contains w_i. By induction hypothesis, $\alpha(v, \{v, w_i\}) = \beta(T_{w_i})$. Since $\mathcal{T}$ is a search-tree, the sets $\beta(v)$, $\alpha(v, \{u, v\})$ and $\beta(T_{w_1}), \ldots, \beta(T_{w_p})$ induce a partition of E, thus $\alpha(v, \{u, v\}) = E \setminus \beta(T_v)$. Since $\mathcal{T}$ is monotone, $\alpha(u, \{u, v\}) = E \setminus \alpha(v, \{u, v\}) = \beta(T_v)$ which finishes the proof.

Lemma 5. *Let G be a connected graph, $q \geq 0$ and $k \geq 2$, $(iii) \Rightarrow (iv)$.*

iii. There is a monotone q-search-tree of width k on G;
iv. There is a q-tree decomposition of width at most $k - 1$ on G.

Proof. Let $\mathcal{T} = (T, \alpha, \beta, r)$ be a q-search-tree of width k.

We claim that $\Theta = (T, \mathcal{X}, r)$ with $\mathcal{X} = \{\chi(v) \mid v \text{ node of } T\}$ is a tree decomposition of width at most $k - 1$. Since G is connected and $|E| > 0$, condition ii. of a tree decomposition implies condition i.

Let $\{x, y\} \in E$ be an edge of G. Since $\mathcal{T}$ is monotone, for every edge $\{u, v\}$ of T, $\{x, y\}$ belongs to either $\alpha(u, \{u, v\})$ or $\alpha(v, \{u, v\})$. Suppose $\{x, y\} \in \alpha(u, \{u, v\})$, by lemma 4, $\{x, y\} \in \beta(T_v)$ with T_v the connected component of $T \setminus \{u, v\}$ that contains v. The edge $\{x, y\}$ thus belongs to at least one $\beta(w)$ for some node w of T_v. By definition of $\chi(w)$, $\{x, y\} \subseteq \chi(w)$.

Let u, v, w be three nodes of T with v on the path $\{u, u', \ldots, v, \ldots, w', w\}$ from u to w. Let T_u (resp., T_w) be the connected component of $T \setminus \{u, u'\}$ (resp., $T \setminus \{w, w'\}$) that contains u (resp., w). Let T_u^v (resp., T_w^v) be the connected component of $T \setminus v$ that contains u (resp., w).

Let $u_1 = u', \ldots, u_p$ be the neighbours of u in T and $x \in \chi(u)$. Either there is an edge of G incident to x in $\beta(u)$, or there exist $1 < i \leq p$ such that there is an edge incident to x in $\alpha(u, \{u, u_i\})$. By lemma 4, there is an edge incident to x in $\beta(T_{u_i}) \subseteq \beta(T_u)$.

Suppose that $x \in \chi(u) \cap \chi(w)$. There exists an edge incident to x in $\beta(T_u^v) \supseteq \beta(T_u)$ and an edge incident to x in $\beta(T_w^v) \supseteq \beta(T_w)$. By lemma 4, we get that $x \in \delta(\mu_v)$. Thus, $x \in \chi(v)$. This proves that Θ is a tree-decomposition. Moreover, by construction, $\mathbf{w}(\Theta) = \mathbf{w}(T, \alpha, \beta) - 1$. Since both $\mathcal{T}$ and Θ use the same underlying three, Θ is a q-tree decomposition of width at most $k - 1$.

References

1. Barát, J.: Directed Path-width and Monotonicity in Digraph Searching. Graphs and Combinatorics 22(2), 161–172 (2006)
2. Barrière, L., Flocchini, P., Fraigniaud, P., Santoro, N.: Capture of an intruder by mobile agents. In: SPAA 2002. Proceedings of the 14th Annual ACM-SIAM Symposium on Parallel Algorithms and Architectures, pp. 200–209 (2002)
3. Berwanger, D., Dawar, A., Hunter, P.W., Kreutzer, S.: Dag-width and Parity Games. In: Durand, B., Thomas, W. (eds.) STACS 2006. LNCS, vol. 3884, pp. 524–536. Springer, Heidelberg (2006)
4. Bienstock, D., Seymour, P.D.: Monotonicity in graph searching. Journal Algorithms 12(2), 239–245 (1991)
5. Dendris, N.D., Kirousis, L.M., Thilikos, D.M.: Fugitive-search games on graphs and related parameters. Theoretical Computer Science 172(1–2), 233–254 (1997)
6. Ellis, J.A., Sudborough, I.H., Turner, J.S.: The Vertex Separation and Search Number of a Graph. Information and Computation 113(1), 50–79 (1994)
7. Fomin, F.V., Fraigniaud, P., Nisse, N.: Nondeterministic Graph Searching: From Pathwidth to Treewidth. In: Jedrzejowicz, J., Szepietowski, A. (eds.) MFCS 2005. LNCS, vol. 3618, pp. 364–375. Springer, Heidelberg (2005)
8. Fraigniaud, P., Nisse, N.: Monotony Properties of Connected Visible Graph Searching. In: Fomin, F.V. (ed.) WG 2006. LNCS, vol. 4271, pp. 229–240. Springer, Heidelberg (2006)
9. Johnson, T., Robertson, N., Seymour, P.D., Thomas, R.: Directed Tree-Width. Journal of Combinatorial Theory Series B 82(1), 138–155 (2001)
10. Kirousis, L.M., Papadimitriou, C.H.: Searching and pebbling. Theoretical Computer Science 47(2), 205–218 (1986)
11. LaPaugh, A.S.: Recontamination does not help to search a graph. Journal of the ACM 40(2), 224–245 (1993)
12. Megiddo, N., Hakimi, S.L., Garey, M.R., Johnson, D.S., Papadimitriou, C.H.: The complexity of searching a graph. Journal of the ACM 35(1), 18–44 (1988)
13. Obdrzálek, J.: DAG-width: connectivity measure for directed graphs. In: SODA 2006. Proceedings of the 17th Annual ACM-SIAM Symposium on Discrete Algorithms, pp. 814–821 (2006)
14. Parsons, T.D.: Pursuit-evasion in a graph. Theory and Applications of Graphs, 426–441 (1976)
15. Robertson, N., Seymour, P.D.: Graph Minors. II. Algorithmic aspects of tree-width. Journal of Algorithms 7(3), 309–322 (1986)
16. Robertson, N., Seymour, P.D.: Graph Minors. X. Obstructions to Tree-Decomposition. Journal of Combinatorial Theory Series B 52(2), 153–190 (1991)
17. Seymour, P.D., Thomas, R.: Graph Searching and a Min-Max Theorem for Tree-Width. Journal of Combinatorial Theory Series B 58(1), 22–33 (1993)
18. Yang, B., Dyer, D., Alspach, B.: Sweeping Graphs with Large Clique Number. In: Fleischer, R., Trippen, G. (eds.) ISAAC 2004. LNCS, vol. 3341, pp. 908–920. Springer, Heidelberg (2004)

Tree-Width and Optimization in Bounded Degree Graphs

Vadim Lozin[1,*] and Martin Milanič[2]

[1] Mathematics Institute, University of Warwick, Coventry CV4 7AL, UK
V.Lozin@warwick.ac.uk
[2] Universität Bielefeld, Technische Fakultät, D-33501 Bielefeld, Germany
mmilanic@cebitec.uni-bielefeld.de

Abstract. It is well known that boundedness of tree-width implies polynomial-time solvability of many algorithmic graph problems. The converse statement is generally not true, i.e., polynomial-time solvability
does not necessarily imply boundedness of tree-width. However, in graphs
of bounded vertex degree, for some problems, the two concepts behave
in a more consistent way. In the present paper, we study this phenom-
enon with respect to three important graph problems – dominating set,
independent dominating set and induced matching – and obtain several results toward revealing the equivalency between boundedness of the
tree-width and polynomial-time solvability of these problems in bounded
degree graphs.

Keywords: Tree-width; Hereditary class of graphs; Dominating set; Induced Matching.

1 Introduction

It is well known that boundedness of tree-width implies polynomial-time solvability of many algorithmic graph problems that are NP-hard in general [2], such
as INDEPENDENT SET, DOMINATING SET, INDEPENDENT DOMINATING SET, IN-
DUCED MATCHING, etc. These problems can also be solved in polynomial time
in some classes of graphs that are not necessarily of bounded tree-width. For
instance, INDEPENDENT SET admits a polynomial-time solution in the class of
claw-free graphs [19], DOMINATING SET and INDEPENDENT DOMINATING SET
are solvable for (claw, net)-free graphs [6] and convex bipartite graphs [8], and
INDUCED MATCHING for co-comparability graphs [10]. An interesting observation is that the tree-width of claw-free graphs remains unbounded even with the
additional restriction to graphs of bounded vertex degree, while for the other
listed examples this is not the case (see more about tree-width of bounded degree graphs in [16,17]). This observation raises the question of the relationship
between boundedness of the tree-width and polynomial-time solvability of some
optimization problems in bounded degree graphs. In the present paper, we study

---

* This author gratefully acknowledges the support of DIMAP – Centre for Discrete
Mathematics and its Applications at the University of Warwick.

A. Brandstädt, D. Kratsch, and H. Müller (Eds.): WG 2007, LNCS 4769, pp. 45–54, 2007.
© Springer-Verlag Berlin Heidelberg 2007

this question with respect to DOMINATING SET, INDEPENDENT DOMINATING SET, and INDUCED MATCHING problems. All our results refer to graph classes that are *hereditary* in the sense that whenever a graph belongs to a class, all induced subgraphs of the graph belong to the same class. Any hereditary class can be described by a unique set of minimal graphs that do not belong to the class, so-called *forbidden induced subgraphs*. The class of graphs containing no induced subgraphs from a set M will be denoted $Free(M)$. Our objective in this paper is to reveal restrictions on the set M that would imply either boundedness of tree-width and polynomial-time solvability of the above problems, or unboundedness of tree-width and NP-hardness of the problems. To this end, we first identify in Section 2 two types of restrictions under which the tree-width is unbounded and the three problems in question are NP-hard. Then in Section 3 by violating those restrictions we discover several areas where the tree-width is bounded and hence the problems are solvable in polynomial time. In the rest of the present section we give necessary definitions and notations.

Most notations we use are customary: $V(G)$ and $E(G)$ denote the vertex set and the edge set of a graph G, respectively. The degree of a vertex v is the number of edges incident with v. By $\Delta(G)$ we denote the maximum vertex degree in G. The *chordality* of a graph G is the size of a largest chordless cycle in G. Given a subset of vertices $U \in V(G)$, we denote by $G - U$ the subgraph induced by $V(G) - U$. As usual, C_n and P_n denote the chordless cycle and the chordless path on n vertices, respectively. For a graph $G = (V, E)$, the *line graph* of G, denoted $L(G)$, is the graph whose vertex set is E, and whose two vertices are adjacent if and only if they share a common vertex as edges of G.

A *dominating set* in a graph is a subset of vertices such that every vertex outside the subset has a neighbor in it. The DOMINATING SET problem is that of finding in a graph a dominating set of minimum cardinality.

A dominating set is *independent* if no two vertices of the set are connected by an edge. The INDEPENDENT DOMINATING SET problem is to find in a graph an independent dominating set of minimum cardinality.

A *matching* in a graph is a subset of edges no two of which have a vertex in common. A matching is *induced* if no two endpoints of different edges of the matching are connected by an edge. The INDUCED MATCHING problem asks to find in a graph an induced matching of maximum size.

2   Classes of Graphs with Unbounded Tree-Width and NP-Hard Problems

Let us denote by H_i the graph on the left of Figure 1 and by $\mathcal{S}_k$ the class of all $(C_3, \ldots, C_k, H_1, \ldots, H_k)$-free graphs of vertex degree at most 3.

The following lemma combines four results proved in [5,13,14,18].

Lemma 1. *For any $k \geq 3$, in the class $\mathcal{S}_k$ the tree-width is unbounded and the following problems are NP-hard:* DOMINATING SET, INDEPENDENT DOMINATING SET, INDUCED MATCHING.

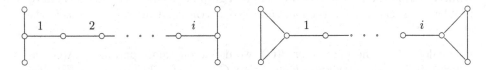

Fig. 1. Graphs H_i (left) and L_i (right)

To state the main result of this section, let us associate with every graph G the following parameter: $\kappa(G)$ is the maximum k such that $G \in \mathcal{S}_k$. If G belongs to no class $\mathcal{S}_k$, we define $\kappa(G)$ to be 0, and if G belongs to all classes $\mathcal{S}_k$, then $\kappa(G)$ is defined to be ∞. Also, for a set of graphs M, we define $\kappa(M) = \sup\{\kappa(G) : G \in M\}$. With these definitions in mind, we can now prove the following result.

Theorem 1. *Let X be the class of M-free graphs of vertex degree at most 3. If $\kappa(M) < \infty$, then in the class X the tree-width is unbounded and the following problems are NP-hard:* DOMINATING SET, INDEPENDENT DOMINATING SET, INDUCED MATCHING.

Proof. To prove the theorem, we will show that there is a k such that $\mathcal{S}_k \subseteq X$. Denote $k := \kappa(M) + 1$ and let G belong to $\mathcal{S}_k$. Assume that G does not belong to X. Then G contains a graph $A \in M$ as an induced subgraph. From the choice of G we know that A belongs to $\mathcal{S}_k$, but then $k \leq \kappa(A) \leq \kappa(M) < k$, a contradiction. Therefore, $G \in X$ and hence, $\mathcal{S}_k \subseteq X$.   □

In what follows, we present another theorem of a similar nature. Again, we start with a preparatory lemma. Denote the class of line graphs of graphs in $\mathcal{S}_k$ by $\mathcal{T}_k$.

Lemma 2. *For any $k \geq 3$, in the class $\mathcal{T}_k$ the tree-width is unbounded and the following problems are NP-hard:* DOMINATING SET, INDEPENDENT DOMINATING SET, INDUCED MATCHING.

Proof. For the tree-width, the statement of this lemma is a consequence of Lemma 1 and the following two facts: first, the relationship connecting the tree-width of a graph G with the clique-width of its line graph

$$(\mathrm{tw}(G) + 1)/4 \leq \mathrm{cw}(L(G)) \leq 2\mathrm{tw}(G) + 2$$

proved in [11], and second, the fact that the clique-width of a graph is bounded above by a function of its tree-width [7].

   To prove the lemma for the dominating set problem, we observe that finding a minimum dominating set in the line graph of G is equivalent to finding a minimum edge dominating set in G. It was shown in [1] that the edge dominating set problem is NP-hard in the class $\mathcal{S}_k$ for any fixed k. Together with Lemma 1 this implies the desired conclusion for the dominating set problem. For the independent dominating set problem this conclusion follows from the fact that in the class of line graphs this problem is equivalent to the dominating set problem [21]. Finally, to prove the lemma for the induced matching problem, we use the reduction from the P_3-FACTOR problem in a graph G to the INDUCED MATCHING problem in $L(G)$ [12] and the fact that P_3-FACTOR is NP-hard for graphs in the class $\mathcal{S}_k$ [1].   □

By analogy with the parameter $\kappa(G)$, we define one more parameter $\lambda(G)$, as follows: $\lambda(G)$ is the maximum k such that $G \in \mathcal{T}_k$. If G belongs to no $\mathcal{T}_k$, then $\lambda(G) := 0$, and if G belongs to every $\mathcal{T}_k$, then $\lambda(G) := \infty$. For a set of graphs M, we define $\lambda(M) = \sup\{\lambda(G) : G \in M\}$. The following theorem is a direct analogue of Theorem 1.

Theorem 2. *Let X be the class of M-free graphs of vertex degree at most 3. If $\lambda(M) < \infty$, then in the class X the tree-width is unbounded and the following problems are NP-hard:* DOMINATING SET, INDEPENDENT DOMINATING SET, INDUCED MATCHING.

3   Classes of Graphs with Bounded Tree-Width

Let $\Delta \geq 3$ be a fixed integer and let M be a set of graphs. The results of the previous section suggest that the tree-width of M-free graphs of vertex degree at most Δ is bounded only if

$$\kappa(M) \text{ is unbounded and } \lambda(M) \text{ is unbounded.} \tag{1}$$

In the present section, we identify several areas where condition (1) is sufficient for boundedness of the tree-width. First of all, let us reveal three major ways to push $\kappa(M)$ to infinity.

One of the possible ways to unbound $\kappa(M)$ is to include in M a graph G with $\kappa(G) = \infty$. According to the definition, in order $\kappa(G)$ to be infinite, G must belong to every class $\mathcal{S}_k$. It is not difficult to see that this is possible only if every connected component of G is of the form $S_{i,j,k}$ represented in Figure 2 (left). Let us denote the class of all such graphs by $\mathcal{S}$. More formally, $\mathcal{S} := \bigcap_{k \geq 3} \mathcal{S}_k$. Any other way to push $\kappa(M)$ to infinity requires the inclusion in M of infinitely many graphs. We distinguish two particular ways of achieving this goal: $M \supseteq \{H_k, H_{k+1}, \ldots\}$ and $M \supseteq \{C_k, C_{k+1}, \ldots\}$ for a constant k.

Translating the above three conditions to the language of line graphs, we obtain three respective ways to unbound $\lambda(M)$. In the first one, we include in M a graph G with $\lambda(G) = \infty$, i.e., the line graph of a graph in $\mathcal{S}$. Let us denote the class of all such graphs by $\mathcal{T}$. In other words, $\mathcal{T}$ is the class of graphs every connected component of which has the form $T_{i,j,k}$ represented in Figure 2, or equivalently, $\mathcal{T} = \bigcap_{k \geq 3} \mathcal{T}_k$. Also, $\lambda(M)$ is unbounded if $M \supseteq \{L_k, L_{k+1}, \ldots\}$, where L_i is the line graph of H_{i+1} (see Figure 1). Finally, $\lambda(M)$ is unbounded if $M \supseteq \{C_k, C_{k+1}, \ldots\}$, since the line graph of a cycle is the cycle itself.

In the rest of this section, we consider all possible combinations of the above ways to unbound $\kappa(M)$ and $\lambda(M)$, and show that each of the combinations leads to a class of graphs of bounded tree-width.

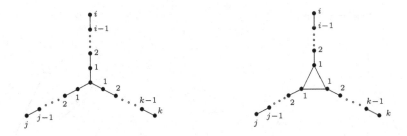

Fig. 2. Graphs $S_{i,j,k}$ (left) and $T_{i,j,k}$ (right)

The above discussion suggests the following nine ways of restricting the set M:

(i) $M \cap S \neq \emptyset$ and $M \cap T \neq \emptyset$,
(ii) $M \cap S \neq \emptyset$ and $M \supseteq \{L_k, L_{k+1}, \ldots\}$ for some $k \geq 1$,
(iii) $M \cap S \neq \emptyset$ and $M \supseteq \{C_k, C_{k+1}, \ldots\}$ for some $k \geq 3$,
(iv) $M \cap T \neq \emptyset$ and $M \supseteq \{H_k, H_{k+1}, \ldots\}$ for some $k \geq 1$,
(v) $M \supseteq \{H_k, L_k, H_{k+1}, L_{k+1}, \ldots\}$ for some $k \geq 1$,
(vi) $M \supseteq \{H_k, C_k, H_{k+1}, C_{k+1}, \ldots\}$ for some $k \geq 3$,
(vii) $M \cap T \neq \emptyset$ and $M \supseteq \{C_k, C_{k+1}, \ldots\}$ for some $k \geq 3$,
(viii) $M \supseteq \{C_k, L_k, C_{k+1}, L_{k+1}, \ldots\}$ for some $k \geq 3$,
(ix) $M \supseteq \{C_k, C_{k+1}, \ldots\}$ for some $k \geq 3$.

We can immediately disregard conditions (iii), (vi), (vii) and (viii), as they are dominated by (ix). We discuss the remaining five cases individually.

(1) Excluding a graph from S and a graph from T. It was shown in [16] that exclusion of a graph from S and a graph from T results in a class in which every graph of bounded vertex degree has bounded tree-width. Since we will make use of this result later on in this section, we restate it here.

Theorem 3 ([16]). *For any positive integer Δ and any two graphs $S \in S$ and $T \in T$ there is an integer N such that every graph of vertex degree at most Δ with no induced subgraphs isomorphic to S or T has tree-width at most N.*

(2) Excluding large cycles. Excluding large cycles bounds the tree-width of bounded-degree graphs in $Free(M)$, as the tree-width of a graph is upper-bounded by a function of its maximum degree and chordality [4].

(3) Excluding large graphs of the form H_i and L_i. For a fixed positive integer k, let C_k denote the class of $(H_k, L_k, H_{k+1}, L_{k+1}, \ldots)$-free graphs. These graphs form a generalization of graphs without long induced paths, for which many structural results are known (see for instance [3,9,15]). An obvious property of connected graphs without long induced paths is that they have small diameter.[1]

---

[1] The *diameter* of a connected graph G is the maximum length of a shortest path connecting two vertices of G.

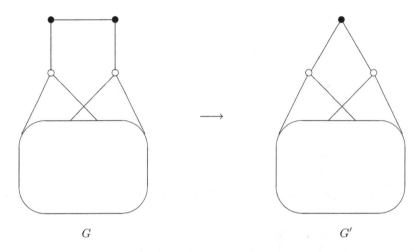

Fig. 3. Breaking a P_4-handle

We will show that connected graphs in $\mathcal{C}_k$ are not far from having small diameter. More precisely, we will prove that any connected graph G from $\mathcal{C}_k$ can be transformed into a graph $\tilde{G} \in \mathcal{C}_{k+1}$ of bounded diameter. Informally, our transformation is inverse to a sequence of edge subdivisions. To describe it formally, we introduce the notion of P_4-handles. A P_4-*handle* in a graph G is an induced P_4 whose two midpoints have degree two in G. We will call a graph P_4-*handle-free* if it has no P_4-handles. Our transformation simply consists of contracting middle edges of P_4-handles as long as possible. (Figure 3 shows one such contraction.) Clearly, it can be applied to any graph G, and it results in a P_4-handle-free graph $\tilde{G}$. Note that up to isomorphism, $\tilde{G}$ is uniquely defined, and the graph G can be obtained from $\tilde{G}$ by applying a sequence of edge subdivisions.

We now show that the diameter of P_4-handle-free graphs in our class is bounded.

Lemma 3. *Let $G \in \mathcal{C}_k$ be a connected P_4-handle-free graph. Then $diam(G) \leq 2k + 18$.*

Proof. The proof is by contradiction. Assume that G contains two vertices a and b at distance at least $2k + 19$, and let $P = (v_1, \ldots, v_s)$ be a shortest a-b path. By assumption, P has at least $2k + 20$ vertices.

As G has no P_4 handles, at least one of v_2 and v_3 has a neighbor, say x, outside P. Similarly, at least one of v_{k+10} and v_{k+11} has a neighbor y outside P, and at least one of v_{2k+18} and v_{2k+19} has a neighbor z outside P.

Since P is a shortest a-b path, the neighborhood of x on P is contained in the set $\{v_1, \ldots, v_5\}$. Similarly, the neighborhood of y on P is contained in the set $\{v_{k+8}, \ldots, v_{k+13}\}$, and the neighborhood of z on P is contained in the set $\{v_{2k+16}, \ldots, v_{2k+20}\}$. In particular, this implies that the vertices x, y and z are distinct. By the minimality of P, the vertices x, y and z are pairwise nonadjacent.

Let v_r be the neighbor of x on P with the largest value of r, and let v_l be the neighbor of z on P with the smallest value of l. To avoid a large graph isomorphic to an H_i or L_i, we conclude without loss of generality that x is adjacent to v_{r-1} and z is nonadjacent to v_{l+1}. Similarly, let us denote by v_R the neighbor of y on P with the largest value of R, and by v_L the neighbor of y on P with the smallest value of L. Then, it follows that y is nonadjacent to v_{L+1}, and adjacent to v_{R-1}. However, this implies that $R - L \geq 3$, which contradicts the minimality of P.   □

With this result in mind, it is now easy to prove the following theorem.

Theorem 4. *For any positive integers k and Δ, there is an N such that the tree-width of every $(H_k, L_k, H_{k+1}, L_{k+1}, \ldots)$-free graph of maximum degree at most Δ is at most N.*

Proof. It is known (see e.g. [18]) that subdivision of an edge does not change the tree-width of the graph. Therefore, the tree-width of $\tilde{G}$ equals to that of G, and hence, to prove the theorem it is enough to show that the tree-width is bounded in the class $X_{k,\Delta} := \{\tilde{G} : G \in \mathcal{C}_k, \Delta(G) \leq \Delta\}$.

It is not difficult to see that $X_{k,\Delta}$ is a subclass of $\mathcal{C}_{k+1}$, and that vertex degrees of graphs in $X_{k,\Delta}$ are also bounded by Δ. Since all graphs $\tilde{G}$ are P_4-handle-free, it follows from Lemma 3 that the diameter of connected graphs in $X_{k,\Delta}$ does not exceed $2k + 20$. As there are only finitely many connected graphs of bounded degree and bounded diameter, the tree-width of graphs in $X_{k,\Delta}$ is bounded above by a constant depending only on k and Δ.   □

(4) Excluding a graph from $\mathcal{S}$, and large graphs of the form L_i. We start with an auxiliary lemma.

Lemma 4. *For any positive integers k and Δ, there is an integer $\rho = \rho(k, \Delta)$ such that any connected $(L_k, L_{k+1}, \ldots)$-free graph G of maximum vertex degree at most Δ contains an induced triangle-free subgraph with at least $|V(G)| - \rho$ vertices.*

Proof. We first show that for any two induced copies T and T' of a triangle in G, the distance between them does not exceed k. Suppose by contradiction that a shortest path P joining a triangle T to another triangle T' consists of $r \geq k+1$ edges. Let us write $P = (v_0, v_1, \ldots, v_{r-1}, v_r)$ where $v_0 \in V(T)$, $v_r \in V(T')$.

Without loss of generality, we may assume that v_0 is the only neighbor of v_1 in T, since otherwise we can re-define T to be a triangle induced by v_1 together with any of its two neighbors in T. Similarly, we may assume that v_r is the only neighbor of v_{r-1} in T'. But now the two triangles T and T' together with the vertices of P connecting them induce a graph L_i with $i \geq k$. Contradiction.

To conclude the proof, assume that G contains an induced triangle T, and let v be a vertex of T. According to the above discussion, the distance from v to a vertex of any other triangle in G (if any) is at most k. Since G is a connected graph of maximum degree at most Δ, there is a constant $\rho = \rho(k, \Delta)$ bounding the number of vertices of G of distance at most k from v. Deletion of all these vertices leaves an induced subgraph of G which is triangle-free.   □

We also recall the following result from [16].

Lemma 5 ([16]). *For a class of graphs X and an integer ρ, let $[X]_\rho$ denote the class of graphs G such that $G - U$ belongs to X for some subset $U \subseteq V(G)$ of cardinality at most ρ. If X is a class of graphs of bounded tree-width, then so is $[X]_\rho$.*

Combining Lemma 4 with Lemma 5 and Theorem 3, we obtain the following conclusion.

Theorem 5. *For any positive integers k and Δ and any graph $S \in \mathcal{S}$, there is an integer N such that the tree-width of every $(S, L_k, L_{k+1}, \ldots)$-free graph of maximum degree at most Δ is at most N.*

(5) **Excluding a graph from $\mathcal{T}$, and large graphs of the form H_i.** This case is similar to the previous one. In particular, the following lemma can be proved by analogy with Lemma 4.

Lemma 6. *For any positive integers k and Δ, there is an integer $\rho = \rho(k, \Delta)$ such that any connected $(H_k, H_{k+1}, \ldots)$-free graph G of maximum vertex degree at most Δ contains an induced claw-free subgraph with at least $|V(G)| - \rho$ vertices.*

Combining Lemma 6 with Lemma 5 and Theorem 3, we obtain the following conclusion.

Theorem 6. *For any positive integers k and Δ and any graph $T \in \mathcal{T}$ there is an integer N such that the tree-width of every $(T, H_k, H_{k+1}, \ldots)$-free graph of maximum degree at most Δ is at most N.*

The results of this section are summarized in the following theorem.

Theorem 7. *Let Δ be a positive integer and let X be a class of M-free graphs of vertex degree at most Δ. If at least one of the following conditions holds:*

(1) there is a $k \geq 3$ such that $M \supseteq \{C_k, C_{k+1}, \ldots\}$, or
(2) there is a $k \geq 1$ such that $M \supseteq \{H_k, L_k, H_{k+1}, L_{k+1}, \ldots\}$, or
(3) $M \cap \mathcal{S} \neq \emptyset$ and there is a $k \geq 1$ such that $M \supseteq \{L_k, L_{k+1}, \ldots\}$, or
(4) $M \cap \mathcal{T} \neq \emptyset$ and there is a $k \geq 1$ such that $M \supseteq \{H_k, H_{k+1}, \ldots\}$, or
(5) $M \cap \mathcal{S} \neq \emptyset$ and $M \cap \mathcal{T} \neq \emptyset$,

then the tree-width of graphs in X is bounded.

Corollary 1. *Let Δ be a positive integer and let X be a class of M-free graphs of vertex degree at most Δ. If at least one of the conditions stated in Theorem 7 holds, then the following problems are solvable in polynomial time for graphs in X:* DOMINATING SET, INDEPENDENT DOMINATING SET, INDUCED MATCHING.

Theorem 7 captures many well-known graph classes, once restricted to graphs of bounded degree. Examples include chordal graphs of bounded degree (condition (1) with $k = 4$), chordal bipartite graphs of bounded degree ((1) with $k = 5$), distance-hereditary graphs of bounded degree ((1) with $k = 5$), weakly chordal graphs of bounded degree ((1) with $k = 5$), AT-free graphs of bounded degree ((1) with $k = 6$), and their subclasses co-comparability graphs of bounded degree and permutation graphs of bounded degree, (bull, fork)-free graphs of bounded degree ((4), (5)), (claw, net)-free graphs of bounded degree ((4), (5)), and P_k-free graphs of bounded degree, for any fixed k ((1)–(5)). In all these classes, boundedness of vertex degrees is a necessary condition for bounded tree-width.

4 Conclusion

In this paper, we presented several results suggesting the idea that polynomial-time solvability of DOMINATING SET, INDEPENDENT DOMINATING SET and INDUCED MATCHING problems in bounded degree graphs is equivalent to bound-edness of tree-width of those graphs. In particular, we revealed two large areas containing graph classes where these problems are NP-hard and the tree-width is unbounded. We also detected several areas where the tree-width is bounded and hence the problems are polynomial-time solvable. There is still a gap of unexplored classes of graphs. Closing this gap is a challenging research problem.

Another interesting topic to investigate is to characterize the class of problems that have a similar relationship with the tree-width in bounded degree graphs. Two possible candidates for this class are the LONGEST CYCLE and LONGEST PATH problems.

Finally, it would be interesting to identify other classes of graphs where polynomial-time solvability and boundedness of tree-width behave consistently. In this respect, a promising direction is given by graphs with bounded expansion (a common generalization of graphs of bounded degree and minor-closed graph classes) introduced recently in [20].

Acknowledgment

We are grateful to the anonymous referees for their helpful comments.

References

1. Alekseev, V.E., Boliac, R., Korobitsyn, D.V., Lozin, V.V.: NP-hard graph problems and boundary classes of graphs. Theoretical Computer Science (in press)
2. Arnborg, S., Proskurowski, A.: Linear time algorithms for NP-hard problems restricted to partial k-trees. Discrete Appl. Math. 23, 11–24 (1989)
3. Bacsó, G., Tuza, Z.: A characterization of graphs without long induced paths. J. Graph Theory 14, 455–464 (1990)
4. Bodlaender, H.L., Thilikos, D.M.: Treewidth for graphs with small chordality. Discrete Appl. Math. 79, 45–61 (1997)

5. Boliac, R., Lozin, V.: Independent domination in finitely defined classes of graphs. Theoret. Comput. Sci. 301, 271–284 (2003)
6. Brandstädt, A., Dragan, F.: On linear and circular structure of (claw, net)-free graphs. Discrete Appl. Math. 129, 285–303 (2003)
7. Courcelle, B., Olariu, S.: Upper bounds to the clique-width of a graph. Discrete Appl. Math. 101, 77–114 (2000)
8. Damaschke, P., Müller, H., Kratsch, D.: Domination in convex and chordal bipartite graphs. Inform. Process. Lett. 36, 231–236 (1990)
9. Dong, J.: On the diameter of i-center in a graph without long induced paths. J. Graph Theory 30, 235–241 (1999)
10. Golumbic, M.C., Lewenstein, M.: New results on induced matchings. Discrete Appl. Math. 101, 157–165 (2000)
11. Gurski, F., Wanke, E.: Line graphs of bounded clique-width. Discrete Math. (2007), doi:10.1016/j.disc.2007.01.020
12. Kobler, D., Rotics, U.: Finding maximum induced matchings in subclasses of claw-free and P_5-free graphs, and in graphs with matching and induced matching of equal maximum size. Algorithmica 37, 327–346 (2003)
13. Korobitsyn, D.V.: On the complexity of determining the domination number in monogenic classes of graphs. Diskret. Mat. 2(3), 90–96 (1990) (in Russian, translation in Discrete Math. and Appl. 2(2), 191–199 (1992))
14. Lozin, V.V.: On maximum induced matchings in bipartite graphs. Inform. Process. Lett. 81, 7–11 (2002)
15. Lozin, V., Rautenbach, D.: Some results on graphs without long induced paths. Inform. Process. Lett. 88, 167–171 (2003)
16. Lozin, V., Rautenbach, D.: On the band-, tree- and clique-width of graphs with bounded vertex degree. SIAM J. Discrete Math. 18, 195–206 (2004)
17. Lozin, V., Rautenbach, D.: Chordal bipartite graphs of bounded tree- and clique-width. Discrete Math. 283, 151–158 (2004)
18. Lozin, V., Rautenbach, D.: The tree- and clique-width of bipartite graphs in special classes. Australasian J. Combinatorics 34, 57–67 (2006)
19. Minty, G.J.: On maximal independent sets of vertices in claw-free graphs. J. Combin. Theory Ser. B 28, 284–304 (1980)
20. Nešetřil, J., de Mendez, P.O.: Grad and classes with bounded expansion I. Decompositions. Eur. J. of Combinatorics (2007), doi:10.1016/j.ejc.2006.07.013
21. Yannakakis, M., Gavril, F.: Edge dominating sets in graphs. SIAM J. Appl. Math. 38, 364–372 (1980)

On Restrictions of Balanced 2-Interval Graphs

Philippe Gambette[1] and Stéphane Vialette[2]

[1] LIAFA - Univ. Paris VII
gambette@liafa.jussieu.fr
[2] LRI - Univ. Paris XI
vialette@lri.fr

Abstract. The class of 2-interval graphs has been introduced for modelling scheduling and allocation problems, and more recently for specific bioinformatics problems. Some of those applications imply restrictions on the 2-interval graphs, and justify the introduction of a hierarchy of subclasses of 2-interval graphs that generalize line graphs: balanced 2-interval graphs, unit 2-interval graphs, and (x,x)-interval graphs. We provide instances that show that all inclusions are strict. We extend the NP-completeness proof of recognizing 2-interval graphs to the recognition of balanced 2-interval graphs. Finally we give hints on the complexity of unit 2-interval graphs recognition, by studying relationships with other graph classes: proper circular-arc, quasi-line graphs, $K_{1,5}$-free graphs, ...

Keywords: 2-interval graphs, graph classes, line graphs, quasi-line graphs, claw-free graphs, circular interval graphs, bioinformatics, scheduling.

1 2-Interval Graphs and Restrictions

The interval number of a graph, and the classes of k-interval graphs have been introduced as a generalization of the class of interval graphs by McGuigan [McG77] in the context of scheduling and allocation problems. Recently, bioinformatics problems have renewed interest in the class of 2-interval graphs (each vertex is associated to a pair of disjoint intervals and edges denote intersection between two such pairs). Indeed, a pair of intervals can model two associated tasks in scheduling [BYHN+06], but also two similar segments of DNA in the context of DNA comparison [JMT92], or two complementary segments of RNA for RNA secondary structure prediction and comparison [Via04].

RNA (ribonucleic acid) are polymers of nucleotides linked in a chain through phosphodiester bonds. Unlike DNA, RNAs are usually single stranded, but many RNA molecules have *secondary structure* in which intramolecular loops are formed by complementary base pairing. RNA secondary structure is generally divided into helices (contiguous base pairs), and various kinds of loops (unpaired nucleotides surrounded by *helices*). The structural stability and function of non-coding RNA (ncRNA) genes are largely determined by the formation of stable secondary structures through complementary bases, and hence ncRNA genes

A. Brandstädt, D. Kratsch, and H. Müller (Eds.): WG 2007, LNCS 4769, pp. 55–65, 2007.

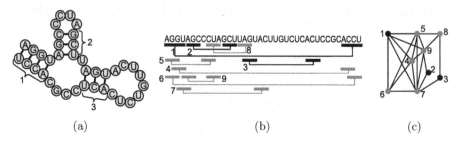

Fig. 1. Helices in a RNA secondary structure (a) can be modeled as a set of balanced 2-intervals among all 2-intervals corresponding to complementary and inverted pairs of letter sequences (b), or as an independent subset in the balanced associated 2-interval graph (c).

across different species are most similar in the pattern of nucleotide complementarity rather than in the genomic sequence. This motivates the use of 2-intervals for modelling RNA secondary structures: each helix of the structure is modeled by a 2-interval. Moreover, the fact that these 2-intervals are usually required to be disjoint in the structure naturally suggests the use of 2-interval graphs. Furthermore, aiming at better modelling RNA secondary structures, it was suggested in [CHLV05] to focus on *balanced 2-interval sets* (each 2-interval is composed of two equal length intervals) and their associated intersection graphs referred as *balanced 2-interval graphs*. Indeed, helices in RNA secondary structures are most of the time composed of equal length contiguous base pairs parts. To the best of our knowledge, nothing is known on the class of balanced 2-interval graphs.

Sharper restrictions have also been introduced in scheduling, where it is possible to consider tasks which all have the same duration, that is 2-interval whose intervals have the same length [BYHN+06,Kar05]. This motivates the study of the classes of unit 2-interval graphs, and (x, x)-interval graphs. In this paper, we consider these subclasses of interval graphs, and in particular we address the problem of recognizing them.

A graph $G = (V, E)$ of order n is a 2-interval graph if it is the intersection graph of a set of n unions of two disjoint intervals on the real line, that is each vertex corresponds to a union of two disjoint intervals $I^k = I_l^k \cup I_r^k$, $k \in [\![1, n]\!]$ (l for "*left*" and r for "*right*"), and there is an edge between I^j and I^k iff $I^j \cap I^k \neq \emptyset$. Note that for the sake of simplicity we use the same letter to denote a vertex and its corresponding 2-interval. A set of 2-intervals corresponding to a graph G is called a realization of G. The set of all intervals, $\bigcup_{k=1}^{n} \{I_l^k, I_r^k\}$, is called the ground set of G (or the ground set of $\{I^1, \ldots, I^n\}$).

The class of 2-interval graphs is a generalization of interval graphs, and also contains all circular-arc graphs (intersection graphs of arcs of a circle), outerplanar graphs (have a planar embedding with all vertices around one of the faces [KW99]), cubic graphs (maximum degree 3 [GW80]), and line graphs (intersection graphs of edges of a graph).

Unfortunately, most classical graph combinatorial problems turn out to be NP-complete for 2-interval graphs: recognition [WS84], maximum independent

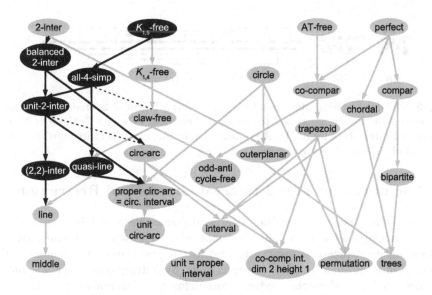

Fig. 2. Graph classes related to 2-interval graphs and its restrictions. A class pointing towards another strictly contains it, and the dashed lines mean that there is no inclusion relationship between the two. Dark classes correspond to classes not yet present in the ISGCI Database [BLS+].

set [BNR96,Via01], coloration [Via01], ... Surprisingly enough, the complexity of the maximum clique problem for 2-interval graphs is still open (although it has been recently proven to be NP-complete for 3-interval graphs [BHLR07]).

For practical application, restricted 2-interval graphs are needed. A 2-interval graph is *balanced* if it has a 2-interval realization in which each 2-interval is composed of two intervals of the same length [CHLV05], *unit* if it has a 2-interval realization in which all intervals of the ground set have length 1 [BFV04], and is called a (x, x)-*interval graph* if it has a 2-interval realization in which all intervals of the ground set are open, have integer endpoints, and length x [BYHN+06,Kar05]. In the following sections, we will study those restrictions of 2-interval graphs, and their position in the hierarchy of graph classes illustrated in Figure 2.

Note that all (x, x)-interval graphs are unit 2-interval graphs, and that all unit 2-interval graphs are balanced 2-interval graphs. We can also notice that $(1, 1)$-interval graphs are exactly line graphs: each interval of length 1 of the ground set can be considered as the vertex of a root graph and each 2-interval as an edge in the root graph. This implies for example that the coloration problem is also NP-complete for $(2, 2)$-interval graphs and wider classes of graphs. It is also known that the complexity of the maximum independent set problem is NP-complete on $(2, 2)$-interval graphs [BNR96]. Recognition of $(1, 2)$-union graphs, a related class (restriction of *multitrack interval graphs*), was also recently proven NP-complete [HK06].

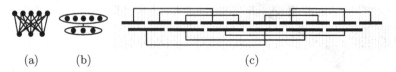

Fig. 3. The complete bipartite graph $K_{5,3}$ (a,b) has a balanced 2-interval realization (c): vertices of S_5 are associated to balanced 2-intervals of length 7, those of S_3 to balanced 2-intervals of length 11. Any realization of this graph is contiguous, *i.e.*, the union of all 2-intervals is an interval.

2   Useful Gadgets for 2-Interval Graphs and Restrictions

For proving hardness of recognizing 2-interval graphs, West and Shmoys considered in [WS84] the complete bipartite graph $K_{5,3}$ as a useful 2-interval gadget. Indeed, all its realizations are contiguous: for any realization, the union of all intervals in its ground set is an interval. Thus, by putting edges between some vertices of a $K_{5,3}$ and another vertex v, we can force one interval of the 2-interval v (or just one extremity of this interval) to be blocked inside the realization of $K_{5,3}$. It is easy to see that $K_{5,3}$ has a balanced 2-interval realization, for example the one in Figure 3.

However, $K_{5,3}$ is not a unit 2-interval graph. Indeed, each 2-interval $I = I_l \cup I_r$ corresponding to a degree 5 vertex intersect 5 disjoint 2-intervals, and hence one of I_l or I_r intersect at least 3 intervals, which is impossible for unit intervals. Therefore, we introduce the new gadget $K_{4,4} - e$ which is a $(2, 2)$-interval graph with only contiguous realizations (the proof is omitted).

3   Balanced 2-Interval Graphs

We show in this section that the class of balanced 2-interval graphs is strictly included in the class of 2-interval graphs, and strictly contains circular-arc graphs. Moreover, we prove that recognizing balanced 2-interval graphs is as hard as recognizing (general) 2-interval graphs.

Property 1. The class of balanced 2-interval graphs is strictly included in the class of 2-interval graphs.

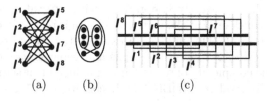

Fig. 4. The graph $K_{4,4} - e$ (a), a nicer representation (b), and a 2-interval realization with open intervals of length 2 (c)

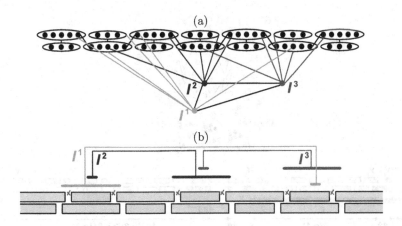

Fig. 5. An example of unbalanced 2-interval graph (a) : any realization groups intervals of the seven $K_{5,3}$ in a block, and the chain of seven blocks creates six "holes" between them, which make it impossible to balance the lengths of the three 2-intervals I^1, I^2, and I^3

Proof. We build a 2-interval graph that has no balanced 2-interval realization. Let's consider a chain of gadgets $K_{5,3}$ (introduced in previous section) to which we add three vertices I^1, I^2, and I^3 as illustrated in Figure 5. In any realization, the presence of holes showed by crosses in the Figure gives the following inequalities for any realization: $l(I_l^2) < l(I_l^1)$, $l(I_l^3) < l(I_r^2)$, and $l(I_r^1) < l(I_r^3)$ (or if the realization of the chain of $K_{5,3}$ appears in the symmetrical order: $l(I_l^1) < l(I_l^3)$, $l(I_r^3) < l(I_l^2)$, and $l(I_r^2) < l(I_r^1)$). If this realization was balanced, then we would have $l(I_l^1) = l(I_r^1) < l(I_r^3) = l(I_l^3) < l(I_r^2) = l(I_l^2)$ (or the symetrical equality): impossible! So this graph has no balanced 2-interval realization although it has a 2-interval generalization.

Theorem 1. *Recognizing balanced 2-interval graphs is NP-complete.*

Proof. We just adapt the proof of West and Shmoys [WS84,GW95]: reduce the problem of Hamiltonian cycle in a 3-regular triangle-free graph to balanced 2-interval recognition.

Let $G = (V, E)$ be a 3-regular triangle-free graph. We build a graph G' which has a 2-interval realization (a special one, very specific, called H-representation and which we prove to be balanced) iff G has a Hamiltonian cycle. The construction of G', illustrated in Figure 6(a) is almost identical to the one by West and Shmoys, so we just prove that G' has a balanced realization, shown in Figure 6 (b), by computing lengths for each interval to ensure it. All $K_{5,3}$ have a balanced realization as shown in section 1 of total length 79, in particular H_3. We can thus affect length 83 to the intervals of v_0. The intervals of the other v_i can have length 3, and their $M(v_i)$ length 79, so through the computation illustrated in

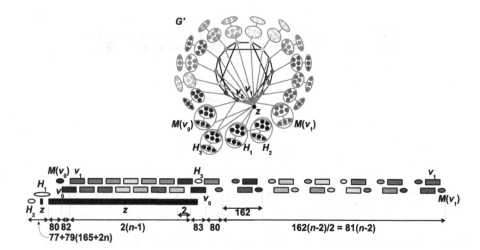

Fig. 6. There is a balanced 2-interval of G' (which has been dilated in the drawing to remain readable) iff there is an H-representation (that is a realization where the left intervals of all 2-intervals are contiguous) for its induced subgraph G iff there is a Hamiltonian cycle in G

Figure 6, intervals of z can have length $80 + 82 + 2(n-1) + 3$, that is $163 + 2n$. We dilate H_1 until a hole between two consecutive intervals of its S_3 can contain an interval of z, that is until the hole has length $165 + 2n$: so after this dilating, H_1 has length $79(165 + 2n)$. Finally if G has a Hamiltonian cycle, then we have found a balanced 2-interval realization of G of total length $13,273 + 241n$.

It is known that circular-arc graphs are 2-interval graphs, they are also balanced 2-interval.

Property 2. The class of circular-arc graphs is strictly included in the class of balanced 2-interval graphs.

Proof. The transformation is simple: if we have a circular-arc representation of a graph $G = (V, E)$, then we choose some point P of the circle. We partition V in $V_1 \cup V_2$, where P intersects all the arcs corresponding to vertices of V_1 and none of the arcs of the vertices of V_2. Then we cut the circle at point P to map it to a line segment: every arc of V_2 becomes an interval, and every arc of V_1 becomes a 2-interval. To obtain a balanced realization we just cut in half the intervals of V_2 to obtain two intervals of equal length for each. And for each 2-interval $[g(I_l), d(I_l)] \cup [g(I_r), d(I_r)]$ of V_1, as both intervals are located on one of the extremities of the realization, we can increase the length of the shortest so that it reaches the length of the longest without changing intersections with the other intervals. The inclusion is strict because $K_{2,3}$ is a balanced 2-interval graph (as a subgraph of $K_{5,3}$ for example) but is not a circular-arc graph (we can find two C_4 in $K_{2,3}$, and only one can be realized with a circular-arc representation).

4   Unit 2-Interval and (x,x)-Interval Graphs

Property 3. Let $x \in \mathbb{N}, x \geq 2$. The class of (x, x)-interval graphs is strictly included in the class of $(x + 1, x + 1)$-interval graphs.

Proof. We first prove that an interval graph with a representation where all intervals have length k (and integer open bounds) has a representation where all intervals have length $k + 1$.

We use the following algorithm. Let S be initialized as the set of all intervals of length k, and let T be initially the empty set. As long as S is not empty, let $I = [a, b]$ be the left-most interval of S, remove from S each interval $[\alpha, \beta]$ such that $\alpha < b$ (including I), add $[\alpha, \beta + 1]$ to T, and translate by $+1$ all the remaining intervals in S. When S is empty, the intersection graph of T, where all intervals have length $k + 1$ is the same as the intersection graph for the original S.

We also build for each $x \geq 2$ a $(x+1, x+1)$-interval graph which is not a (x, x)-interval graph. We consider the bipartite graph K_{2x} and a perfect matching $\{(v_i, v'_i), i \in [\![1, x]\!]\}$. We call K'_x the graph obtained from K_{2x} with the following transformations, illustrated in Figure 7(a): remove edges (v_i, v'_i) of the perfect matching, add four graphs $K_{4,4} - e$ called X_1, X_2, X_3, X_4 (for each X_i, we call v^i_l and v^i_r the vertices of degree 3), link v^2_r and v^3_l, link all v_i to v^1_r and v^4_l, link all v'_i to v^2_l and v^3_r, and finally add a vertex a (resp. b) linked to all v_i, v'_i, and to two adjacent vertices of X_1 (resp. X_4) of degree 4. We illustrate in Figure 7(b) that K'_x has a realization with intervals of length $x + 1$. We can prove by induction on x that K'_x has no realization with intervals of length x: it is rather technical, so we just give the idea. Any realization of K'_x forces the block X_2 to share an extremity with the block X_3, so each 2-interval v'_i has one interval intersecting the other extremity of X_2, and the other intersecting the other extremity of X_3. Then constraints on the position of vertices v_i force their intervals to appear as two "stairways" as shown in Figure 7(b). So v^1_r must contain x extremities of intervals which have to be different, so it must have length $x + 1$.

The complexity of recognizing unit 2-interval graphs and (x, x)-interval graphs remains open, however the following shows a relationship between those complexities.

Lemma 1. $\{unit\ 2\text{-}interval\ graphs\} = \bigcup_{x \in \mathbb{N}^*} \{(x, x)\text{-}interval\ graphs\}.$

Proof. The $\supset$ part is trivial. To prove $\subset$, let $G = (V, E)$ be a unit 2-interval graph. Then it has a realization with $|V| = n$ 2-intervals, that is $2n$ intervals of the ground set. So we consider the interval graph of the ground set, which is a unit interval graph. There is a linear time algorithm based on breadth-first search to compute a realization of such a graph where interval endpoints are rational, with denominator $2n$ [CKN+95]. So by dilating by a factor $2n$ such a realization, we obtain a realization of G where intervals of the ground set have length $2n$.

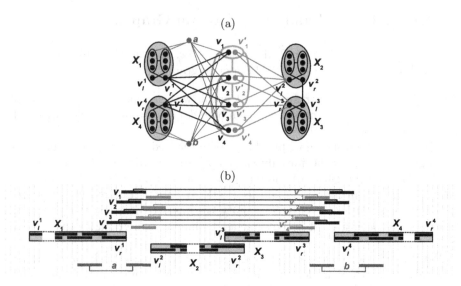

Fig. 7. The graph K_4' (a) is (5,5)-interval but not (4,4)-interval

Theorem 2. *If recognizing (x, x)-interval graphs is polynomial for any integer x then recognizing unit 2-interval graphs is polynomial.*

5   Investigating the Complexity of Unit 2-Interval Graphs

In this section we show that all proper circular-arc graphs (circular-arc graphs such that no arc is included in another in the representation) are unit 2-interval graphs, and we study a class of graphs which generalizes quasi-line graphs and contains unit 2-interval graphs.

Property 4. The class of proper circular-arc graphs is strictly included in the class of unit 2-interval graphs.

Proof. As in the proof of Property 2, we cut the circle of the representation of a proper circular-arc graph G to get a proper interval realization, which we transform into a unit interval realization [Rob69], which provides a unit 2-interval representation of G. To complete the proof, we notice that the domino (two cycles C_4 having an edge in common) is a unit 2-interval graph but not a circular-arc graph.

Quasi-line graphs are those graphs whose vertices are bisimplicial, *i.e.*, the closed neighborhood of each vertex can be partitioned into two cliques. They have been introduced as a generalization of line graphs and a useful subclass of claw-free graphs [Ben81,FFR97,CS05,KR07]. Following the example of quasi-line graphs that generalize line graphs, we introduce here a new class of graphs for generalizing unit 2-interval graphs. Let $k \in \mathbb{N}^*$. A graph $G = (V, E)$ is *all-k-simplicial* if the neighborhood of each vertex $v \in V$ can be partitioned into at most k cliques (note that quasi-line graphs are exactly all-2-simplicial graphs).

Property 5. The class of unit 2-interval graphs is strictly included in the class of all-4-simplicial graphs.

Proof. The inclusion is trivial. We show that it is strict. Consider the following graph which is all-4-simplicial but not unit 2-interval: start with the cycle C_4, call its vertices v_i, $i \in [1, 4]$, add four $K_{4,4} - e$ gadgets called X_i, and for each i we connect the vertex v_i to two connected vertices of degree 4 in X_i. This graph is certainly all-4-simplicial. But if we try to build a 2-interval realization of this graph, then each of the 2-intervals v_k has an interval trapped into the block X_k. So each 2-interval v_k has only one interval to realize the intersections with the other v_i: this is impossible as we have to realize a C_4 which has no interval representation.

Property 6. The class of claw-free graphs is not included in the class of all-4-simplicial graphs.

Proof. The Kneser Graph $KG(7, 2)$ is triangle-free, not 4-colorable [Lov78]. The graph obtained by adding an isolated vertex v and then taking the complement graph, i.e., $\overline{KG(7, 2) \uplus \{v\}}$, is claw-free as $KG(7, 2)$ is triangle-free. And if it was all-4-simplicial, then the neighborhood of v in $\overline{KG(7, 2) \uplus \{v\}}$, that is $\overline{KG(7, 2)}$, would be a union of at most four cliques, so $KG(7, 2)$ would be 4-colorable: impossible so this graph is claw-free but not all-4-simplicial.

Property 7. The class of all-k-simplicial graphs is strictly included in the class of $K_{1,k+1}$-free graphs.

Proof. If a graph G contains $K_{1,k+1}$, then it has a vertex with $k + 1$ independent neighbors, and hence G is not all-k-simplicial. The wheel W_{2k+1} is a simple example of $K_{1,k+1}$-free graph in which the center can not have its neighborhood (a C_{2k+1}) partitioned into k cliques or less.

Unfortunately, all-k-simplicial graphs do not have a nice structure which could help unit 2-interval graph recognition.

Theorem 3. *Recognizing all-k-simplicial graphs is NP-complete for $k \geq 3$.*

Proof. We reduce from GRAPH k-COLORABILITY, which is known to be NP-complete for $k \geq 3$ [Kar72]. Let $G = (V, E)$ be a graph, and let G' be the complement graph of G to which we add a universal vertex v. We claim that G is k-colorable iff G' is all-k-simplicial. If G is k-colorable, then the non-neighborhood of any vertex in G is k-colorable, so the neighborhood of any vertex in $\overline{G}$ is a union of at most k cliques. And the neighborhood of v is also a union of at most k cliques, so G' is all-k-simplicial. Conversely, if G' is all-k-simplicial, then in particular the neighborhood of v is a union of at most k cliques. Let's partition it into k vertex-disjoint cliques $X_1, \ldots, X_k$. Then, coloring G such that two vertices have the same color iff they are in the same X_i leads to a valid k-coloring of G.

6   Conclusion

Motivated by practical applications in scheduling and computational biology, we focused in this paper on balanced 2-interval graphs and unit 2-intervals graphs. Also, we introduced two natural new classes: (x, x)-interval graphs and all-k-simplicial graphs.

We mention here some directions for future works. First, the complexity of recognizing unit 2-interval graphs and (x, x)-interval graphs remains open. Second, the relationships between quasi-line graphs and subclasses of balanced 2-intervals graphs still have to be investigated. Last, since most problems remain NP-hard for balanced 2-interval graphs, there is a natural interest in investigating the complexity and approximation of classical optimization problems on unit 2-interval graphs and (x, x)-interval graphs.

Acknowledgments

We are grateful to Vincent Limouzy in particular for bringing to our attention the class of quasi-line graphs, and Michael Rao and Michel Habib for useful discussions.

References

Ben81.   Rebea, A.B.: Étude des stables dans les graphes quasi-adjoints. PhD thesis, Université de Grenoble (1981)

BFV04.   Blin, G., Fertin, G., Vialette, S.: New results for the 2-interval pattern problem. In: Sahinalp, S.C., Muthukrishnan, S.M., Dogrusoz, U. (eds.) CPM 2004. LNCS, vol. 3109, Springer, Heidelberg (2004)

BHLR07.   Butman, A., Hermelin, D., Lewenstein, M., Rawitz, D.: Optimization problems in multiple-interval graphs. In: Proceedings of SODA 2007, pp. 268–277 (2007)

BLS+.   Brandstädt, A., Le, V.B., Szymczak, T., Siegemund, F., de Ridder, H.N., Knorr, S., Rzehak, M., Mowitz, M., Ryabova, N.: ISGCI: Information System on Graph Class Inclusions.
            http://wwwteo.informatik.uni-rostock.de/isgci/classes.cgi

BNR96.   Bafna, V., Narayanan, B.O., Ravi, R.: Nonoverlapping local alignments (weighted independent sets of axis-parallel rectangles). Discrete Applied Math. 71(1), 41–54 (1996)

BYHN+06.   Bar-Yehuda, R., Halldórson, M.M., Naor, J., Shachnai, H., Shapira, I.: Scheduling split intervals. SIAM Journal on Computing 36(1), 1–15 (2006)

CHLV05.   Crochemore, M., Hermelin, D., Landau, G.M., Vialette, S.: Approximating the 2-interval pattern problem. In: Brodal, G.S., Leonardi, S. (eds.) ESA 2005. LNCS, vol. 3669, pp. 426–437. Springer, Heidelberg (2005)

CKN+95.   Corneil, D.G., Kim, H., Natarajan, S., Olariu, S., Sprague, A.P.: Simple linear time recognition of unit interval graphs. Information Processing Letters 55, 99–104 (1995)

CS05.   Chudnovsky, M., Seymour, P.: The structure of claw-free graphs. In: Surveys in Combinatorics. London. Math. Soc. Lecture Notes, vol. 327, pp. 153–172. Cambridge University Press, Cambridge (2005)

FFR97.   Faudree, R., Flandrin, E., Ryjáček, Z.: Claw-free graphs - a survey. Discrete Mathematics 164, 87–147 (1997)

GW80.   Griggs, J.R., West, D.B.: Extremal values of the interval number of a graph. SIAM Journal on Algebraic and Discrete Methods 1, 1–7 (1980)

GW95.   Gyárfás, A., West, D.B.: Multitrack interval graphs. Congress Numerantium 109, 109–116 (1995)

HK06.   Halldórsson, M.M., Karlsson, R.K.: Strip graphs: Recognition and scheduling. In: Fomin, F.V. (ed.) WG 2006. LNCS, vol. 4271, pp. 137–146. Springer, Heidelberg (2006)

JMT92.   Joseph, D., Meidanis, J., Tiwari, P.: Determining DNA sequence similarity using maximum independent set algorithms for interval graphs. In: Nurmi, O., Ukkonen, E. (eds.) SWAT 1992. LNCS, vol. 621, pp. 326–337. Springer, Heidelberg (1992)

Kar72.   Karp, R.M.: Reducibility among combinatorial problems, pp. 85–103. Plenum Press, New York (1972)

Kar05.   Karlsson, R.: A survey of split intervals and related graphs, Manuscript (2005)

KR07.   King, A., Reed, B.: Bounding χ in terms of ω ans δ for quasi-line graphs. Article in preparation (2007)

KW99.   Kostochka, A.V., West, D.B.: Every outerplanar graph is the union of two interval graphs. Congress Numerantium 139, 5–8 (1999)

Lov78.   Lovász, L.: Kneser's conjecture, chromatic number, and homotopy. Journal of Combinatorial Theory Series A 25, 319–324 (1978)

McG77.   McGuigan, R.: Presentation at NSF-CBMS Conference at Colby College (1977)

Rob69.   Roberts, F.S.: Indifference graphs. In: Proof Techniques in Graph Theory, Proceedings of the Second Ann Arbor Graph Theory Conference, pp. 139–146 (1969)

Via01.   Vialette, S.: Aspects algorithmiques de la prédiction des structures secondaires d'ARN. PhD thesis, Université Paris 7 (2001)

Via04.   Vialette, S.: On the computational complexity of 2-interval pattern matching. Theoretical Computer Science 312(2-3), 223–249 (2004)

WS84.   West, D.B., Shmoys, D.B.: Recognizing graphs with fixed interval number is NP-complete. Discrete Applied Math. 8, 295–305 (1984)

Graph Operations Characterizing Rank-Width and Balanced Graph Expressions*

Bruno Courcelle and Mamadou Moustapha Kanté

LaBRI, Université Bordeaux 1, CNRS, 351 cours de la libération 33405
Talence Cedex, France
{bruno.courcelle,mamadou.kante}@labri.fr

Abstract. Graph complexity measures like *tree-width, clique-width, NLC-width* and *rank-width* are important because they yield *Fixed Parameter Tractable* algorithms. Rank-width is based on ranks of adjacency matrices of graphs over $GF(2)$. We propose here algebraic operations on graphs that characterize rank-width. For algorithmic purposes, it is important to represent graphs by balanced terms. We give a unique theorem that generalizes several "balancing theorems" for tree-width and clique-width. New results are obtained for rank-width and a variant of clique-width, called *m-clique-width*.

1 Introduction

Graph complexity measures like *tree-width* [17], *clique-width* [7], *NLC-width* [18] and *rank-width* [16] are important parameters for the construction of polynomial algorithms. Every graph property expressible by a formula of MS (*Monadic Second-Order*) logic has a *Fixed Parameter Linear* algorithm if tree-width is taken as parameter and a *Fixed Parameter Cubic* algorithm if clique-width (equivalently rank-width) is taken as parameter. These results are proved in the books by Downey and Fellows [11] and by Flum and Grohe [12] for tree-width, by Courcelle and al. [6] with help of results by Oum and Seymour [15,16] for rank-width and clique-width.

Clique-width and rank-width are equivalent in the sense that the same classes of undirected graphs have bounded clique-width and bounded rank-width. Clique-width has the advantage of having a definition in terms of very simple graph operations. Furthermore this definition is the basis of the construction of algorithms for checking MS graph properties in linear time in the size of the algebraic expressions defining the input graphs. Rank-width has the advantage of a good behavior with respect to vertex-minor inclusion, so that the class of graphs of rank-width at most k is characterized by finitely many excluded vertex-minors. Furthermore, the cubic-time algorithm that constructs for a given graph an algebraic expression of clique-width at most $2^{3k} - 1$ if the graph has clique-width at most k, is based on the decomposition underlying rank-width.

---

* Research supported by the french ANR-project Graph decompositions and Algorithms (GRAAL). B. Courcelle is member of Institut Universitaire de France.

A. Brandstädt, D. Kratsch, and H. Müller (Eds.): WG 2007, LNCS 4769, pp. 66–75, 2007.
© Springer-Verlag Berlin Heidelberg 2007

In this article we propose algebraic operations on graphs that characterize rank-width as follows:

a graph G has rank-width at most k if and only if (iff for short) it is the value of a term in $T(R_k, C_k)$

where R_k is a finite set of graph operations, C_k a finite set of constants, both depending on k.

In a few words, the operations are based on coloring vertices by sets of colors $\subseteq [k] := \{1, 2, \ldots, k\}$, like in the variant of clique-width called *m-clique-width* (see definitions of Sect. 2 and [7,8]), but vertex colors are manipulated by linear transformations on the $GF(2)$ vector space $\{0, 1\}^k$ rather than with set union over subsets of $\{1, \ldots, k\}$. Furthermore, edges are created between two disjoint graphs by means of bilinear forms, taking the vectors of colors as arguments. It is thus somewhat natural that they can generate (exactly) the set of graphs of rank-width at most k since rank-width is based on ranks of $GF(2)$ matrices.

The operation that replaces anywhere a vertex color a by the color b, and the one that adds edges between any vertex colored by a and any vertex colored by b are typical examples of quantifier-free transformations. Quantifier-free transformations modify logical structures by redefining certain relations by quantifier-free formulas (see [10,3] for graph algebras).

For algorithmic purposes, it is useful and sometimes crucial to represent graphs by a-balanced binary terms, i.e, trees of height at most $a(\log(n)+1)$ where n is the number of nodes and a is a constant. This is the case for instance, of the labeling schemes considered in [9,8]. Another practical use of balanced terms is the design of parallel algorithms. This is considered for example by Bodlaender to design parallel algorithms to construct minimum-width tree-decompositions of graphs or to solve some NP-complete problems [1,2].

Therefore it is quite natural to ask whether, every graph of "width" k admits an a-balanced binary "decomposition" of width $f(k)$ for some fixed function f. It is known that every graph of tree-width k admits a 2-balanced binary tree-decomposition of width at most $3k + 2$ [1] and every graph of m-clique-width k admits a 6-balanced m-clique-width expression of width at most $2k$ [8]. We investigate the problem of a unified framework. We prove a general theorem covering several particular cases saying that every term in $T(F, C)$ representing a *binary structure* is equivalent to a 3-balanced one in $T(F', C')$, where (F', C') is a *binary signature* and $(F, C) \subseteq (F', C')$. For that we introduce a kind of generalization of the notion of associative and commutative operation, called *flexibility*.

The main results of this article are: an algebraic characterization of rank-width, a unified framework for "balancing theorems" with application to rank-width, clique-width, NLC-width and m-clique-width.

2   Notations and Definitions

We denote by $[k]$ the set $\{1, \ldots, k\}$. Graphs are finite, simple, loop-free, undirected unless otherwise specified. A graph G is defined as $< V_G, edg_G >$ where

$edg_G \subseteq V_G \times V_G$ is the symmetric adjacency relation. Without loss of generality we assume that V_G is always linearly ordered. This order will be used to represent edg_G by a square matrix over $GF(2)$.

A *sub-cubic tree* is a tree such that the degree of each node is at most 3. All logarithms are in base 2.

Let t be a rooted tree and $a \in \mathbb{R}$. We say that t is a-balanced if the *height* of t, i.e., the maximal distance of a leaf to the root, is at most $a(\log(n) + 1)$ where n is the number of nodes of t.[1]

Let F be a set of binary functions and C be a set of constants. We denote by $T(F, C)$ the set of well-formed terms built with $F \cup C$. They will be discussed as colored directed and rooted ordered trees in the usual way. A *context* is a term in $T(F, C \cup \{u\})$ having a single occurrence of the variable u (a nullary symbol). We denote by $Cxt(F, C)$ the set of contexts. We denote by Id the particular context u. Let s be a context and t be a term or a context, we denote by $s[t/u]$ the result of the substitution of t for u in s.

We define two binary operations on terms and contexts: $s \circ s' = s[s'/u]$, belonging to $Cxt(F, C)$ for s, s' in $Cxt(F, C)$ and $s \bullet t = s[t/u]$, belonging to $T(F, C)$ for s in $Cxt(F, C)$ and t in $T(F, C)$.

We now recall the definition of *rank-width*, a graph complexity measure introduced by Oum and Seymour in their investigations on recognition algorithms for graphs of bounded *clique-width* [16]. For an (R, C)-matrix $M = (m_{ij} \mid i \in R, j \in C)$ over a field F, if $X \subseteq R$, $Y \subseteq C$, we let $M[X, Y]$ denote the sub-matrix $(m_{ij} \mid i \in X, j \in Y)$. For a graph G, we let A_G be its adjacency (V_G, V_G)-matrix over $GF(2)$.

Cut-rank Functions. Let $G = \langle V_G, edg_G \rangle$ be a graph. We define the *cut-rank* function ρ_G of G by letting $\rho_G(X) = rk(A_G[X, V_G \backslash X])$ for $X \subseteq V_G$, where rk is the matrix rank function. We let $\rho_G(\emptyset) = \rho_G(V_G) = 0$.

Rank-width. A *layout* of a graph G is a pair (T, f) of a sub-cubic tree T and a bijective function $f : V_G \to \{t \mid t$ is a node of degree 1 in $T\}$.

For an edge e of T, the connected components of $T \backslash e$ induce a bipartition of the set of nodes of degree 1 of T, hence a bipartition (X_e, Y_e) of the set of vertices of G. The width of an edge e of a layout (T, f) is $\rho_G(X_e) = \rho_G(Y_e)$. The width of a layout (T, f) is the maximum width over all edges of T. The *rank-width* of G, denoted by $rwd(G)$, is the minimum width over all layouts of G.

The notions of rank-width and of clique-width are equivalent in the sense that a class of graphs has bounded rank-width iff it has bounded clique-width. Oum has given in [15] an $O(n^3)$-time approximation algorithm that reports that a graph has rank-width at least $k + 1$ or outputs a layout of width at most $3k - 1$. This has been improved in [13] which gives a cubic-time algorithm that outputs a layout of width k if the graph has rank-width k. But if we want to solve problems definable in MS on graphs of bounded rank-width, we need to transform the layout into a clique-width expression (see [16]) and, after that, to use techniques

---

[1] This definition is meaningful in the case $n = 1$.

by Courcelle and al. [6]. In this paper, we propose an algebraic characterization of rank-width, which will allow us to solve MS definable problems without transforming the layout into a clique-width expression. This is important because the transformation of a layout of width k may give a $(2^{k+1} - 1)$-clique-width expression. The exponent 2^{k+1} is part of the large size of constants in FPT algorithms.

Proposition 1. *[16,8,7,14] For every undirected graph G,*

(1) $rwd(G) \leq cwd(G) \leq 2^{rwd(G)+1} - 1$ (3) $mcwd(G) \leq twd(G) + 3$

(2) $mcwd(G) \leq cwd(G) \leq 2^{mcwd(G)+1}$ (4) $rwd(G) \leq 4 \times twd(G) + 2$.

Here twd, cwd and $mcwd$ denote respectively *tree-width* [17], clique-width [7] and *m-clique-width* (we recall below the definition of m-clique-width [8]).

M-clique-width. Let L be a finite set of colors. A *multi-colored graph* is a triple $< V_G, edg_G, \delta_G >$ consisting of a graph $< V_G, edg_G >$ and a mapping δ_G associating with each x in V_G the set of its colors, a subset of L. A vertex may have zero, one or several colors.

The following constants will be used: for $A \subseteq L$ we let **A** be a constant denoting the graph G with single vertex x and $\delta_G(x) = A$. We write $\mathbf{A}(x)$ if we need to specify the vertex x. The following binary operations will be used: for $R \subseteq L \times L$, for *recolorings* $g, h : L \to 2^L$ and for multi-colored graphs G and H we define $K = G \otimes_{R,g,h} H$ if G and H are disjoint (otherwise we replace H by a disjoint copy) where

$$V_K = V_G \cup V_H,$$
$$edg_K = edg_G \cup edg_H \cup \{xy \mid x \in V_G, y \in V_H, R \cap (\delta_G(x) \times \delta_H(y)) \neq \emptyset\},$$
$$\delta_K(x) = (g \circ \delta_G)(x) = \{a \mid a \in g(b), b \in \delta_G(x)\} \text{ if } x \in V_G,$$
$$\delta_K(x) = (h \circ \delta_H)(x) \text{ if } x \in V_H .$$

As in the operations by Wanke [18] these operations add edges between two disjoint graphs, that are the two arguments of (many) binary operations. This is a difference with clique-width where a single binary operation is used, and $\eta_{i,j}$ applied to $G \oplus H$ may add edges to G and to H.

We let F_L be the set of all binary operations $\otimes_{R,g,h}$ and C_L be the set of constants $\{\mathbf{A} \mid A \subseteq L\}$. Every term t in $T(F_L, C_L)$ denotes a multi-colored graph $val(t)$ with colors in L, and every multi-colored graph G is the value of such a term for large enough L. To simplify the notation, we will write F_k and C_k if $L = [k]$. We let $mcwd(G)$ be the minimum k such that G is the value of a term $t \in T(F_k, C_k)$ and call this number the *m-clique-width* of G.

3   Vectorial Colorings and Rank-Width

Handling multiple colorings of vertices with k colors is clearly the same thing as handling colorings with colors in $\{0,1\}^k$. Let $k \geq 1$ and $\mathbb{B} = \{0,1\}$. A $\mathbb{B}^k$-coloring of a graph G is a mapping $\gamma : V_G \to \mathbb{B}^k$ with no constraint on the values of γ for neighbor vertices. We consider that $x \in V_G$ has color i (among others) iff $\gamma(x)[i]$ (the i-th component of $\gamma(x)$) is 1. A $\mathbb{B}^k$-colored graph is a triple $G =< V_G, edg_G, \gamma_G >$ where γ_G is a $\mathbb{B}^k$-coloring of $< V_G, edg_G >$. The empty $\mathbb{B}^k$-colored graph is denoted by $\emptyset_k$. (This constant can be eliminated from expressions by Remark 1). We define some operations on these graphs.

A mapping $h : \mathbb{B}^k \to \mathbb{B}^\ell$ is *linear* if for some $(k \times \ell)$-matrix and all row-vectors $u \in \mathbb{B}^k$ we have $h(u) = u \cdot N$. We say that h is described by N. A mapping $f : \mathbb{B}^k \times \mathbb{B}^\ell \to \mathbb{B}$ is said *bilinear* if for some $(k \times \ell)$-matrix and all row-vectors $u \in \mathbb{B}^k$, $v \in \mathbb{B}^\ell$ we have $f(u,v) = u \cdot M \cdot v^T$ where v^T indicates *transposition* of the row-vector v(we say that f is described by M).

With a $\mathbb{B}^k$-colored graph $G =< V_G, edg_G, \gamma_G >$ we associate the (V_G, V_G)-adjacency (symmetric) matrix A_G and the $(V_G, [k])$-*color matrix* Γ_G, the row vectors of which are the vectors $\gamma_G(x)$ in $\mathbb{B}^k$ for x in V_G. We define the *color-rank* of G as the rank of Γ_G and we denote it by $crk(G)$. Clearly, $crk(G) \le k$ if G is $\mathbb{B}^k$-colored.[2,3]

Linear Recolorings. For $h : \mathbb{B}^k \to \mathbb{B}^\ell$ a linear mapping and G a $\mathbb{B}^k$-colored graph, we let $Recol_h(G) = H = < V_G, edg_G, \gamma_H >$ where $\gamma_H = h \circ \gamma_G$. Hence $\Gamma_H = \Gamma_G \cdot N$ and H is a $\mathbb{B}^\ell$-colored graph. If h and h' are linear recolorings, described respectively by N and N', then $h \circ h'$ is linear and is described by $N' \cdot N$.

Bilinear Product of Graphs. Let $f : \mathbb{B}^k \times \mathbb{B}^\ell \to \{0,1\}$ be a bilinear mapping, let $g : \mathbb{B}^k \to \mathbb{B}^m$ and $h : \mathbb{B}^\ell \to \mathbb{B}^m$ be arbitrary linear mappings. For G, $\mathbb{B}^k$-colored and H, $\mathbb{B}^\ell$-colored, we let $K = G \otimes_{f,g,h} H$ be defined as follows, where, as usual, we assume $V_G \cap V_H = \emptyset$:

$$V_K = V_G \cup V_H,$$
$$edg_K = edg_G \cup edg_H \cup \{xy \mid x \in V_G, y \in V_H, \gamma_G(x) \cdot M \cdot (\gamma_H(y))^T = 1\},$$
$$\gamma_K(x) = \gamma_G(x) \cdot N \text{ if } x \in V_G, \qquad \gamma_K(x) = \gamma_H(x) \cdot P \text{ if } x \in V_H,$$

where f, g, h are described respectively by M, N, P. Hence K is a $\mathbb{B}^m$-colored graph. We order the graph $K = G \otimes_{f,g,h} H$ by preserving the orderings of V_G and V_H and letting $x < y$ for $x \in V_G$ and $y \in V_H$. We will use the notation $\otimes_{M,N,P}$ instead of $\otimes_{f,g,h}$.

Constants. We will use **1** to denote the graph with a single vertex with its $\mathbb{B}^1$-coloring by (1). In order to avoid the use of recolorings, and to deal only with constants and binary operations, we will also use constants for the graphs $Recol_h(\mathbf{1})$ where h ranges over linear recolorings defined by 1-row matrices $N = u \in \mathbb{B}^k$. Such constants will be denoted by **u**. We use C_k to denote the set of constants **u** for $u \in \mathbb{B}^\ell$, $\ell \le k$.

---

[2] The color-rank of G should not be confused with its *rank*. All ranks are relative to $GF(2)$.

[3] A graph $G =< V_G, edg_G >$ is made canonically into a $\mathbb{B}^k$-colored graph for each k, with $\gamma_G(x) = (0, \ldots, 0)$ for each x.

Remark 1. We have

$$G \otimes_{M,N,P} H = H \otimes_{M^T,P,N} G, \quad Recol_Q(G) \otimes_{M,N,P} Recol_{Q'}(H) = G \otimes_{QMQ'^T,QN,Q'P} H,$$

$$G \otimes_{M,N,P} \emptyset_k = Recol_N(G), \qquad\qquad Recol_Q(G \otimes_{M,N,P} H) = G \otimes_{M,NQ,PQ} H .$$

where M^T denotes the transposition of the matrix M. We let R_n be the set of linear recolorings and bilinear products. We denote by $val(t)$ the graph defined by a term $t \in T(R_n, C_n)$. This graph is the value of the term in the corresponding algebra. We can assume with Remark 1 that a term t in $T(R_n, C_n)$ is written with the binary operations $\otimes_{M,N,P}$ and the constants $\mathbf{u}$ where $u \in \mathbb{B}^1 \cup \ldots \cup \mathbb{B}^n$.

Proposition 2.   *1. The operations $Recol_N$ are quantifier-free operations.*
 2. The operations $\otimes_{M,N,P}$ are expressible in terms of $\oplus$ and quantifier-free operations.

Corollary 1. *For each n, every* MS *graph property of a graph G can be decided in time $O(|t|)$, if G is the value of a given term $t \in T(R_n, C_n)$.*

Theorem 1. *A graph G has rank-width at most n iff it is the value of a term in $T(R_n, C_n)$.*

For the "If" direction, we let G be defined by a term t in $T(R_n, C_n)$ (t has its root colored by a binary operation $\otimes_{M,N,P}$). We take the syntactic tree of t as a layout of G. It is sufficient to prove the claim below to prove that the rank-width of this layout is at most n.

Claim 2. *([4]) If $t = c \bullet t'$, $t' \in T(R_n, C_n)$, $c \in Cxt(R_n, C_n) - \{Id\}$, $G = val(t)$, $H = val(t')$ then we have: $A_G[V_H, V_G - V_H] = \Gamma_H \cdot B$ and $\Gamma_{G[V_H]} = \Gamma_H \cdot C$ for some matrices B and C, and, $rk(A_G[V_H, V_G - V_H]) \le n$.*

For the converse, we prove some technical lemmas. We write $G = H \otimes_M K$ instead of $H \otimes_{M,N,P} K$ if we do not care about the coloring of G but only of its vertices and edges.

Lemma 1. *([4]) Let G be a graph with a bipartition (V_1, V_2) of V_G. Let $m = rk(A_G[V_1, V_2])$. Then $G = H \otimes_M K$ where M is a nonsingular $(m \times m)$-matrix, for some $\mathbb{B}^m$-colorings H and K of $G[V_1]$ and $G[V_2]$ respectively.*

Proposition 3. *([4]) Assume $G = H \otimes_A K$ with A of dimension $p \times q$ of rank k. Let M be a $(k \times k)$-sub-matrix of rank k of A. Then we have N of dimension $p \times k$, P of dimension $q \times k$ such that $A = N \cdot M \cdot P^T$ and $G = Recol_N(H) \otimes_M Recol_P(K)$.*

Lemma 2. *([4]) Let G be a graph, let H, K, L be induced subgraphs such that (V_H, V_K, V_L) is a 3-partition of V_G, with each component not empty. Let $h = \rho_G(V_H)$, $k = \rho_G(V_K)$, $\ell = \rho_G(V_L)$. There exist matrices of appropriate dimensions such that*

$$G = (H \otimes_{M,N_1,N_2} K) \otimes_P L.$$

We can thus prove the following proposition (the "only if" direction of Theorem 1).

Proposition 4. *([4]) Every graph of rank-width at most n is the value of a term in $T(R_n, C_n)$.*

4   A General Framework for Establishing Balancing Theorems

It is known that every graph of tree-width k has a 2-balanced binary tree-decomposition of width at most $3k + 2$ [1] and every graph of m-clique-width k has a 6-balanced m-clique-width expression of width at most $2k$ [8]. We will propose a general framework for establishing *balancing theorems*. This will allow us to prove similar theorems for rank-width, clique-width and NLC-width. Our general framework combines two ideas.

The first idea, coming from [9] consists in introducing binary operations ∘ and • on terms and contexts representing respectively the composition of the unary functions associated with two contexts and the evaluation of such a function for an argument defined by a term. We use a result of [9] showing that every term t in $T(F, C)$ can be replaced by an equivalent 3-balanced *special term* t^b written with ∘, • and the constant Id (the trivial context defining the identity). This construction makes no assumption on the algebraic properties of the signature (F, C).

The second idea introduces a kind of generalization of the notion of an associative and commutative operation. It concerns a subsignature (F, C) of (F', C'). Roughly speaking if f in F is not associative, hence if we do not have $f(x, f(y, z)) = f(f(x, y), z)$ for $f \in F$ then we require that $f(x, f(y, z)) = f'(f(x, y), z)$ for some $f' \in F'$. We say that (F', C') is (F, C)-flexible if this condition and similar ones hold. This condition makes it possible to eliminate from a term written with F, C, ∘, • and Id the operations ∘, • and the constant Id and somehow, to express them in terms of operations of F'.

The idea is to associate with a context $c \in Cxt(F, C)$ an object m_c denoted by a term $\tilde{c}$ in $T(F', C')$ and a function $f^c \in F'$ such that $c \bullet t$ is equivalent to $f^c(\tilde{c}, t)$. For two contexts c and c' we have (by the condition of flexibility) an operation $f^{c,c'}$ in F' such that $\widetilde{c \circ c'}$ is equivalent to $f^{c,c'}(\tilde{c}, \tilde{c'})$. It follows that a special term t over F, C can be transformed into an equivalent term in $T(F', C')$ of no larger height.

By combining the two constructions, we can transform a term $t \in T(F, C)$ into an equivalent 3-balanced term in $T(F', C')$.

In our applications to graph operations we will apply this to a signature (F, C) using k colors (e.g. (R_k, C_k) corresponding to rank-width at most k) and prove that some finite $(F', C') \supseteq (F, C)$ is (F, C)-flexible. This technique also applies to branch-width and tree-width [5].

Let $\mathcal{S}$ be a countable set whose elements are called *sorts*. A *binary $\mathcal{S}$-signature* is a pair (F, C) where F is a set of binary function symbols, each of them having a type $s_1 \times s_2 \to s$ where $s_1, s_2, s \in \mathcal{S}$, and C is a set of nullary symbols, each of them having a type s in $\mathcal{S}$. A nullary symbol is called a *constant*. We say that a binary $\mathcal{T}$-signature (F, C) is a *sub-signature* of (F', C') if $\mathcal{T} \subseteq \mathcal{S}$, $F \subseteq F'$, $C \subseteq C'$ and the types of the elements of F and C are the same for (F', C') and for (F, C). Let $\sigma : F \cup C \to \mathcal{S}$ where $\sigma(f) = s$ if f is a constant of type s or a binary function of type $s_1 \times s_2 \to s$. We define the type of a term $t \in T(F, C)$ as $\sigma(r_t)$, where r_t is its first symbol (the one at the root of its syntactic tree).

Special Terms. We let $S = T(F \cup \{\circ, \bullet\}, C \cup \{Id\})$. We let S_c and S_t be the least subsets of S such that:

$$S_t := S_c \bullet S_t \ \cup \ f(S_t, S_t) \ \cup \ b$$
$$S_c := S_c \circ S_c \ \cup \ f(S_t, S_c) \ \cup \ f(S_c, S_t) \ \cup \ f(S_t, Id) \ \cup \ f(Id, S_t)$$

with rules for each f in F, each b in C. We denote them by $SPE_t(F, C)$ and $SPE_c(F, C)$ if we need to specify F and C. Note that $Id \notin S_t \cup S_c$. The notions of context and the operations $\circ$ and $\bullet$ extend in presence of sorts. We have actually several operations $\circ$, $\bullet$ and several constants Id depending on sorts, but we will overlook this technical point.

For terms t in $SPE_t(F, C) \cup SPE_c(F, C)$ we denote by $|t|_{FC}$ the number of occurrences of symbols from $F \cup C$, by $|t|_0$ the number of occurrences of $\circ$ and $\bullet$, and, by $|t|_{Id}$ the number of occurrences of Id.

Every term t in $SPE_t(F, C)$ evaluates into a term $Eval(t)$ in $T(F, C)$ and every term c in $SPE_c(F, C)$ evaluates into a context $Eval(c)$ in $Cxt(F, C) - \{Id\}$.

A more careful proof than the one of [9, Theorem 1] gives the following result.

Theorem 3. *([4]) For every term t in $T(F, C) - C$ one can construct a term t^b in $SPE_t(F, C)$ such that $|t^b|_{FC} = |t|_{FC} = |t|$, $Eval(t^b) = t$ and $ht(t^b) \le 3 \log(|t| - 1)$. This term can be constructed in time $O(n \log(n))$ if $n = |t|$.*

Comb-term. Let $X_{n+1} = \{x_1, \ldots, x_{n+1}\}$. A *comb-term* is a term in $T(F, X_{n+1})$ of the form $q = f_1(x_1, f_2(x_2, \ldots, f_n(x_n, x_{n+1})) \ldots)$. It contains no constant. We denote it also by $q(x_1, \ldots, x_n, x_{n+1})$ in order to specify the list of variables, in the order in which they occur.

Commutativity. A binary S-signature (F, C) is *commutative* with respect to a class of algebras $\mathscr{C}$ (that will be implicitly assumed in most cases) if for every $f \in F$ there exists a function $\tilde{f}$ in F such that

$$\tilde{f}_M(x, y) = f_M(y, x) \tag{1}$$

for all $M \in \mathscr{C}$, all $x, y \in D_M$.

Comb-decomposition. The *comb-decomposition* of a term $t \in T(F, C) - C$ is the unique writing of t as $q(t_1, \ldots, t_n, b)$ where $q(x_1, \ldots, x_{n+1})$ is a comb-term, $b \in C$ and $t_i \in T(F, C)$.

The following definition makes sense only if F is commutative. Let $c \in Cxt(F, C) - \{Id\}$. Let us define by structural induction on c a comb-term $q(x_1, \ldots, x_n, u)$ for some n, and a sequence $(t_1, \ldots, t_n)$ of terms in $T(F, C)$ such that $c \simeq q(t_1, \ldots, t_n, u)$ and $\simeq$ denotes the equivalence of terms with respect to the intended class $\mathscr{C}$ of algebras (for which F is commutative).

We define $Comb(c)$ and $seq(c)$ as follows:

1. $Comb(c) = f(x_1, u)$ and $seq(c) = (t)$ if $c = f(t, Id)$.
2. $Comb(c) = Comb(c')$ and $seq(c) = seq(c')$ if $c = f(c_1, t)$ and $c' = \tilde{f}(t, c_1)$.
3. $Comb(c) = f(x_1, q(x_2, \ldots, x_{n+1}, u))$ and $seq(c) = (t) \cdot seq(c')$ if $c = f(t, c_1)$, $c_1 \ne Id$ and $Comb(c') = q(x_1, \ldots, x_n, u)$.

These definitions actually extend to contexts defined as terms in $SPE_c(F, C)$. We need only add one clause to *(1)-(3)*:

(4) If $c = c' \circ c''$ (so that $c' \neq Id$, $c'' \neq Id$) if $Comb(c') = q'(x_1, \ldots, x_p, u)$ and $Comb(c'') = q''(x_1, \ldots, x_n, u)$ then we define

$$Comb(c) \text{ as } q'(x_1, \ldots, x_p, q''(x_{p+1}, \ldots, x_{n+p}, u))$$
$$\text{and } seq(c) \text{ as } seq(c') \cdot seq(c'').$$

In the following, we will extend the equivalence relation $\simeq$ by letting $Eval(t) \simeq t$ and $Eval(c) \simeq c$ for terms in $SPE_t(F, C) \cup SPE_c(F, C)$.

Flexibility. We let (F', C') and (F, C) be two binary signatures such that $(F, C) \subseteq (F', C')$. We let $\mathscr{C}$ be a set of (F', C')-algebras. All equivalences of terms and contexts denoted by $\simeq$ will be considered with respect to $\mathscr{C}$. We say that (F', C') is (F, C)-*flexible* if the following conditions hold:

1. F and F' are commutative.
2. There exist three mappings: $q \mapsto \hat{q}$, $q \mapsto f^q$ and $(q, q') \mapsto f^{q,q'}$ which satisfy the following properties:
   (2.1) For every comb-term $q(x_1, \ldots, x_n, u)$ over F with $n \geq 2$, $\hat{q}$ is a comb-term $\hat{q}(x_1, \ldots, x_n)$ over F'.
   (2.2) If $q(x_1, u)$ is the comb-term $g(x_1, u)$ then $\hat{q} = x_1$ and $f^q = g$.
   (2.3) For every q as in (2.1), we have $f^q \in F'$ and $q \simeq f^q(\hat{q}, u)$.
   (2.4) For every two comb-terms as in (2.1) or (2.2) $q(x_1, \ldots, x_p, u)$ and q' $(x_1, \ldots, x_n, u)$ we have $f^{q,q'} \in F'$ and

$$\widehat{q''} \simeq f^{q,q'}(\hat{q}(x_1, \ldots, x_p), \widehat{q'}(x_{p+1}, \ldots, x_{p+n}))$$

   where $q'' = q(x_1, \ldots, x_p, q'(x_{p+1}, \ldots, x_{p+n}, u))$.

If q is a comb-term as in (2.2), Property (2.3) also holds from the definitions of $\hat{q}$ and f^q.

Proposition 5. *([4]) If (F', C') is (F, C)-flexible, then for every term t in $SPE_t(F, C)$ one can define a term $\tilde{t}$ in $T(F', C')$ that is equivalent to t and such that $|\tilde{t}|_{F'C'} = |t|_{FC}$ and $ht(\tilde{t}) \leq ht(t)$.*

Combining Theorem 3 and Proposition 5 we get the following theorem:

Theorem 4. *([4]) Let (F', C') be an (F, C)-flexible S-signature. Every term t in $T(F, C)$ of size n is equivalent to a 3-balanced term t' in $T(F', C')$. This term can be constructed in time $O(n \log(n))$, if we assume that $\hat{q}$, f^q, $f^{q,q'}$ can be constructed in time $O(\max\{|q|, |q'|\})$.*

We can apply this theorem to m-clique-width, rank-width, clique-width and NLC-width. It will suffice to check the flexibility condition for appropriate super-signatures of the signatures that define m-clique-width, rank-width, clique-width and NLC-width.

Theorem 5. *([4,8])*

1. *Every graph of m-clique-width k is the value of a 3-balanced term of m-clique-width at most $2k$.*
2. *Every graph of rank-width k is the value of a 3-balanced term of rank-width at most $2k$.*
3. *Every graph of clique-width or NLC-width k is the value of a 3-balanced clique-width expression of clique-width or NLC-width at most $k \times 2^{k+1}$.*

References

1. Bodlaender, H.L.: NC-algorithms for graphs with small tree-width. In: van Leeuwen, J. (ed.) Graph-Theoretic Concepts in Computer Science. LNCS, vol. 344, pp. 1–10. Springer, Heidelberg (1989)
2. Bodlaender, H.L., Hagerup, T.: Parallel algorithms with optimal speedup for bounded tree-width. SIAM J. Comput. 27, 1725–1746 (1998)
3. Blumensath, A., Courcelle, B.: Recognizability, hypergraph operations and logical types. Information and Computation 204, 853–919 (2006)
4. Courcelle, B., Kanté, M.M.: Multiple Colorings : graph operations characterizing rank-width and balanced graph expressions. Available on http://www.labri.fr/perso/courcell/courcelle_kante07.pdf
5. Courcelle, B., Kanté, M.M.: Balanced Graph Expressions, manuscript (2007)
6. Courcelle, B., Makowsky, J.A., Rotics, U.: Linear time solvable optimization problems on graphs of bounded clique-width. Theory of Computing Systems 33, 125–150 (2000)
7. Courcelle, B., Olariu, S.: Upper bounds to the clique-width of graphs. Discrete Applied Mathematics 101, 77–114 (2000)
8. Courcelle, B., Twigg, A.: Compact Forbidden-set Routing. In: Thomas, W., Weil, P. (eds.) STACS 2007. LNCS, vol. 4393, pp. 37–48. Springer, Heidelberg (2007)
9. Courcelle, B., Vanicat, R.: Query efficient implementation of graphs of bounded clique-width. Discrete Applied Mathematics 131, 129–150 (2003)
10. Courcelle, B., Weil, P.: The recognizability of sets of graphs is a robust property. Theor. Comput. Sci. 342, 173–228 (2005)
11. Downey, R., Fellows, M.: Parameterized complexity. Springer, Heidelberg (1999)
12. Flum, J., Grohe, M.: Parameterized Complexity Theory. Springer, Heidelberg (2006)
13. Hliněný, P., Oum, S.: Finding Branch-decompositions and Rank-decompositions. In: Algorithms–ESA 2007. LNCS, vol. 4698, pp. 163–174. Springer, Heidelberg (October 2007)
14. Kanté, M.M.: Vertex-minor reductions can simulate edge contractions. Discrete Applied Mathematics 155(17), 2328–2340 (2007)
15. Oum, S.: Approximating rank-width and Clique-width Quickly. In: Kratsch, D. (ed.) WG 2005. LNCS, vol. 3787, pp. 49–58. Springer, Heidelberg (2005)
16. Oum, S., Seymour, P.: Approximating clique-width and branch-width. J. Combin Theory, Ser B 96, 514–528 (2006)
17. Robertson, N., Seymour, P.: Graph minors V: excluding a planar graph. J. Combin. Theory (B) 41, 92–114 (1986)
18. Wanke, E.: k-NLC graphs and Polynomial algorithms. Discrete Applied Mathematics 54, 251–266 (1994)

The Clique-Width of Tree-Power and Leaf-Power Graphs

(Extended Abstract)

Frank Gurski and Egon Wanke

Heinrich-Heine-University Düsseldorf
Institute of Computer Science
D-40225 Düsseldorf, Germany
{gurski-wg07,wanke-wg07}@acs.uni-duesseldorf.de

Abstract. The *k-power graph* of a graph G is the graph in which two vertices are adjacent if and only if there is a path between them in G of length at most k. We show that (1.) the k-power graph of a tree has NLC-width at most $k+2$ and clique-width at most $k+2+\max(\lfloor \frac{k}{2} \rfloor - 1, 0)$, (2.) the k-leaf-power graph of a tree has NLC-width at most k and clique-width at most $k+\max(\lfloor \frac{k}{2} \rfloor - 2, 0)$, and (3.) the k-power graph of a graph of tree-width l has NLC-width at most $(k+1)^{l+1} - 1$ and clique-width at most $2 \cdot (k+1)^{l+1} - 2$.

Keywords: tree powers, leaf powers, tree-width, clique-width, NLC-width, strictly chordal.

1 Introduction

The clique-width of a graph G is the smallest integer k such that G can be defined by operations on vertex-labeled graphs using k labels [CO00]. These operations are the vertex disjoint union, the addition of edges between vertices controlled by a label pair, and the relabeling of vertices. The NLC-width[1] of a graph G is defined similarly in terms of closely related operations [Wan94]. The only essential difference between the composition mechanisms of clique-width bounded graphs and NLC-width bounded graphs is the addition of edges. In an NLC-width composition the addition of edges is combined with the union operation. Every graph of clique-width at most k has NLC-width at most k and every graph of NLC-width at most k has clique-width at most $2k$ [Joh98]. Both concepts are useful, because it is sometimes much more comfortable to use NLC-width expressions instead of clique-width expressions and vice versa, respectively. The concept of clique-width generalizes the well-known concept of tree-width defined in [RS86] by the existence of a tree-decomposition. Clique-width bounded graphs and tree-width bounded graphs are particularly interesting from an algorithmic point of view. Many NP-complete graph problems can be solved in polynomial

---

[1] The abbreviation NLC results from the *node label controlled* embedding mechanism originally defined for graph grammars.

A. Brandstädt, D. Kratsch, and H. Müller (Eds.): WG 2007, LNCS 4769, pp. 76–85, 2007.

time for graphs of bounded clique-width [CMR00] and for graphs of bounded tree-width [CM93], respectively.

A well known concept in graph theory is the concept of graph powers [BLS99]. The k-*power graph* G^k of a graph G is a graph with the same vertex set as G. Two vertices in G^k are adjacent if and only if there is a path between them in G of length at most k. Determining whether a given graph is a k-power graph is NP-complete for every fixed integer $k \geq 2$ [MS94].

In this paper, we consider power graphs of graphs of bounded tree-width and leaf-power graphs. A *tree-power graph* is the power graph of a tree, a *leaf-power graph* is the subgraph of a tree-power graph induced by the leaves of the tree. Obviously even leaf-power graphs have unbounded tree-width. In order to obtain a tree structured decomposition we consider the clique-width of these graphs.

Todinca has stated in [Tod03] that the k-power graph of a graph of clique-width l has clique-width at most $2 \cdot l \cdot k^l$. Since trees have NLC-width and clique-width 3, its follows that a k-tree-power graph has NLC-width and clique-width at most $6 \cdot k^3$. We prove in this paper that k-tree-power graphs have NLC-width at most $k + 2$ and clique-width at most $k + 2 + \max(\lfloor \frac{k}{2} \rfloor - 1, 0)$. We also show that k-leaf-power graphs have NLC-width at most k and clique-width at most $k + \max(\lfloor \frac{k}{2} \rfloor - 2, 0)$.

Corneil and Rotics have shown in [CR05] that every graph G of tree-width l has clique-width at most $3 \cdot 2^{l-1}$. This result implies that G^k has NLC-width and clique-width at most $3 \cdot 2^l \cdot k^{3 \cdot 2^{l-1}}$. We improve this bound and show that the k-power graph of a graph of tree-width l has NLC-width at most $(k + 1)^{l+1} - 1$ and clique-width at most $2 \cdot (k + 1)^{l+1} - 2$.

Since strictly chordal graphs are 4-leaf-power graphs, see [KLY06], our results imply that strictly chordal graphs have NLC-width and clique-width at most 4.

2   Preliminaries

Let $[k] := \{1, \ldots, k\}$ be the set of all integers between 1 and k. We work with finite undirected vertex labeled *graphs* $G = (V_G, E_G, \mathrm{lab}_G)$, where V_G is a finite set of *vertices* labeled by some mapping $\mathrm{lab}_G : V_G \to [k]$ and $E_G \subseteq \{\{u, v\} \mid u, v \in V_G, u \neq v\}$ is a finite set of *edges*. The labeled graph consisting of a single vertex labeled by $a \in [k]$ is denoted by $\bullet_a$.

The notion of clique-width is defined by Courcelle and Olariu in [CO00].

Definition 1 (Clique-width, [CO00]). *Let k be some positive integer. The class CW_k of labeled graphs is recursively defined as follows.*

1. *The single vertex graph $\bullet_a$ for some $a \in [k]$ is in CW_k.*
2. *Let $G = (V_G, E_G, \mathrm{lab}_G) \in CW_k$ and $J = (V_J, E_J, \mathrm{lab}_J) \in CW_k$ be two vertex disjoint labeled graphs, then*

$$G \oplus J := (V', E', lab')$$

defined by $V' := V_G \cup V_J$, $E' := E_G \cup E_J$, and

$$lab'(u) := \begin{cases} lab_G(u) & \text{if } u \in V_G \\ lab_J(u) & \text{if } u \in V_J \end{cases}$$

is in CW_k.

1. *Let* $a, b \in [k]$ *be two distinct integers and* $G = (V_G, E_G, lab_G) \in CW_k$ *be a labeled graph, then*
   (a) $\rho_{a \rightarrow b}(G) := (V_G, E_G, lab')$ *defined by*

   $$lab'(u) := \begin{cases} lab_G(u) & \text{if } lab_G(u) \neq a \\ b & \text{if } lab_G(u) = a \end{cases}$$

   is in CW_k *and*
   (b) $\eta_{a,b}(G) := (V_G, E', lab_G)$ *defined by*

   $$E' := E_G \cup \{\{u, v\} \mid u, v \in V_G, \ u \neq v, \ lab_G(u) = a, \ lab_G(v) = b\}$$

   is in CW_k.

The notion of NLC-width is defined by Wanke in [Wan94].

Definition 2 (NLC-width, [Wan94]). *Let* k *be some positive integer. The class* NLC_k *of labeled graphs is recursively defined as follows.*

1. *The single vertex graph* $\bullet_a$ *for some* $a \in [k]$ *is in* NLC_k.
2. *Let* $G = (V_G, E_G, lab_G) \in NLC_k$ *and* $R : [k] \rightarrow [k]$ *be a function, then*

   $$\circ_R(G) := (V_G, E_G, lab')$$

   defined by $lab'(u) := R(lab_G(u))$ *is in* NLC_k.
3. *Let* $G = (V_G, E_G, lab_G) \in NLC_k$ *and* $J = (V_J, E_J, lab_J) \in NLC_k$ *be two vertex disjoint labeled graphs and* $S \subseteq [k]^2$ *be a set of label pairs, then*

   $$G \times_S J := (V', E', lab')$$

   defined by $V' := V_G \cup V_J$,

   $$E' := E_G \cup E_J \cup \{\{u, v\} \mid u \in V_G, \ v \in V_J, \ (lab_G(u), lab_J(v)) \in S\},$$

   and

   $$lab'(u) := \begin{cases} lab_G(u) & \text{if } u \in V_G \\ lab_J(u) & \text{if } u \in V_J \end{cases}$$

   is in NLC_k.

The *clique-width* (*NLC-width*) of a labeled graph G is the least integer k such that $G \in CW_k$ ($G \in NLC_k$, respectively). An expression built with the operations $\bullet_a, \oplus, \rho_{a \rightarrow b}, \eta_{a,b}$ for integers $a, b \in [k]$ is called a *clique-width* k*-expression*. An expression built with the operations $\bullet_a, \times_S, \circ_R$ for $a \in [k]$, $S \subseteq [k]^2$, and $R : [k] \rightarrow [k]$ is called an *NLC-width* k*-expression*. The graph defined by an expression X is denoted by val(X).

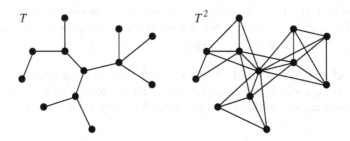

Fig. 1. The figure shows a tree T and its 2-tree-power graph T^2

3   Clique-Width of Tree-Power Graphs

Any k-power graph of a tree is a chordal graph, see [LS95]. Determining whether a given graph G is the k-power graph of a tree can be done in linear time for every fixed integer k [CKL06].

Theorem 1. *The k-power graph of a tree has NLC-width at most $k + 2$ and clique-width at most $k + 2 + \max(\lfloor \frac{k}{2} \rfloor - 1, 0)$.*

Proof (Sketch). Let T be a tree with root u. Then for every vertex v of T we define an NLC-width $(k + 2)$-expression X_v. This expression will define the k-power graph of the subtree of T with root v. The set of labels is the set of integers[2] between 0 and $k + 1$. In $\mathrm{val}(X_v)$ a vertex has label $r \leq k$ if its distance to v is r. That is, the label of v is 0, the labels of the children of v are 1, and so on. All vertices with a distance to v greater than k have label $k + 1$.

Let $v_1, \ldots, v_m$ be the children of vertex v in T. Then

$$X_v := \begin{cases} \bullet_0 & \text{if } m = 0 \\ \circ_R(X_{v_1}) \times_S \bullet_0 & \text{if } m = 1 \\ \circ_R(X_{v_1}) \times_S \ldots \times_S \circ_R(X_{v_m}) \times_S \bullet_0 & \text{if } m > 1 \end{cases}$$

where

$$S := \{(a, b) \mid a, b \in \{0, \ldots, k\} \text{ and } a + b \leq k\}$$

and

$$R(a) := \begin{cases} a + 1 \text{ if } 0 \leq a \leq k \\ k + 1 \text{ if } a = k + 1 \end{cases}.$$

Then X_u defines the k-power graph of tree T and uses at most $k + 2$ labels.

To define a clique-width expression X_v, the edges between the vertices of two subtrees can be inserted with additional auxiliary labels. These labels are necessary, because we can not insert edges between equal labeled vertices. We only need to relabel the labels $1, \ldots, \lfloor \frac{k}{2} \rfloor$ in one of the two combined graphs. After

---

[2] In the definition of the NLC-width and clique-width, the vertex labels are always positive integers. It is easy to see that the labels we use for the vertices can be changed to conform the definition without to increase the number of used labels.

all succeeding edge insertions between the two subtrees the auxiliary labels can be relabeled back to $1, \ldots, \lfloor \frac{k}{2} \rfloor$. Since label 0 can be used as an auxiliary label, we only need $\lfloor \frac{k}{2} \rfloor - 1$ auxiliary labels at all.   □

By the results of [CKL06] and Theorem 1, it follows that for every k-tree power graph G, an NLC-width $(k+2)$-expression and a clique-width $(k+2+\max(\lfloor \frac{k}{2} \rfloor - 1, 0))$-expression can be found in linear time for every fixed integer k.

4   Clique-Width of Leaf-Power Graphs

The notion of a leaf-power graph was introduced in [NRT02] motivated by the reconstruction of phylogenetic trees as a certain case of tree-power graphs. The *k-leaf-power* graph T^k of a tree T is a graph whose vertices are the leaves of T. Two vertices in T^k are adjacent if and only if there is a path between them in T of length at most k. Figure 2 shows an example of a 3-leaf-power graph.

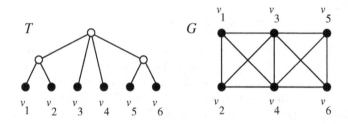

Fig. 2. A tree T and its 3-leaf-power graph G

Every k-leaf-power graph is strongly chordal, see [BL06]. A graph is a 2-leaf-power graph if and only if it is the disjoint union of cliques, thus they have NLC-width 1 and clique-width at most 2. A graph is a 3-leaf-power graph if and only if it is obtained from a tree T by substituting the vertices of T by cliques, see [BL06]. That is, the NLC-width and clique-width of 3-leaf-power graphs is at most 3. 3-leaf-power and 4-leaf-power graphs can be recognized in linear time, see [BL06] and [BLS05]. For the recognition of k-leaf-power graphs with $k \geq 5$ no polynomial time algorithms are known.

In [Rau06, DGHN05] a forbidden subgraph characterization for prime 4-leaf-power graphs is given. All strictly chordal graphs, i.e. chordal graphs whose clique hypergraph is a strict hypertree, are k-leaf-power graphs for $k \geq 4$. The corresponding trees can be found in linear time, see [KLY06].

We next give a bound on the NLC-width and clique-width of k-leaf-power graphs for $k \geq 2$.

Theorem 2. *The k-leaf-power graph of a tree has NLC-width at most k and clique-width at most $k + \max(\lfloor \frac{k}{2} \rfloor - 2, 0)$.*

Proof (Sketch). Let T be a tree with root u. Then for every inner vertex v of T we define an NLC-width k-expression X_v. This expression will define the k-leaf-power graph of the subtree of T with root v. The set of labels is the set of integers between 1 and k. In $\text{val}(X_v)$ a vertex u has label $r < k$ if its distance to v is r. All vertices with a distance to v greater than or equal to k have label k.

Let $v_1, \ldots, v_m$ be the non-leaf children of inner vertex v in T and let l be the number of children of v that are leaves. A basic observation for leaf-power graphs is that the leaf children of v induce a complete graph of T^k, for $k \geq 2$. Let C_l be an NLC-width 1-expression for a complete graph on l vertices. Then

$$X_v := \begin{cases} C_l & \text{if } m = 0 \text{ and } l > 0 \\ \circ_R(X_{v_1}) \times_S C_l & \text{if } m = 1 \text{ and } l > 0 \\ \circ_R(X_{v_1}) \times_S \ldots \times_S \circ_R(X_{v_m}) \times_S C_l & \text{if } m > 1 \text{ and } l > 0 \\ \circ_R(X_{v_1}) & \text{if } m = 1 \text{ and } l = 0 \\ \circ_R(X_{v_1}) \times_S \ldots \times_S \circ_R(X_{v_m}) & \text{if } m > 1 \text{ and } l = 0 \end{cases}$$

where

$$S := \{(a, b) \mid a, b \in \{1, \ldots, k - 1\} \text{ and } a + b \leq k\}$$

and

$$R(a) := \begin{cases} a + 1 & \text{if } 1 \leq a \leq k - 1 \\ k & \text{if } a = k \end{cases}.$$

Then X_u defines the k-leaf-power graph of T and uses at most k labels.

To define a clique-width expression X_v, C_l is a clique-width 2-expression that defines a clique with l vertices all labeled by label 1. The edges between the leaves of two subtrees can be inserted similarly as in the proof of Theorem 1 with additional auxiliary labels. Now we only need to relabel the labels $2, \ldots, \lfloor \frac{k}{2} \rfloor$ in one of the two combined graphs. Since label 1 can be used as auxiliary label, we only need $\lfloor \frac{k}{2} \rfloor - 2$ auxiliary labels at all.   □

In [Tod03] the set of clique-width operations was extended by operations of the form $\eta_{a,a}$ in order to define clique-width expressions that allow to insert edges between equal labeled vertices. The set of k-labeled graphs which can be defined by these extended clique-width operations using k labels is denoted by CW'_k. A simple observation shows that $\text{CW}_k \subseteq \text{CW}'_k \subseteq \text{CW}_{2k}$. It is easy to see that every k-tree-power graph is in CW'_{k+2} and every k-leaf-power graph is in CW'_k.

5   Clique-Width of k-Power Graphs of Graphs of Bounded Tree-Width

Now we consider the clique-width of power graphs of graphs of bounded tree-width.

Theorem 3. *The k-power graph of a graph of tree-width l has NLC-width at most $(k + 1)^{l+1} - 1$.*

To prove Theorem 3, we need the following notations. The tree-width of a graph is defined by Robertson and Seymour in [RS86]. A graph of tree-width l is also known as partial l-tree, see Rose [Ros74]. A *partial l-tree* is a subgraph of an l-tree which is defined as follows:

1. A complete graph with l vertices is an l-tree.
2. If G is an l-tree, then the result obtained by adding a new vertex with edges to l vertices of a complete subgraph of G is an l-tree.

Let G be a graph with n vertices. Let $o = (v_1, \ldots, v_n)$ be an order of the n vertices of G, i.e. every vertex of G appears in sequence o exactly once. Let $N^+(G, o, i)$ and $N^-(G, o, i)$ for $i = 1, \ldots, n$ be the set of neighbors v_j of vertex v_i with $i < j$ and $j < i$, respectively. That is,

$$N^+(G, o, i) := \{v_j \mid \{v_i, v_j\} \in E_G \ \wedge \ i < j\}$$

and

$$N^-(G, o, i) := \{v_j \mid \{v_j, v_i\} \in E_G \ \wedge \ j < i\}.$$

A vertex order $o = (v_1, \ldots, v_n)$ for graph G is called a *perfect elimination order* (PEO) if the vertices of $N^+(G, o, i)$ for $i = 1, \ldots, n$ induce a complete subgraph of G.

Proof (of Theorem 3, Sketch). Let G be a partial l-tree that is a subgraph of an l-tree $\hat{G}$, such that G has the same vertex set as $\hat{G}$. By the recursive definition of l-trees, we know that each l-tree $\hat{G}$ has a PEO. Let $o := (v_1, \ldots, v_n)$ be a PEO of $\hat{G}$. Then, the vertices of the sets $N^+(\hat{G}, o, 1), \ldots, N^+(\hat{G}, o, n-l)$ induce l vertex complete subgraphs of $\hat{G}$. The vertices of each $N^+(\hat{G}, o, i)$ for $n - l < i \leq n$ induce an $(n - i)$ vertex complete subgraph of $\hat{G}$.

Each l-tree $\hat{G}$ is $(l + 1)$-colorable, because we can assign to v_i any color not used by the vertices of $N^+(\hat{G}, o, i)$ for $i = 1, \ldots, n$. Let $\text{col} : V_{\hat{G}} \to [l + 1]$ be an $(l + 1)$-coloring of l-tree $\hat{G}$, that is, $\text{col}(v_i) \neq \text{col}(v_j)$ for all edges $\{v_i, v_j\} \in E_{\hat{G}}$. The mapping col is also an $(l + 1)$-coloring of partial l-tree G.

We define C_{v_i}, $1 \leq i \leq n$, to be the set of all colors of the vertices from $N^+(G, o, i)$, i.e.,

$$C_{v_i} := \{\text{col}(v_j) \mid \{v_i, v_j\} \in E_G \wedge i < j\}.$$

The structure of l-tree $\hat{G}$ with respect to PEO o can be characterized by the following tree $T := (V_T, E_T)$ defined by

$$V_T := V_{\hat{G}}$$
$$E_T := \{\{v_i, v_j\} \in E_{\hat{G}} \mid i < j \ \wedge \ \forall j', i < j' < j \ : \ \{v_i, v_{j'}\} \notin E_{\hat{G}}\}.$$

Graph T is a tree, because for each vertex v_i, $i < n$, tree T has exactly one edge $\{v_i, v_j\} \in E_{\hat{G}}$ with $i < j$. Let v_n be the root of T.

We recursively define for $i = 1, \ldots, n$ an NLC-width expression X_i which defines the k-power graph of the subgraph of G induced by the vertices v_j, $1 \leq j \leq i$. The vertices will be labeled by sets[2] of pairs (c, d), where $c \in [l+1]$ is a *color* and $d \in [k]$ is a *distance*. In a vertex label, for every color $c \in [l+1]$ there is at most one pair (c, d) for some $d \in [k]$. This implies that we have at most $(k+1)^{l+1}$ different vertex labels. A color c is either not used, or used together with exactly one of k distances. We will see later that there is one label which is not used at all, this is the label $\{(1, 1), \ldots, (l+1, 1)\}$.

For a label $C \subseteq [l+1] \times [k]$ and a positive integer r let

$$C^{+r} := \{(c, d+r) \mid (c, d) \in C \text{ and } d + r \leq k\}.$$

For two labels $C_1, C_2 \subseteq [l+1] \times [k]$ let

$$C_1 \overset{\min}{\cup} C_2$$

be the set of all pairs $(c, d) \in C_1 \cup C_2$ such that there is no pair $(c, d') \in C_1 \cup C_2$ with $d' < d$.

We first define an NLC-width expression X_n for a subgraph of G^k with vertex set V_{G^k}. After that, we will show how the missing edges are added by inserting additional pairs in the relations S used by the edge insertion operations $\times_S$ of X_n.

Expression X_i for $i = 1, \ldots, n$ is defined as follows.

1. If $N^-(T, o, i) = \emptyset$, i.e. if vertex v_i is a leaf in tree T, then we define

$$X_i := \bullet_{\{(c,1) \mid c \in C_{v_i}\}}.$$

2. If $N^-(T, o, i) = \{v_{j_1}, \ldots, v_{j_m}\}$, i.e. if $v_{j_1}, \ldots, v_{j_m}$ are the sons of vertex v_i in tree T, then we define

$$
\begin{aligned}
X_{i,1} &:= X_{j_1} \\
X_{i,2} &:= X_{i,1} \times_{S_1} X_{j_2} \\
&\ \ \vdots \\
X_{i,m} &:= X_{i,m-1} \times_{S_{m-1}} X_{j_m} \\
X_i &:= \circ_R(\bullet_{D_i} \times_{S_m} X_{i,m}),
\end{aligned}
$$

where all $S_1, \ldots, S_{m-1}$ are (so far) empty and

$$S_m := \{(D_i, C) \mid C \subseteq [l+1] \times [k] \wedge (\mathrm{col}(v_i), d) \in C \text{ for some } d\}.$$

The label D_i is defined as follows. Consider all vertices of $\mathrm{val}(X_{i,m})$ whose labels have a pair with color $\mathrm{col}(v_i)$. These are exactly the vertices which will be connected with the vertex v_i of $\mathrm{val}(\bullet_{D_i})$ by $\times_{S_m}$. Every of these labels has exactly one pair with color $\mathrm{col}(v_i)$. Let $C_1, \ldots, C_r$ be the labels of these vertices and let $(\mathrm{col}(v_i), d_1), \ldots, (\mathrm{col}(v_i), d_r)$ be the corresponding pairs with color $\mathrm{col}(v_i)$. Then D_i is the set

$$D_i := \{(c,1) \mid c \in C_{v_i}\} \; \overset{\min}{\underset{1 \le j \le r}{\bigcup}} \; (C_j - \{(\mathrm{col}(v_i), d_j)\})^{+d_j}$$

Operation $\circ_R$ relabels every label C_s, $1 \le s \le r$, as defined above into

$$R(C_s) := C_s - \{(\mathrm{col}(v_i), d_s)\} \; \overset{\min}{\bigcup} \; \{(c, 1 + d_s) \mid c \in C_{v_i}\}$$

$$\overset{\min}{\underset{1 \le j \le r, j \ne s}{\bigcup}} \; (C_j - \{(\mathrm{col}(v_i), d_j)\})^{+(d_j + d_s)}$$

Note that operation $\circ_R$ does not change label D_i of v_i, because it has no pair with color $\mathrm{col}(v_i)$.

Let $p = (v_{i_1}, \ldots, v_{i_m})$ be a path in the partial l-tree G. Path p is called a *high-end-path* if $i_j < i_m$ for $j = 1, \ldots, m - 1$. The graph defined by NLC-width expression X_n above has an edge between two vertices v_i, v_j, $i < j$, if and only if there is a high-end-path from v_i to v_j in G of length at most k. Additionally[3], if u and u' are two equal labeled vertices of $\mathrm{val}(X_i)$, $1 \le i \le n$, then there is a high-end-path from u to some vertex v_j, $j > i$, of length $k' \le k$ in G if and only if there is a high-end-path from u' to v_j of length k' in G.

Now we extend the sets S of all edge insertion operations $\times_S$ of X_n as follows. For every subexpression $Y \times_S Z$ of X_n such that graph $\mathrm{val}(Y)$ defined by Y has a vertex u, graph $\mathrm{val}(Z)$ defined by Z has a vertex w, and there is a path p of length $k' \le k$ between u and w in G, we add the label pair $(\mathrm{lab}_{\mathrm{val}(Y)}(u), \mathrm{lab}_{\mathrm{val}(Z)}(w))$ to S. The new expression is denoted by $\widetilde{X_n}$.

The graph $\mathrm{val}(\widetilde{X_n})$ defined by the new expression $\widetilde{X_n}$ certainly has all edges of the k-power graph G^k. The following observation shows that it has only edges of G^k. Let v_j be the vertex of the path p with the largest index j with respect to the perfect elimination order o. Then, the first part of p from u to v_j is a high-end-path of length $k_1 \le k'$ and the last part of p from w back to v_j is a high-end-path of length $k_2 \le k'$ such that $k_1 + k_2 = k'$. By the definition of the vertex labeling, every vertex of $\mathrm{val}(Y)$ labeled by $\mathrm{lab}_{\mathrm{val}(Y)}(u)$ has a high-end-path to v_j of length k_1 in G, and every vertex of $\mathrm{val}(Z)$ labeled by $\mathrm{lab}_{\mathrm{val}(Z)}(w)$ has a high-end-path to v_j of length k_2 in G. Thus, all vertices of $\mathrm{val}(Y)$ labeled by $\mathrm{lab}_{\mathrm{val}(Y)}(u)$ can be connected with all vertices of $\mathrm{val}(Z)$ labeled by $\mathrm{lab}_{\mathrm{val}(Z)}(w)$, because there are paths between them of length $k' \le k$ in G. □

Since every graph of NLC-width k has clique-width at most $2k$, we get the following corollary.

Corollary 1. *The k-power graph of a graph of tree-width l has clique-width at most $2 \cdot (k + 1)^{l+1} - 2$.*

---

[3] The proof of this fact is omitted due to of space restrictions for this abstract.

References

[BL06]   Brandstädt, A., Le, V.B.: Structure and linear time recognition of 3-leaf powers. Information Processing Letters 98, 113–138 (2006)

[BLS99]   Brandstädt, A., Le, V.B., Spinrad, J.P.: Graph Classes: A Survey. In: SIAM Monographs on Discrete Mathematics and Applications, SIAM, Philadelphia (1999)

[BLS05]   Brandstädt, A., Le, V.B., Sritharan, R.: Structure and linear time recognition of 4-leaf powers. Manuscript (2005)

[CKL06]   Chang, M.-S., Ko, M.-T., Lu, H.-I.: Linear-time algorithms for tree root problems. In: Arge, L., Freivalds, R. (eds.) SWAT 2006. LNCS, vol. 4059, pp. 411–422. Springer, Heidelberg (2006)

[CM93]   Courcelle, B., Mosbah, M.: Monadic second-order evaluations on tree-decomposable graphs. Theoretical Computer Science 109, 49–82 (1993)

[CMR00]   Courcelle, B., Makowsky, J.A., Rotics, U.: Linear time solvable optimization problems on graphs of bounded clique-width. Theory of Computing Systems 33(2), 125–150 (2000)

[CO00]   Courcelle, B., Olariu, S.: Upper bounds to the clique width of graphs. Discrete Applied Mathematics 101, 77–114 (2000)

[CR05]   Corneil, D.G., Rotics, U.: On the relationship between clique-width and treewidth. SIAM Journal on Computing 4, 825–847 (2005)

[DGHN05]   Dom, M., Guo, J., Hüffner, F., Niedermeier, R.: Extending the tractability border for closest leaf powers. In: Kratsch, D. (ed.) WG 2005. LNCS, vol. 3787, pp. 397–408. Springer, Heidelberg (2005)

[Joh98]   Johansson, Ö.: Clique-decomposition, NLC-decomposition, and modular decomposition - relationships and results for random graphs. Congressus Numerantium 132, 39–60 (1998)

[KLY06]   Kennedy, W., Lin, G., Yan, G.: Strictly chordal graphs are leaf powers. Journal of Discrete Algorithms 4(4), 511–525 (2006)

[LS95]   Lin, Y.L., Skiena, S.S.: Algorithms for square roots of graphs. SIAM Journal on Discrete Mathematics 8(1), 99–118 (1995)

[MS94]   Motwani, R., Sudan, M.: Computing roots of graphs is hard. Discrete Applied Mathematics 54(1), 81–88 (1994)

[NRT02]   Nishimura, N., Ragde, P., Thilikos, D.M.: On graph powers for leaf-labeled trees. Journal of Algorithms 56, 69–108 (2002)

[Rau06]   Rautenbach, D.: Some remarks about leaf roots. Discrete Mathematics 306, 1456–1461 (2006)

[Ros74]   Rose, D.J.: On simple characterizations of k-trees. Discrete Mathematics 7, 317–322 (1974)

[RS86]   Robertson, N., Seymour, P.D.: Graph minors II. Algorithmic aspects of tree width. Journal of Algorithms 7, 309–322 (1986)

[Tod03]   Todinca, I.: Coloring powers of graphs of bounded clique-width. In: Bodlaender, H.L. (ed.) WG 2003. LNCS, vol. 2880, pp. 370–382. Springer, Heidelberg (2003)

[Wan94]   Wanke, E.: k-NLC graphs and polynomial algorithms. Discrete Applied Mathematics 54, 251–266 (1994)

NLC-2 Graph Recognition and Isomorphism*

Vincent Limouzy[1], Fabien de Montgolfier[1], and Michaël Rao[1]

LIAFA - Univ. Paris Diderot
{limouzy,fm,rao}@liafa.jussieu.fr

Abstract. NLC-width is a variant of clique-width with many application in graph algorithmic. This paper is devoted to graphs of NLC-width two. After giving new structural properties of the class, we propose a $O(n^2m)$-time algorithm, improving Johansson's algorithm [14]. Moreover, our alogrithm is simple to understand. The above properties and algorithm allow us to propose a robust $O(n^2m)$-time isomorphism algorithm for NLC-2 graphs. As far as we know, it is the first polynomial-time algorithm.

1 Introduction

NLC-width is a graph parameter introduced by Wanke [16]. This notion is tightly related to clique-width introduced by Courcelle *et al.* [2]. Both parameters were introduced to generalise the well known tree-width. The motivation on research about such *width* parameter is that, when the width (NLC-, clique- or tree-width) is bounded by a constant, then many NP-complete problems can be solved in polynomial (even linear) time, if the decomposition is provided.

Such parameters give insights on graph structural properties. Unfortunately, finding the minimum NLC-width of the graph was shown to be NP-hard by Gurski *et al.* [12]. Some results however are known. Let NLC-k be the class of graph of NLC width bounded by k. NLC-1 is exactly the class of cographs. Probe-cographs, bi-cographs and weak-bisplit graphs [9] belong to NLC-2. Johansson [14] proved that recognising NLC-2 graphs is polynomial and provided an $O(n^4 \log(n))$ recognition algorithm. Complexity for recognition of NLC-k, $k \geq 3$, is still unknown.

In this paper we improve Johansson's result down to $O(n^2m)$. Our approach relies on graph decompositions. We establish the tight links that exist between NLC-2 graphs and the so-called modular decomposition, split decomposition, and bi-join decomposition.

NLC-2 can be defined as a graph colouring problem. Unlike NLC-k classes, for $k \geq 3$, *recolouring* is useless for prime NLC-2 graphs. That allow us to propose a canonical decomposition of bi-coloured NLC-2 graphs, defined as certain bi-coloured split operations. This decomposition can be computed in $O(nm)$ time if the colouring is provided. If a graph is *prime*, there using split and bi-join

---

* Research supported by the ANR project *Graph Decompositions and Algorithms* (GRAAL) and by INRIA project-team GANG.

A. Brandstädt, D. Kratsch, and H. Müller (Eds.): WG 2007, LNCS 4769, pp. 86–98, 2007.

decompositions, we show that there is at most $O(n)$ colourings to check. Finally, modular decomposition properties allow to reduce NLC-2 graph decomposition to prime NLC-2 graph decomposition. Section 3 explains this $O(n^2 m)$-time decomposition algorithm.

In Section 4 is proposed an isomorphism algorithm. Using modular, split and bi-join decompositions and the canonical NLC-2 decomposition, isomorphism between two NLC-2 graphs can be tested in $O(n^2 m)$ time.

2   Preliminaries

A graph $G = (V, E)$ is pair of a set of *vertices* V and a set of *edges* E. For a graph G, $V(G)$ denote its set of vertices, $E(G)$ its set of edges, $n(G) = |V(G)|$ and $m(G) = |E(G)|$ (or V, E, n and m if the graph is clear in the context). $N(x) = \{y \in V : \{x, y\} \in E\}$ denotes the *neighbourhood* of the vertex x, and $N[x] = N(v) \cup \{v\}$. For $W \subseteq V$, $G[W] = (W, E \cap W^2)$ denote the *graph induced by* W. Let A and B be two disjoint subsets of V. Then we note $A \oplus B$ if for all $(a, b) \in A \times B$, then $\{a, b\} \in E$, and we note $A \oslash B$ if for all $(a, b) \in A \times B$, then $\{a, b\} \notin E$. Two graphs $G = (V, E)$ and $G' = (V', E')$ are *isomorphic* (noted $G \simeq G'$) if there is a bijection $\varphi : V \to V'$ such that $\{x, y\} \in E \Leftrightarrow \{\varphi(x), \varphi(y)\} \in E'$, for all $u, v \in V$.

A *k-labelling* (or *labelling*) is a function $l : V \to \{1, \dots, k\}$. A *k-labelled graph* is a pair of a graph $G = (V, E)$ and a k-labelling l on V. It is denoted by (G, l) or by (V, E, l). Two labelled graphs (V, E, l) and (V', E', l') are isomorphic if there is a bijection $\varphi : V \to V'$ such that $\{u, v\} \in E \Leftrightarrow \{\varphi(x), \varphi(y)\} \in E'$ and $l(u) = l'(\varphi(u))$ for all $u, v \in V$. Let k be a positive integer. The class of *NLC-k graphs* is defined recursively by the following operations.

- For all $i \in \{1, \dots, k\}$, $\cdot(i)$ is in NLC-k, where $\cdot(i)$ is the graph with one vertex labelled i.
- Let $G_1 = (V_1, E_1, l_1)$ and $G_2 = (V_2, E_2, l_2)$ be NLC-k and let $S \subseteq \{1, \dots, k\}^2$. Then $G_1 \times_S G_2$ is in NLC-k, where $G_1 \times_S G_2 = (V, E, l)$ with $V = V_1 \cup V_2$,

$$E = E_1 \cup E_2 \cup \{\{u, v\} : (u, v) \in V_1 \times V_2 \text{ and } (l_1(u), l_2(v)) \in S\}$$

$$\text{and for all } u \in V, \, l(u) = \begin{cases} l_1(u) \text{ if } u \in V_1 \\ l_2(u) \text{ if } u \in V_2. \end{cases}$$

- Let $R : \{1, \dots, k\} \to \{1, \dots, k\}$ and $G = (V, E, l)$ be NLC-k. Then $\rho_R(G)$ is in NLC-k, where $\rho_R(G) = (V, E, l')$ such that $l'(u) = R(l(u))$ for all $u \in V$.

A graph is NLC-k if there is a k-labelling of G such that (G, l) is in NLC-k. A k-labelled graph is *NLC-k ρ-free* if it can be constructed without the ρ_R operation.

Modules and modular decomposition. A *module* in a graph is a non-empty subset $X \subseteq V$ such that for all $u \in V \setminus X$, then either $N(u) \cap X = \emptyset$ or $X \subseteq N(u)$. A module is *trivial* if $|X| \in \{1, |V|\}$. A graph is *prime* (w.r.t. modular decomposition) if all its modules are trivial. Two sets X and X' *overlap* if $X \cap X', X \setminus X'$

and $X' \setminus X$ are non-empty. A module X is *strong* if there is no module X' such that X and X' overlap. Let $\mathcal{M}'(G)$ be the set of modules of G, let $\mathcal{M}(G)$ be the set of its strong modules, and let $\mathcal{P}(G) = \{M_1, \ldots, M_k\}$ be the maximal (w.r.t. inclusion) members of $\mathcal{M}(G) \setminus \{V\}$.

Theorem 1. *[11] Let $G = (V, E)$ be a graph such that $|V| \geq 2$. Then:*

- *if G is not connected, then $\mathcal{P}(G)$ is the set of connected components of G,*
- *if $\overline{G}$ is not connected, then $\mathcal{P}(G)$ is the set of connected components of $\overline{G}$,*
- *if G and $\overline{G}$ are connected, then $\mathcal{P}(G)$ is a partition of V and is formed with the maximal members of $\mathcal{M}' \setminus \{V\}$.*

$\mathcal{P}(G)$ is a partition of V, and G can be decomposed into $G[M_1], \ldots, G[M_k]$, where $\mathcal{P}(G) = \{M_1, \ldots, M_k\}$. The *characteristic graph* G^* of a graph G is the graph of vertex set $\mathcal{P}(G)$ and two $P, P' \in \mathcal{P}(G)$ are adjacent if there is an edge between P and P' in G (and so there is no non-edges since P and P' are two modules). The recursive decomposition of a graph by this operation gives the *modular decomposition* of the graph, and can be represented by a rooted tree, called the *modular decomposition tree*. It can be computed in linear time [15]. The nodes of the modular decomposition tree are exactly the strong modules, so in the following we make no distinction between the modular decomposition of G and $\mathcal{M}(G)$. Note that $|\mathcal{M}(G)| \leq 2 \times n - 1$. For $M \in \mathcal{M}(G)$, let $G_M = G[M]$ and G_M^* its characteristic graph.

Lemma 1. *[14] Let G be a graph. G is NLC-k if and only if every characteristic graph in the modular decomposition of G is NLC-k.*

Moreover, a NLC-k expression for G can be easily constructed from the modular decomposition and from NLC-k expressions of prime graphs. On prime graphs, NLC-2 recognition is easier:

Lemma 2. *[14] Let G be a prime graph. Then G is NLC-2 if and only if there is a 2-labelling l such that (G, l) is NLC-2 ρ-Free.*

Bi-partitive family. A *bipartition* of V is a pair $\{X, Y\}$ such that $X \cap Y = \emptyset$, $X \cup Y = V$ and X and Y are both non empty. Two bipartitions $\{X, Y\}$ and $\{X', Y'\}$ *overlap* if $X \cap Y$, $X \cap Y'$, $X' \cap Y$ and $X' \cap Y'$ are non empty. A family $\mathcal{F}$ of bipartitions of V is *bipartitive* if (1) for all $v \in V$, $\{\{v\}, V \setminus \{v\}\} \in \mathcal{F}$ and (2) for all $\{X, Y\}$ and $\{X', Y'\}$ in $\mathcal{F}$ such that $\{X, Y\}$ and $\{X', Y'\}$ overlap, then $\{X \cap X', Y \cup Y'\}$, $\{X \cap Y', Y \cup X'\}$, $\{Y \cap X', X \cup Y'\}$, $\{Y \cap Y', X \cup X'\}$ and $\{X \Delta X', X \Delta Y'\}$ are in $\mathcal{F}$ (where $X \Delta Y = (X \setminus Y) \cup (Y \setminus X)$). Bipartitive families are very close to partitive families [1], which generalise properties of modules in a graph.

A member $\{X, Y\}$ of a bipartitive family $\mathcal{F}$ is *strong* if there is no $\{X', Y'\}$ such that $\{X, Y\}$ and $\{X', Y'\}$ overlap. Let T be a tree. For an edge e in the tree, $\{C_e^1, C_e^2\}$ denote the bipartition of leaves of T such that two leaves are in the same set if and only if the path between them avoids e. Similarly, for an internal node α, $\{C_\alpha^1, \ldots, C_\alpha^{d(\alpha)}\}$ denote the partition of leaves of T such that two leaves are in the same set if and only if the path between them avoid α.

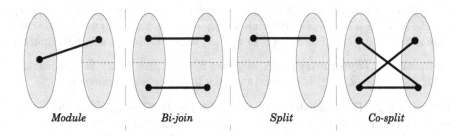

Fig. 1. A module, a bi-join, a split and a co-split

Theorem 2. *[4] Let $\mathcal{F}$ be a bipartitive family on V. Then there is an unique unrooted tree T, called the* representative tree *of $\mathcal{F}$, such that the set of leaves of T is V, the internal nodes of T are labelled* **degenerate** *or* **prime**, *and*
- *for every edge e of T, $\{C_e^1, C_e^2\}$ is a strong member of $\mathcal{F}$, and there is no other strong member in $\mathcal{F}$,*
- *for every node α labelled* **degenerate**, *and for every $\emptyset \subsetneq I \subsetneq \{1, \ldots, d(\alpha)\}$, $\{\cup_{i \in I} C_\alpha^i, V \setminus \cup_{i \in I} C_\alpha^i\}$ is in $\mathcal{F}$, and there is no other member in $\mathcal{F}$.*

A *split* in a graph $G = (V, E)$ is a bipartition $\{X, Y\}$ of V such that the set of vertices in X having a neighbour in Y have the same neighbourhood in Y (*i.e.*, for all $u, v \in X$ such that $N(u) \cap Y \neq \emptyset$ and $N(v) \cap Y \neq \emptyset$, then $N(u) \cap Y = N(v) \cap Y$). A *co-split* in a graph G is a split in $\overline{G}$. The family of split in a connected graph is a bipartitive family [3]. The split decomposition tree is the representative tree of the family of splits, and can be computed in linear time [5]. Let α be an internal node of the split decomposition tree of a connected graph G. For all $i \in \{1, \ldots, d(\alpha)\}$ let $v_i \in C_\alpha^i$ such that $N(v_i) \setminus C_\alpha^i \neq \emptyset$. Since G is connected, such a v_i always exists. $G[\{v_1, \ldots, v_{d(\alpha)}\}]$ denote the *characteristic graph* of α. The characteristic graph of a **degenerate** node is a complete graph or a star [3].

A *bi-join* in a graph is a bipartition $\{X, Y\}$ such that for all $u, v \in X$, $\{N(u) \cap Y, Y \setminus N(u)\} = \{N(v) \cap Y, Y \setminus N(v)\}$. The family of bi-joins in a graph is bipartitive. The *bi-join decomposition tree* is the representative tree of the family of bi-joins, and can be computed in linear time [7,8]. Let α be an internal node of the bi-join decomposition tree of a graph G. For all $i \in \{1, \ldots, d(\alpha)\}$ let $v_i \in C_\alpha^i$. $G[\{v_1, \ldots, v_{d(\alpha)}\}]$ denote the *characteristic graph* of α. The characteristic graph of a **degenerate** node is a complete bipartite graph or a disjoint union of two complete graphs [7,8].

3   Recognition of NLC-2 Graphs

3.1   NLC-2 ρ-Free Canonical Decomposition

In this section, $G = (V, E, l)$ is a 2-labelled graph such that every mono-coloured module (*i.e.* a module M such that $\forall v, v' \in M$, $l(v) = l(v')$) has size 1. A couple

(X,Y) is a *cut* if $X \cup Y = V$, $X \cap Y = \emptyset$, $X \neq \emptyset$ and $Y \neq \emptyset$. Let $S \subseteq \{1,2\} \times \{1,2\}$. A cut (X,Y) is a S-*cut* of G if for all $u \in X$ and $v \in Y$, then $\{u,v\} \in E$ if and only if $(l(u), l(v)) \in S$. For $S \subseteq \{1,2\} \times \{1,2\}$ let $\mathcal{F}_S(G)$ be the set of S-cut of G.

Definition 1 (Symmetry). *We say that* $S \in \{1,2\} \times \{1,2\}$ *is* symmetric *if* $(1,2) \in S \iff (2,1) \in S$, *otherwise we say that* S *is* non-symmetric.

Definition 2 (Degenerate property). *A family* $\mathcal{F}$ *of cuts has the* degenerate *property if there is a partition* $\mathcal{P}$ *of* V *such that for all* $\emptyset \subsetneq \mathcal{X} \subsetneq \mathcal{P}$, $(\bigcup_{X \in \mathcal{X}} X, \bigcup_{Y \in \mathcal{P} \setminus \mathcal{X}} Y)$ *is in* $\mathcal{F}$, *and there is no others cut in* $\mathcal{F}$.

Lemma 3. *For every symmetric* $S \subseteq \{1,2\} \times \{1,2\}$, $\mathcal{F}_S(G)$ *has the degenerate property.*

Proof. The family $\mathcal{F}_{\{\}}(G)$ has the degenerate property since (X,Y) is a $\{\}$-cut if and only if there is no edges between X and Y ($\mathcal{P}$ is exactly the set of connected components). For $W \subseteq V$, let $G|W = (V, E\Delta W^2, l)$. For $i \in \{1,2\}$ let $V_i = \{v \in V : l(v) = i\}$. Let $G_1 = G|V_1$, $G_2 = G|V_2$ and $G_{12} = (G|V_1)|V_2$.

- $\mathcal{F}_{\{(1,1)\}}(G) = \mathcal{F}_{\{\}}(G_1)$, $\mathcal{F}_{\{(2,2)\}}(G) = \mathcal{F}_{\{\}}(G_2)$, $\mathcal{F}_{\{(1,1),(2,2)\}}(G) = \mathcal{F}_{\{\}}(G_{12})$,
- $\mathcal{F}_{\{(1,1),(1,2),(2,1),(2,2)\}}(G) = \mathcal{F}_{\{\}}(\overline{G})$, $\mathcal{F}_{\{(1,2),(2,1),(2,2)\}}(G) = \mathcal{F}_{\{\}}(\overline{G_1})$,
  $\mathcal{F}_{\{(1,1),(1,2),(2,1)\}}(G) = \mathcal{F}_{\{\}}(\overline{G_2})$, $\mathcal{F}_{\{(1,2),(2,1)\}}(G) = \mathcal{F}_{\{\}}(\overline{G_{12}})$.

Thus for every symmetric $S \subseteq \{1,2\} \times \{1,2\}$, $\mathcal{F}_S(G)$ has the degenerate property.

Definition 3 (Linear property). *A family* $\mathcal{F}$ *of cuts has the* linear *property if for all* (X,Y) *and* (X',Y') *in* $\mathcal{F}$, *either* $X \subseteq X'$ *or* $X' \subseteq X$.

Lemma 4. *For every non-symmetric* $S \subseteq \{1,2\} \times \{1,2\}$, $\mathcal{F}_S(G)$ *has the linear property.*

Proof. Case $S = \{(1,2)\}$: suppose that $X \setminus X'$ and $X' \setminus X$ are both non-empty. Then if $u \in X \setminus X'$ is labelled 1 and $v \in X' \setminus X$ is labelled 2, u and v has to be adjacent and non-adjacent, contradiction. Thus $X \setminus X'$ and $X' \setminus X$ are mono-coloured. Now suppose w.l.o.g. that all vertices in $X\Delta X'$ are labelled 1. Then $X\Delta X'$ is adjacent to all vertices labelled 2 in $Y \cap Y'$ and non adjacent to all vertices labelled 1 in $Y \cap Y'$. Moreover $X\Delta X'$ is non adjacent to all vertices in $X \cap X'$. Thus $X\Delta X'$ is a mono-coloured module, and $|X\Delta X'| \geq 2$. Contradiction. For others non-symmetric S, we bring back to case $\{(1,2)\}$ like in the proof of lemma 3.

For $S \subseteq \{1,2\} \times \{1,2\}$, let $\mathcal{P}_S(G)$ denote the unique partition of V such that (1) for all $(X,Y) \in \mathcal{F}_S(G)$ and $P \in \mathcal{P}_S(G)$, $P \subseteq X$ or $P \subseteq Y$, and (2) for all $P, P' \in \mathcal{P}$, $P \neq P'$, there is a $(X,Y) \in \mathcal{F}_S(G)$ such that $P \subseteq X$ and $P' \subseteq Y$, or $P \subseteq Y$ and $P' \subseteq X$. For a non-symmetric $S \in \{1,2\} \times \{1,2\}$, let $\mathcal{P}'_S(G) = (P_1, \ldots, P_k)$ denote the unique ordering of elements in $\mathcal{P}_S(G)$ such that for all $(X,Y) \in \mathcal{F}_S(G)$, there is a l such that $X = \cup_{i \in \{1,\ldots,l\}} P_i$.

Lemma 5. *If* G *is in NLC-2 ρ-Free, then there is a* $S \subseteq \{1,2\} \times \{1,2\}$ *such that* $\mathcal{F}_S(G)$ *is non-empty.*

Proof. If G is NLC-2 ρ-Free, then there is a $S \subseteq \{1,2\} \times \{1,2\}$, and two graphs G_1 and G_2 such that $G = G_1 \times_S G_2$. Thus $(V(G_1), V(G_2)) \in \mathcal{F}_S(G)$ and $\mathcal{F}_S(G)$ is non empty.

Lemma 6. *Let $G = (V, E, l)$ 2-labelled graph and let $S \subseteq \{1,2\} \times \{1,2\}$. If G is NLC-2 ρ-Free and has no mono-coloured non-trivial module, then for all $P \in \mathcal{P}_S(G)$, $G[P]$ has no mono-coloured non-trivial module.*

Proof. If M is a mono-coloured module of $G[P]$, then M is a mono-coloured module of G. Contradiction.

Lemma 7. *Let $G = (V, E, l)$ 2-labelled graph and let $S \subseteq \{1,2\} \times \{1,2\}$. Then G is NLC-2 ρ-Free if and only if for all $P \in \mathcal{P}_S(G)$, $G[P]$ is NLC-2 ρ-Free.*

Proof. The "only if" is immediate. Now suppose that for all $P \in \mathcal{P}_S(G)$, $G[P]$ is NLC-2 ρ-Free. If S is symmetric, let $\mathcal{P}_S(G) = \{P_1, \ldots, P_{|\mathcal{P}_S(G)|}\}$. Then $G = ((G[P_1] \times_S G[P_2]) \times_S \ldots \times_S G[P_{|\mathcal{P}_S(G)|}]$, and G is NLC-2 ρ-Free. Otherwise, if S is non-symmetric, let $\mathcal{P}'_S(G) = (P_1, \ldots, P_{|\mathcal{P}_S(G)|})$. Then $G = ((G[P_1] \times_S G[P_2]) \times_S \ldots \times_S G[P_{|\mathcal{P}_S(G)|}]$, and G is NLC-2 ρ-Free.

The *NLC-2 ρ-Free decomposition tree* of a 2-labelled graph G is a rooted tree such that the leaves are the vertices of G, and the internal nodes are labelled by $\times_S$, with $S \subseteq \{1,2\} \times \{1,2\}$. An internal node is **degenerated** if S is symmetric, and **linear** if S is non-symmetric. By lemmas 5, 6 and 7, G is NLC-2 ρ-Free if and only if it has a NLC-2 ρ-Free decomposition tree. This decomposition tree is not unique. But we can define a *canonical decomposition tree* if we fix a total order on the subsets of $\{1,2\} \times \{1,2\}$ (for example, the lexicographic order). If two graphs are isomorphic, then they have the same canonical decomposition tree. Algorithm 1 computes the canonical decomposition tree of a 2-labelled prime graph, or fails if G is not NLC-2 ρ-Free.

Input. A 2-labelled graph $G = (V, E, l)$
Output. A NLC-2 ρ-Free decomposition tree, or fail if G is not NLC-2 ρ-Free

1 **if** $|V| = 1$ **then return** *the leaf* $\cdot (l(v))$ (where $V = \{v\}$)
2 Let $\mathcal{S}$ be the set of subsets of $\{1,2\} \times \{1,2\}$ and σ be the lexicographic order of $\mathcal{S}$
3 **foreach** $S \in \mathcal{S}$ *w.r.t.* σ **do**
4   Compute $\mathcal{P}_S(G)$, and $\mathcal{P}'_S(G)$ if S is non-symmetric (see algorithm 2)
5   **if** $|\mathcal{P}_S(G)| > 1$ **then**
6   Create a new $\times_S$ node β
7   **foreach** $P \in \mathcal{P}_S(G)$ *(w.r.t. $\mathcal{P}'_S(G)$ if S is non-symmetric)* **do**
8   make NLC-2 ρ-Free decomposition tree of $G[P]$ be a child of β.
9   **return** *the tree rooted at* β
10 **fail with** *Not NLC-2 ρ-Free*

Algorithm 1. Computation of the NLC-2 ρ-Free canonical decomposition tree

Algorithm 2 computes $\mathcal{P}_S$ and $\mathcal{P}'_S$ for a 2-labelled prime graph G and $S \subseteq \{1,2\} \times \{1,2\}$ in linear time. We need some additional definitions for this algorithm and its proof of correctness. A *bipartite graph* is a triplet (X, Y, E) such that $E \subseteq X \times Y$. The *bi-complement* of a bipartite graph (X, Y, E) is the bipartite graph $(X, Y, (X \times Y) \setminus E)$. A *bipartite trigraph (BT)* is a bipartite graph with two types of edges: the *join* edges and the *mixed* edges. It is denoted by $\mathcal{B} = (X, Y, E_j, E_m)$ where E_j are the set of *join* edges, and E_m the set of *mixed* edges. A *BT-module* in a BT is a $M \subseteq X$ or $M \subseteq Y$ such that M is a module in (X, Y, E_j) and there is no *mixed* edges between M and $(X \cup Y) \setminus M$. For $v \in X \cup Y$, let $N_j(v) = \{u \in X \cup Y : \{u, v\} \in E_j\}$ and $N_m(v) = \{u \in X \cup Y : \{u, v\} \in E_m\}$. Let $d_j(v) = |N_j(v)|$ and $d_m(v) = |N_m(v)|$. A *semi-join* in a BT (X, Y, E_j, E_m) is a cut (A, B) of $X \cup Y$, such that there is no edges between $A \cap Y$ and $B \cap X$, and there is only *join* edges between $A \cap X$ and $B \cap Y$.

In algorithm 2, $\mathcal{B}$ is obtained from the graph G. Vertices of X correspond to subsets of vertices labelled 1 in G, and vertices of Y correspond to subsets of vertices labelled 2. There is a *join* edge between M and M' in $\mathcal{B}$ if $M \textcircled{1} M'$ in G, and there is a *mixed* edge between $M \in X$ and $M' \in Y$ in $\mathcal{B}$ if there is at least an edge and a non-edge between M and M' in G. Such a graph $\mathcal{B}$ can easily be built in linear time from a given graph G. It suffices to consider a list and an array bounded by the number of component in G with the same colour. The following lemmas are close to observations in [9], but deal with BT instead of bipartite graphs.

Lemma 8. *Let $G = (X, Y, E_j, E_m)$ be a BT such that every BT-module has size 1. Let $(x_1, \ldots, x_{|X|})$ be X sorted by $(d_j(x), d_m(x))$ in lexicographic decreasing order. If (A, B) is a semi-join of G, then there is a $k \in \{0, \ldots, |X|\}$ such that $A \cap X = \{x_1, \ldots, x_k\}$.*

Input. A 2-labelled graph G, and $S \subseteq \{1,2\} \times \{1,2\}$
Output. $\mathcal{P}_S$ if S is symmetric, $\mathcal{P}'_S$ if S is non-symmetric
1 $V_i \leftarrow \{v : v \in V$ and $l(v) = i\}$;
2 **if** $(1,1) \in S$ **then** $\mathcal{C}_1 \leftarrow$ co-connected components of $G[V_1]$;
3 **else** $\mathcal{C}_1 \leftarrow$ connected components of $G[V_1]$;
4 **if** $(2,2) \in S$ **then** $\mathcal{C}_2 \leftarrow$ co-connected components of $G[V_2]$;
5 **else** $\mathcal{C}_2 \leftarrow$ connected components of $G[V_2]$;
6 $\mathcal{B} = (\mathcal{C}_1, \mathcal{C}_2, E_j, E_m) \leftarrow$ the bipartite trigraph between the elements of $\mathcal{C}_1$ and $\mathcal{C}_2$;
7 **if** $S \cap \{(1,2), (2,1)\} = \emptyset$ **then**
8 $\quad$ **return** *connected components of* $(\mathcal{C}_1, \mathcal{C}_2, E_j \cup E_m)$
9 **else if** $S \cap \{(1,2), (2,1)\} = \{(1,2), (2,1)\}$ **then**
10 $\quad$ **return** *connected components of the bi-complement of* $(\mathcal{C}_1, \mathcal{C}_2, E_j)$
11 **else** Search all semi-joins of $\mathcal{B}$ (using lemmas 8 and 9) ;

Algorithm 2. Computation of $\mathcal{P}_S$ and $\mathcal{P}'_S$

Lemma 9. *Let $k \in \{0, \ldots, |X|\}$ and $k' \in \{0, \ldots, |Y|\}$. Then $(A, (X \cup Y) \setminus A)$, where $A = \{x_1, \ldots, x_k, y_1, \ldots, y_{k'}\}$, is a semi-join of G if and only if $\sum_{i=1}^{k} d_j(x_i) - \sum_{i=1}^{k'} d_j(y_i) = k \times (|Y| - k')$ and $\sum_{i=1}^{k} d_m(x_i) - \sum_{i=1}^{k'} d_m(y_i) = 0$.*

Theorem 3. *Algorithm 2 is correct and runs in linear time.*

Proof. **Correctness:** Suppose that (A, B) is a S-cut. If $(1,1) \notin S$, then there is no edge between $A \cap V_1$ and $B \cap V_1$, thus (A, B) cannot cut a component C_1 (and similarly for $(1,1) \in S$, and for C_2). Now we work on the BT $\mathcal{B} = (C_1, C_2, E_j, E_m)$. If $S \cap \{(1,2), (2,1)\} = \emptyset$, then S-cuts correspond exactly to connected components of $\mathcal{B}$, and if $S \cap \{(1,2), (2,1)\} = \{(1,2), (2,1)\}$ then S-cuts correspond exactly to connected components of the BT of $\overline{G}$, which is $(C_1, C_2, (C_1 \times C_2) \setminus (E_j \cup E_m), E_m)$. Finally, if S is non-symmetric, S-cuts correspond to semi-joins of $\mathcal{B}$.

Complexity: It is well admitted that we can perform a BFS on a graph or its complement in linear time [13,6]. The instructions on lines [2-5,8] can be done with a BFS on a graph or its complement. It is easy to see that we can do a BFS on the bi-complement in linear time (like a BFS on a complement graph, with two vertex lists for X and Y), so instruction line 10 can be done in linear time. Finally, the operations at line 11 are done in linear time.

These results can be summarized as:

Theorem 4. *Algorithm 1 computes the canonical NLC-2 ρ-Free decomposition tree of a 2-labelled graph in $O(nm)$ time.*

3.2 NLC-2 Decomposition of a Prime Graph

In this section, G is an unlabelled prime (w.r.t. modular decomposition) graph, with $|V| \geq 3$.

Definition 4 (2-*bimodule*). *A bipartition $\{X, Y\}$ of V is a 2-bimodule if X can be partitioned into X_1 and X_2, and Y into Y_1 and Y_2 such that for all $(i, j) \in \{1, 2\} \times \{1, 2\}$, then either $X_i \circledcirc Y_j$ or $X_i \textcircled{1} Y_j$. It is easy to see that if $\{X, Y\}$ is a 2-bimodule if and only if $\{X, Y\}$ is a split, a co-split or a bi-join. Moreover, if $\min(|X|, |Y|) > 1$ then $\{X, Y\}$ cannot be both of them in the same time (since G is prime).*

Let $l : V \to \{1, 2\}$ be a 2-labelling. Then $s(l)$ denote the 2-labelling on V such that for all $v \in V$, $s(l)(v) = 1$ if and only if $l(v) = 2$.

Definition 5 (Labelling induced by a 2-bimodule). *Let $\{X, Y\}$ be a 2-bimodule. We define the labelling $l : V \to \{1, 2\}$ of G induced by $\{X, Y\}$. If $|X| = |Y| = 1$, then $l(x) = 1$ and $l(y) = 2$, where $X = \{x\}$ and $Y = \{y\}$. If $|X| = 1$, then $l(v) = 1$ iff $v \in N[x]$. Similarly if $|Y| = 1$, then $l(v) = 1$ iff $v \in N[y]$. Now we suppose $\min(|X|, |Y|) > 1$. If $\{X, Y\}$ is a split, then the set of vertices in X with a neighbour Y and the set of vertices in Y with a neighbour* in X is labelled 1, others vertices are labelled 2. If $\{X,Y\}$ is a co-split, then a labelling of G induced by $\{X,Y\}$ is a labelling of $\overline{G}$ induced by the split $\{X,Y\}$. Finally if $\{X,Y\}$ is a bi-join, l is such that $\{v \in X : l(v) = 1\}$ is a join with $\{v \in Y : l(v) = 1\}$ and $\{v \in X : l(v) = 2\}$ is a join with $\{v \in Y : l(v) = 2\}$. Note that if $\{X,Y\}$ is a bi-join, then there is two possibles labelling l_1 and l_2, with $l_1 = s(l_2)$. If $\{X,Y\}$ is a 2-bimodule of G and l a labelling induced by $\{X,Y\}$, then every mono-coloured module has size 1 (since G is prime and $|V| \geq 3$).

Definition 6 (Good 2-bimodule). A 2-bimodule $\{X,Y\}$ is good if the graph G with the labelling induced by $\{X,Y\}$ is NLC-2 ρ-Free. The following proposition comes immediately from lemma 2.

Proposition 1. G is NLC-2 if and only if G has a good 2-bimodule.

Lemma 10. If G has a good 2-bimodule $\{X,Y\}$ which is a split, then G has a good 2-bimodule which is a strong split.

Proof. There is a node α in the split decomposition tree and we have $\emptyset \subsetneq I \subsetneq \{1,\ldots,d(\alpha)\}$ such that $\{X,Y\} = \{\cup_{i \in I} C^i_\alpha, \cup_{i \notin I} C^i_\alpha\}$. Let $l : V \to \{1,2\}$ be the labelling of G induced by $\{X,Y\}$. For all $i \in \{1,\ldots,d(\alpha)\}$, $(G[C^i_\alpha], l|_{C^i_\alpha})$ is NLC-2 ρ-Free (where $l|_W$ is the function l restricted at W).

Let l' be the 2-labelling of V such that for all i, and $v \in C^i_\alpha$, $l(v) = 1$ if and only if v has a neighbour outside of C^i_α. For all i, either $l|_{C^i_\alpha} = l'|_{C^i_\alpha}$, or $\forall v \in C^i_\alpha$, $l(v) = 2$. Then for all i, $(G[C^i_\alpha], l'|_{C^i_\alpha})$ is NLC-2 ρ-Free, and thus (G, l') is NLC-2 ρ-Free. Since there is a dominating vertex in the characteristic graph of α, there is a j such that the labelling induced by the strong split $\{C^j_\alpha, V \setminus C^j_\alpha\}$ is l'. Thus the strong split $\{C^j_\alpha, V \setminus C^j_\alpha\}$ is good.

Previous lemma on $\overline{G}$ say that if G has a good 2-bimodule $\{X,Y\}$ which is a co-split, then G has a good 2-bimodule which is a strong co-split. The following lemma is similar to Lemma 10.

Lemma 11. If G has a good 2-bimodule $\{X,Y\}$ which is a bi-join, then G has a good 2-bimodule which is a strong bi-join.

Input. A graph G
Result. Yes iff G is NLC-2
$S \leftarrow$ the set of strong splits, co-splits and bi-joins of G ;
foreach $\{X,Y\} \in S$ **do**
  $\quad l \leftarrow$ the labelling of G induced by $\{X,Y\}$;
  $\quad$ **if** $(G[X], G[Y], l)$ is NLC-2 ρ-Free **then return** Yes ;
return No ;

Algorithm 3. Recognition of prime NLC-2 graphs

Theorem 5. Algorithm 3 recognises prime NLC-2 graphs, and its time complexity is $O(n^2 m)$.

Proof. Trivially if the algorithm return Yes, then G is NLC-2. On the other hand, by proposition 1, and lemmas 10 and 11, if G is NLC-2, then it has a good strong 2-bimodule and the algorithm returns Yes.

The set S can be computed using algorithms for computing split decomposition on G and $\overline{G}$, and bi-join decomposition on G. Note that it is not required to use a linear time algorithm for split decomposition [5]: some simpler algorithms run in $O(n^2m)$ [3,10]. [7,8] show that bi-join decomposition can be computed in linear time, using a reduction to modular decomposition. But there also, modular decomposition algorithms simpler than [15] may be used. The set S has $O(n)$ elements. Testing if a 2-bimodule is good takes $O(nm)$ using algorithm 1. So total running time is $O(n^2m)$.

3.3 NLC-2 Decomposition

Using lemma 1, modular decomposition and algorithm 3, we get:

Theorem 6. *NLC-2 graphs can be recognised in $O(n^2m)$, and a NLC-2 expression can be generated in the same time.*

4 Graph Isomorphism on NLC-2 Graphs

4.1 Graph Isomorphism on NLC-2 ρ-Free Prime Graphs

Proposition 2. *Consider a symmetric $S \in \{1,2\} \times \{1,2\}$. Two graphs G and H are isomorphic if and only if there is a bijection π between $\mathcal{P}_S(G)$ and $\mathcal{P}_S(H)$ such that for all $P \in \mathcal{P}_S(G)$, $G[P]$ is isomorphic to $H[\pi(P)]$.*

Proposition 3. *Let a non-symmetric $S \in \{1,2\} \times \{1,2\}$ and let G and H be two graphs. Let $\mathcal{P}'_S(G) = (P_1, \ldots, P_k)$ and $\mathcal{P}'_S(H) = (P'_1, \ldots, P'_{k'})$ then G and H are isomorphic if and only if $k = k'$ and for all $i \in \{1, \ldots, k\}$, $G[P_i]$ is isomorphic to $H[P'_i]$.*

These two propositions are direct consequences of the linear and degenerate properties of S-cuts. Then two NLC-2 ρ-Free 2-labelled graphs G and H are isomorphic if and only if there is an isomorphism between their canonical NLC-2 ρ-Free decomposition tree which respects the order of children of linear nodes. This isomorphism can be tested in linear time, thus isomorphism of NLC-2 ρ-Free graphs can be done in $O(nm)$ time.

4.2 Graph Isomorphism on Prime NLC-2 Graphs

Theorem 7. *Algorithm 4 test isomorphism between two prime NLC-2 graphs in time $O(n^2m)$.*

Proof. If the algorithm returns "yes", then trivially $G \simeq H$. On the other hand suppose that $G \simeq H$ and let $\pi : V(G) \to V(H)$ be a bijection such that $\{u,v\} \in E(G)$ iff $(\pi(u), \pi(v)) \in E(H)$. Then $\{X', Y'\}$ with $X' = \pi(X)$ and $Y' = \pi(Y)$

is a good 2-bimodule if H. If $\min(|X|, |Y|) > 1$ and $\{X', Y'\}$ is a bi-join, then by definition there is two labelling induced by $\{X, Y\}$, and $(G, l) \simeq (H, l')$ or $(G, l) \simeq (H, s(l'))$. Otherwise the labelling is unique and $(G, l) \simeq (H, l')$.

Input. Two prime NLC-2 graphs G and H
Result. Yes if $G \simeq H$, No otherwise
$\mathcal{S} \leftarrow$ the set of strong splits, co-splits and bi-joins of G ;
$\mathcal{S}' \leftarrow$ the set of strong splits, co-splits and bi-joins of H ;
if *there is no good 2-bimodule in* $\mathcal{S}$ **then fail with** "G is not NLC-2";
$\{X, Y\} \leftarrow$ a good 2-bimodule in $\mathcal{S}$;
$l \leftarrow$ the labelling of G induced by $\{X, Y\}$;
foreach $\{X', Y'\} \in \mathcal{S}'$ *such that* $\{X', Y'\}$ *is good* **do**
   $l' \leftarrow$ the labelling of H induced by $\{X', Y'\}$;
   if $|X| > 1$ **and** $|Y| > 1$ **and** $\{X, Y\}$ *is a bi-join* **then**
       if $(G, l) \simeq (H, l')$ **or** $(G, l) \simeq (H, s(l'))$ **then return** *Yes* ;
   else if $(G, l) \simeq (H, l')$ **then return** *Yes* ;
return *No* ;

Algorithm 4. Isomorphism for prime NLC-2 graphs

The sets $\mathcal{S}$ and $\mathcal{S}'$ can be computed in $O(n^2)$ time using linear time algorithms for computing split decomposition on G and $\overline{G}$, and bi-join decomposition on G. The sets $\mathcal{S}$ and $\mathcal{S}'$ have $O(n)$ elements. Test if a 2-bimodule is good take $O(nm)$ using algorithm 1, and test if two 2-labelled prime graphs are isomorphic take also $O(nm)$. Thus the total running time is $O(n^2m)$.

4.3   Graph Isomorphism on NLC-2 Graphs

It is easy to show that graph isomorphism on prime NLC-2 graphs with an additional labels into $\{1, \ldots, q\}$ can be done in $O(n^2m)$ time. For that, we add the additional label of v at the leaf corresponding to v in the NLC-2 ρ-Free decomposition tree.

We show that we can do graph isomorphism on NLC-2 graphs in time $O(n^2m)$, using the modular decomposition and algorithm 4. Let $\mathcal{M}(G)$ and $\mathcal{M}(H)$ be the modular decomposition of G and H. For $M \in \mathcal{M}(G)$, let G_M be $G[M]$, and for $M \in \mathcal{M}(H)$, let H_M be $H[M]$. Let G_M^* be the characteristic graph of G_M (note that $|V(G_M^*)|$ is the number of children of M in the modular decomposition tree). Let $\mathcal{M}_{(i,*)} = \{M \in \mathcal{M}(G) \cup \mathcal{M}(H) : |M| = i\}$, let $\mathcal{M}_{(*,j)} = \{M \in \mathcal{M}(G) \cup \mathcal{M}(H) : |V(G_M^*)| = j\}$ and let $\mathcal{M}_{(i,j)} = \mathcal{M}_{(i,*)} \cap \mathcal{M}_{(*,j)}$. Note that $\sum_{j=1}^{n}(|\mathcal{M}_{(*,j)}| \times j)$ is the number of vertices in G plus the number of edges in the modular decomposition tree, and thus is at most $3n - 2$.

Theorem 8. *Algorithm 5 tests isomorphism between two NLC-2 graphs in time* $O(n^2m)$.

Proof. The correctness comes from the fact that at each step, for all $M, M' \in \mathcal{M}(G) \cup \mathcal{M}(H)$ such that $l(M)$ and $l(M')$ are set, G_M and $G_{M'}$ are isomorphic if and only if $l(M) = l(M')$. The total time $f(n, m)$ of this algorithm is ("big O" is omitted): $f(n, m) \leq \sum_i \sum_j \left(j^2 m |\mathcal{M}_{(i,j)}|^2\right) \leq m \sum_j \left(j^2 \sum_i \left(|\mathcal{M}_{(i,j)}|^2\right)\right) \leq m \sum_j \left(j^2 |\mathcal{M}_{(*,j)}|^2\right) \leq m \sum_j \left(\left(j |\mathcal{M}_{(*,j)}|\right)^2\right) \leq n^2 m$.

Input. Two NLC-2 graphs G and H
Result. Yes if $G \simeq H$, No otherwise
for *every* $M \in \mathcal{M}(G) \cup \mathcal{M}(H)$ *such that* $|M| = 1$ **do** $l(M) \leftarrow 1$;
for *i from* 2 *to* n **do**
> **for** *j from* 2 *to* i **do**
> > Compute the partition $\mathcal{P}$ of $\mathcal{M}_{(i,j)}$ such that M and M' are in the same class of $\mathcal{P}$ if and only if $(G^*_M, l) \simeq (G^*_{M'}, l)$. ;
> > **foreach** $P \in \mathcal{P}$ **do**
> > > $a \leftarrow$ a new label (an integer not in $\mathrm{Img}(l)$) ;
> > > For all $M \in P$, $l(M) \leftarrow a$;

Algorithm 5. Isomorphism on NLC-2 graphs

References

1. Chein, M., Habib, M., Maurer, M.C.: Partitive hypergraphs. Discrete Math. 37(1), 35–50 (1981)
2. Courcelle, B., Engelfriet, J., Rozenberg, G.: Handle-rewriting hypergraph grammars. J. Comput. Syst. Sci. 46(2), 218–270 (1993)
3. Cunningham, W.H.: Decomposition of directed graphs. SIAM J. Algebraic Discrete Methods 3(2), 214–228 (1982)
4. Cunningham, W.H., Edmonds, J.: A combinatorial decomposition theory. Canad. J. Math. 32, 734–765 (1980)
5. Dahlhaus, E.: Parallel algorithms for hierarchical clustering and applications to split decomposition and parity graph recognition. J. Algorithms 36(2), 205–240 (2000)
6. Dahlhaus, E., Gustedt, J., McConnell, R.M.: Partially complemented representations of digraphs. Discrete Math. Theor. Comput. Sci. 5(1), 147–168 (2002)
7. de Montgolfier, F., Rao, M.: The bi-join decomposition. In: ICGT. ENDM, vol. 22, pp. 173–177 (2005)
8. de Montgolfier, F., Rao, M.: Bipartitives families and the bi-join decomposition. Technical report (2005), https://hal.archives-ouvertes.fr/hal-00132862
9. Fouquet, J.-L., Giakoumakis, V., Vanherpe, J.-M.: Bipartite graphs totally decomposable by canonical decomposition. Internat. J. Found. Comput. Sci. 10(4), 513–533 (1999)
10. Gabor, C.P., Supowit, K.J., Hsu, W.-L.: Recognizing circle graphs in polynomial time. J. ACM 36(3), 435–473 (1989)
11. Gallai, T.: Transitiv orientierbare Graphen. Acta Math. Acad. Sci. Hungar. 18, 25–66 (1967)
12. Gurski, F., Wanke, E.: Minimizing NLC-width is NP-Complete. In: Kratsch, D. (ed.) WG 2005. LNCS, vol. 3787, pp. 69–80. Springer, Heidelberg (2005)

13. Habib, M., Paul, C., Viennot, L.: Partition refinement techniques: An interesting algorithmic tool kit. Internat. J. Found. Comput. Sci. 10(2), 147–170 (1999)
14. Johansson, Ö.: NLC_2-decomposition in polynomial time. Internat. J. Found. Comput. Sci. 11(3), 373–395 (2000)
15. McConnell, R.M., Spinrad, J.P.: Modular decomposition and transitive orientation. Discrete Math. 201(1-3), 189–241 (1999)
16. Wanke, E.: k-NLC Graphs and Polynomial Algorithms. Discrete Appl. Math. 54(2-3), 251–266 (1994)

A Characterisation of the Minimal Triangulations of Permutation Graphs

Daniel Meister

Institutt for Informatikk, Universitetet i Bergen, 5020 Bergen, Norway

Abstract. A minimal triangulation of a graph is a chordal graph obtained from adding an inclusion-minimal set of edges to the graph. For permutation graphs, i.e., graphs that are both comparability and cocomparability graphs, it is known that minimal triangulations are interval graphs. We (negatively) answer the question whether every interval graph is a minimal triangulation of a permutation graph. We give a non-trivial characterisation of the class of interval graphs that are minimal triangulations of permutation graphs and obtain as a surprising result that only "a few" interval graphs are minimal triangulations of permutation graphs.

1 Introduction

For some graph classes, the class of minimal triangulations is known. An early result shows that minimal triangulations of cographs are trivially perfect graphs [3]. Since cographs are exactly the P_4-free graphs, and trivially perfect graphs are exactly the P_4-free chordal graphs, it follows that every minimal triangulation of a cograph is a cograph. This result was extended by Parra and Scheffler: for $k \leq 5$, a graph is P_k-free if and only if every of its minimal triangulations is P_k-free [14]. The cograph result was generalised also in another direction, in particular to permutation graphs:

Theorem 1 ([2]). *Minimal triangulations of permutation graphs are interval graphs.*

Later, it was shown that this result even holds for cocomparability graphs [10] and AT-free graphs [13]. By the following characterisation, the class of AT-free graphs is the largest class of graphs containing only graphs whose minimal triangulations are interval graphs: a graph is AT-free if and only if it has only minimal triangulations that are interval graphs [13], [14]. So, for some graph classes, minimal triangulations reflect structural properties of the base graphs: minimal triangulations of P_k-free graphs for $k \leq 5$ are P_k-free, minimal triangulations of cocomparability graphs and AT-free graphs are cocomparability graphs and AT-free graphs, respectively. In all these cases, the base graph class contains the class of minimal triangulations.

The case of permutation graphs is different. The class of interval graphs is not contained in the class of permutation graphs, and vice versa. A minimal triangulation of a permutation graph can be a graph that is not a permutation graph.

A. Brandstädt, D. Kratsch, and H. Müller (Eds.): WG 2007, LNCS 4769, pp. 99–108, 2007.

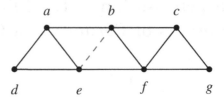

Fig. 1. Depicted is a permutation graph G, and adding edge be gives an interval graph that is a minimal triangulation of G. This interval graph is not a permutation graph.

An example is given in Figure 1. On the other hand, not every interval graph can be a minimal triangulation of a permutation graph. Such an example is depicted in Figure 4. Thus, the class of minimal triangulations of permutation graphs is a non-trivial subclass of the class of interval graphs, and the question arises how this class can be characterised. In this paper, we exactly address this problem. We give a characterisation of the class of minimal triangulations of permutation graphs, which even results in a recognition algorithm. This algorithm assumes the input graph to be given by an interval model and decides then in time linear in the number of vertices, independent of the number of edges.

Our characterisation is based on minimal separators and potential maximal cliques. A potential maximal clique is a set of vertices that is a maximal clique in a minimal triangulation [4]. We define a graph over the set of potential maximal cliques of a graph in Section 3, which we call the *connectors graph*, and show a connection to minimal triangulations. In Sections 4 and 5, we investigate the structure of these graphs for chordal and permutation graphs. It turns out that every graph with two isolated vertices can be the connectors graph of a chordal, even an interval graph and that connectors graphs of permutation graphs have nice bipartite substructures. In Section 6, we give the final characterisation of the minimal triangulations of permutation graphs using connectors graphs, establishing a connection between minimal triangulations of permutation graphs and bipartite graphs. This result has two interesting implications. The one is that only a few interval graphs are minimal triangulations of permutation graphs. The other is the astounding observation that permutation graphs are cocomparability and comparability graphs, and minimal triangulations of permutation graphs are cocomparability graphs with a comparability (bipartite) structural property.

Due to space restrictions, proofs are omitted.

2   Graph Preliminaries, Chordal Graphs and Minimal Triangulations

We only consider simple finite undirected graphs. A *graph* is a pair $G = (V, E)$ with V the vertex set and E the edge set. Edges of G are 2-elementary subsets of V, and they are denoted as uv where u and v are vertices of G. In this case, u and v are called *adjacent*; otherwise, u and v are *non-adjacent*. Let $H = (W, F)$ be a graph. We say that G is a *subgraph* of H, if $V \subseteq W$ and $E \subseteq F$. If G is not a proper subgraph of another subgraph of H on the vertex set of G, then G is called an *induced* subgraph of H, denoted as $G \sqsubseteq H$. For a set A of vertices of G, $G[A]$ denotes the subgraph of G induced by A, i.e., $G[A] = (A, X)$ for (A, X) an induced subgraph of G. Induced paths and cycles and connected components are defined as usual. For a set $S \subseteq V$, $G \setminus S$ denotes the subgraph of G induced by $V \setminus S$. Let a and b be vertices of G. A set S of vertices of G is an a, b-*separator* of G, if a and b are contained in different connected components of $G \setminus S$. If no proper subset of S is an a, b-separator of G, S is a *minimal a, b-separator* of G. A *minimal separator* of G is a minimal a, b-separator of G for some vertices a and b.

A graph is called *chordal*, if it does not contain an induced cycle of length greater than 3. Chordal graphs are exactly the graphs whose minimal separators all are cliques [6]. Let $G = (V, E)$ be a graph. A tree T is a *clique-tree* for G, if it has a vertex for every maximal clique of G and for every pair C', C'' of maximal cliques of G, every clique corresponding to a vertex on the path in T between the vertices corresponding to C' and C'' contains the vertices in $C' \cap C''$. A graph is chordal if and only if it has a clique-tree [5], [7], [16].

Theorem 2 ([1]). *Let $G = (V, E)$ be a chordal graph, and let T be a clique-tree for G. Then, a set S of vertices of G is a minimal separator of G if and only if there are maximal cliques C' and C'' of G that are neighbours with respect to T such that $S = C' \cap C''$.*

A graph H on the vertex set of G is a *triangulation* of G, if H is chordal and contains G as a subgraph. If no proper subgraph of H is a triangulation of G then H is a *minimal triangulation* of G. An easy polynomial-time test for minimal triangulation is based on the following result: Let G and H be two graphs on the same vertex set. Then, H is a minimal triangulation of G if and only if G is a subgraph of H, H is chordal and $H - e$ is not chordal for every edge e of H not contained in G [15].

Theorem 3 ([11]). *Let $G = (V, E)$ be a graph. A set S of vertices of G is a minimal separator of G if and only if S is a minimal separator of a minimal triangulation of G.*

Interval graphs are defined as the intersection graphs of families of closed intervals of the real line in the following sense: given a family of closed intervals of the real line, the obtained interval graph has a vertex for every interval, and two vertices are adjacent if and only if the corresponding intervals have a non-empty intersection. Such a family of intervals is also called an *interval model* for an interval graph. Alternatively, a graph is an interval graph if and only if it has a clique-tree that is a path [8], and such an alignment of the maximal cliques is called *consecutive clique arrangement*.

3   An Invariant for Minimal Triangulations

In some sense, it is well understood how minimal triangulations are related to the corresponding base graphs. Several characterisation results are known. But only a few is known about structures of minimal triangulations that are "inherited" from the base graph. We identify a structural graph property that is inherited by its minimal triangulations. Let $G = (V, E)$ be a graph. A set C of vertices of G is called a *potential maximal clique* of G, if C is a maximal clique in a minimal triangulation of G. By definition, the potential maximal cliques of a chordal graph are the maximal cliques of the chordal graph. Potential maximal cliques were introduced by Bouchitté and Todinca [4]. Potential maximal cliques contain minimal separators of the graph.

Definition 1. *Let $G = (V, E)$ be a graph. A potential maximal clique C of G is called a* **connector** *if and only if there is a* **plug pair** *for C, which is a pair $[a, b]$ of vertices in C such that there are two minimal separators S' and S'' of G contained in C such that $a \in S'$ and $b \in S''$ and there is no minimal separator of G contained in C that contains a and b.*

For a plug pair, we write $[a, b]$ instead of (a, b) to emphasise that there is no inherent order of the vertices. For a given graph, not every pair of vertices constitutes a plug pair.

Lemma 1. *Let $G = (V, E)$ be a graph. Let $[a, b]$ be a plug pair for a potential maximal clique C of G. Then, a and b are adjacent in G.*

Based on connectors and plug pairs, we are interested in the question how connectors are related to each other. For expressing such a relation, we define the so-called connectors graph using plug pairs. Let $G = (V, E)$ be a graph. The *connectors graph* of G, denoted as $\mathrm{con}(G)$, is defined as follows:

(vc) $\mathrm{con}(G)$ contains a vertex for every potential maximal clique of G

(ec) uv is an edge of $\mathrm{con}(G)$ if and only if the potential maximal cliques A_u and A_v are connectors and there are vertices a, b, c such that $[a, b]$ and $[b, c]$ are plug pairs for A_u and A_v, respectively.

The vertices of a connectors graph are associated with the corresponding potential maximal cliques, which can be considered as the *names* of the vertices. We are interested in the relationship between the connectors graphs of a graph and its minimal triangulations.

Bouchitté and Todinca showed a strong correspondence between the minimal separators of a graph and of its minimal triangulations [4]. In fact, this result is an extention of Theorem 3.

Theorem 4 ([4]). *Let $G = (V, E)$ be a graph and let H be a minimal triangulation of G. Let C be a maximal clique of H. Then, a set S of vertices is a minimal separator of H contained in C if and only if S is a minimal separator of G contained in C.*

Corollary 1. *Let $G = (V, E)$ be a graph, and let H be a minimal triangulation of G. Then, $\mathrm{con}(H) \sqsubseteq \mathrm{con}(G)$.*

4   Connectors Graphs of Chordal Graphs

Mainly, we are interested in the construction and structural properties of connectors graphs of chordal graphs. First, we show that the connectors graph of a chordal graph can be constructed from a clique-tree. Here, it is important to verify that the necessary properties are invariant with respect to clique-trees.

Lemma 2. *Let $G = (V, E)$ be a chordal graph. Let C be a maximal clique of G and let $[a, b]$ be a pair of vertices of G.*

1. *If $[a, b]$ is a plug pair for C then for every clique-tree T for G, there are two maximal cliques C' and C'' of G such that C' and C'' are neighbours of C with respect to T and $a \in C \cap C'$ and $b \in C \cap C''$ and no maximal clique of G except for C contains a and b.*
2. *Let T be a clique-tree for G. If there are two maximal cliques C' and C'' of G such that C' and C'' are neighbours of C with respect to T and $a \in C \cap C'$ and $b \in C \cap C''$ and no maximal clique of G except for C contains a and b, then $[a, b]$ is a plug pair for C.*

The main result of this section shows that connectors graphs of interval graphs can have any structure. We prove this statement by giving a concrete construction. A vertex of a graph is called *isolated*, if it does not have any neighbour.

Theorem 5. *Let $G = (V, E)$ be a graph with two isolated vertices. Then, there is an interval graph whose connectors graph is (isomorphic to) G.*

We see that connectors graphs of well-known chordal graph classes have no special property. This is an important property of connectors graphs, since they are made for analysing minimal triangulations of non-chordal graphs. So, any structural property of the connectors graphs of a considered graph class may yield a non-trivial result about the structure of the minimal triangulations. To complete Theorem 5, note that every chordal graph that is not complete has two maximal cliques that are not connectors.

5   Connectors Graphs of Permutation Graphs

Let M be a finite set of size n, and let σ_1 and σ_2 be two orderings over M. The graph on M defined by σ_1 and σ_2 has a vertex for every element of M, and two vertices of M are adjacent if and only if the corresponding elements are ordered differently by σ_1 and σ_2. Considering σ_1 and σ_2 as binary relations, the edge set of the defined graph corresponds to the complement of the intersection, $\sigma_1 \cap \sigma_2$. Graphs defined in this way are called *permutation graphs*, and we denote the graph defined by the pair (σ_1, σ_2) as $G(\sigma_1, \sigma_2)$. With a permutation graph, we associate a so-called *permutation diagram*: on two horizontal lines, mark points for each vertex and label the points with the names of the vertices in order defined by σ_1 for the upper line and σ_2 for the lower line. Connect the two points with the same label by a line segment. It is an easy property that two vertices of the permutation graph are adjacent if and only if the corresponding line segments of the permutation diagram intersect. For further information on permutation graphs and diagrams, we refer to the book by Golumbic [9]. Our results are best understood using permutation diagrams as representation models for permutation graphs.

For studying potential maximal cliques of permutation graphs, minimal separation lines are the best means. Let $G = G(\sigma_1, \sigma_2)$ be a permutation graph on n vertices. A *separation line* s of G is a pair (a, e) where $a, e \in \{\frac{1}{2}, 1\frac{1}{2}, \ldots, n\frac{1}{2}\}$. Separation lines for permutation graphs were introduced by Bodlaender, Kloks, Kratsch [2] (under the name *scanlines*). Let $s_1 = (a_1, e_1)$ and $s_2 = (a_2, e_2)$ be two separation lines of G. If $a_1 \leq a_2$ and $e_1 \leq e_2$, we also write $s_1 \leq s_2$; if $s_1 \leq s_2$ and $s_1 \neq s_2$, we write $s_1 < s_2$. Let $\sigma_1 = \langle x_1, \ldots, x_n \rangle$ and $\sigma_2 = \langle x_{\pi(1)}, \ldots, x_{\pi(n)} \rangle$ for some permutation π over $\{1, \ldots, n\}$. Separation line (a, e) *crosses* vertex x_i, if either $i < a$ or $\pi^{-1}(i) < e$. The set of vertices crossed by s is denoted as $\text{int}(s)$. If $i < a$ and x_i is not crossed by s, then x_i is *to the left of* s; if $i > a$ and x_i is not crossed by s, then x_i is *to the right of* s. For s_1 and s_2 and vertex x, if $s_1 < s_2$ and x is to the left of s_2 and to the right of s_1, we say that x is *between s_1 and s_2*. If $\{x_i : i \in \{a - \frac{1}{2}, a\frac{1}{2}, \pi(e - \frac{1}{2}), \pi(e\frac{1}{2})\}\} \cap \text{int}(s) = \emptyset$, s is called a *minimal separation line* of G. Note that the first set may contain between one and four vertices. If s_1 and s_2 are minimal separation lines of G and $s_1 < s_2$ and there is no minimal separation line t of G such that $s_1 < t < s_2$, then s_2 is called a *successor* of s_1.

Theorem 6 ([12]). *Let $G = G(\sigma_1, \sigma_2)$ be a permutation graph. Let C be a set of vertices of G. Then, C is a potential maximal clique of G if and only if there is a pair of minimal separation lines of G, s_1 and s_2, such that s_2 is a successor of s_1 and C is equal to the union of $\text{int}(s_1) \cup \text{int}(s_2)$ and the set of vertices between s_1 and s_2. Separation lines s_1 and s_2 are unique for C.*

In this section, every potential maximal clique of a permutation graph is associated with a pair of minimal separation lines that define the potential maximal clique in the sense of Theorem 6. Let $G = G(\sigma_1, \sigma_2)$ be a permutation graph. Let s_1 and s_2 be minimal separation lines of G where s_2 is a successor of s_1. Then, we define $L(s_1, s_2) =_{\text{def}} \text{int}(s_1) \setminus \text{int}(s_2)$ and $R(s_1, s_2) =_{\text{def}} \text{int}(s_2) \setminus \text{int}(s_1)$.

Lemma 3. *Let $G = G(\sigma_1, \sigma_2)$ be a permutation graph. Let C be a potential maximal clique of G defined by the pair (s_1, s_2) of minimal separation lines of G where s_2 is a successor of s_1. Then, $[a, b]$ is a plug pair for C if and only if $a \in L(s_1, s_2)$ and $b \in R(s_1, s_2)$ or $b \in L(s_1, s_2)$ and $a \in R(s_1, s_2)$.*

Using Lemma 3, the connectors graph of a permutation graph can be generated easily. The same lemma and the definition of permutation diagrams justify the following definition.

Definition 2. *Let $G = G(\sigma_1, \sigma_2)$ be a permutation graph. Let C be a potential maximal clique of G defined by the pair (s_1, s_2) of minimal separation lines of G where s_2 is a successor of s_1. Let C be a connector, and let $[a, b]$ be a plug*

Fig. 2. The figure shows the two possible situations for the vertices of a plug pair $[a, b]$ in a connector of a permutation graph. The vertical lines represent the two minimal separation lines defining the connector. The left hand figure illustrates the case when vertices a and b meet at the top whereas a and b meet at the bottom in the right hand figure.

pair for C where we assume $a \in L(s_1, s_2)$ and $b \in R(s_1, s_2)$. We say that a and b **meet at the top** *if and only if $b \prec_{\sigma_1} a$. Otherwise, we say that a and b* **meet at the bottom**.

From Lemma 1, we know that the vertices of a plug pair are adjacent. Taking into account this fact, the definition of "meeting at the top/bottom" has a clear geometric representation. It is illustrated in Figure 2. We show now that "meeting at the top/bottom" is a property of the connector, not only of a single plug pair.

Lemma 4. *Let $G = G(\sigma_1, \sigma_2)$ be a permutation graph. Let C be a connector of G. Then, there is a plug pair $[a, b]$ for C such that a and b meet at the top if and only if for every plug pair $[c, d]$ for C, c and d meet at the top.*

Lemma 4 motivates the following definition for connectors.

Definition 3. *Let $G = G(\sigma_1, \sigma_2)$ be a permutation graph. Let C be a connector of G. We say that C is* **oriented to the top** *if and only if there is a plug pair $[a, b]$ for C such that a and b meet at the top. Otherwise, we say that C is* **oriented to the bottom**.

Let $G = G(\sigma_1, \sigma_2)$ be a permutation graph. For two connectors A and B of G, we write $A \parallel B$, if both A and B are oriented to the top or to the bottom; otherwise, we write $A \perp B$. The following lemma presents the crucial property from which we will derive our main result in this section. Let A and B be potential maximal cliques of G defined by pairs (s_1, s_2) and (t_1, t_2) of minimal separation lines of G, respectively. Let s_2 and t_2 be successors of s_1 and t_1, respectively. We say that A and B are *parallel*, if $s_1 < s_2 \leq t_1 < t_2$ or $t_1 < t_2 \leq s_1 < s_2$.

Lemma 5. *Let $G = G(\sigma_1, \sigma_2)$ be a permutation graph. Let A and B be parallel connectors of G. Let $[a, b]$ be a plug pair for A and let $[b, c]$ be a plug pair for B. Then, $A \perp B$.*

It would be nice, if the prerequisite of parallel connectors in Lemma 5 was not necessary. But a simple example shows that the statement does not hold for non-parallel connectors in general (Figure 3).

Our main result for permutation graphs:

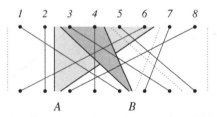

Fig. 3. Depicted is a permutation diagram of a permutation graph on eight vertices. The dotted and the thick line segments represent the minimal separation lines.

Theorem 7. *Let* $G = G(\sigma_1, \sigma_2)$ *be a permutation graph. Let* S *be a set of vertices of the connectors graph* $\mathrm{con}(G)$ *of* G *that correspond to pairwise parallel potential maximal cliques. Then,* S *induces a bipartite subgraph of* $\mathrm{con}(G)$.

Note that Theorem 7 does not make a statement about the structure of the whole connectors graph.

6   Characterising Minimal Triangulations of Permutation Graphs

It is known that minimal triangulations of permutation graphs are interval graphs (Theorem 1). Figure 4 depicts an interval graph H that is not a minimal triangulation of a permutation graph. So, which interval graphs are minimal triangulations of permutation graphs? We give the first characterisation of the class of interval graphs that are minimal triangulations of permutation graphs. The proof consists of two parts: combining the results of the previous sections, we can exclude a large number of interval graphs from this class, and for showing that the remaining interval graphs are minimal triangulations of permutation graphs, we construct a witness (a permutation graph). The proof also relies on the following theorem, that characterises the set of minimal triangulations of a single permutation graph.

Theorem 8 ([12]). *Let* $G = G(\sigma_1, \sigma_2)$ *be a permutation graph, and let* H *be a chordal graph on the vertex set of* G. *Then,* H *is a minimal triangulation of* G *if and only if the set of maximal cliques of* G *is a maximal set of pairwise parallel potential maximal cliques of* G.

Theorem 9. *Let* $H = (V, E)$ *be an interval graph. Then,* H *is a minimal triangulation of a permutation graph if and only if* $\mathrm{con}(H)$ *is bipartite.*

Theorem 9 implies a fast algorithm for recognising the class of minimal triangulations of permutation graphs. We can also use the algorithm of the proof of Theorem 9 to construct a permutation graph witnessing that an input interval graph is indeed a minimal triangulation of a permutation graph. Our algorithms

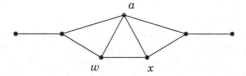

Fig. 4. The depicted graph H is an interval graph and minimal triangulation of the graphs H, $H-aw$, $H-ax$, $H-aw, ax$. None of the four graphs is a permutation graph, which can be verified easily, since a chordless path on six vertices has an almost unique representation in the permutation diagram.

require the input graph to be given by an interval model. Then, the algorithms take time only proportional in the number of vertices. We call such a running time *vertex-linear*.

Corollary 2. *There is a vertex-linear-time algorithm deciding the following question: given an interval graph H represented by an interval model, is H a minimal triangulation of a permutation graph. In the positive case, the algorithm outputs a permutation graph G represented by a permutation diagram and a set of pairwise non-intersecting minimal separation lines proving that H is a minimal triangulation of G.*

7   Final Remarks

We gave a characterisation of the minimal triangulations of permutation graphs. This characterisation is based on connectors graphs, that were introduced here, and shows a connection between interval graphs that are minimal triangulations of permutation graphs and bipartite graphs. This result also implies a statement about the number of interval graphs that are minimal triangulations of permutation graphs. Even though there is not much known about connectors graphs of interval graphs, it seems quite natural to assume that interval graphs are equally distributed among the connectors graphs. And if we accept that the class of bipartite graphs is only a small, nevertheless important, graph class, we can conclude that only "a few" interval graphs can be minimal triangulations of permutation graphs.

Finding a characterisation of the class of minimal triangulations of permutation graphs was a non-trivial and interesting task, since the class of permutation graphs does not contain the class of minimal triangulations of permutation graphs. Other graph classes with this property are the class of P_6-free graphs [14] and the class of gem-free graphs. The gem is a minimal triangulation of the C_5. The most natural such graph class, however, is the class of comparability graphs. The question is whether our techniques can be applied to these or similar graph classes to obtain results about the classes of their minimal triangulations. One step could be to identify structural properties that are preserved during the filling process.

References

1. Blair, J.R.S., Peyton, B.: An introduction to chordal graphs and clique trees. In: Graph Theory and Sparse Matrix Computation, pp. 1–29 (1993)
2. Bodlaender, H.L., Kloks, T., Kratsch, D.: Treewidth and Pathwidth of Permutation Graphs. SIAM Journal on Discrete Mathematics 8, 606–616 (1995)
3. Bodlaender, H.L., Möhring, R.H.: The pathwidth and treewidth of cographs. SIAM Journal on Discrete Mathematics 6, 181–188 (1993)
4. Bouchitté, V., Todinca, I.: Treewidth and Minimum Fill-in: Grouping the Minimal Separators. SIAM Journal on Computing 31, 212–232 (2001)
5. Buneman, P.: A Characterisation of Rigid Circuit Graphs. Discrete Mathematics 9, 205–212 (1974)
6. Dirac, G.A.: On rigid circuit graphs. Abhandlungen aus dem Mathematischen Seminar der Universität Hamburg 25, 71–76 (1962)
7. Gavril, F.: The Intersection Graphs of Subtrees in Trees Are Exactly the Chordal Graphs. Journal of Combinatorial Theory (B) 16, 47–56 (1974)
8. Gilmore, P.C., Hoffman, A.J.: A Characterization of Comparability Graphs and of Interval Graphs. Canadian Journal of Mathematics 16, 539–548 (1964)
9. Golumbic, M.C.: Algorithmic Graph Theory and Perfect Graphs. Academic Press, New York (1980)
10. Habib, M., Möhring, R.H.: Treewidth of cocomparability graphs and a new order-theoretic parameter. Technical report 336/1992, Fachbereich 3 Mathematik, Technische Universität Berlin (1992)
11. Kloks, T., Kratsch, D., Spinrad, J.: On treewidth and minimum fill-in of asteroidal triple-free graphs. Theoretical Computer Science 175, 309–335 (1997)
12. Meister, D.: Computing Treewidth and Minimum Fill-in for Permutation Graphs in Linear Time. In: Kratsch, D. (ed.) WG 2005. LNCS, vol. 3787, pp. 91–102. Springer, Heidelberg (2005)
13. Möhring, R.H.: Triangulating graphs without asteroidal triples. Discrete Applied Mathematics 64, 281–287 (1996)
14. Parra, A., Scheffler, P.: Characterizations and algorithmic applications of chordal graph embeddings. Discrete Applied Mathematics 79, 171–188 (1997)
15. Rose, D.J., Tarjan, R.E., Lueker, G.S.: Algorithmic aspects of vertex elimination on graphs. SIAM Jounal on Computing 5, 266–283 (1976)
16. Walter, J.R.: Representations of Chordal Graphs as Subtrees of a Tree. Journal of Graph Theory 2, 265–267 (1978)

The 3-Steiner Root Problem

Maw-Shang Chang[1,*] and Ming-Tat Ko[2]

[1] Department of Computer Science and Information Engineering
National Chung Cheng University, Chiayi 621, Taiwan, R.O.C.
mschang@cs.ccu.edu.tw
[2] Institute of Information Science
Academia Sinica, Taipei 115, Taiwan, R.O.C.
mtko@iis.sinica.edu.tw

Abstract. For a graph G and a positive integer k, the k-power of G is the graph G^k with $V(G)$ as its vertex set and $\{(u,v)|u,v \in V(G),\ d_G(u,v) \le k\}$ as its edge set where $d_G(u,v)$ is the distance between u and v in graph G. The k-*Steiner root problem* on a graph G asks for a tree T with $V(G) \subseteq V(T)$ and G is the subgraph of T^k induced by $V(G)$. If such a tree T exists, we call it a k-*Steiner root* of G. This paper gives a linear time algorithm for the 3-Steiner root problem. Consider an unrooted tree T with leaves one-to-one labeled by the elements of a set V. The k-*leaf power* of T is a graph, denoted T_L^k, with $T_L^k = (V, E)$, where $E = \{(u,v) \mid u,v \in V \text{ and } d_T(u,v) \le k\}$. We call T a k-*leaf root* of T_L^k. The k-leaf power recognition problem is to decide whether a graph has such a k-leaf root. The complexity of this problem is still open for $k \ge 5$ [6]. It can be solved in polynomial time if the $(k-2)$-Steiner root problem can be solved in polynomial time [6]. Our result implies that the k-leaf power recognition problem can be solved in linear time for $k = 5$.

Keywords: Tree power, tree root, Steiner root, leaf power, efficient algorithm.

1 Introduction

For a graph G, let $V(G)$ and $E(G)$ be the vertex and edge sets of G, respectively. For $x, y \in V(G)$, the distance between x and y in G, denoted by $d_G(x,y)$, is the length of a shortest path from x to y in G. The k-*power* of a graph G is the graph G^k with $V(G^k) = V(G)$ and $E(G^k) = \{(u,v)|u,v \in V(G), d_G(u,v) \le k\}$. Reversely, G is called a k-*root* of G^k. Deciding whether a graph G is a power of some other graph H is called the *graph root problem*. When H is required to be a tree and the power k is specified, the problem is called the k-*tree root problem* and H is called the k-*tree root* of G. Kearney and Corneil showed that the problem can be solved in $O(n^3)$ time [14]. The result was improved to $O(n+m)$ by Chang, Ko, and Lu [2].

Consider an unrooted tree T with leaf set V. The k-*leaf power* of T is a graph, denoted as T_L^k, with $V(T_L^k) = V$ and $E(T_L^k) = \{(u,v) \mid u,v \in V \text{ and } d_T(u,v) \le k\}$.

* Partially supported by the National Science Council of Taiwan, grant NSC 96-2221-E-194 -045 -MY3.

A. Brandstädt, D. Kratsch, and H. Müller (Eds.): WG 2007, LNCS 4769, pp. 109–120, 2007.
© Springer-Verlag Berlin Heidelberg 2007

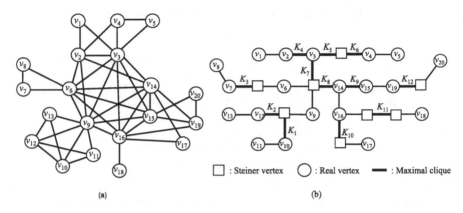

Fig. 1. (a)A graph G, (b)A 3-Steiner root T of G

We call T a k-leaf root of T_L^k. The k-leaf power recognition problem is to decide whether a graph has such a k-leaf root. While requiring that all internal nodes of a k-leaf root have degrees at least three, the k-leaf power recognition problem becomes the k-*phylogenetic root problem* and such a k-leaf root is called a k-*phylogenetic root* of G. These two problems were studied in [1,4,5,15,10,11,12,13]. Nishimura et. al. showed that the k-leaf power recognition problem can be solved in $O(|V|^3)$ time for $k \leq 4$ [15]. Lin et. al. showed that the k-phylogenetic root problem can be solved in $O(|V|+|E|)$ time for $k \leq 4$ [13]. Under the assumption that the maximum degree of the phylogenetic root is bounded from above by a constant, there is a linear-time algorithm that determines whether a graph has a k-phylogenetic root for arbitrary k [4,5]. The complexity of the k-leaf power recognition problem is still open for $k \geq 5$ [7]. Special cases of the problem with $k = 5$ were studied in [10,12].

A tree T is called a Steiner tree of vertex set V if $V \subseteq V(T)$. Given a graph G and positive integer k, the k-*Steiner root problem* asks for a Steiner tree T of $V(G)$ such that G is the subgraph of T^k induced by $V(G)$. If such a tree T exists, it is called a k-*Steiner root* of G. For a graph G, denote the set of k-Steiner roots of G by $\mathcal{SR}_k(G)$. A graph and a 3-Steiner root of the graph are shown in Figure 1(a) and (b) respectively. This problem was first studied in [13], and can be solved in linear time for $k \leq 3$ [1,13]. But it remains open for $k \geq 3$.

This paper gives a linear time algorithm for the 3-Steiner root problem. The k-leaf power recognition problem can be solved in polynomial time if the $(k-2)$-Steiner root problem can be solved in polynomial time [6]. Our result implies that the k-leaf power recognition problem can be solved in linear time for $k = 5$.

2   Preliminaries

For any set S, let $|S|$ denote the cardinality of S. All graphs in this paper are undirected and simple, and have no self-loops. Let $N_G(v)$ denote the set of neighbors of v in graph G and $N_G[v]$ denote $N_G(v) \cup \{v\}$. For a subset $W \subseteq V$, $N_G(W)$ denotes $\bigcup_{x \in W} N_G(x) \setminus W$. For a subset U of $V(G)$, define $G - U$ to be the subgraph of G induced by $V(G) \setminus U$, i.e., $G - U = G[V(G) \setminus U]$. For $X \subset V(G)$ and $y \in V(G) \setminus X$, the distance between X and y, denoted by $d_G(y, X)$, is the minimum among distances between y and vertices in X. For two subsets X and Y of $V(G)$, the distance $d_G(X, Y)$ in graph G between X and Y is the minimum distance $d_G(x, y)$ in G between a vertex $x \in X$ and another vertex $y \in Y$. A separator of a connected graph G is a subset S of $V(G)$ such that $G - S$ has at least two components. A separator S is a u-v separator if vertices u and v are in different connected components of $G - S$. A separator (u-v separator resp.) S is minimal if no proper subset of S is a separator (u-v separator resp.). A minimal vertex separator is a minimal u-v separator for some u and v. Note that a minimal vertex separator is not necessarily a minimal separator, as a minimal u-v separator may contain a minimal x-y separator for some other pair of vertices x, y. A minimal vertex separator S of G is of type 1, type 2, or type 3 if $|S| = 1$, $|S| = 2$, or $|S| \geq 3$, respectively. For a subset X of $V(G)$, $\Delta(X)$ denote the set of minimal vertex separators in G contained in $X \subseteq V(G)$.

Definition 1. *The* eccentricity $\epsilon(v)$ *of a vertex* v *in a connected graph* G *is* $\max_{u \in V(G)} d_G(u, v)$. *The* radius $\gamma(G)$ *is the minimum eccentricity of the vertices. The* diameter $\delta(G)$ *is the maximum eccentricity of the vertices. A vertex* v *is a* central vertex *if* $\epsilon(v) = \gamma(G)$, *and the* center *of* G, *denoted by* $C(G)$, *is the set of all central vertices.*

Note that $\delta(G)$, the diameter of G, is the length of the longest induced path in G. The center of a tree is consisting of either a vertex or two adjacent vertices [8]. Furthermore, the center of a tree T consists of a vertex if and only if $\delta(T)$ is even and two adjacent vertices if and only if $\delta(T)$ is odd. A tree T is called a vertex if $\delta(T) = 0$, an edge if $\delta(T) = 1$, a star if $\delta(T) = 2$, and a conjoined star if $\delta(T) = 3$. In a star T, let a be the central vertex of T. Then $E(T) = \{(a, x) | x \in V(T) \setminus a\}$. The star T is represented by $(a, V(T) \setminus a)$. In a conjoined star T, let a and b be the central vertices of T, $S_a = N_T(a) \setminus b$, $S_b = N_T(b) \setminus a$. Then S_a and S_b are disjoint subsets of vertices such that $S_a \cup S_b = V(T) \setminus \{a, b\}$ and $E(T) = \{(a, b)\} \cup \{(a, x) | x \in S_a\} \cup \{(b, y) | y \in S_b\}$. The conjoined star is represented by (a, S_a, b, S_b). A branch at a vertex u of a tree T is a maximal subtree containing u as a leaf. Thus the number of branches at u is the degree of u.

A graph G is chordal if it contains no induced subgraph which is a cycle of size greater than three. In other words, G is chordal if every cycle of length at least four has a chord, i.e., an edge between two nonconsecutive vertices of the cycle. Since any induced subgraph of a chordal graph is still a chordal graph and any power of a tree is a chordal graph, any graph having Steiner roots is chordal. Namely, for a graph G, if $\mathcal{SR}_k(G) \neq \emptyset$ for some $k \geq 0$, then G is chordal. Thus, in the rest of paper, we only consider chordal graphs. Chordal graphs have been extensively studied [9].

3   Sketch of the Algorithm

Let T be a k-*Steiner root* of graph G. We call the vertices and edges of T *tree vertices* and *tree edges*, respectively. The vertices in $V(G)$ and $V(T) \setminus V(G)$ are called *real* vertices and *Steiner* vertices, respectively. For a set S of tree vertices, we use $Real(S)$ to denote the set of real vertices in S. Let X and Y be subsets of $V(T)$ and $V(G)$, respectively. We use $T[X]$ to denote the subgraph (a forest) of T induced by X. We use $T(Y)$ to denote the minimal subtree of T containing Y. Actually, $T(Y)$ is the subtree T' of T satisfying the conditions that $Real(V(T')) = Y$ and all leaves of T' are real vertices. Let S be any minimal vertex separator of G and K be any maximal clique of G. If $T(S)$ is a star, its topology is represented by $(a, S \setminus a)$, where a is the central vertex of $T(S)$. When a is a Steiner vertex, the topology is denoted as (β, S), where β denotes a generic Steiner vertex. If $T(K)$ is a conjoined star, let its two central vertices of be a and b, where a and b can be Steiner vertices. Let $S_a = Real(N_T(a) \setminus \{b\})$ and $S_b = Real(N_T(b) \setminus \{a\})$. Then $S_a \cup S_b \cup Real(\{a, b\}) = K$. The topology of $T(K)$ is represented by (a, S_a, b, S_b). We refer to edge (a, b) as the central edges of $T(K)$. Please refer to Figure 1(b) for examples. A bold edge in tree T of Figure 1(b) is the central edge of $T(K)$ of some maximal clique K of G. For instance, $K_8 = \{v_3, v_6, v_9, v_{14}, v_{15}, v_{16}\}$, $T(K_8)$ is a conjoined stars, and the central edge of $T(K_8)$ is (v_{14}, β) where β is a Steiner vertex.

In the following, α and β are generic Steiner vertices. A minimal vertex separator $S \subset K$ is *maximal with respect to* K if S is not properly contained in any other minimal vertex separator contained in K. A minimal vertex separator is *maximal* if it is maximal with respect to every maximal clique containing it.

Lemma 1. *Suppose G is connected and has a 3-Steiner root T. Let S and K be a minimal vertex separator and a maximal clique of G, respectively. Let $\mathcal{N}_1$, $\mathcal{N}_2$, and $\mathcal{N}_3$ denote the set of minimal vertex separators of type 1, type 2, and type 3, contained in K. Let $\mathcal{N}_2^a$ denote the set of the minimal vertex separators in $\mathcal{N}_2$ containing vertex a. The following observations on topologies of $T(S)$ and $T(K)$ are correct:*

1. *$T(S)$ is a star, an edge or a vertex; and $T(K)$ is a conjoined star, a star, or an edge.*
2. *If S is of type 2 and S is not maximal, then $T(S)$ is an edge.*
3. *If $T(K)$ is a star, then all minimal vertex separators contained in K are of type 1 or type 2. Besides there is a vertex $a \in K$ such that $\mathcal{N}_2 = \mathcal{N}_2^a$.*
4. *If S is of type 3, then $T(S)$ is a star.*
5. *If $S \subset K$ and $T(S)$ is a star, then $T(K)$ is a conjoined star and the central vertex of $T(S)$ is one of the central vertices of $T(K)$.*
6. *For any 3-Steiner root T of G there is at most one minimal vertex separator S of type 2 in $\Delta(K)$ with $T(S)$ not an edge.*
7. *If $\mathcal{N}_2$ is nonempty, then, either there are two vertices $a, b \in K$ such that $\mathcal{N}_2^a \neq \emptyset$, $\mathcal{N}_2^b \neq \emptyset$ $\mathcal{N}_2^a \cap \mathcal{N}_2^b = \emptyset$ and $\mathcal{N}_2^a \cup \mathcal{N}_2^b = \mathcal{N}_2$, or there is a vertex $a \in K$ such that $\mathcal{N}_2 = \mathcal{N}_2^a$.*

8. *A maximal clique of G contains at most two minimal vertex separators of type 3. If K contains two minimal vertex separators A and B of type 3, then $K = A \cup B$, $|A \cap B| \leq 2$, and for every $S' \in \Delta(K) \setminus \{A, B\}$, $S' \subset A$ or $S' \subset B$.*

Two graphs G_1 and G_2 share a vertex set X if $V(G_1) \cap V(G_2) = X$ and $G_1[X] = G_2[X]$. For two graphs G_1 and G_2 sharing a vertex set X, the *join graph* of G_1 and G_2 at X, denoted as $G_1 \bigoplus_X G_2$, or $G_1 \bigoplus G_2$ if X is clear, is the graph with $V(G_1) \cup V(G_2)$ as the vertex set and with $E(G_1) \cup E(G_2)$ as the edge set. Note that the subgraph of $G_1 \bigoplus_X G_2$ induced by $V(G_1)$ ($V(G_2)$ resp.) is exactly G_1 (G_2 resp.).

Definition 2. *Suppose G_1 and G_2 share a clique S and both G_1 and G_2 have 3-Steiner roots. Let T_1 and T_2 be 3-Steiner roots of G_1 and G_2, respectively. We say that T_1 and T_2 are* compatible *if and only if one of the following four conditions is satisfied:*

- *$|S| = 1$.*
- *$|S| = 2$ and both $T_1(S)$ and $T_2(S)$ are an edge.*
- *Both $T_1(S)$ and $T_2(S)$ are stars and the central vertices of $T_1(S)$ and $T_2(S)$ are a common vertex in S.*
- *Both $T_1(S)$ and $T_2(S)$ are stars and both central vertices of $T_1(S)$ and $T_2(S)$ are Steiner vertices.*

For two compatible 3-Steiner roots T_1 and T_2 of G_1 and G_2 sharing a clique S, respectively, define the join Steiner tree *of T_1 and T_2 at S*, denoted as $T_1 \bigodot_S T_2$ or as $T_1 \bigodot T_2$ if S is clear, as follows: First identify the two central vertices of $T_1(S)$ and $T_2(S)$ if both $T_1(S)$ and $T_2(S)$ are stars and both central vertices of $T_1(S)$ and $T_2(S)$ are Steiner vertices. Now T_1 and T_2 share a vertex set $V(T_1(S)) = V(T_2(S))$. Then let $T_1 \bigodot_S T_2 = T_1 \bigoplus_{V(T_1(S))} T_2$.

Immediately following the above definition we have the following lemma.

Lemma 2. *Suppose G_1 and G_2 share a clique S and both G_1 and G_2 have 3-Steiner roots. Let T_1 and T_2 be 3-Steiner roots of G_1 and G_2, respectively. If T_1 and T_2 are compatible and $T = T_1 \bigodot_S T_2$ has the property that $d_T(V(G_1) \setminus S, V(G_2) \setminus S) > 3$, then T is a 3-Steiner root of $G_1 \bigoplus_S G_2$.*

Let $G = (V, E)$ be a connected chordal graph and $\bar{K}$ be a maximal clique of G. A *decomposition tree* of G with respect to $\bar{K}$, denoted by $\mathcal{D}(G, \bar{K})$, is recursively defined as follows. Every node H of a decomposition tree is associated with a 5-tuple $(H_G, H_V, H_K, H_C, H_S)$. Decomposition tree $\mathcal{D}(G, \bar{K})$ is a rooted tree rooted at node R with $R_G = G$, $R_V = V$, $R_K = \bar{K}$, $R_C = V$, and $R_S = \emptyset$. Node H has no child if H_G is a clique. Suppose H_G is not a clique and $H_G - H_K$ has k components. Let $H_G[C_1], H_G[C_2], \ldots, H_G[C_k]$ be the k components of $H_G - H_K$. Let $S_i = N_G(C_i)$ and $V_i = C_i \cup S_i$ for $1 \leq i \leq k$. Clearly S_i is a proper subset of H_K and is a minimal vertex separator of G for $1 \leq i \leq k$. Let K_i be a maximal clique of G such that $S_i \subset K_i$ and $K_i \subseteq V_i$ for $1 \leq i \leq k$.

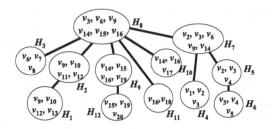

Fig. 2. A decomposition tree $\mathcal{D}(G, \bar{K}\})$ of graph G shown in Figure 1(a) where $\bar{K} = \{v_3, v_6, v_9, v_{14}, v_{15}, v_{16}\}$. Each circle represents a node of the decomposition tree. The set of vertices in the circle of a node H is H_K, a maximal clique of G. For example, $(H_1)_K = \{v_9, v_{10}, v_{12}, v_{13}\}$.

Then H has k child nodes $H_1, H_2, \ldots, H_k$ with $(H_i)_V = V_i$, $(H_i)_G = G[V_i]$, $(H_i)_K = K_i$, $(H_i)_S = S_i$, and $(H_i)_C = C_i$ for $1 \leq i \leq k$. Figure 2 shows a decomposition tree of graph G shown in Figure 1(a). Since G is a chordal graph, each node H of $\mathcal{D}(G, \bar{K})$ corresponds to a maximal clique H_K of G. Each edge $(H, \hat{H})$ of $\mathcal{D}(G, \bar{K})$, where $\hat{H}$ is a child node of H, corresponds to a minimal vertex separator $\hat{H}_S = N_G(\hat{H}_C) = H_K \cap \hat{H}_K$ of G. Let $\mathcal{D}_H$ be the subtree of $\mathcal{D}(G, \bar{K})$ rooted at node H and $V(\mathcal{D}_H)$ be the set of nodes of $\mathcal{D}_H$. Then $H_V = \bigcup_{\hat{H} \in V(\mathcal{D}_H)} \hat{H}_K$. In fact, a decomposition tree of a chordal graph is a special clique tree.

Now we are ready to describe the algorithm for determining whether a connected chordal graph having a 3-Steiner root. The algorithm arbitrarily selects a maximal clique $\bar{K}$ of G and constructs a decomposition tree $\mathcal{D}(G, \bar{K})$ of G with respect to $\bar{K}$. Then for each node H of $\mathcal{D}(G, \bar{K})$ perform the following operations in the ordering from leaves up to the root in a dynamic programming way: Firstly compute a set $\mathcal{P}(H)$ of topologies of H_K. Next compute a set $\mathcal{T}(H)$ of 3-Steiner roots of H_G. Initially $\mathcal{T}(H)$ is empty. For each topology Γ in $\mathcal{P}(H)$ the algorithm tries to constructs a 3-Steiner root Υ of H_G such that $\Upsilon(H_K) = \Gamma$. If it constructs such a tree successfully, includes it in $\mathcal{T}(H)$. If $\mathcal{T}(H)$ is empty after all topologies in $\mathcal{P}(H)$ are considered, then G has no 3-Steiner root and stop. If $\mathcal{T}(R)$, where R is the root of $\mathcal{D}(G, \bar{K})$, is not empty after all nodes of $\mathcal{D}(G, \bar{K})$ are considered, then G has a 3-Steiner root and any tree in $\mathcal{T}(R)$ is a 3-Steiner root of G. Otherwise G has no 3-Steiner root.

Suppose T is a 3-Steiner root of G, H is a node of $\mathcal{D}(G, \bar{K})$, and $\hat{H}$ is a child node of H. Obviously $\hat{H}_G$ and $G[H_K]$ share $\hat{H}_S$ and $T(\hat{H}_V)$ and $T(H_K)$ are compatible. Suppose $\hat{H}_1, \hat{H}_2, \ldots, \hat{H}_h$ are child nodes of H. Then $T(H_V) = T(H_K) \odot_{(\hat{H}_1)_S} T((\hat{H}_1)_V) \cdots \odot_{(\hat{H}_h)_S} T((\hat{H}_h)_V)$. Given a topology Γ of H_K, Algorithm 1 builts a 3-Steiner root Υ of H_G with that $\Upsilon(H_K) = \Gamma$ by selecting a tree Π from $T(\hat{H}_i)$ for each child node $\hat{H}_i$ and joining them with Γ by $\odot$ operation. Please refer to Algorithm 1 for details.

The ideas behind the algorithm are two heuristic rules: Firstly, the algorithm chooses those topologies Γ of H_K to be included in $\mathcal{P}(H)$ satisfying the condition that the vertices of H_K are away from each other in Γ as far as possible according to the decomposition tree. Secondly, when building a 3-Steiner root Υ of H_G with $\Upsilon(H_K) = \Gamma$ Algorithm 1 builds one satisfying the condition that for each child node $\hat{H}$ of H the distance $d_\Upsilon(\hat{H}_C, H_K \setminus \hat{H}_S)$ is the largest among all 3-Steiner roots of H_G with $\Upsilon(\hat{H}) = \Gamma(\hat{H})$. When choosing a tree Π from $\mathcal{T}(\hat{H})$ of a child node $\hat{H}$ of H we select one satisfying the conditions that $\Pi(\hat{H}_S) = \Gamma(\hat{H}_S)$ and the distance $d_T(\hat{H}_C, H_K \setminus \hat{H}_S)$ between $\hat{H}_C$ and $H_K \setminus \hat{H}_S$ in tree $T = \Gamma \odot_{\hat{H}_S} \Pi$ is the largest. These two heuristic rules are inspired by the following observations.

For a node H with $|H_S| = 1$, define $\lambda^*(H)$ to be the maximum distance $d_\Upsilon(v, H_C)$, where v is the only vertex in H_S, among all 3-Steiner roots Υ of H_G and define $\Upsilon^*(H)$ to be a 3-Steiner root Υ of H_G with $d_\Upsilon(v, H_C) = \lambda(H)$. Notice that $1 \leq \lambda^*(H) \leq 3$. It is straightforward to verify that if G has a 3-Steiner root then G has a 3-Steiner root T such that $T(H_V) = \Upsilon^*(H)$.

For a node H with $|H_S| = 2$, either $T(H_S)$ is a star with a Steiner central vertex or $T(H_S)$ is an edge in any 3-Steiner root T of G. Suppose $|H_S| = 2$. Let $\Upsilon_{star}(H)$ be a 3-Steiner root of H_G such that both vertices in H_S are leaves of $\Upsilon_{star}(H)$ and the distance $\lambda_{star}(H) = d_{\Upsilon_{star}(H)}(H_C, H_S)$ between H_S and H_C in tree $\Upsilon_{star}(H)$ is the maximum among all distances between H_S and H_C in all 3-Steiner roots of H_G. We observe that G has a 3-Steiner root T with $T(H_V) = \Upsilon_{star}(H)$ if G has a 3-Steiner root T' with $T'(H_S)$ being a star. Assume that $H_S = \{a, b\}$ and H^* is the parent node of H. Suppose T is a 3-Steiner root of G with that $T(H_S)$ is an edge. If b is a leaf of $T(H_V)$, and $d_{T(H_V)}(H_S, H_C) = 1$, then for any sibling node H' of H with $H'_S = H_S$, a is a leaf of $T(H'_V)$ and $d_{T(H'_V)}(H'_S, H'_C) = 2$. Besides, $d_{T(H^*_K)}(H_S, H^*_K \setminus H_S) > 1$. Similarly, if $d_{T(H^*_K)}(H_S, H^*_K \setminus H_S) = 1$ and b is a central vertex and a is a leaf of $T(H^*_K)$, then for all child node H' of H^*, b is a leaf and $d_{T(H'_V)}(H'_S, H'_C) > 1$.

Now consider a node H with $|H_S| > 2$. For a topology $\mathcal{S}$ of H_S, define $\Upsilon(H, \mathcal{S})$ be a 3-Steiner root Π of H_G with that $\Pi(H_S) = \mathcal{S}$ and the distance $d_\Pi(H_S, H_C)$ between H_S and H_C in Π is maximum among distances between H_S and H_C in all 3-Steiner roots Π' of H_G with $\Pi'(H_S) = \mathcal{S}$. It is straightforward to verify that if G has a 3-Steiner root T with $T(H_S) = \mathcal{S}$, then G has a 3-Steiner root T such that $T(H_V) = \Upsilon(H, \mathcal{S})$.

The correctness of the algorithm is based upon the above observations.

Now we describe the algorithm for obtaining a set $\mathcal{P}(H)$ of potential topologies of H_K for a node H of $\mathcal{D}(G, \bar{K})$. For simplicity of notations, let $\mathcal{N}_1$, $\mathcal{N}_2$, and $\mathcal{N}_3$ denote the set of minimal vertex separators of type 1, type 2, and type 3 contained in H_K, respectively, and let $\mathcal{N}_2^a$ denote the set of the minimal vertex separators in $\mathcal{N}_2$ containing vertex a. For a node H with $|H_S| = 1$, define $\lambda(H)$ to be the maximum distance $d_\Pi(H_S, H_C)$ among all 3-Steiner roots Π of H_G in $\mathcal{T}(H)$ and define $\Upsilon(H)$ to be a 3-Steiner root Π of H_G in $\mathcal{T}(H)$ with $d_\Pi(v, H_C) = \lambda(H)$. In the following, assume that G has a 3-Steiner root and α and β are generic Steiner vertices. The containment relation between maximal cliques and minimal vertex separators are important in solving the problem. We classify H_K into 7 types according to the minimal vertex separators contained in H_K.

1. H_K is a type (a) clique, i.e., H_K contains two minimal separators A and B of type 3 and $A \cap B = \{a, b\}$. Then, let $\mathcal{P}(H) = \{(a, A \setminus \{a, b\}, b, B \setminus \{a, b\}), (b, A \setminus \{a, b\}, a, B \setminus \{a, b\})\}$.

2. H_K is a type (b) clique, i.e., H_K contains two minimal separators A and B of type 3 and $A \cap B = \{a\}$. Then, let $\mathcal{P}(H) = \{(a, A \setminus \{a\}, \beta, B \setminus \{a\}), (\beta, A \setminus \{a\}, a, B \setminus \{a\})\}$. For example, $(H_8)_K$ of the decomposition tree in Figure 2 is a type (b) clique.

3. H_K is a type (c) clique, i.e., H_K contains two minimal separators A and B of type 3 and $A \cap B = \emptyset$. Then, either $\mathcal{P}(H) = \{(\alpha, A, \beta, B)\}$.

4. H_K is a type (d) clique, i.e., H_K contains exactly one minimal vertex separator A of type 3. For example, $(H_7)_K$ of the decomposition tree in Figure 2 is a type (d) clique. In this case $T(H_K)$ is a conjoined stars for any 3-Steiner root of G. Let $B = \bigcup_{S \in \Delta(H_K), S \nsubseteq A} S$. We can verify that $|A \cap B| \leq 1$. We first determine a set C_A of vertices in A that are possible to be the central vertex of $T(A)$ for some 3-Steiner root T of G. Suppose $A \subset K'$ for some maximal clique $K' \neq H_K$ such that K' contains two minimal vertex separators A and B' of type 3. If there exists K' with $A \cap B' = \emptyset$, then $C_A = \{\beta\}$; else if there exists K' with $A \cap B' = \{a'\}$, then $C_A = \{a', \beta\}$; else if there exists K' with $A \cap B' = \{a', b'\}$, then $C_A = \{a', b'\}$; else if there is a minimal vertex separator $S = \{a', b'\} \subset A$, then $C_A = \{a', b'\}$; else all maximal cliques K' containing A contains exactly one minimal vertex separator of type 3, i.e., A, and all minimal vertex separators properly contained in A are of type 1, we let $C_A = \{\beta\}$. In all the above cases, we have $|C_A| \leq 2$. Consider the following three cases:

   (a) $A \cap B = \{a\}$. In this case, there is a minimal vertex separator S' of type 2 such that $S' \subset B$ and $a \in S'$. Suppose B contains exactly one minimal vertex separator of type 2, namely $S' = \{a, b\}$, and $H_K = A \cup S'$. Then for a 3-Steiner root T of G with $T(S')$ being a star, $T(H_K) = (a, A \setminus \{a\}, \beta, \{b\})$. For a 3-Steiner root T of G with $T(S')$ being an edge, a is not a central vertex of $T(A)$ but is the central vertex of $T(H_K)$ other than the central vertex of $T(A)$ and we have $T(H_K) = (x, A \setminus \{x, a\}, a, H_K \setminus A)$ where $x \in C_A \setminus \{a\}$. Thus $\mathcal{P}(H) = \{(a, A \setminus \{a\}, \beta, \{b\})\} \cup \{(x, A \setminus \{x, a\}, a, H_K \setminus A) \mid x \in C_A \setminus \{a\}\}$.

   On the other hand, suppose B contains more than one minimal vertex separator of type 2 or $|H_K| > |A| + 1$. For any 3-Steiner root T of G, $T(S')$ is an edge for any S' contained in B. We see that a is not the central vertex of $T(A)$ for any 3-Steiner root T of G and a must be the central vertex of $T(H_K)$ other than the central vertex of $T(A)$ for any 3-Steiner root T of G. Let $\mathcal{P}(H) = \{(x, A \setminus \{x, a\}, a, H_K \setminus A) \mid x \in C_A\}$.

   (b) $A \cap B = \emptyset$ and B contains a minimal vertex separator S' of type 2. Then $H_K \setminus A = S'$ and A does not contain any minimal vertex separator of type 2. Let $\mathcal{P}(H) = \{(\alpha, A, \beta, H_K \setminus A)\}$.

   (c) $A \cap B = \emptyset$ and all minimal vertex separators contained in B are of type 1. Let $\mathcal{P}(H) = \{(x, A \setminus \{x\}, \beta, H_K \setminus A) \mid x \in C_A\}$.

5. H_K is a type (e) clique, i.e., $\mathcal{N}_3 = \emptyset$ and $\mathcal{N}_2$ is a disjoint union of nonempty sets $\mathcal{N}_2^a$ and $\mathcal{N}_2^b$ where $\{a, b\} \subset H_K$ and $a \neq b$. Let $A = (\bigcup_{S \in \mathcal{N}_2^a} S) \setminus \{a\}$. There are three subcases:

   (a) $|\mathcal{N}_2^a| \geq 2$ and $|\mathcal{N}_2^b| \geq 2$. Then $\mathcal{P}(H) = \{(a, A, b, H_K \setminus (A \cup \{a, b\}))\}$.

   (b) $|\mathcal{N}_2^a| \geq 2$ and $\mathcal{N}_2^b = \{S = \{b, c\}\}$. Then $\mathcal{P}(H) = \{(a, A, b, H_K \setminus (A \cup \{a, b\})), (a, A, c, H_K \setminus (A \cup \{a, c\}))\}$

   (c) $|\mathcal{N}_2^a| = 1$ and $|\mathcal{N}_2^b| = 1$. Let $\mathcal{N}_2^a = \{\{a, b\}\}$ and $\mathcal{N}_2^b = \{\{c, d\}\}$. If $H_K = \{a, b, c, d\}$, then $\mathcal{P}(H) = \{(\alpha, \{a, b\}, \beta, \{c, d\}), (a, \{b\}, c, H_K \setminus \{a, b, c\}), (a, \{b\}, d, H_K \setminus \{a, b, d\}), (b, \{a\}, c, H_K \setminus \{a, b, c\}), (b, \{a\}, d, H_K \setminus \{a, b, d\})\}$; else $\mathcal{P}(H) = \{(a, \{b\}, c, H_K \setminus \{a, b, c\}), (a, \{b\}, d, H_K \setminus \{a, b, d\}), (b, \{a\}, c, H_K \setminus \{a, b, c\}), (b, \{a\}, d, H_K \setminus \{a, b, d\})\}$.

6. H_K is a type (f) clique, i.e., $\mathcal{N}_3 = \emptyset$ and there is a vertex $a \in H_K$ such that $\mathcal{N}_2 = \mathcal{N}_2^a$. Let $S_a = \bigcup_{S \in \mathcal{N}_2^a} S$. There are two cases:

   (a) $|\mathcal{N}_2^a| \geq 2$. We have several cases:

      i. There exists a child node $\hat{H}$ of H with $|\hat{H}_S| = 2$ satisfying one of the following two conditions hold: (i) $\Pi(\hat{H}_S)$ is a star for all trees $\Pi \in \mathcal{T}(\hat{H})$; and (ii) there does not exists a tree $\Pi \in \mathcal{T}(\hat{H})$ with that $\Pi(\hat{H}_S)$ is an edge, a is a leaf of Π, and $d_{\Pi}(\hat{H}_S, \hat{H}_C) > 1$. Then assume $\hat{H}_S = \{a, x\}$ and let $\mathcal{P}(H) = \{(a, H_K \setminus \{a, x\}, \beta, \{x\}), (a, \{x\}, \beta, H_K \setminus \{a, x\})\}$.

      ii. The above case does not hold and there exists a child node $\hat{H}$ of H with $|\hat{H}_S| = 1$ such that $\lambda(\hat{H}) < 3$. Assume $\hat{H}_S = \{x\}$. If $x \in S_a$ and $a \neq x$, let $\mathcal{P}(H) = \{(a, H_K \setminus \{a, x\}, \beta, \{x\})\}$. Otherwise $x \notin S_a$, let $\mathcal{P}(H) = \{(a, S_a \setminus \{a\}, \beta, H_K \setminus S_a)\}$.

      iii. $|H_S| = 2$ and the above two cases do not hold. Assume $H_S = \{a, b\}$. In a 3-Steiner root T of G either $T(H_S)$ is an edge or $T(H_S)$ is a star. Let $\mathcal{P}(H) = \{(a, S_a \setminus \{a\}, \beta, H_K \setminus S_a), (a, H_K \setminus \{a, b\}, \beta, \{b\})\}$.

      iv. $|H_S| = 1$ and the above three cases do not hold. Assume $H_S = \{x\}$. If $x \in S_a$ and $a \neq x$, let $\mathcal{P}(H) = \{(a, H_K \setminus \{a, x\}, \beta, \{x\}), (a, S_a \setminus \{a\}, \beta, H_K \setminus S_a)\}$. Otherwise $x \notin S_a$, let $\mathcal{P}(H) = \{(a, S_a \setminus \{a\}, \beta, H_K \setminus S_a)\}$.

   (b) $|\mathcal{N}_2^a| = 1$ and $\mathcal{N}_2^a = \{S = \{a, b\}\}$. Let $\mathcal{P}(H) = \{(\alpha, S, \beta, H_K \setminus S), (a, \{b\}, \beta, H_K \setminus S), (b, \{a\}, \beta, H_K \setminus S)\}$.

7. H_K is a type (g) clique, i.e., $\mathcal{N}_3 = \emptyset$ and $\mathcal{N}_2 = \emptyset$. We claim that if G has a 3-Steiner root, then there is a 3-Steiner root T of G such that $T(K) = (\alpha, V_a, \beta, V_b)$ where $V_a \cap V_b = \emptyset$ and $V_a \cup V_b = H_K$. Assume $H_S = \{u\}$. If $\lambda(\hat{H}) > 1$ for all child nodes, then let $\mathcal{P}(H) = \{(\alpha, \{u\}, \beta, H_K \setminus \{u\})\}$. Otherwise there exists exactly one child node $\hat{H}$ of H with $\lambda(\hat{H}) = 1$. Assume $\hat{H}_S = \{v\}$. Let $\mathcal{P}(H) = \{(\alpha, H_K \setminus \{v\}, \beta, \{v\})\}$.

Algorithm 1. Computing a 3-Steiner root Υ of H_G with $\Upsilon(H_K) = \Gamma$

Input: A topology Γ of H_K.

Output: A 3-Steiner root Υ of H_G with $\Upsilon(H_K) = \Gamma$ such that G has a 3-Steiner root T with $T(H_V) = \Upsilon$.

For a node H with $|H_S| = 1$, define $\lambda(H)$ to be the maximum distance $d_\Pi(v, H_C)$, where v is the only vertex in H_S, among all 3-Steiner roots Π of H_G in $\mathcal{T}(H)$ and define $\Upsilon(H)$ to be a 3-Steiner root Π of H_G in $\mathcal{T}(H)$ with $d_\Pi(v, H_C) = \lambda(H)$.

Initially let $\Upsilon = \Gamma$;

for each child node $\hat{H}$ of H with $\hat{H}_S$ being of type 3 or of type 1 **do**
  If $\hat{H}_S$ is of type 1, let Π be $\Upsilon(\hat{H})$; otherwise let Π be a tree in $\mathcal{T}(\hat{H})$ with $\Pi(\hat{H}_S) = \Gamma(\hat{H}_S)$.
  let $\Upsilon = \Upsilon \odot_{\hat{H}_S} \Pi$;
end for

Let $\Delta_H = \Delta(H_K) \setminus \Delta(H_S)$ where $\Delta(H_S)$ is the set of minimal vertex separators contained in H_S including H_S.

for each separator S of type 2 in Δ_H **do**
  let $S = \{a, b\}$ and let $\mathcal{H}$ be the set of child nodes $\hat{H}$ with $\hat{H}_S = S$.
  if $\Gamma(S)$ is a star **then**
    for each node $\hat{H} \in \mathcal{H}$ **do**
      Let Π be a tree in $\mathcal{T}(\hat{H})$ with $\Pi(S)$ being a star and both central vertices of $\Pi(\hat{H}_K)$ being Steiner vertices if such a tree exists and let Π be any tree in $\mathcal{T}(\hat{H})$ with $\Pi(S)$ being a star otherwise.
      let $\Upsilon = \Upsilon \odot_{\hat{H}_S} \Pi$
    end for
  else
    $\Gamma(H_S)$ is an edge. Without loss of generality assume that b is a central vertex and a is a leaf of Γ.
    if $d_\Gamma(S, H_K \setminus S) > 1$ **then**
      for each node $\hat{H} \in \mathcal{H}$ **do**
        let Π be a tree $\Pi \in \mathcal{T}(\hat{H})$ with that $\Pi(S)$ is an edge, a is a leaf of Π, and $d_\Pi(S, \hat{H}_C) > 1$ if such a tree exists; and let Π be a tree in $\mathcal{T}(\hat{H})$ with that b is a leaf of Π and $\Pi(S)$ is an edge otherwise.
        let $\Upsilon = \Upsilon \odot_{\hat{H}_S} \Pi$.
      end for
    else
      $d_\Gamma(S, H_K \setminus S) = 1$
      for each node $\hat{H} \in \mathcal{H}$ **do**
        Let Π be a tree in $\mathcal{T}(\hat{H})$ with that b is a leaf.
        let $\Upsilon = \Upsilon \odot_{\hat{H}_S} \Pi$.
      end for
    end if
  end if
end for
if Υ is a 3-Steiner root of H_G **then**
  output Υ;
else
  output a null tree.
end if

4   Complexity

We use n and m for the numbers of vertices and edges of a graph, respectively. A chordal graph can be recognized in $O(n+m)$ time. A clique tree of a chordal graph can be constructed in $O(n+m)$ time. A decomposition tree $\mathcal{D}(G, \bar{K})$ of a chordal graph G can be constructed from a clique tree of G in $O(n+m)$ time. There are $O(n)$ maximal cliques and $O(n)$ minimal vertex separators in a chordal graph. Every node of a decomposition tree corresponds to a maximal cliques. Hence the number of nodes of a decomposition tree is $O(n)$. Every edge of a decomposition tree corresponds to a minimal vertex separator. It is easy to verify that only constant number of topologies of H_K in $\mathcal{P}(H)$ for every node H of $\mathcal{D}(G, \bar{K})$. For each topology $\Gamma \in \mathcal{P}(H)$, we run Algorithm 1 once to construct the a 3-Steiner root Υ of H_G if it exists. The running time of Algorithm 1 is $O(|H_K| + \sum_{1 \le i \le h} |(\hat{H}_i)_S|)$ where $\hat{H}_1, \hat{H}_2, \ldots, \hat{H}_h$ are all child nodes of H. Since we run Algorithm 1 $|\mathcal{P}(H)|$ times for each node H of $\mathcal{D}(G, \bar{K})$, the total running time of Algorithm 1 is $\sum_{H \in V(\mathcal{D}(G, \bar{K}))} O(|\mathcal{P}(H)| \cdot (|H_K| + \sum_i |(\hat{H}_i)_S|))$ $= \sum_{H \in V(\mathcal{D}(G, \bar{K}))} O(|\mathcal{P}(H)| \cdot 2 \cdot |H_K|) = O(n+m)$. The total running time for computing $\mathcal{P}(H)$ for each node H of $\mathcal{D}(G, \bar{K})$ is $O(n+m)$. It can be analyzed in the same way as we did for the total running time of Algorithm 1. Thus we have the following theorem:

Theorem 1. *A graph having a 3-Steiner root can be recognized in $O(n+m)$ time.*

References

1. Brandstädt, A., Van Bang Le: Structure and linear time recognition of 3-leaf powers. Information Processing Letters 98, 133–138 (2006)
2. Chang, M.-S., Ko, M.-T., Lu, H.-I.: Linear time algorithms for tree root problems. In: Arge, L., Freivalds, R. (eds.) SWAT 2006. LNCS, vol. 4059, pp. 411–422. Springer, Heidelberg (2006)
3. Sunil Chandran, L.: A linear time algorithm for enumerating all the maximum and minimal separators of a chordal graph. In: Wang, J. (ed.) COCOON 2001. LNCS, vol. 2108, pp. 308–317. Springer, Heidelberg (2001)
4. Chen, Z.-Z., Jiang, T., Lin, G.-H.: Computing phylogenetic roots with bounded degrees and errors. SIAM J. Comput. 32, 864–879 (2003)
5. Chen, Z.-Z., Tsukiji, T.: Computing bounded-degree phylogenetic roots of disconnected graphs. Journal of Algorithms 59, 125–148 (2006)
6. Dom, M., Guo, J., Hüffner, F., Niedermeier, R.: Extending the tractability border for closest leaf powers. In: Kratsch, D. (ed.) WG 2005. LNCS, vol. 3787, pp. 397–408. Springer, Heidelberg (2005)
7. Dom, M., Guo, J., Hüffner, F., Niedermeier, R.: Error compensation in leaf power problems. Algorithmica 44, 363–381 (2006)
8. Harary, F.: Graph Theory. Addison-Wesley Publishing Company, Reading (1969)
9. Ho, C.-W., Lee, R.C.T.: Counting clique trees and computing perfect elimination schemes in parallel. Infom. Process. Lett. 31, 61–68 (1989)

10. Kennedy, W., Lin, G.: 5th Phylogenetic Root Construction for Strictly Chordal Graphs. In: Deng, X., Du, D.-Z. (eds.) ISAAC 2005. LNCS, vol. 3827, pp. 738–747. Springer, Heidelberg (2005)

11. Kennedy, W., Lin, G., Yan, G.: Strictly chordal graphs are leaf powers. Journal of Discrete Algorithms (to appear)

12. Kong, H., Yan, G.Y.: Algorithm for phylogenetic 5-root problem. Optimization Methods and Software (submitted for publication)

13. Lin, G.-H., Kearney, P.E., Jiang, T.: Phylogenetic k-root and Steiner k-root. In: Lee, D.T., Teng, S.-H. (eds.) ISAAC 2000. LNCS, vol. 1969, pp. 539–551. Springer, Heidelberg (2000)

14. Kearney, P.E., Corneil, D.G.: Tree powers. J. of Algorithms 29, 111–131 (1998)

15. Nishimura, N., Ragde, P., Thilikos, D.M.: On graph powers for leaf-labeled trees. J. of Algorithms 42, 69–108 (2002)

On Finding Graph Clusterings with Maximum Modularity[*]

Ulrik Brandes[1], Daniel Delling[2], Marco Gaertler[2], Robert Görke[2], Martin Hoefer[1], Zoran Nikoloski[3], and Dorothea Wagner[2]

[1] Department of Computer and Information Science, University of Konstanz
{Ulrik.Brandes,Martin.Hoefer}@uni-konstanz.de
[2] Faculty of Informatics, Universität Karlsruhe (TH)
{delling,gaertler,rgoerke,wagner}@informatik.uni-karlsruhe.de
[3] Department of Applied Mathematics, Faculty of Mathematics and Physics,
Charles University, Prague
nikoloski@kam.mff.cuni.cz

Abstract. Modularity is a recently introduced quality measure for graph clusterings. It has immediately received considerable attention in several disciplines, and in particular in the complex systems literature, although its properties are not well understood. We study the problem of finding clusterings with maximum modularity, thus providing theoretical foundations for past and present work based on this measure. More precisely, we prove the conjectured hardness of maximizing modularity both in the general case and with the restriction to cuts, and give an Integer Linear Programming formulation. This is complemented by first insights into the behavior and performance of the commonly applied greedy agglomaration approach.

1 Introduction

Graph clustering is a fundamental graph-theoretic problem in data and, more specifically, network analysis [1]. Studied for decades and applied in many settings, it is currently popular as the problem of partitioning networks into communities. In this line of research, a novel graph clustering index called *modularity* has been proposed recently [2]. The rapidly growing interest in this measure prompted a series of follow-up studies on various applications and possible adjustments (see, e.g., [3,4,5,6]). Moreover, an array of heuristic algorithms has been proposed to optimize modularity. These are based on a greedy agglomeration [7], on spectral division [8,9], simulated annealing [10], or extremal optimization [11] to name but a few prominent examples. While these studies often provide subjective plausibility arguments in favor of the resulting partitions, we

---

[*] This work was partially supported by the DFG under grants BR 2158/2-3, WA 654/14-3, Research Training Group 1042 "Explorative Analysis and Visualization of Large Information Spaces" and the EU under grant DELIS (contract no. 001907).

A. Brandstädt, D. Kratsch, and H. Müller (Eds.): WG 2007, LNCS 4769, pp. 121–132, 2007.

know of only one attempt to characterize properties of clusterings with maximum modularity [3]. In particular, none of the proposed algorithms has been shown to produce partitions that are optimal with respect to modularity.

In this paper we study the problem of finding clusterings with maximum modularity, thus providing theoretical foundations for past and present work based on this measure. More precisely, we proof the conjectured hardness of maximizing modularity both in the general case and the restriction to cuts, and give an integer linear programming formulation to facilitate optimization without enumeration of all clusterings. Since the most commonly employed heuristic to optimize modularity is based on greedy agglomeration, we investigate its worst-case behavior. In fact, we give a graph family for which the greedy approach yields an approximation factor no better than two. In addition, our examples indicate that the quality of greedy clusterings may heavily depend on the tie-breaking strategy utilized. In fact, in the worst case, no approximation factor can be provided. These performance studies are concluded by partitioning some previously considered networks optimally, which does yield further insight.

This paper is organized as follows. Section 2 contains brief preliminaries, formulations of modularity and an ILP formulation of the problem. Basic and counterintuitive properties of modularity are observed in Sect. 3. Our $\mathcal{NP}$-completeness proofs are given in Sect. 4, followed by an analysis of the greedy approach in Sect. 5. Our work is concluded by characterizations of revisited examples from previous work in Sect. 6 and a brief discussion in Sect. 7.

2   Preliminaries

Throughout this paper, we will use the notation of [1]. More precisely, we assume that $G = (V, E)$ is an undirected connected graph with $n := |V|$ vertices, $m := |E|$ edges. Denote by $\mathcal{C} = \{C_1, \ldots, C_k\}$ a partition of V. We call $\mathcal{C}$ a *clustering* of G and the C_i, which are required to be non-empty, *clusters*; $\mathcal{C}$ is called *trivial* if either $k = 1$ or $k = n$. We denote the set of all possible clusterings of a graph G with $\mathcal{A}(G)$. In the following, we often identify a cluster C_i with the induced subgraph of G, i.e., the graph $G[C_i] := (C_i, E(C_i))$, where $E(C_i) := \{\{v, w\} \in E : v, w \in C_i\}$. Then $E(\mathcal{C}) := \bigcup_{i=1}^{k} E(C_i)$ is the set of *intra-cluster edges* and $E \setminus E(\mathcal{C})$ the set of *inter-cluster edges*. The number of intra-cluster edges is denoted by $m(\mathcal{C})$ and the number of inter-cluster edges by $\overline{m}(\mathcal{C})$. The set of edges that connect C_i and C_j is denoted by $E(C_i, C_j)$.

2.1   Definition of Modularity

Modularity is a quality index for clusterings. Given a simple graph $G = (V, E)$, we follow [2] and define the *modularity* $\mathsf{q}(\mathcal{C})$ of a clustering $\mathcal{C}$ as

$$\mathsf{q}(\mathcal{C}) = \sum_{C \in \mathcal{C}} \left[\frac{|E(C)|}{m} - \left(\frac{\sum_{v \in C} \deg(v)}{2m} \right)^2 \right] . \tag{1}$$

This formula reveals an inherent trade-off: to maximize the first term, many edges should be contained in clusters, whereas the minimization of the second term is achieved by splitting the graph into many clusters with small total degrees. Note that the first term $|E(\mathcal{C})|/m$ is also known as *coverage* [1].

2.2 Maximizing Modularity Via Integer Linear Programming

The problem of maximizing modularity can be cast into a very simple and intuitive integer linear program (ILP). Given a graph $G = (V, E)$ with $n :=$ $|V|$ nodes, we define n^2 decision variables $X_{uv} \in \{0, 1\}$, one for every pair of nodes $u, v \in V$. The key idea is that these variables can be interpreted as an equivalence relation (over V) and thus form a clustering. In order to ensure consistency, we need the following constraints, which guarantee

reflexivity $\forall u: X_{uu} = 1$,

symmetry $\forall u, v: X_{uv} = X_{vu}$, and

transitivity $\forall u, v, w: \begin{cases} X_{uv} + X_{vw} - 2 \cdot X_{uw} \le 1 \\ X_{uw} + X_{uv} - 2 \cdot X_{vw} \le 1 \\ X_{vw} + X_{uw} - 2 \cdot X_{uv} \le 1 \end{cases}$.

The objective function of modularity then becomes

$$\frac{1}{2m} \sum_{(u,v) \in V^2} \left(E_{uv} - \frac{\deg(u) \deg(v)}{2m} \right) X_{uv} \ ,$$

$$\text{with } E_{uv} = \begin{cases} 1 & , \text{ if } (u, v) \in E \\ 0 & , \text{ otherwise} \end{cases} .$$

Note that this ILP can be simplified by pruning redundant variables and constraints, leaving only $\binom{n}{2}$ variables and $\binom{n}{3}$ constraints.

3   Fundamental Observations

In the following, we identify basic structural properties that clusterings with maximum modularity fulfill. We first focus on the range of modularity, for which Lemma 1 gives the lower and upper bound.

Lemma 1. *Let G be an undirected and unweighted graph and $\mathcal{C} \in \mathcal{A}(G)$. Then $-1/2 \le q(\mathcal{C}) \le 1$ holds.*

Lemma 1 is proven by minimizing modularity, for details see [12]. As a result, any bipartite graph $K_{a,b}$ with the canonic clustering $\mathcal{C} = \{C_a, C_b\}$ yields the minimum modularity of $-1/2$. The upper bound is obvious from our reformulation in Equation (1), and has been observed previously [3,4,13]. It can only be attained in the specific case of a graph with no edges, where coverage is commonly defined to be 1. The following four results strongly characterize the rough structure of a clustering with maximum modularity.

Corollary 1. *Isolated nodes have no impact on modularity.*

Corollary 1 directly follows from the fact that modularity depends on edges and degrees, thus, an isolated node does not contribute, regardless of its association to a cluster.

Lemma 2. *A clustering with maximum modularity has no cluster that consists of a single node with degree 1.*

Lemma 3. *There is always a clustering with maximum modularity, in which each cluster consists of a connected subgraph.*

The proofs of Lemmas 2 and 3 can be found in [12] and are straightforward. Both rely on the fact that a strict increase in modularity is possible, if they are violated.

Corollary 2. *A clustering of maximum modularity does not include disconnected clusters.*

Corollary 2 follows from Lemma 3 and Equation (1). In the following, we exclude isolated nodes from further consideration, i. e., all nodes are assumed to be of degree greater than zero. Thus, the search for an optimum can be restricted to clusterings, in which clusters are connected subgraphs and there are no clusters consisting of nodes with degree 1.

3.1   Counterintuitive Behavior

In the last section, we listed some intuitive and desirable properties like connectivity within clusters for clusterings of maximum modularity. However, due to the enforced balance between coverage and the sums of squared cluster degrees, counter-intuitive situations arise. These are non-locality, scaling behavior, and sensitivity to satellites.

Non-Locality. At a first view, modularity seems to be a local quality measure. Recalling Equation (1), each cluster contributes separately. However, the example presented in Figures 1(a) and 1(b) exhibit a typical non-local behavior. In these figures, clusters are represented by colors. By adding an additional node connected to the leftmost node, the optimal clustering is altered completely. According to Lemma 2 the additional node has to be clustered together with the leftmost node. This leads to a shift of the leftmost white node from the white cluster to the black cluster, although locally its neighborhood structure has not changed.

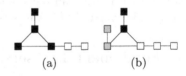

(a)   (b)

Fig. 1. Non-local behavior. Clusters are represented by colors.

Sensitivity to Satellites. A *clique with leaves* is a graph of $2n$ nodes that consists of a clique K_n and n *leaf* nodes of degree one, such that each node of the clique is connected to exactly one leaf node. For a clique, the trivial clustering with $k = 1$ has maximum modularity. For a clique with leaves, however, the optimal clustering changes to $k = n$ clusters, in which each cluster consists of a connected pair of leaf and clique nodes. Figure 2(a) shows such an example.

Scaling Behavior. Figures 2(a) and 2(b) display the
scaling behavior of modularity. By simply doubling
the graph presented in Figure 2(a), the optimal clustering is altered completely. While in Figure 2(a) we
obtain three clusters each consisting of the minor K_2,
the clustering with maximum modularity of the graph
in Figure 2(b) consists of two clusters, each being a
graph equal to the one in Figure 2(a). This behavior is

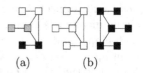

(a)   (b)

Fig. 2. Scaling behavior.
Clustering by colors.

in line with the previous observations in [3,5], where it was observed that size and
structure of clusters in the optimum clustering depend on the total number of
links in the network. Hence, clusters that are identified in smaller graphs might
be combined to a larger cluster in a optimum clustering of a larger graph. The
formulation of Equation 1 mathematically explains this observation as modularity optimization strives to optimize the trade-off between coverage and degree
sums. This provides a rigorous understanding of the observations made in [3,5].

4 $\mathcal{NP}$-Completeness

It has been conjectured that maximizing modularity is hard [7], but no formal
proof was provided to date. We next show that decision version of modularity
maximization is indeed $\mathcal{NP}$-complete.

Problem 1 (MODULARITY). *Given a graph G and a number K, is there a
clustering $\mathcal{C}$ of G, for which $q(\mathcal{C}) \geq K$?*

Note that we may ignore the fact that, in principle, K could be a real number
in the range $[0, 1]$, because $4m^2 \cdot q(\mathcal{C})$ is integer for every partition $\mathcal{C}$ of G and
polynomially bounded in the size of G. Our hardness result for MODULARITY is
based on a transformation from the following decision problem.

Problem 2 (3-PARTITION). *Given $3k$ positive integer numbers $a_1, \ldots, a_{3k}$ such
that the sum $\sum_{i=1}^{3k} a_i = kb$ and $b/4 < a_i < b/2$ for an integer b and for all
$i = 1, \ldots, 3k$, is there a partition of these numbers into k sets, such that the
numbers in each set sum up to b?*

We show that an instance $A = \{a_1, \ldots, a_{3k}\}$ of 3-PARTITION can be transformed
into an instance $(G(A), K(A))$ of MODULARITY, such that $G(A)$ has a clustering
with modularity at least $K(A)$, if and only if $a_1, \ldots, a_{3k}$ can be partitioned into
k sets of sum $b = 1/k \cdot \sum_{i=1}^{k} a_i$ each. It is crucial that 3-PARTITION is *strongly*
$\mathcal{NP}$-complete [14], i.e. the problem remains $\mathcal{NP}$-complete even if the input is
represented in unary coding. This implies that no algorithm can decide the
problem in time polynomial even in the sum of the input values, unless $\mathcal{P} = \mathcal{NP}$.
More importantly, our transformation need only be pseudo-polynomial.

The reduction is defined as follows. Given an instance A of 3-PARTITION, construct a graph $G(A)$ with k cliques (completely connected subgraphs) $H_1, \ldots, H_k$
of size $a = \sum_{i=1}^{3k} a_i$ each. For each element $a_i \in A$ we introduce a single *element*

node, and connect it to a_i nodes in each of the k cliques in such a way that each clique member is connected to exactly one element node. It is easy to see that each clique node then has degree a and the element node corresponding to element $a_i \in A$ has degree ka_i. The number of edges in $G(A)$ is $m = k/2 \cdot a(a+1)$. Note that the size of $G(A)$ is polynomial in the unary coding size of A, so that our transformation is indeed pseudo-polynomial. Before specifying bound $K(A)$ for the instance of MODULARITY, we will show three properties of maximum modularity clusterings of $G(A)$. Together these properties establish the desired characterization of solutions for 3-PARTITION by solutions for MODULARITY.

Lemma 4. *In a maximum modularity clustering of $G(A)$, none of the cliques $H_1, \ldots, H_k$ is split.*

Lemma 5. *In a maximum modularity clustering of $G(A)$, every cluster contains at most one of the cliques $H_1, \ldots, H_k$.*

The proofs of Lemmas 4 and 5 can be found in [12]. Both are based on the fact that modularity can be increased by a modification of the clustering, if either Lemma is violated. Next, we observe that the optimum clustering places at most one clique completely into a single cluster.

The previous two lemmas show that any clustering can be strictly improved to a clustering that contains k *clique clusters*, such that each one completely contains one of the cliques $H_1, \ldots, H_k$ (possibly plus some additional element nodes). In particular, this must hold for the optimum clustering as well. Now that we know how the cliques are clustered we turn to the element nodes. As they are not directly connected, it is never optimal to create a cluster consisting only of element nodes. Splitting such a cluster into singleton clusters, one for each element node, reduces the squared degree sums but keeps the edge coverage at the same value. Hence, such a split yields a clustering with strictly higher modularity. The next lemma shows that we can further strictly improve the modularity of a clustering with a singleton cluster of an element node by joining it with one of the clique clusters.

Lemma 6. *In a maximum modularity clustering of $G(A)$, there is no cluster composed of element nodes only.*

Closely following the proofs of the previous two lemmas, we obtain the proof of Lemma 6 in [12]. We have shown that for the graphs $G(A)$ the clustering of maximum modularity consists of exactly k clique clusters, and each element node belongs to exactly one of the clique clusters. Combining the above results, we now state our main result:

Theorem 3. MODULARITY *is strongly $\mathcal{NP}$-complete.*

Proof. For a given clustering $\mathcal{C}$ of $G(A)$ we can check in polynomial time whether $\mathsf{q}(\mathcal{C}) \geq K(A)$, so clearly MODULARITY $\in \mathcal{NP}$. For $\mathcal{NP}$-completeness we transform an instance $A = \{a_1, \ldots, a_{3k}\}$ of 3-PARTITION into an instance $(G(A), K(A))$ of MODULARITY. We have already outlined the construction of the graph $G(A)$

above. For the correct parameter $K(A)$ we consider a clustering in $G(A)$ with the properties derived in the previous lemmas, i.e., a clustering with exactly k clique clusters. Any such clustering yields exactly $(k-1)a$ inter-cluster edges, so the edge coverage is given by

$$\sum_{C \in \mathcal{C}} \frac{|E(C)|}{m} = \frac{m - (k-1)a}{m} = 1 - \frac{2(k-1)a}{ka(a+1)} = 1 - \frac{2k-2}{k(a+1)}.$$

Hence, the clustering $\mathcal{C} = (C_1, \ldots, C_k)$ with maximum modularity must minimize $\deg(C_1)^2 + \deg(C_2)^2 + \ldots + \deg(C_k)^2$. This requires a distribution of the element nodes between the clusters which is as even as possible with respect to the sum of degrees per cluster. In the optimum case we can assign each cluster element nodes corresponding to elements that sum to $b = 1/k \cdot a$. In this case the sum of degrees of element nodes in each clique cluster is equal to $k \cdot 1/k \cdot a = a$. This yields $\deg(C_i) = a^2 + a$ for each clique cluster C_i, $i = 1, \ldots, k$, and gives

$$\deg(C_1)^2 + \ldots + \deg(C_k)^2 \geq k(a^2 + a)^2 = ka^2(a+1)^2.$$

Equality holds only in the case, in which an assignment of b to each cluster is possible. Hence, if there is a clustering $\mathcal{C}$ with $\mathsf{q}(\mathcal{C})$ of at least

$$K(A) = 1 - \frac{2k-2}{k(a+1)} - \frac{ka^2(a+1)^2}{k^2a^2(a+1)^2} = \frac{(k-1)(a-1)}{k(a+1)}$$

then we know that this clustering must split the element nodes perfectly to the k clique clusters. As each element node is contained in exactly one cluster, this yields a solution for the instance of 3-PARTITION. With this choice of $K(A)$ the instance $(G(A), K(A))$ of MODULARITY is satisfiable only if the instance A of 3-PARTITION is satisfiable.

Otherwise, suppose the instance for 3-PARTITION is satisfiable. Then there is a partition into k sets such that the sum over each set is $1/k \cdot a$. If we cluster the corresponding graph by joining the element nodes of each set with a different clique, we get a clustering of modularity $K(A)$. This shows that the instance $(G(A), K(A))$ of MODULARITY is satisfiable if the instance A of 3-PARTITION is satisfiable. This completes the reduction and proves the theorem.   □

This result naturally holds also for the straightforward generalization of maximizing modularity in weighted graphs [15]. Instead of using the numbers of edges the definition of modularity employs the sum of edge weights for edges within clusters, between clusters and in the total graph.

4.1   Special Case: Modularity with a Bounded Number of Clusters

A common clustering approach is based on iteratively identifying cuts, see for example [16,17,18]. The general problem being $\mathcal{NP}$-complete, we now complete our hardness results by proving that the restricted optimization problem is hard as well. More precisely, we consider the two problems of computing the clustering with maximum modularity that splits the graph into exactly or at most two clusters. Although these are two different problems, our hardness result holds for both versions, hence, we define the problem cumulatively.

Problem 4 (2-MODULARITY). *Given a graph G and a number K, is there a clustering C of G into exactly/at most 2 clusters, for which $q(C) \geq K$?*

Our proof uses a reduction similar to the one for showing the hardness of the "MinDisAgree[2]" problem of correlation clustering [19]. The reduction is from MINIMUM BISECTION FOR CUBIC GRAPHS (MB3).

Problem 5 (MINIMUM BISECTION FOR CUBIC GRAPHS). *Given a 3-regular graph G with n nodes and an integer c, is there a clustering into two clusters of $n/2$ nodes each such that it cuts at most c edges?*

This problem has been shown to be strongly $\mathcal{NP}$-complete in [20]. We construct an instance of 2-MODULARITY from an instance of MB3 as follows. For each node v from the graph $G = (V, E)$ we attach $n - 1$ new nodes and construct an n-clique. We denote these cliques as $cliq(v)$ and refer to them as *node clique* for $v \in V$. Hence, in total we construct n different new cliques, and after this transformation each node from the original graph has degree $n + 2$. Note that a cubic graph with n nodes has exactly $1.5n$ edges. In our adjusted graph there are exactly $m = (n(n - 1) + 3)n/2$ edges.

We will show that an optimum clustering C^* of 2-MODULARITY in the adjusted graph has exactly two clusters. Furthermore, such a clustering corresponds to a minimum bisection of the underlying MB3 instance. In particular, we give a bound K such that the MB3 instance has a bisection cut of size at most c if and only if the corresponding graph has 2-modularity at least K. We begin by noting that there is always a clustering C with $q(C) > 0$. Hence, C^* must have exactly two clusters, as no more than two clusters are allowed. This serves to show that our proof works for both versions of 2-modularity, in which at most or exactly two clusters must be found.

Lemma 7. *For every graph constructed from a MB3 instance, there exists a clustering $C = \{C_1, C_2\}$ such that $q(C) > 0$. In particular, the clustering C^* has two clusters.*

The proof of Lemma 7 can be found in [12]. Next, we show that in an optimum clustering, all the nodes of one node clique $cliq(v)$ are located in one cluster. The proof is also published in [12]

Lemma 8. *For every node $v \in V$ a cluster $C \in C^*$ exists, such that $cliq(v) \subseteq C$.*

The final lemma before defining the appropriate input parameter K for the 2-MODULARITY and thus proving the correspondence between the two problems shows that the clusters in the optimum clusterings have the same size. The proof can be found in [12].

Lemma 9. *In C^*, each cluster contains exactly $n/2$ complete node cliques.*

Finally, we can state theorem about the complexity of 2-MODULARITY:

Theorem 6. 2-MODULARITY *is strongly* $\mathcal{NP}$-*complete.*

Proof. Let (G, c) be an instance of MINIMUM BISECTION FOR CUBIC GRAPHS, then we construct a new graph G' as stated above and define $K := 1/2 - c/m$.

As we have shown in Lemma 9 that each cluster of $\mathcal{C}^*$ that is an optimum clustering of G' with respect to 2-MODULARITY has exactly $n/2$ complete node cliques, the sum of degrees in the clusters is exactly m. Thus, it is easy to see that if the clustering $\mathcal{C}^*$ meets the following inequality

$$\mathsf{q}\,(\mathcal{C}^*) \geq 1 - \frac{c}{m} - \frac{2m^2}{4m^2} = \frac{1}{2} - \frac{c}{m} = K \ ,$$

then the number of inter-cluster edges can be at most c. Thus the clustering $\mathcal{C}^*$ induces a balanced cut in G with at most c cut edges.   □

This proof is particularly interesting as it highlights that maximizing modularity in general is hard due to the hardness of minimizing the squared degree sums on the one hand, whereas in the case of two clusters this is due to the hardness of minimizing the edge cut.

5   The Greedy Algorithm

In contrast to the abovementioned iterative cutting strategy, another commonly used approach to find clusterings with good quality scores is based on greedy agglomeration.In the case of modularity, this approach is particularly widespread [7]. It starts with the singleton clustering and iteratively merges those two clusters that yield a clustering with the best modularity, i. e., the largest increase or the smallest decrease is chosen. After $n - 1$ merges the clustering that achieved the highest modularity is returned. Note that $n - 1$ is an upper bound on the number of iterations and that one can terminate the algorithm as soon as no further increase in modularity is possible. This is due to a property called *single-peakedness*, proven in [7].

Since it is $\mathcal{NP}$-hard to maximize modularity in general graphs, it is unlikely that this greedy algorithm is optimal. In fact, we sketch a graph family, where the above greedy algorithm has an approximation factor of 2, asymptotically (Theorem 8). While the former result relies on a deterministic procedure of the algorithm, in the following we even point out instances where a specific way of breaking ties of equally attractive merges yield a clustering with modularity of 0, while the optimum clustering has a strictly positive score (Theorem 7).

Modularity is defined such that it takes values in the interval $[-1/2, 1]$ for any graph and any clustering (Lemma 1). In particular the modularity of a trivial clustering placing all vertices into a single cluster has a value of 0. We exploit this technical peculiarity to show that the greedy algorithm has an unbounded approximation ratio.

Theorem 7. *There is no finite approximation factor for the greedy algorithm for finding clusterings with maximum modularity.*

The full proof can be found in [12]. The key observation is a worst-case scenario in the sense that greedy is in each iteration supposed to pick exactly the "worst" merge choice of several equivalently attractive alternatives. As mentioned earlier, this negative result is due to the formulation of modularity, which yields values from the interval $[-1/2, 1]$. For instance, a linear remapping of the range of modularity to the interval $[0, 1]$, the greedy algorithm yields a value of $1/3$ compared to the new optimum score of $2/3$. In this case the approximation factor would be 2. Next, we provide a weaker lower bound for a different class of graphs, but making no assumptions on random choices of the algorithm.

Theorem 8. *The approximation factor of the greedy algorithm for finding clusterings with maximum modularity is no better than 2.*

The founding idea of the proof of Theorem 8 is a special graph family which is constructed by attaching a path to each node of a clique. An example is given in Figure 3. We show that greedy algorithm always yields n clusters, each of which includes a vertex v and the attached path. The clustering with maximal modularity, however, seperates the clique from the paths. The approximation factor asymptotically approaches 2 for n going to infinity with paths of length $1/2\sqrt{n}$ attached to a clique of size n. See [12] for details.

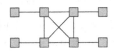

Fig. 3. Clique with attached paths

6   Examples Revisited

Applying our results about maximizing modularity gained so far, we revisit two example networks that were used in related work [21,2,8]. More precisely, we compare published greedy solutions with respective optima, thus revealing two peculiarities of modularity. First, we illustrate a behavioral pattern of the greedy merge strategy and, second, we relativize the quality of the greedy approach.

The first instance, Figure 4, is the karate club network of Zachary originally introduced in [21] and used for demonstration in [2]. The real-world partition of the club is given by the shape of the nodes, while the colors indicate the clustering calculated by the greedy algorithm and blocks refer to a optimum clustering maximizing modularity, that has been obtained by solving the above ILP. The corresponding scores of modularity are 0.431 for the optimum clustering, 0.397 for the greedy clustering, and 0.383 for the clustering given by the split. Observe the following peculiarity: Due to the attempt to balance the squared sum of degrees (over the clusters), a node with large degree (white square) and one with small degree (white circle) are merged relatively soon. Using the same argument, such a cluster will unlikely be merged with another one. As a result, a cluster rarely has only one node, but relative small clusters still occur, featuring skew distribution of node degrees.

The second instance, Figure 5, is a network of books on politics, compiled by V. Krebs and used for demonstration in [8]. Nodes represent books on American

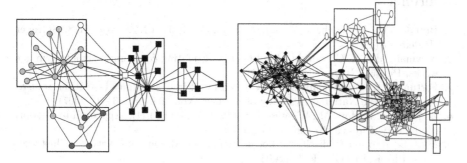

Fig. 4. The network of a Karate club introduced by Zachary [21]

Fig. 5. The networks of books on politics compiled by V. Krebs

politics and edges join pairs of books that are frequently purchased together bought from `Amazon.com`. The optimum clustering maximizing modularity is given by the shapes of nodes, the colors of nodes indicate a clustering calculated by the greedy algorithm and the blocks show a clustering calculated by Geometric MST Clustering (GMC) which is introduced in [22] using the geometric mean of *coverage* and *performance*, both of which are quality indices discussed in the same paper. The corresponding scores of modularity are 0.527 for the optimum clustering, 0.502 for the greedy clustering, and 0.510 for the GMC clustering. A key observation is that GMC outperforms the greedy algorithm although it does not consider modularity in its calculations. Moreover, the comparison of the structure of the calculated clusterings reveals that several clusterings close to the optimum one still have relative large modularity score. Thus, the good performance of the greedy approach comes as no surprise.

7   Conclusion

We provide the first formal assessments of a popular clustering index known as modularity. We have settled the open question about the complexity status of modularity maximization by proving its $\mathcal{NP}$-completeness in the strong sense. We show that this even holds for the restricted version with a bound of two on the number of clusters. This justifies the further investigation of approximation algorithms and heuristics, such as the widespread greedy approach. For the latter we prove a first lower bound on the approximation factor. Our analysis of the greedy algorithm also includes a brief comparison with the optimum clustering which is calculated via ILP on some real-world instances, thus encouraging a reconsideration of previous results. For the future we plan an extended analysis and the development of a clustering algorithm with provable performance guarantees. The special properties of the measure, its popularity in application domains and the absence of fundamental theoretical insights hitherto, render further mathematically rigorous treatment of modularity necessary.

References

1. Brandes, U., Erlebach, T. (eds.): Network Analysis. LNCS, vol. 3418. Springer, Heidelberg (2005)
2. Newman, M.E.J., Girvan, M.: Finding and evaluating community structure in networks. Physical Review E 69 (2004)
3. Fortunato, S., Barthelemy, M.: Resolution Limit in Community Detection. In: Proceedings of the National Academy of Sciences, vol. 104, pp. 36–41 (2007)
4. Ziv, E., Middendorf, M., Wiggins, C.: Information-Theoretic Approach to Network Modularity. Physical Review E 71 (2005)
5. Muff, S., Rao, F., Caflisch, A.: Local Modularity Measure for Network Clusterizations. Physical Review E 72 (2005)
6. Gaertler, M., Görke, R., Wagner, D.: Significance-Driven Graph Clustering. In: AAIM 2007. LNCS, vol. 4508, pp. 11–26. Springer, Heidelberg (2007)
7. Clauset, A., Newman, M.E.J., Moore, C.: Finding community structure in very large networks. Physical Review E 70 (2004)
8. Newman, M.: Modularity and Community Structure in Networks. In: Proceedings of the National Academy of Sciences, pp. 8577–8582 (2005)
9. White, S., Smyth, P.: A Spectral Clustering Approach to Finding Communities in Graph. In: SIAM Data Mining Conference (2005)
10. Reichardt, J., Bornholdt, S.: Statistical Mechanics of Community Detection. Physical Review E 74 (2006)
11. Duch, J., Arenas, A.: Community Detection in Complex Networks using Extremal Optimization. Physical Review E 72 (2005)
12. Brandes, U., Delling, D., Gaertler, M., Görke, R., Höfer, M., Nikoloski, Z., Wagner, D.: On Modularity Clustering. IEEE Transactions on Knowledge and Data Engineering (to appear, 2007)
13. Danon, L., Díaz-Guilera, A., Duch, J., Arenas, A.: Comparing community structure identification. Journal of Statistical Mechanics (2005)
14. Garey, M.R., Johnson, D.S.: Computers and Intractability. A Guide to the Theory of $\mathcal{NP}$-Completeness. W.H. Freeman and Company, New York (1979)
15. Newman, M.: Analysis of Weighted Networks. Technical report, Cornell University, Santa Fe Institute, University of Michigan (2004)
16. Alpert, C.J., Kahng, A.B.: Recent Directions in Netlist Partitioning: A Survey. Integration: The VLSI Journal 19, 1–81 (1995)
17. Hartuv, E., Shamir, R.: A Clustering Algorithm based on Graph Connectivity. Information Processing Letters 76, 175–181 (2000)
18. Vempala, S., Kannan, R., Vetta, A.: On Clusterings - Good, Bad and Spectral. In: FOCS 2000. Proceedings of the 41st Annual IEEE Symposium on Foundations of Computer Science, pp. 367–378 (2000)
19. Giotis, I., Guruswami, V.: Correlation Clustering with a Fixed Number of Clusters. In: SODA 2006. Proceedings of the 17th Annual ACM–SIAM Symposium on Discrete Algorithms, New York, NY, USA, pp. 1167–1176 (2006)
20. Bui, T.N., Chaudhuri, S., Leighton, F.T., Sipser, M.: Graph bisection algorithms with good average case behavior. Combinatorica 7, 171–191 (1987)
21. Zachary, W.W.: An Information Flow Model for Conflict and Fission in Small Groups. Journal of Anthropological Research 33, 452–473 (1977)
22. Brandes, U., Gaertler, M., Wagner, D.: Experiments on Graph Clustering Algorithms. In: Di Battista, G., Zwick, U. (eds.) ESA 2003. LNCS, vol. 2832, pp. 568–579. Springer, Heidelberg (2003)

On Minimum Area Planar Upward Drawings of Directed Trees and Other Families of Directed Acyclic Graphs

Fabrizio Frati

Dipartimento di Informatica e Automazione – Università di Roma Tre
frati@dia.uniroma3.it

Abstract. It has been shown in [9] that there exist planar digraphs that require exponential area in every upward *straight-line* planar drawing. On the other hand, upward *poly-line* planar drawings of planar graphs can be realized in $\Theta(n^2)$ area [9]. In this paper we consider families of DAGs that naturally arise in practice, like DAGs whose underlying graph is a tree (*directed trees*), is a bipartite graph (*directed bipartite graphs*), or is an outerplanar graph (*directed outerplanar graphs*). Concerning *directed trees*, we show that optimal $\Theta(n \log n)$ area upward straight-line/poly-line planar drawings can be constructed. However, we prove that if the order of the neighbors of each node is assigned, then exponential area is required for straight-line upward drawings and quadratic area is required for poly-line upward drawings, results surprisingly and sharply contrasting with the area bounds for planar upward drawings of undirected trees. After having established tight bounds on the area requirements of planar upward drawings of several families of directed trees, we show how the results obtained for trees can be exploited to determine asymptotic optimal values for the area occupation of planar upward drawings of *directed bipartite graphs* and *directed outerplanar graphs*.

1   Introduction

Upward drawings of directed acyclic digraphs (*DAGs* for short) have several applications in the visualization of hierarchical structures, as PERT diagrams, subroutine-call charts, Hasse diagrams, and is-a relationships, and hence they have been intensively studied from a theoretical point of view. It is known that testing the *upward planarity* of a graph is an NP-complete problem if the graph has a variable embedding [13], while it is polynomially solvable if the embedding of the graph is *fixed* [2], if the underlying graph is supposed to be an *outerplanar graph* ([15]), if the digraph has a *single source* ([14]), or if it's a *bipartite DAG* ([7]). Di Battista and Tamassia ([8]) showed that a graph is upward planar if and only if it's a subgraph of an *st-planar* graph. Moreover, some families of DAGs are always upward planar, like the *series-parallel* digraphs and the digraphs whose underlying graph is a *tree*.

Concerning algorithms for obtaining upward drawings of DAGs in small area, Di Battista et al. have shown in [9] that every upward planar embedding can

A. Brandstädt, D. Kratsch, and H. Müller (Eds.): WG 2007, LNCS 4769, pp. 133–144, 2007.
© Springer-Verlag Berlin Heidelberg 2007

be drawn with upward *poly-line* edges in optimal $\Theta(n^2)$ area, while there exist graphs that require exponential area in any planar *straight-line* upward drawing. Hence, it is natural to restrict the attention to interesting families of DAGs, searching for better area bounds. This research direction has been taken by Bertolazzi et al. in [1], where it is shown that *series-parallel* digraphs admit upward planar straight-line drawings in $\Theta(n^2)$ area, while exponential area is generally required if the embedding is chosen in advance.

In this paper we study classes of DAGs that commonly arise in practice, as DAGs whose underlying graph is a tree (*directed trees*), is a bipartite graph (*directed bipartite graphs*), or is an outerplanar graph (*directed outerplanar graphs*). All of such digraph classes exhibit simple and strong structural properties that allow to create planar upward drawings with less constraints and in a easier way with respect to general digraphs. Consequently, we are able to construct straight-line planar upward drawings of directed trees in $\Theta(n \log n)$ area, and to get $\Theta(n)$ area straight-line planar upward drawings for some sub-classes of directed trees. Surprisingly, we prove that when constraints are imposed on the drawings by forcing an ordering of the neighbors of each vertex, then again exponential area is required for constructing straight-line planar upward drawings and quadratic area is required for constructing poly-line planar upward drawings. Such negative results contrast with the fact that sub-quadratic area is sufficient for constructing straight-line order-preserving upward planar drawings of undirected trees ([3]). Furthermore, we prove that the lower bounds obtained for directed trees extend also to directed bipartite graphs and directed outerplanar graphs.

More in detail, we provide the following results: (i) straight-line and poly-line planar upward drawings of directed trees can be constructed in optimal $\Theta(n \log n)$ area (Sec. 3); (ii) straight-line order-preserving planar upward drawings of directed trees require (and can be constructed in) exponential area (Sec. 4); (iii) poly-line order-preserving planar upward drawings of directed trees require (and can be constructed in) quadratic area (Sec. 4); (iv) *directed binary trees* have the same area requirements of general directed trees (Sec. 5); (v) *directed caterpillars* and *directed spider trees* admit linear area straight-line drawings (Sec. 5); (vi) straight-line planar upward drawings of directed bipartite graphs require (and can be constructed in) exponential area (Sec. 5); (vii) poly-line planar upward drawings of directed bipartite graphs require (and can be constructed in) quadratic area (Sec. 5); (viii) straight-line outerplanar upward drawings of directed outerplanar graphs require (and can be constructed in) exponential area (Sec. 5); and

Table 1. A table summarizing the results on minimum area upward drawings of directed trees. Straight-line and poly-line non-order-preserving drawings are in the same columns, since they have the same area bounds. Constants b and c are greater than 1.

	Straight-line / Poly-line				Straight-line Order-Pres.				Poly-line Order-Pres.			
	UB	ref.	LB	ref.	UB	ref.	LB	ref	UB	ref.	LB	ref.
Dir. Trees	$O(n \log n)$	Th. 1	$\Omega(n \log n)$	Th. 1	$O(c^n)$	[12]	$\Omega(b^n)$	Th. 2	$O(n^2)$	[9]	$\Omega(n^2)$	Th. 3
Dir. Binary Trees	$O(n \log n)$	Th. 1	$\Omega(n \log n)$	Th. 1	$O(c^n)$	[12]	$\Omega(b^n)$	Th. 2	$O(n^2)$	[9]	$\Omega(n^2)$	Th. 3
Dir. Caterpillars	$O(n)$	Th. 4	$\Omega(n)$	trivial	$O(c^n)$	[12]	$\Omega(b^n)$	Th. 2	$O(n^2)$	[9]	$\Omega(n^2)$	Th. 3
Dir. Spider Trees	$O(n)$	Th. 5	$\Omega(n)$	trivial	$O(n)$	Th. 5	$\Omega(n)$	trivial	$O(n)$	Th. 5	$\Omega(n)$	trivial

(ix) poly-line planar upward drawings of directed outerplanar graphs require (and can be constructed in) quadratic area (Sec. 5). Table 1 summarizes the area requirements of planar upward drawings of directed trees, directed binary trees, directed caterpillars, and directed spider trees.

2   Preliminaries

We assume familiarity with graphs and their drawings (see also [5]).

A *grid drawing* of a graph is a mapping of each vertex to a distinct point of the plane with integer coordinates and of each edge to a Jordan curve between the endpoints of the edge. A *poly-line drawing* is such that the edges are sequences of rectilinear segments, with bends having integer coordinates. A *straight-line drawing* is such that all edges are rectilinear segments. A *planar drawing* is such that no two edges intersect. An *upward drawing* of a *digraph* is a planar drawing with each directed edge represented by a curve monotonically increasing in the vertical direction. In the following when we refer to upward drawings we always mean planar upward grid drawings. The graph obtained from a digraph G by considering its edges without orientation is called the *underlying graph* of G. An *embedding* of a graph is a circular ordering of the edges incident on each vertex. A drawing is *order-preserving* if the order of the edges incident on each vertex is the same as the one of an embedding specified in advance. The *bounding box* $B(\Gamma)$ of a drawing Γ is the smallest rectangle with sides parallel to the axes that covers Γ completely. We denote by $l(\Gamma)$, by $r(\Gamma)$, by $t(\Gamma)$, and by $b(\Gamma)$ the *left* side, the *right* side, the *top* side, the *bottom* side of $B(\Gamma)$, respectively. The *height* (*width*) of Γ is the height (width) of its bounding box plus one. The *area* of Γ is the height of Γ multiplied by its width. We denote by $y(v)$ the y-coordinate of a vertex v that is drawn on the plane.

An *outerplanar graph* is a graph that has a planar embedding in which all vertices are incident to the same face. Such an embedding is called *outerplanar embedding*. A *bipartite graph* is a graph G that has the vertices partitioned into two subsets such that G has edges only between vertices of different subsets. A *caterpillar* C is a tree such that the removal from C of all the leaves and of their incident edges turns C in a path. A *spider tree* is a tree having only one vertex of degree greater than two.

3   Upward Drawings of Trees

In this section we show that directed trees admit straight-line upward drawings in $\Theta(n \log n)$ area and that such an area is necessary in the worst case, even if bends are allowed on the edges. Concerning the lower bound, Crescenzi et al. in [4] showed a non-directed rooted binary tree T that requires $\Omega(n \log n)$ area in any *strictly upward* grid drawing. Now T can be turned in a directed binary tree T' by directing its edges away from the root. Since an upward drawing of T' is a strictly upward drawing of T, the lower bound on the area requirement of upward drawings of directed trees follows.

Now we show that every directed tree has an $O(n \log n)$ area straight-line upward drawing. This is done by means of an algorithm that consider a directed tree T, removes from T a path called *spine*, recursively draws each disconnected subtree, and finally puts the drawings of the subtrees together with a drawing of the spine, obtaining a drawing of T. This *divide et impera* strategy has been intensively used in algorithms for drawing undirected trees and outerplanar graphs ([3,11,10,6]). Let us describe the algorithm more formally.

Preprocessing: The input is a directed tree T with n nodes. We derive a non-directed rooted tree T' from T by removing the orientations from the edges of T and by choosing a node r in T as root of T'.

Divide: Let T^* be the current non-directed rooted tree and let r^* be its root (at the first step the current tree is T' rooted at r).

If the number of nodes in T^* is greater than one, then select a spine $S^* = (v_0, v_1, \ldots, v_k)$ in T^* with the following properties: (i) $v_0 = r^*$, (ii) for $1 \leq i \leq k$, v_i is the root of the heaviest (i.e. with the greatest number of nodes) subtree among the subtrees rooted at the children of v_{i-1}, and (iii) each edge (v_{i-1}, v_i) is directed from v_i to v_{i-1} in T, for $1 \leq i < k$, and (iv) edge (v_{k-1}, v_k) is directed from v_{k-1} to v_k in T, or v_k is a leaf. Remove from T^* the nodes of S^*, but for v_k, disconnecting T^* in several non-directed subtrees. We classify such subtrees into sets $T^*(\uparrow, v_i)$ and $T^*(\downarrow, v_i)$, with $0 \leq i < k$, so that a tree rooted at a vertex v goes into set $T^*(\uparrow, v_i)$ (resp. $T^*(\downarrow, v_i)$) if in the directed tree T there is an edge directed from v to v_i (resp. there is an edge directed from v_i to v). Notice that each set could contain several trees. We denote by $T^*(v_k)$ the tree rooted at v_k and by $r(T^*)$ the root of a non-directed tree T^*.

Impera: Assume that in the *Divide* step a tree T^* has been disconnected in a spine S^*, in a subtree $T^*(v_k)$, and in several subtrees in $T^*(\uparrow, v_i)$ and in $T^*(\downarrow, v_i)$, with $0 \leq i < k$. Introduce again the direction on the edges of T^*, obtaining a directed tree $T(v_k)$ from $T^*(v_k)$, obtaining a set of directed trees $T(\uparrow, v_i)$ from the trees in $T^*(\uparrow, v_i)$, and obtaining a set of directed trees $T(\downarrow, v_i)$ from the trees in $T^*(\downarrow, v_i)$. Assume to have for each of such directed trees a drawing with the following properties: $(\mathbf{P_1})$ the drawing is planar, upward, and straight-line; $(\mathbf{P_2})$ the root of the tree is placed on the left side of the bounding box of the drawing; and $(\mathbf{P_3})$ no node of the tree is placed in the drawing below and on the same vertical line of the root of the tree.

Notice that such a drawing can be trivially constructed for a tree with at most one node. Now we show how to construct a drawing Γ satisfying properties P_1, P_2, and P_3 for the directed tree $\overline{T}$ obtained from T^* by introducing again the directions on the edges. Notice that, in the last *Impera* step, Γ will be a drawing of the whole directed tree T. We distinguish two cases:

k = 1: Place the drawings of the trees in $T(\downarrow, v_0)$ stacked one above the other at one unit of vertical distance, with the left side of their bounding boxes on the same vertical line l, obtaining a drawing Γ'. Place v_0 one unit to the left of l and one unit below $b(\Gamma')$. Place the drawings of the trees in $T(\uparrow, v_0)$ stacked one above the other at one unit of vertical distance, with the left side of their bounding boxes on l, and so that the highest horizontal line intersecting a drawing of a tree in $T(\uparrow, v_0)$ is one unit below v_0, obtaining a drawing Γ''. If (v_0, v_1) is directed from v_0 to v_1, then place the drawing of $T(v_1)$ so that the left side of its bounding box is on the same vertical line of v_0 and so that the bottom side of its bounding box is one unit above $t(\Gamma'')$ (see Fig. 1.a). Otherwise, that is v_1 is a leaf and (v_0, v_1) is directed from v_1 to v_0, place v_1 on the same vertical line of v_0 and one unit below $b(\Gamma'')$.

$k \geq 2$: Place the drawings of the trees in $T(\downarrow, v_0)$ stacked one above the other at one unit of vertical distance, with the left side of their bounding boxes on the same vertical line l, obtaining a drawing Γ'. Place v_0 two units to the left of l and one unit below $b(\Gamma')$. Place the drawings of the trees in $T(\uparrow, v_0)$ stacked one above the other at one unit of vertical distance, with the left side of their bounding boxes on l, and so that the highest horizontal line intersecting a drawing of a tree in $T(\uparrow, v_0)$ is one unit below v_0, obtaining a drawing Γ_0. For $i = 1, 2, \ldots, k - 2$, place the drawings of the trees in $T(\downarrow, v_i)$ stacked one above the other at one unit of vertical distance, with the left side of their bounding boxes on l, and so that the highest horizontal line intersecting a drawing of a tree in $T(\downarrow, v_i)$ is one unit below $b(\Gamma_{i-1})$, obtaining a drawing Γ'. Place v_i one unit to the left of l and and one unit below $b(\Gamma')$. Place the drawings of the trees in $T(\uparrow, v_i)$ stacked one above the other at one unit of vertical distance, with the left side of their bounding boxes on l, and so that the highest horizontal line intersecting a drawing of a tree in $T(\uparrow, v_i)$ is one unit below v_i, obtaining a drawing Γ_i. Let W be the maximum between the width of the drawing of $T(v_k)$ minus 1 and the maximum width of a drawing of a tree in $T(\uparrow, v_i)$ or in $T(\downarrow, v_i)$ plus 2, with $0 \leq i < k$. Let l' be the vertical line W units to the right of v_0. Mirror the drawings of the trees in $T(\uparrow, v_{k-1})$ with respect to a vertical line and place them stacked one above the other at one unit of vertical distance, with the right side of their mirrored bounding boxes one unit to the left of l' and so that the highest horizontal line intersecting a drawing of a tree in $T(\uparrow, v_{k-1})$ is one unit below $b(\Gamma_{k-2})$. Mirror the drawings of the trees in $T(\downarrow, v_{k-1})$ with respect to a vertical line and place them stacked one above the other at one unit of vertical distance, with the right side of their mirrored bounding boxes one unit to the left of l', and so that the lowest horizontal line intersecting a drawing of a tree in $T(\downarrow, v_{k-1})$ is one unit above $t(\Gamma_{k-2})$. Place v_{k-1} on l' one unit below v_{k-2}, obtaining a drawing Γ_{k-1}. Finally, if edge (v_{k-1}, v_k) is directed from v_{k-1} to v_k, mirror the drawing of $T(v_k)$ with respect to a vertical line and place it with the right side of its mirrored bounding box on l' so that the bottom side of its bounding box is one unit above $t(\Gamma_{k-1})$; otherwise, that is v_k is a leaf and edge (v_{k-1}, v_k) is directed from v_k to v_{k-1}, place v_k on l' one unit below $b(\Gamma_{k-1})$.

The planarity and the upwardness of the final drawing Γ of T can be easily verified. Concerning the area requirements of Γ, the *height* of Γ is $O(n)$, since there is at least one node of the tree for each horizontal line intersecting Γ. Denote by $w(T(\uparrow, v_i))$, by $w(T(\downarrow, v_i))$, by $w(T(v_i))$, and by $w(n)$ the width of the drawing of a tree in $T(\uparrow, v_i)$, of a tree in $T(\downarrow, v_i)$, of a tree $T(v_i)$, and of

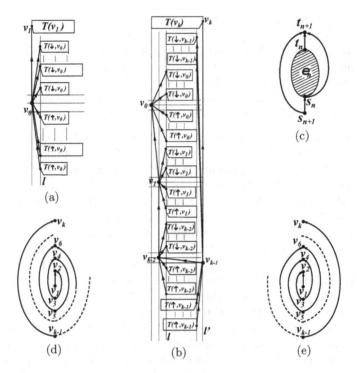

Fig. 1. (a) and (b) *Impera* step of the algorithm for obtaining straight-line non-order preserving upward drawings of trees, in the case $k = 1$ and $k \geq 2$. (c) Embedding $\mathcal{E}_{n+1}$ of the series-parallel digraph presented in [1]. (d) A clockwise coil. (e) A counter-clockwise coil.

an n-nodes tree constructed by the above described algorithm, respectively. In case $k = 1$ we have $w(T) = \max\{w(T(v_1)), 1 + w(T(\uparrow, v_0)), 1 + w(T(\downarrow, v_0))\}$, and in case $k \geq 2$ we have $w(T) = \max_{0 \leq i < k}\{w(T(v_k)), 3 + w(T(\uparrow, v_i)), 3 + w(T(\downarrow, v_i))\}$. By the definition of S, each tree in $T(\uparrow, v_i)$ and each tree in $T(\downarrow, v_i)$ has at most $n/2$ nodes, and $T(v_k)$ has at most $n - k$ nodes. It follows that $w(n) = \max\{w(n-1), 3 + w(n/2)\}$, that easily solves to $w(n) = O(\log n)$. So we have the following:

Theorem 1. *Every n-nodes directed tree admits an upward straight-line drawing in optimal $\Theta(n \log n)$ area.*

4   Upward Drawings of Trees with Fixed Embedding

We discuss the area requirement of order-preserving upward drawings of directed trees. Garg and Tamassia ([12]) proved that any upward planar embedding can be realized with straight-line edges in exponential area. Hence, exponential area straight-line upward drawings of embedded directed trees are feasible.

Now we prove the claimed exponential lower bound. Bertolazzi et al. showed in [1] an embedding $\mathcal{E}_n$ of a $2n$-vertex *series-parallel* digraph requiring $\Omega(4^n)$ area in any order-preserving upward straight-line drawing. Such an embedding is recursively defined as follows: $\mathcal{E}_0$ consists of a single edge (s_0, t_0); $\mathcal{E}_{n+1}$ is obtained from $\mathcal{E}_n$ by adding (i) two new nodes s_{n+1} and t_{n+1}, (ii) an edge from s_{n+1} to s_n, (iii) an edge from t_n to t_{n+1}, (iv) an edge from s_n to t_{n+1} on the right of $\mathcal{E}_n$, and (v) an edge from s_{n+1} to t_{n+1} on the left of $\mathcal{E}_n$ (see Fig. 1.c).

We define a *clockwise coil* S to be an upward planar drawing of a directed path $P = (v_1, v_2, \ldots, v_k)$ that respects three properties: **property (i)** the edges (v_i, v_{i+1}) of P, with i odd (with i even), are directed from v_i to v_{i+1} (resp. from v_{i+1} to v_i), **property (ii)** $y(v_i) < y(v_j)$ $(y(v_i) > y(v_j))$, for every i odd (resp. for every i even) and every j such that $j < i$, and **property (iii)** for i odd (for i even) every vertex v_j, with $j < i$, is contained in the region $R(v_i, v_{i+1})$ delimited by the edge (v_i, v_{i+1}) and by the horizontal half-lines starting at v_i and at v_{i+1} and directed toward increasing x-coordinates (resp. toward decreasing x-coordinates) (see Fig. 1.d). A *counter-clockwise coil* is defined analogously, with *odd* replaced by *even* and vice-versa in property (iii) (see Fig. 1.e). We have:

Lemma 1. *A straight-line n-vertex clockwise or counter-clockwise coil S requires $\Omega(2^n)$ area.*

Proof. Consider any straight-line clockwise coil S. We show that adding segments (v_i, v_{i+2}), for $i = 1, 2 \ldots, n-2$, augments S in a planar drawing S'. Namely, we prove that a segment (v_i, v_{i+2}) does not intersect (a) any segment (v_j, v_{j+1}) of S, with $j \leq i$, (b) segment (v_{i+1}, v_{i+2}) of S, (c) segment (v_{i+2}, v_{i+3}) of S, (d) any segment (v_j, v_{j+1}) of S, with $j > i+2$, and (e) any segment (v_j, v_{j+2}), with $j \neq i$ added to S.

(a) Suppose i is odd (is even). By property (ii) no vertex v_j of S, with $j < i+2$ and $j \neq i$, lies in the open half-plane $\mathcal{H}$ below (resp. above) the horizontal line passing through v_i. Moreover, v_{i+2} is contained in $\mathcal{H}$. Hence, (v_i, v_{i+2}) does not create crossings with any segment (v_j, v_{j+1}) of S, with $j \leq i$. (b) Since they are adjacent, (v_i, v_{i+2}) and (v_{i+1}, v_{i+2}) cross only if they overlap. But in such a case (v_i, v_{i+1}) and (v_{i+1}, v_{i+2}) overlap, too. However, this is not possible by the supposed planarity of S. (c) By property (iii) v_i is contained inside $R(v_{i+2}, v_{i+3})$. Hence (v_i, v_{i+2}) is internal to $R(v_{i+2}, v_{i+3})$ and can not cross (v_{i+2}, v_{i+3}) that is on the border of $R(v_{i+2}, v_{i+3})$. (d) By property (iii) v_i and v_{i+2} are contained inside $R(v_j, v_{j+1})$, so (v_i, v_{i+2}) is internal to $R(v_{i+2}, v_{i+3})$ and can not cross (v_{i+2}, v_{i+3}) that is on the border of $R(v_{i+2}, v_{i+3})$. (e) It's easy to see that segments (v_i, v_{i+2}), for $i = 1, 2 \ldots, n-2$, form a directed path with increasing y-coordinate and so they don't cross each other.

Now one can observe that S' is an upward drawing of $\mathcal{E}_{n/2}$ (see [1] and the beginning of the section). Hence, an n-vertex straight-line clockwise coil S requires the same area of a straight-line drawing of $\mathcal{E}_{n/2}$, that is $\Omega(4^{n/2}) = \Omega(2^n)$. If S is a counter-clockwise straight-line coil a straightforward modification of the previous proof shows that S requires $\Omega(2^n)$ area. $\square$

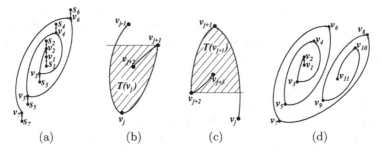

Fig. 2. (a) An upward drawing of T^* with embedding $\mathcal{E}^*$. (b) $T(v_j)$. (c) $T(v_{j+1})$. (d) An upward drawing of P^*. Notice that $(v_1, v_2, \ldots, v_7)$ is a counter-clockwise coil, while $(v_{11}, v_{10}, \ldots, v_7)$ is a clockwise coil.

Now let T^* be a tree composed by an $n/2$-nodes path $P^* = (v_1, v_2, \ldots, v_{n/2})$ and by $n/2$ leaves s_i, $1 \leq i \leq n/2$, such that s_i adjacent to v_i, with n even and $n/2$ odd. Edges (v_i, v_{i+1}), with i odd (with i even), are directed from v_i to v_{i+1} (resp. from v_{i+1} to v_i). Edges (v_i, s_i), with i odd (with i even), are directed from s_i to v_i (resp. from v_i to s_i). We fix for T^* an embedding $\mathcal{E}^*$ such that for each node v_i, $2 \leq i \leq n/2$, the clockwise order of the edges incident in v_i is $[s_i, v_{i-1}, v_{i+1}]$ (see Fig. 2.a). We claim the following:

Lemma 2. *Every upward drawing Γ^* of T^* with embedding $\mathcal{E}^*$ contains a clockwise or a counter-clockwise coil of at least $n/4$ nodes.*

Proof. Observe that, by the embedding constraints of $\mathcal{E}^*$ and by the upwardness of Γ^*, path P^* turns in clockwise direction at every edge (v_{i-1}, v_i), for $i = 2, 3, \ldots, n/2$, i. e. considering the half-lines t_1 and t_2 starting at v_i and tangent to the curves representing edges (v_{i-2}, v_{i-1}) and (v_{i_1}, v_i), respectively, the angle described by a clockwise movement that leads t_1 to overlap with t_2 is less than π. Let j be the highest index such that the drawing S_1^* of the subpath $(v_1, v_2, \ldots, v_j)$ of P^* is a counter-clockwise coil. If $j \geq n/4$ or if such a j doesn't exist, i.e. P^* is entirely drawn as a counter-clockwise coil, the lemma follows. Otherwise, we claim that the drawing S_2^* of the subpath $(v_{n/2}, v_{n/2-1}, \ldots, v_{j+1}, v_j)$ of P^* is a clockwise coil. Property (i) follows from the upwardness of Γ^*. Consider three vertices v_{i-1}, v_i, and v_{i+1} that are consecutive in P^*. Let v_t be the one between v_{i-1} and v_{i+1} such that $|y(v_i) - y(v_t)|$ is minimum. Denote by $T(v_i)$, with $i = j, j+1, \ldots, n/2-1$ the triangle with curved edges delimited by (v_i, v_{i-1}), by (v_i, v_{i+1}), and by the horizontal line through v_t. Assume j is odd. Since $(v_1, v_2, \ldots, v_j, v_{j+1})$ is not a coil, then $y(v_{j-1}) \geq y(v_{j+1})$. Since (v_{j+1}, v_{j+2}) turns in clockwise direction with respect to (v_j, v_{j+1}), the planarity and the upwardness of Γ^* imply that v_{j+2} is inside $T(v_j)$, and so $y(v_{j+2}) > y(v_j)$ (see Fig. 2.b). Since (v_{j+2}, v_{j+3}) turns in clockwise direction with respect to (v_{j+1}, v_{j+2}), the planarity and the upwardness of Γ^* imply that v_{j+3} is inside $T(v_{j+1})$, and so $y(v_{j+3}) > y(v_{j+1})$ (see Fig. 2.c). Proceeding in the same way, it follows that, for all $i = j, j+1, \ldots, n/2-2$, $y(v_{i+2}) > y(v_i)$

$(y(v_{i+2}) > y(v_i))$ with i odd (resp. with i even). Hence, property (ii) is satisfied by S_2^*. Further, property (iii) is satisfied by S_2^*, since every vertex v_k, with $k \geq i+2$ is contained inside $T(v_i)$ and, consequently, inside $R(v_i, v_{i+1})$, that encloses $T(v_i)$. If j is even an analogous proof shows that S_2^* is a clockwise coil. Finally, since $j < n/4$, S_2^* contains at least $n/2 - j > n/4$ nodes.   □

Theorem 2. *There exists an n-nodes embedded directed tree requiring $\Omega(b^n)$ area, with b greater than 1, in any upward straight-line order-preserving drawing.*

Proof. Consider T^* and its embedding $\mathcal{E}^*$ described in this section. By Lemma 2 every upward drawing of T^* with embedding $\mathcal{E}^*$ contains a coil of at least $n/4$ nodes that, by Lemma 1, requires $\Omega(2^{n/4}) = \Omega((\sqrt[4]{2})^n) = \Omega(b^n)$, with $b = \sqrt[4]{2}$.
□

Now we turn to poly-line drawings. Di Battista et al. have shown in [9] that every upward planar embedding can be drawn with poly-line edges in $O(n^2)$ area. It follows that quadratic area poly-line upward drawings of embedded directed trees are feasible. Concerning the lower bound, we have the following:

Lemma 3. *An n-vertex poly-line clockwise or counter-clockwise coil S requires $\Omega(n^2)$ area.*

Proof. By property (ii) vertex v_i, with i odd, has y-coordinate less than the one of every vertex v_j, with $j < i$. This implies that $n/2$ vertices v_i such that i is odd occupy $n/2$ distinct horizontal lines and so the height of S is $\Omega(n)$. Concerning the width of S, suppose w.l.o.g. to draw S starting from a drawing Γ_1 of v_1, and then iteratively constructing a drawing Γ_i by adding vertex v_i and edge (v_{i-1}, v_i) to Γ_{i-1}, for $i = 2, \ldots, n$. We claim that the width of Γ_i is at least the width of Γ_{i-1} plus one. Suppose that the width of Γ_i is equal to the width of Γ_{i-1}. Then edge (v_{i-1}, v_i) can not be on the left or on the right of Γ_{i-1} and so property (iii) can not be satisfied. It follows that the width of S is $\Omega(n)$.   □

Hence, we can again consider directed tree T^* with fixed embedding $\mathcal{E}^*$. By Lemma 2 every upward drawing of T^* with embedding $\mathcal{E}^*$ contains a clockwise or a counter-clockwise coil S of at least $n/4$ nodes. By Lemma 3 $\Omega(n^2)$ area is required for S.

Theorem 3. *There exists an n-nodes directed tree T^* and an embedding of T^* requiring $\Omega(n^2)$ area in any upward poly-line order-preserving drawing.*

5   Upward Drawings of Some Families of DAGs

In the first part of this section we study the area requirement of planar upward drawings of some families of directed trees, like *directed binary trees*, *directed caterpillars*, and *directed spider trees*, searching for better area bounds with respect to those obtained for general trees. In the second part of this section we show that the results obtained for directed trees can be exploited to obtain area bounds for several others families of DAGs, like *directed bipartite graphs* and *directed outerplanar graphs.* The proofs of the theorems claimed in this section are omitted, for reasons of space.

Concerning *directed binary trees,* one can observe that the lower bounds on the area requirement of planar upward drawings of directed trees presented in Sections 3 and 4 are obtained by considering directed *binary* trees. Hence such lower bounds are still valid here. Moreover, the algorithms for drawing directed trees clearly apply also to directed binary trees, hence the optimal bounds on the area requirement of planar upward drawings of directed binary trees are the same of the ones of general trees.

Analogously, concerning *directed caterpillars,* we notice that the lower bound on the area requirement of order-preserving upward drawings of directed trees presented in Section 4 was obtained by considering a directed caterpillar. Hence such a lower bound is still valid here. On the other hand, for non-order-preserving drawings one can obtain better results with respect to those for general trees, as shown by the following:

Theorem 4. *Every n-nodes directed caterpillar tree admits an upward straight-line drawing in optimal $\Theta(n)$ area.*

For *directed spider trees* linear area is achievable also for order-preserving drawings:

Theorem 5. *Every n-nodes directed spider tree admits an upward order-preserving straight-line drawing in optimal $\Theta(n)$ area.*

Considering families of DAGs richer than directed trees, exponential area is sometimes necessary even without forcing an order of the neighbors of each vertex. In the following we show the inductive construction of an n-vertex directed bipartite graph B_n. Such a digraph contains an $O(n)$ nodes coil in any upward planar drawing, hence it requires exponential area in any straight-line upward drawing and quadratic area in any poly-line upward drawing. Such lower bounds are again matched by the upper bounds in [12,9]. We define B_n as the directed bipartite graph with vertex sets V and U, inductively defined as follows: (i) B_8 has vertices $v_{-2}, v_{-1}, v_1, v_2 \in V$ and $u_{-2}, u_{-1}, u_1, u_2 \in U$, the edges of a directed path $(v_{-2}, u_{-2}, v_{-1}, u_{-1}, v_1, u_1, v_2, u_2)$, and the directed edges (v_1, u_2), (v_{-1}, u_1), (v_{-2}, u_1) and (v_{-1}, u_2) (see Fig. 3.a); (ii) B_n, with n multiple of 4, is done by B_{n-4}, by four new vertices $v_{n/4}, u_{n/4}, v_{-n/4}$, and $u_{-n/4}$ and by eight directed edges $(v_{-n/4}, u_{-n/4})$, $(u_{-n/4}, v_{-n/4+1})$, $(u_{n/4-1}, v_{n/4})$, $(v_{n/4}, u_{n/4})$, $(v_{-n/4+2}, u_{n/4})$, $(v_{-n/4+1}, u_{n/4-1})$, $(v_{-n/4}, u_{n/4-1})$, and $(v_{-n/4+1}, u_{n/4})$ (see Fig. 3.b). An extensive study of the properties of B_n leads to the followings:

Theorem 6. *There exists an n-vertex directed bipartite graph requiring $\Omega(b^n)$ area, with b greater than 1, in any upward straight-line drawing.*

Theorem 7. *There exists an n-vertex directed bipartite graph requiring $\Omega(n^2)$ area in any upward poly-line drawing.*

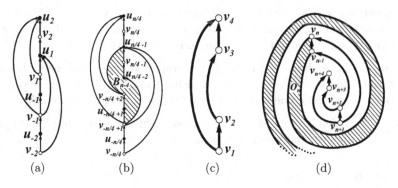

Fig. 3. (a) B_8. (b) B_n. (c) O_4. (d) O_n.

Again using arguments based on the results obtained for directed trees, it can be shown that *directed outerplanar graphs* generally require exponential area in any outerplanar straight-line upward drawing and quadratic area in any poly-line upward drawing. These results are achieved by considering the n-vertex directed outerplanar graph O_n inductively defined as follows: (i) O_4 has four vertices v_1, v_2, v_3, and v_4 and four directed edges (v_1, v_2), (v_1, v_4), (v_2, v_3), and (v_3, v_4) (see Fig. 3.c); (ii) O_{n+4} is composed by O_n, by four new vertices v_{n+1}, v_{n+2}, v_{n+3}, and v_{n+4}, and by six new directed edges (v_{n+1}, v_n), (v_{n+2}, v_{n-1}), (v_{n+1}, v_{n+2}), (v_{n+2}, v_{n+3}), (v_{n+1}, v_{n+4}), and (v_{n+3}, v_{n+4}) (see Fig. 3.d). Studying the properties of upward drawings of O_n the followings can be proved:

Theorem 8. *There exists an n-vertex directed outerplanar graph requiring $\Omega(b^n)$ area, with b greater than 1, in any upward outerplanar straight-line drawing.*

Theorem 9. *There exists an n-vertex directed outerplanar graph requiring $\Omega(n^2)$ area in any upward poly-line drawing.*

6   Conclusions and Open Problems

In this paper we have studied the area requirement of upward drawings of several classes of DAGs that frequently arise in theory and in practice.

We provided tight bounds on the area requirement of straight-/poly-line order-/non-order-preserving upward drawings of general directed trees and of several families of directed trees. However, the following problem is still open:

Problem 1. Which is the minimum area of upward straight/poly-line order/non order-preserving drawings of *complete* and *balanced* trees?

Concerning directed bipartite graphs, we have shown an exponential area lower bound for straight-line upward drawings, but the following is still open:

Problem 2. Which is the minimum area of an upward drawing of a *bipartite DAG*? Bipartite DAGs [7] are those DAGs having a vertex set partitioned into two subsets V_1 and V_2 with each edge directed from a vertex of V_1 to a vertex of V_2. Consequently, bipartite DAGs form a subclass of the digraphs whose underlying graph is bipartite, that was considered in this paper.

Further, we have shown an outerplanar graph requiring exponential area in any straight-line *outerplanar* upward drawing. However, when considering *non-outerplanar* drawings, one could obtain better area bounds, so we ask:

Problem 3. Which is the minimum area of straight-line non-outerplanar upward drawings of directed outerplanar graphs?

Acknowledgments

Thanks to Walter Didimo and Giuseppe Liotta for reporting the problem of obtaining minimum area upward drawings of directed trees. Thanks to Giuseppe Di Battista for very useful discussions.

References

1. Bertolazzi, P., Cohen, R.F., Di Battista, G., Tamassia, R., Tollis, I.G.: How to draw a series-parallel digraph. Int. J. Comput. Geom. Appl. 4(4), 385–402 (1994)
2. Bertolazzi, P., Di Battista, G., Liotta, G., Mannino, C.: Upward drawings of triconnected digraphs. Algorithmica 12(6), 476–497 (1994)
3. Chan, T.M.: A near-linear area bound for drawing binary trees. Algorithmica 34(1), 1–13 (2002)
4. Crescenzi, P., Di Battista, G., Piperno, A.: A note on optimal area algorithms for upward drawings of binary trees. Comput. Geom. 2, 187–200 (1992)
5. Di Battista, G., Eades, P., Tamassia, R., Tollis, I.G.: Graph Drawing. Prentice-Hall, Upper Saddle River, NJ (1999)
6. Di Battista, G., Frati, F.: Small area drawings of outerplanar graphs. In: Graph Drawing, pp. 89–100 (2005)
7. Di Battista, G., Liu, W.P., Rival, I.: Bipartite graphs, upward drawings, and planarity. Inf. Process. Lett. 36(6), 317–322 (1990)
8. Di Battista, G., Tamassia, R.: Algorithms for plane representations of acyclic digraphs. Theor. Comput. Sci. 61, 175–198 (1988)
9. Di Battista, G., Tamassia, R., Tollis, I.G.: Area requirement and symmetry display of planar upward drawings. Disc. & Computat. Geometry 7, 381–401 (1992)
10. Garg, A., Rusu, A.: Area-efficient drawings of outerplanar graphs. In: Graph Drawing, pp. 129–134 (2003)
11. Garg, A., Rusu, A.: Area-efficient order-preserving planar straight-line drawings of ordered trees. Int. J. Comput. Geometry Appl. 13(6), 487–505 (2003)
12. Garg, A., Tamassia, R.: Efficient computation of planar straight-line. In: Graph Drawing (Proc. ALCOM Workshop on Graph Drawing), pp. 298–306 (1994)
13. Garg, A., Tamassia, R.: On the computational complexity of upward and rectilinear planarity testing. SIAM J. Comput. 31(2), 601–625 (2001)
14. Hutton, M.D., Lubiw, A.: Upward planarity testing of single-source acyclic digraphs. SIAM J. Comput. 25(2), 291–311 (1996)
15. Papakostas, A.: Upward planarity testing of outerplanar dags. In: Graph Drawing, pp. 298–306 (1994)

A Very Practical Algorithm
for the Two-Paths Problem
in 3-Connected Planar Graphs

Torben Hagerup

Institut für Informatik, Universität Augsburg, 86135 Augsburg, Germany
hagerup@informatik.uni-augsburg.de

Abstract. A linear-time algorithm that does not need a planar embedding is presented for the problem of computing two vertex-disjoint paths, each with prescribed endpoints, in an undirected 3-connected planar graph.

1 Introduction

The *two-paths problem* (2PP for short) is, given an undirected graph G and four distinct vertices s_1, s_2, t_1, t_2 in G, to decide whether G contains a pair (p, q) of two vertex-disjoint paths, p from s_1 to t_1 and q from s_2 to t_2, and to compute such a pair (p, q) if it exists. Known polynomial-time algorithms [9,10] for the 2PP first reduce a general instance to one in which G is 3-connected and then take one of two routes, depending on whether G is planar. The 2PP in 3-connected planar graphs was shown to be solvable in linear time by Perl and Shiloach [6]. Their algorithm and its correctness proof are fairly involved, but a very simple solution was later given by Woeginger [11]. This note presents a new linear-time algorithm for the 2PP in 3-connected planar graphs, the *symmetric algorithm*, of complexity comparable to Woeginger's algorithm. In contrast with the previous algorithms, the symmetric algorithm does not use a planar embedding of the input graph. If such an embedding is not available, the symmetric algorithm is likely to offer practical advantages in terms of coding effort, program size and running time.

This paragraph situates the problem considered here in a broader context. According to Perl and Shiloach [6,9], Itai reduced the 2PP in general graphs to the 2PP in 3-connected graphs, and the reduction can be made to preserve planarity. The reduction is more complicated than the algorithm presented here, but works in linear time, so that the 2PP can be solved in linear time in planar graphs. In general graphs with n vertices and m edges, the fastest algorithm known for the 2PP works in $O(m\alpha(m, n) + n)$ time, where α is an inverse of the Ackermann function [10]. If we modify the problem by requiring the paths computed to be edge-disjoint rather than vertex-disjoint (let us speak of the 2EPP), the fastest algorithms known for the planar and general cases have the running times mentioned above plus an additive term of $O(n \log n)$ [10]. If we

A. Brandstädt, D. Kratsch, and H. Müller (Eds.): WG 2007, LNCS 4769, pp. 145–150, 2007.

step from undirected to directed graphs, the 2PP and the 2EPP both become NP-complete [1]. In planar directed graphs, the 2PP can be solved in polynomial time [8], whereas it is unknown whether this is the case for the 2EPP (even assuming $P \neq NP$). Let us generalize further by allowing the number k of pairs of vertices to be connected via disjoint paths to be larger than 2. The resulting kPP and kEPP are NP-complete, even in undirected planar graphs, if k is arbitrary and part of the input [3,4]. If k is a constant larger than 2, the kPP and the kEPP can be solved in polynomial time in undirected graphs [5,7], but the known algorithms to do so are based on the computation of tree decompositions and are not practical. In planar directed graphs, an algorithm of Schrijver [8] solves the kPP in polynomial time for constant k.

2   Woeginger's Algorithm

For given vertices s and t, an *s-t path* is a simple path from s to t. Two or more s-t paths are *internally disjoint* if no two of the paths share a vertex other than s and t. The input graph G being *3-connected* means that for each pair (s, t) of distinct vertices in G, G contains 3 internally disjoint s-t paths. This also implies that in every planar embedding of G, every boundary of a face is a simple cycle. When p is a path in G, denote by $G - p$ the graph obtained from G by removing all vertices on p and their incident edges.

Woeginger's algorithm computes a planar embedding of the input graph G and then distinguishes between two cases. In Case 1, the boundary of some face F in the planar embedding contains both s_1 and t_1. Woeginger shows that if the given instance has a solution (p, q), then p can be chosen as one of the two s_1-t_1 paths p_1 and p_2 on the boundary of F. It therefore suffices to test whether $G - p_1$ or $G - p_2$ contains an s_2-t_2 path q.

In Case 2, no face boundary contains both s_1 and t_1. In this case, the given instance always has a solution (p, q). Moreover, Woeginger shows that if p_1, p_2 and p_3 are internally disjoint s_1-t_1 paths, then p can be chosen as one of these, so that one can proceed similarly as in Case 1.

3   The Symmetric Algorithm

Let us return to Case 1 of Woeginger's algorithm. Assume that s_1 and t_1 both lie on the boundary of a face F and let p_1 and p_2 be the s_1-t_1 paths on the boundary of F. It is easy to see and proved in detail by Perl and Shiloach [6, Theorem 4.1] that if each of p_1 and p_2 contains one of s_2 and t_2, then the given instance has no solution. Assume that this is not the case and, without loss of generality, that p_1 contains neither s_2 nor t_2. Then, if q_1, q_2 and q_3 are internally disjoint s_2-t_2 paths in G, p_1 is vertex-disjoint from at least one of q_1, q_2 and q_3. To see this, assume otherwise and, for $i = 1, \ldots, 3$, let v_i be a vertex shared between p_1 and q_i. The graph G' obtained from G by adding a new vertex x

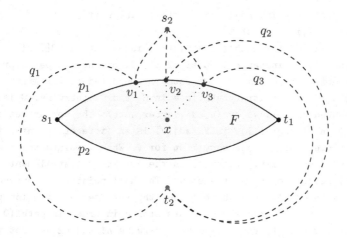

Fig. 1. If q_1, q_2 and q_3 all touch p_1, they furnish a planar embedding of $K_{3,3}$

and edges from x to v_1, v_2 and v_3 is clearly planar, since x can be embedded in F (see Fig. 1). However, the subgraph of G' spanned by the edges on q_1, q_2 and q_3 or incident on x is a subdivision of the complete bipartite graph on the vertex sets $\{s_2, t_2, x\}$ and $\{v_1, v_2, v_3\}$. This contradicts Kuratowski's theorem (which can be found, e.g., in [2]).

The new algorithm can be formulated as follows:

> *Symmetric algorithm:*
> **for** $j = 1, 2$ **do**
> Compute three internally disjoint s_j-t_j paths p_1, p_2 and p_3;
> **for** $i = 1, 2, 3$ **do**
>   **if** $G - p_i$ contains an s_{3-j}-t_{3-j} path q **then**
>   **if** $j = 1$ **then return** (p_i, q) **else return** (q, p_i);
> **return** "No solution exists";

The correctness of the algorithm follows from what was observed above. The part of the computation with $j = 1$ corresponds to Case 2 in Woeginger's algorithm. If the computation continues with $j = 2$, Case 1 obtains, and a solution is found if one exists. Since the paths p_1, p_2 and p_3 can be computed in linear time per path by standard flow-augmentation methods, a linear overall time bound is immediate, and no planar embedding is used.

4   The Symmetric Algorithm in D

This section gives a complete implementation of the symmetric algorithm in a C-like subset of the programming language D (http://www.digitalmars.com/d), chosen here because it allows nested functions.

```
int symmetric(int n,int adj[],int vertex[],int next[],
  int s1,int s2,int t1,int t2,
  out int n1,inout int path1[],out int n2,inout int path2[]) {
// Returns 1 if an instance I=(G,s1,s2,t1,t2) of the two-paths problem
// is solvable, where G is a undirected 3-connected planar graph and
// s1,s2,t1,t2 are arbitrary vertices in G, and 0 otherwise. G is
// represented as follows by (n,adj,vertex,next): The vertex set is
// V={0,...,n-1}. For all v in V, adj[v] is an (integer) pointer to the
// first element of an adjacency list for v. When x points to an element
// of an adjacency list, vertex[x] is the vertex of that element, and
// next[x] points to the next element. The null pointer is represented
// by the integer -1. If I is solvable, the vertices of disjoint paths
// from s1 to t1 and from s2 to t2 are stored in order in path1[0..n1-1]
// and path2[0..n2-1], respectively. Otherwise n1 and n2 are set to 0,
// and path1 and path2 are not changed. The caller must ensure that path1
// and path2 are sufficiently large, e.g., of size n. If G is not
// 3-connected or not planar, the procedure terminates normally, and
// any two paths computed are a correct solution; but the procedure
// may incorrectly indicate that the input instance has no solution.

  int onetry(int s1,int s2,int t1,int t2,
      out int n1,inout int path1[],out int n2,inout int path2[]) {
    int[] avail=new int[n],succ=new int[n];
    // C: int *avail=malloc(sizeof(int)*n),*succ=malloc(sizeof(int)*n);
    int second[3]; int reached=0,k=0,v;

    void dfs1(int u) {
      int v,w,x;
      avail[u]=0;
      if ((v=succ[u])>=0&&avail[v]) { dfs1(v); if (reached) succ[u]=-1; }
      x=adj[u];
      while (x>=0&&!reached) {
        v=vertex[x]; x=next[x];
        if (v==s1) { second[k++]=u; reached=1; }
        else if (avail[v]) {
          if ((w=succ[v])<0) dfs1(v); // v is not on an earlier path
          else if (avail[w]) { avail[v]=0; dfs1(w); } // omit dfs1(v)
          if (reached) { succ[v]=u; avail[v]=1; }
        }
        // continue from t1 if fewer than three paths found
        if (u==t1&&k<3) reached=0;
      }
      if (reached) avail[u]=1;
    }
```

```
    void dfs2(int u) {
      int v,x;
      if (u==s2) reached=1;
      else {
        avail[u]=0;
        x=adj[u];
        while (x>=0&&!reached)
          { if (avail[v=vertex[x]]) dfs2(v); x=next[x]; }
      }
      if (reached) path2[n2++]=u; // record the s2-t2 path
    }


    // Begin onetry
    for (v=0;v<n;v++) { avail[v]=1; succ[v]=-1; }
    if (s1==t1) second[k++]=-1; // record a single trivial s1-t1 path
    else dfs1(t1); // compute three internally disjoint s1-t1 paths
    // The second vertex of the k'th path is second[k-1], for k=1,2,3.
    // The path successor of a vertex v other than s1 is succ[v].
    reached=n1=n2=0;
    while (k) { // search for s2-t2 path disjoint from k'th s1-t1 path
      for (v=0;v<n;v++) avail[v]=1;
      avail[s1]=0; // mark the k'th s1-t1 path as visited
      v=second[--k];
      while (v>=0) { avail[v]=0; v=succ[v]; };
      if (avail[t2]) dfs2(t2);
      if (reached) { // record the successful s1-t1 path and quit
        path1[n1++]=s1;
        v=second[k];
        while (v>=0) { path1[n1++]=v; v=succ[v]; }
        break;
      }
    }
    // C: free(avail); free(succ);
    return reached;
  }


  // Begin symmetric
  if (onetry(s1,s2,t1,t2,n1,path1,n2,path2)) return 1;
  else return onetry(s2,s1,t2,t1,n2,path2,n1,path1);
}
```

References

1. Fortune, S., Hopcroft, J., Wyllie, J.: The directed subgraph homeomorphism problem. Theoret. Comput. Sci. 10, 111–121 (1980)
2. Harary, F.: Graph Theory. Addison-Wesley, Reading, Mass (1969)
3. Kramer, M.R., van Leeuwen, J.: The complexity of wire-routing and finding minimum area layouts for arbitrary VLSI circuits. In: Preparata, F.P. (ed.) Advances in Computing Research, vol. 2, pp. 129–146. JAI Press, Greenwich, Conn (1984)
4. Lynch, J.F.: The equivalence of theorem proving and the interconnection problem. ACM SIGDA Newsletter 5, 31–36 (1975)
5. Perković, L., Reed, B.: An improved algorithm for finding tree decompositions of small width. Internat. J. Foundat. Comput. Sci. 11, 365–371 (2000)
6. Perl, Y., Shiloach, Y.: Finding two disjoint paths between two pairs of vertices in a graph. J. ACM 25, 1–9 (1978)
7. Robertson, N., Seymour, P.D.: Graph Minors. XIII. The disjoint paths problem. J. Comb. Theory, Ser. B 63, 65–110 (1995)
8. Schrijver, A.: Finding k disjoint paths in a directed planar graph. SIAM J. Comput. 23, 780–788 (1994)
9. Shiloach, Y.: A polynomial solution to the undirected two paths problem. J. ACM 27, 445–456 (1980)
10. Tholey, T.: Solving the 2-disjoint paths problem in nearly linear time. Theory Comput. Systems 39, 51–78 (2006)
11. Woeginger, G.: A simple solution to the two paths problem in planar graphs. Inform. Process. Lett. 36, 191–192 (1990)

Approximation Algorithms for Geometric Intersection Graphs*

Klaus Jansen

Institut für Informatik, Universität zu Kiel, Olshausenstr. 40, D-24098 Kiel, Germany
kj@informatik.uni-kiel.de

Abstract. In this paper we describe together with an overview about other results the main ideas of our polynomial time approximation schemes for the maximum weight independent set problem (selecting a set of disjoint disks in the plane of maximum total weight) in disk graphs and for the maximum bisection problem (finding a partition of the vertex set into two subsets of equal cardinality with maximum number of edges between the subsets) in unit-disk graphs.

For a set V of geometric objects, the corresponding geometric intersection graph is the undirected graph $G = (V, E)$ with vertex set V and an set E of edges between two vertices if the corresponding objects intersect. Assume that we are given a set $\mathcal{D} = \{D_1, \ldots, D_n\}$ of n disks in the plane, where D_i has diameter d_i and center $c_i = (x_i, y_i)$. Disk D_i and D_j intersect if $dist(c_i, c_j) \leq (d_i + d_j)/2$, where $dist(p, q)$ denotes the Euclidean distance between points p and q in the plane. A disk graph is the intersection graph of a set of disks. We assume that the input to our problems is the set $\mathcal{D}$ of disks, not only the corresponding intersection graph. This is an important distinction, because determining for a given graph whether it is a disk graph is known to be NP-hard [10], and hence no efficient method is known for computing a disk representation if only the intersection graph is given. Interestingly, every planar graph is a coin graph, i.e. the intersection graph of a set of interior disjoint disks [13]. Therefore, the class of disk graphs properly contains the class of planar graphs. Applications of geometric intersection graphs are frequency assignment in networks [9,15] and map labelling [1].

We are interested in approximation algorithms for NP-hard optimization problems on disk graphs, in particular for the maximum weighted independent set and maximum bisection problem. The goal of the maximum weight independent set problem (MWIS) is to compute, for a given set of disks $\mathcal{D}$ with certain vertex weights $w_1, \ldots, w_n$, a subset $U \subset \mathcal{D}$ of disjoint non-overlapping disks with maximum total weight $w(U) = \sum_{D_i \in U} w_i$. MIS refers to the unweighted version of this problem (i.e. with $w_i = 1$ for each disk $D_i \in \mathcal{D}$). The goal of the maximum bisection problem (MBS) is to partition the disk set into two subsets

* Research of the author was supported in part by the EU Thematic Network APPOL I + II, Approximation and Online Algorithms, IST-1999-14084 and IST-2001-32007.

A. Brandstädt, D. Kratsch, and H. Müller (Eds.): WG 2007, LNCS 4769, pp. 151–153, 2007.
© Springer-Verlag Berlin Heidelberg 2007

of the same cardinality (assuming that the number n of disks is even) and to maximize the number of edges across the partition. For unit disk graphs (intersection graphs of disks with equal diameter) and planar graphs, MIS remains NP-hard [4,7]. Recently, it was proved that MBS is also NP-hard for unit disk graphs and planar graphs [5,14].

For a given set $\mathcal{D}$ of disks in the plane, let $OPT(\mathcal{D})$ denote the maximum value (the weight or the number of edges) of an optimum solution for MWIS and MBS, respectively. An algorithm A is a ρ-approximation algorithm for MWIS or MBS (with $\rho \geq 1$) if it runs in time polynomial in the input size (the size of the representation for $\mathcal{D}$) and always computes a solution of total value at least $(1/\rho)OPT(\mathcal{D})$. A polynomial time approximation scheme is a family of approximation algorithms $\{A_\epsilon | \epsilon > 0\}$ where A_ϵ computes a solution of value at least $1/(1 + \epsilon)OPT(\mathcal{D})$ and runs in polynomial time in the input size for each fixed $\epsilon > 0$.

Hunt et al. [11] gave a PTAS for MWIS on unit disk graphs, and Baker [3] provided a PTAS for MWIS on planar graphs. For intersection graphs of disks with arbitrary diameters, the best previous known approximation algorithm achieves approximation ratio 5 for MIS [16]. In [11,15], the question was raised whether a PTAS exists for disk graphs. As the class of disk graphs contains the class of unit disk graphs and the class of planar graphs, a PTAS for disk graphs generalizes the result for unit disk graphs due to Hunt et al. [11] and the result for planar graphs due to Baker [3]. We resolve this question by presenting a PTAS for MWIS in disk graphs (with given presentation) [6]. The basic idea of the PTAS is as follows. The plane is partitioned into squares on different levels, and some of the disks are removed from the input so that different squares on the same level yield independent subproblems with respect to all disks that are on this level or on a level with disks of smaller diameter. Furthermore, at most a constant number of disks with larger diameter can be disjoint and intersect a square on the current level. Therefore, all such sets can be enumerated in polynomial time for each square, and a dynamic programming approach becomes feasible. In addition to this result, we [6] propose also a PTAS for the minimum weight vertex cover problem (MWVC).

The complexity and approximability status of MBS on planar graphs have been long-standing open problems. Contrary to the polynomial time algorithm of planar maximum cut (finding a partition into two subsets that maximizes the number of edges with endpoints in both subsets) [8], planar maximum bisection has been proven only recently to be NP-hard in exact setting by Jerrum [14]. The proof of this result can be found in [12]. We [12] have proved that there is a PTAS for MBS in planar graphs. This is obtained by combining (via tree-typed dynamic programming) the original Baker's method [3] of dividing the input graph into families of k-outerplanar graphs with our method of finding maximum partitions of bounded treewidth graphs. For dense graphs, Arora, Karger, and Karpinski [2] gave a PTAS for the maximum bisection problem. We [12] have found a PTAS for the maximum bisection problem on unit disk graphs. It is obtained by combining (again via tree-typed dynamic programming) the idea of Hunt et al. [11] of dividing the input graph defined by plane conditions into family of subgraphs with the aforementioned methods in [2] of finding maximum partitions of dense graphs.

References

1. Agarwal, P.K., van Kreveld, M., Suri, S.: Label placement by maximum independent set in rectangles. Computational Geometry 11, 209–218 (1998)
2. Arora, S., Karger, D., Karpinski, M.: Polynomial time approximation schemes for dense instances of NP-hard problems. Journal of Computer and System Science 58, 193–210 (1999)
3. Baker, B.: Approximation algorithms for NP-complete problems on planar graphs. Journal of the ACM 41, 153–180 (1994)
4. Clark, B.N., Colbourn, C.J., Johnson, D.S.: Unit disk graphs. Discrete Mathematics 86, 165–177 (1990)
5. Diaz, J., Kaminski, M.: Max-CUT and MAX-BISECTION are NP-hard on unit disk graphs. Theoretical Computer Science 377, 271–276 (2007)
6. Erlebach, T., Jansen, K., Seidel, E.: Polynomial-time approximation schemes for geometric graphs. SIAM Journal on Computing 34, 1302–1323 (2005)
7. Garey, M.R., Johnson, D.S., Stockmeyer, L.J.: Some simplified NP-complete graph problems. Theoretical Computer Science 1, 237–267 (1976)
8. Hadlock, F.: Finding a maximum cut of a planar graph in polynomial time. SIAM Journal on Computing 4, 221–225 (1975)
9. Hale, W.K.: Frequency assignment: theory and applications. Proceedings of the IEEE 68, 1497–1514 (1980)
10. Hlineny, P., Kratochvil, J.: Representing graphs by disks and balls. Discrete Mathematics 229, 101–124 (2001)
11. Hunt III, H.B., Marathe, M.V., Radhakrishnan, V., Ravi, S.S., Rosenkrantz, D.J., Stearns, R.E.: NC-approximation schemes for NP-and PSPACE-hard problems for geometric graphs. Journal of Algorithms 26, 238–274 (1998)
12. Jansen, K., Karpinski, M., Lingas, A., Seidel, E.: Polynomial time approximation schemes for max-bisection on planar and geometric graphs. SIAM Journal on Computing 35, 110–119 (2005)
13. Koebe, P.: Kontaktprobleme der konformen Abbildung, Berichteüber die Verhandlungen der Sächsischen Akademie der Wissenschaften. Leipzig, Math.-Phys. Klasse 88, 141–164 (1936)
14. Jerrum, M.: private communication (2000)
15. Malesinska, E.: Graph-theoretical models for frequency assignment problems, PhD thesis, Technische Universität Berlin (1997)
16. Marathe, M.V., Breu, H., Hunt III, H.B., Ravi, S.S., Rosenkrantz, D.J.: Simple heuristics for unit disk graphs. Networks 25, 59–68 (1995)

An Equivalent Version of the Caccetta-Häggkvist Conjecture in an Online Load Balancing Problem

Angelo Monti[1], Paolo Penna[2,*], and Riccardo Silvestri[1]

[1] Dipartimento di Informatica, Università degli Studi di Roma "La Sapienza",
via Salaria 113, Roma, Italy
{monti,silver}@di.uniroma1.it
[2] Dipartimento di Informatica ed Applicazioni "R.M. Capocelli",
Università di Salerno, via S. Allende 2, Baronissi (SA), Italy
penna@dia.unisa.it

Abstract. We study the competitive ratio of certain *online* algorithms for a well-studied class of *load balancing* problems. These algorithms are obtained and analyzed according to a method by Crescenzi *et al* (2004). We show that an exact analysis of their competitive ratio on certain "uniform" instances would resolve a fundamental conjecture by Caccetta and Häggkvist (1978). The conjecture is that any digraph on n nodes and minimum outdegree d must contain a directed cycle involving at most $\lceil n/d \rceil$ nodes. Our results are the first relating this conjecture to the competitive analysis of certain algorithms, thus suggesting a new approach to the conjecture itself. We also prove that, on "uniform" instances, the analysis by Crescenzi *et al* (2004) gives only trivial upper bounds, unless we find a counterexample to the conjecture. This is in contrast with other (notable) examples where the same analysis yields optimal (non-trivial) bounds.

Keywords: Caccetta-Häggkvist conjecture, online load balancing, competitive analysis.

1 Introduction

We consider a combinatorial problem which has applications to the construction of *competitive*[1] algorithms for the well-studied class of *online load balancing* problems considered in e.g. [4,3,2,5] (see Section 1.2 for a formal definition). Our work is motivated by a technique from Crescenzi *et al.* [8] in which the simple greedy algorithm is "tuned" on the problem at hand. A rather informal

---

[*] Fully supported by the European Union under the Project IST-15964, Algorithmic Principles for Building Efficient Overlay Computers (AEOLUS).

[1] Intuitively speaking, an online algorithm is c-competitive if there exists a constant b such that the algorithm outputs a solution whose cost is at most $c \cdot opt + b$ where opt is the optimum for the instance considered up to the current time step. In this case, c is the competitive ratio of the algorithm.

A. Brandstädt, D. Kratsch, and H. Müller (Eds.): WG 2007, LNCS 4769, pp. 154–165, 2007.
© Springer-Verlag Berlin Heidelberg 2007

description of this technique is as follows (see Section 1.2 for a more formal description):

> Each online load balancing problem specifies a set of "feasible modifications" of the greedy algorithm and an "easy-to-compute" upper bound $c(\cdot)$ on the competitive ratio. In particular, every such feasible modification M describes a modified version of the greedy algorithm whose competitive ratio on this problem is at most $c(M)$.

This approach has been applied to the *linear* and to the *hierarchical* server topologies studied in [5] where it is rather easy to find an M such that $c(M)$ results in a dramatic improvement over the competitive ratio of the greedy algorithm and matches the lower bound for the problem considered [8]. It is thus natural to try to apply the same technique to more problems.

1.1 Our Contribution

In this work we consider a natural class of *s-uniform* online load balancing problems in which every task can be assigned to some *s-subset* of the n processors (this subset can vary arbitrarily from task to task). The resulting combinatorial problem is to determine (exactly) the *minimum competitive bound* $C(n, s)$ which is the smallest value that the above function $c(\cdot)$ can assume for s-uniform instances. Our major contribution is to show that the minimum competitive bound $C(n, s)$ leads to an *equivalent* version of one of the most fundamental and intriguing conjectures in graph theory (which also accounts for dozens of connections to other basic questions in combinatorics and number theory [14]):

Conjecture 1 (Caccetta-Häggkvist 1978 [7]). *Any digraph on n nodes with minimum outdegree at least d contains a directed cycle of length at most $\lceil n/d \rceil$.*

We indeed prove that, if the above conjecture is true, then $C(n, s) = n/s$. Observe that, there is a trivial upper bound $C(n, s) \leq n/s$ (see Section 1.2). Thus any improvement on the trivial bound would give a counterexample to the conjecture. At the heart of this result is another interesting number associated to the analysis of s-uniform instances which we call the *blind competitive bound* $B(n, s)$. This number is "tightly coupled" with the Caccetta-Häggkvist conjecture since we prove that, for $s \leq \sqrt{n}$,

$$B(n, s) = 1 + n - \lceil n/s \rceil$$

if and only if the conjecture holds. The number $B(n, s)$ is the minimum for $c(\cdot)$ when considering certain modifications M which result in "blind" algorithms that assign tasks without even "looking at the processors": tasks which can be potentially allocated to the same subset of processors are *all* assigned to a predetermined and fixed processor.

Our results can be seen as the hardness of obtaining any non-trivial bound with the method of [8] in the case of s-uniform instances (this is in contrast with other instances considered in [8]). These hardness results are in some sense of a "new type" since they do *not* rely on computational assumptions and they are obtained by relating two (apparently) different problems. We feel one of the main contributions of this work is to connect the analysis of online algorithms to a fundamental conjecture in graph theory and to show that such an analysis is as difficult as solving the latter.

From another point of view, our results suggest a possible way for proving the conjecture by showing a lower bound on the competitive ratio of the online algorithms yielded by certain modifications of greedy. Such bounds have also a practical interest since these algorithms use only *local information* (namely, each task can decide its own allocation by considering only the current load of the processors in its associated subset). Blind algorithms are a notable example since lower bounds are probably easier to prove, while any tight result on the competitive ratio of the best blind algorithm for s-uniform instances would either prove or disprove the conjecture. We stress that the Caccetta-Häggkvist conjecture is considered a central and important problem in combinatorics, graph, and number theory. Thirty years of significant efforts culminated in a large number of deep connections among these areas. They have been the main subject of a recent workshop held at the American Institute of Mathematics dedicated to this conjecture (see Sullivan's paper surveying these results [14]).

Roadmap. In Section 1.2 we introduce online load balancing, the technique in [8], and the related combinatorial problems. In Section 2 we introduce and study blind algorithms, and we relate the blind competitive bound to the Caccetta-Häggkvist conjecture. We apply these results to the minimum competitive bound in Section 3. Finally, we further discuss our results and their implications in Section 4.

1.2   Online Load Balancing, Modified Greedy Algorithms, and Their Analysis

In this section we go back to our initial application that is *online load balancing* of *temporary weighted tasks* in the case of *restricted assignment* with *no preemption*. Here each task t is specified by a *subset S_t of processors* that can execute that task, a *weight W_t*, and a *duration D_t*. Tasks arrive one by one, each task t needs to be allocated upon its arrival to one of the processors in S_t. No task can be reallocated. The duration D_t is *unknown* and the task simply disappears without any prior notice after D_t time units from its arrival. At every time step, a processor has a *load* equal to the sum of the weights of those tasks currently in the system and which have been assigned to it. The goal is to keep, over time, the maximum processor load as low as possible. We are interested in designing online algorithms with a small *competitive ratio c*, that is, the algorithm must guarantee that the load of each processor never exceeds $c \cdot opt + b$, where *opt* is the optimum for the instance and b is a fixed constant.

In general, online algorithms with a "good" competitive ratio are designed "ad-hoc" for a family $\mathcal{F}$ containing all possible subsets of processors that can be associated to any task. A notable example is the hierarchical server topologies by Bar-Noy *et al* [5] where the "combinatorial structure" of $\mathcal{F}$ impacts significantly on the competitive ratio of the algorithms. Moreover, "general purpose" algorithms, such as the greedy one, are in general "far" from the optimal [3,4,5]. The approach in [8] constructs a "modified" version of the greedy algorithm for the problem '$\mathcal{F}$' as follows:

- In an *offline phase*, each $S \in \mathcal{F}$ is mapped into a *non-empty* subset $\mathrm{M}(S) \subseteq S$, for some function $\mathrm{M}(\cdot)$.
- In the *online phase*, each task t is allocated to the currently *least-loaded* processor in $\mathrm{M}(S_t)$.

Notice that we limit ourselves to a subset of available processors. As shown in [8], by carefully choosing M, the modified greedy algorithm avoids allocations that are "too far" from the optimum. The main result in [8] is that the competitive ratio of this algorithm is at most $1 + c_{\mathcal{F}}(\mathrm{M})$, with

$$c_{\mathcal{F}}(\mathrm{M}) := \max_{S \in \mathcal{F}} \frac{|\mathrm{Adversary}_{\mathcal{F}}(S, \mathrm{M})|}{|\mathrm{M}(S)|}, \tag{1}$$

where $\mathrm{Adversary}_{\mathcal{F}}(S, \mathrm{M})$ consists of the *union* of all subsets S' in $\mathcal{F}$ such that $\mathrm{M}(S')$ intersects $\mathrm{M}(S)$. Intuitively, the tasks allocated to $\mathrm{M}(S)$ could have been assigned only to processors in $\mathrm{Adversary}_{\mathcal{F}}(S, \mathrm{M})$.

In this work, we focus on *s-uniform instances*, that is, the case in which $\mathcal{F}$ contains *all s-subsets* of the n processors. This is a natural restriction modeling problems where each task is guaranteed (only) to be assignable to s out of the n processors (though this set can change arbitrarily from task to task). With the minimum competitive bound $C(n, s)$ we ask how small the bound in (1) can be depending on n and s (see Definition 2). Notice that the resulting algorithm uses only *local* information as it assigns a task t by simply considering the current load of (a subset of) the processors that can execute that task. When this subset, which is specified by M, consists of a *single* processor, the corresponding algorithm requires "no information" on the processors' loads. The blind competitive bound $B(n, s)$ is defined as the minimum competitive bound, when restricting to these "blind" algorithms (see Definition 1). This number is a *tight* bound on the competitive ratio of these algorithms and its analysis is fundamental for the minimum competitive bound too. Both numbers initiate the study of online algorithms for load balancing problems which use only *local* information. In our view, one of the main contributions of this work is a stringent connection between the competitive analysis of certain local online load balancing algorithms and the Caccetta-Häggkvist conjecture.

Preliminaries and notation. We are given a family $\mathcal{F}$ of *distinct* subsets of an n-set (the latter, representing the processors). We let $Feas(\mathcal{F})$ be the set of all functions M mapping every subset $S \in \mathcal{F}$ into a *nonempty* subset $\mathrm{M}(S) \subseteq S$. We let

$$\mathrm{Adversary}_{\mathcal{F}}(S, \mathrm{M}) := \bigcup_{S' \in \mathcal{F}: \; \mathrm{M}(S') \cap \mathrm{M}(S) \neq \emptyset} S'.$$

In the sequel, s denotes the cardinality of the sets in $\mathcal{F}$. We will always assume that s and n are positive integers satisfying $2 \leq s \leq n$ (the case $s = 1$ is trivial and not interesting for the application). Observe that $\text{Adversary}_{\mathcal{F}}(S, M)$ contains at most n elements (i.e., the n processors). Thus, the identity function $M_{\text{trivial}}(S) = S$ yields a *trivial upper bound*:

$$c_{\mathcal{F}}(M_{\text{trivial}}) \leq n/s. \tag{2}$$

We typically consider families containing *all* possible s-subsets of an n-set. In this case we write $Feas(n, s)$ and omit the subscript '$\mathcal{F}$'.

2   Blind Algorithms and the Caccetta-Häggkvist Conjecture

A simple (and somewhat naive) class of (online) algorithms assign tasks in a fixed manner without "looking" at the current loads of the processors: every task t is allocated to the processor $p(S_t)$ for some function $p(\cdot)$ (thus ignoring the allocation of all other tasks). These algorithms and their analysis via the upper bound in (1) are captured by the following:

Definition 1. *A* blind algorithm *is a function* M *mapping every s-subset of the n processors into a 1-subset of this s-subset. The* blind competitive ratio *is*

$$B(n, s) := \min_{M \in Blind(n,s)} \{c(M)\},$$

where $Blind(n, s)$ consists of all blind algorithms.

We stress that a simple argument shows that, for blind algorithms, the upper bound in (1) gives a tight analysis:

Fact 2. *The competitive ratio of any blind algorithm* M *is exactly $c(M)$. Hence, $B(n, s)$ is the minimum competitive ratio over all blind algorithms.*

In this section, we show that $B(n, s) = 1 + n - \lceil n/s \rceil$, where the lower bound holds *if and only if* the Caccetta-Häggkvist conjecture (see Conjecture 1) is true. The upper and the lower bounds will follow from the next two lemmata.

Lemma 1. *Let G be any digraph on n nodes with minimum outdegree d and not containing any directed cycle of length at most s. Then there exists* M $\in$ *$Blind(n, s)$ with $c(M) = n - d$, that is, $B(n, s) \leq n - d$.*

Proof. We construct M $\in Blind(n, s)$ as follows. We identify the nodes of G with the n processors. For every s-subset S we search for an $a \in S$ such that in G there is no edge from a to another element in S. Observe that such an element must exist since otherwise we have a directed cycle involving only elements in S, and thus a directed cycle of length at most s. We then set $M(S) := \{a\}$. Observe that, if (a, b) is an edge in G and an s-subset T contains b, then it cannot be the case $M(T) = \{a\}$. This implies that the set $\text{Adversary}(S, M)$ does not contain any node in the outneighborhood of a. Since node a has outdegree at least d, this set has cardinality at most $n - d$. Since $|M(S)| = 1$ for all S, from (1) we obtain $c(M) = \max_S |\text{Adversary}(S, M)| \leq n - d$. ☐

Lemma 2. *Let n, s, and $d \geq s$ be positive integers such that $B(n, s) \leq n - d$. Then there exists a digraph G on at most n nodes with minimum outdegree at least d and not containing a directed cycle of length at most s.*

Proof. Let $G = G(M)$ be the digraph on n nodes containing the edge (a, b) if and only if there exists no S such that $b \in S$ and $M(S) = \{a\}$. By construction the outneighborhood of a contains all but the elements in $\text{Adversary}(S, M)$, that is, its outdegree is $n - |\text{Adversary}(S, M)|$. Since $|\text{Adversary}(S, M)| \leq B(n, s)$, the outdegree of any node a is at least $n - B(n, s) \geq d$. Hence, the graph G has minimum outdegree $d_G \geq d$.

We observe that the subgraph induced by *any* subset of s nodes must contain a *sink*, that is, a node having outdegree 0 in that subgraph: Indeed, for any S, the element a such that $M(S) = \{a\}$ must be a sink. In particular, there is no directed cycle of length s.

Using this fact, we iteratively remove nodes from G and obtain a subgraph G' with $n' \leq n$ nodes, without directed cycles of length at most s, and minimum outdegree equal to the minimum outdegree d of G. Towards this end, we proceed as follows. While we can pick a set C of nodes that form a directed cycle of length at most $s - 1$ in G' (recall that there is no directed cycle of length s), we add to C, one by one, nodes of G' that have an edge directed to the current set of nodes. This process must stop when reaching at most $s - 1$ nodes since otherwise, when C reaches cardinality s, by construction, it does not contain a sink, thus a contradiction. Notice that there is no edge from a node in $G' - C$ to a node in C. We can thus remove the nodes in C from the graph without decreasing its minimum outdegree.

At the end of this process, the graph G' does not contain any directed cycle of length s or smaller and its minimum outdegree is at least $d_G \geq d$. Observe that G' cannot be empty since every removed set C as above must have some outgoing edge (because of $d \geq s \geq |C|$) and this edge cannot be ingoing to the previously removed components. ☐

Lemmata 1 and 2 will give us the upper and the lower bound:

Theorem 3. *For any n and $s \leq \sqrt{n}$, it holds that*

$$B(n, s) = 1 + n - \lceil n/s \rceil$$

where the lower bound holds unless Conjecture 1 is false.

Proof. Let us set $d = \max\{s, \lceil n/s \rceil\}$. By contradiction, assume $B(n, s) \leq n - d$. Lemma 2 implies the existence of a digraph G on $n' \leq n$ nodes with minimum outdegree $d \geq \lceil n'/s \rceil$ and not containing directed cycles of length s or smaller. However, Conjecture 1 implies that G must have a directed cycle of length at most $\lceil n'/d \rceil \leq \lceil n'/\lceil n'/s \rceil \rceil \leq \lceil n'/(n'/s) \rceil = s$, thus a contradiction. Since $B(n,s)$ is integer, it must be $B(n,s) \geq 1 + n - d$. Since $s \leq \sqrt{n}$, we have $d = \lceil n/s \rceil$, which proves the lower bound. In order to prove the upper bound, we consider the following digraph G, first described by Behzad, Chartrand and Wall [6]. We let $[n] = \{0,\ldots,n-1\}$ be the set of nodes. For every node $x \in [n]$, we let its out-neighborhood being the $d-1$ nodes in the interval $[(x+1) \bmod n, (x+d-1) \bmod n]$. By construction, the resulting digraph G has minimum outdegree $d-1$ and, since $d-1 = \lceil n/s \rceil - 1 < n/s$, does not have any directed cycle of length at most s. Lemma 1 thus implies $B(n,s) \leq n - (d-1) = n - \lceil n/s \rceil + 1$, that is the upper bound. $\qquad\square$

Remark 1. Notice that the Caccetta-Häggkvist conjecture is not "interesting" for $d > n/2$ since in this case it is easy to show that a two-cycle must exist, i.e., the conjecture holds. Lemma 2 implies that $B(n,2) = n/2$, for any n. In contrast, proving a tight bound for $B(n,3)$ is the first hard case: It corresponds to the case $d = n/3$ of the conjecture which is one of the most studied [14, Section 2.2].

It is possible to settle (weaker) lower bounds on $B(n,s)$ by using some "approximate" results for the Caccetta-Häggkvist conjecture. It is known that the conjecture holds if we consider some "additive" constant α. That is, a minimum outdegree d guarantees that every digraph on n nodes must have a directed cycle of length at most $n/d + \alpha$. Currently, the best known bound is $\alpha = 73$ by Shen [13]. This type of results imply the following:

Theorem 4. *For any n and $\alpha < s \leq \sqrt{n} + \alpha/2$, it holds that $B(n,s) \geq 1 + n - \lceil n/(s-\alpha) \rceil$.*

Proof. Since $s > \alpha$, we can consider $d = \lceil n/(s-\alpha) \rceil$. By contradiction, assume $B(n,s) \leq n - \lceil n/(s-\alpha) \rceil = n - d$. From $s \leq \sqrt{n} + \alpha/2$, we have $d \geq s$ and thus Lemma 2 implies that there exists a digraph on n nodes with minimum outdegree d and not containing any directed cycle of length s or smaller. Since $n/d + \alpha = n/\lceil n/(s-\alpha) \rceil + \alpha \leq n/(n/(s-\alpha)) + \alpha = (s-\alpha) + \alpha = s$, this graph does not contain a directed cycle of length $n/d + \alpha$ or smaller. This contradicts the definition of α. Since $B(n,s)$ is integer, it must be $B(n,s) \geq 1 + n - d$ and the theorem follows. $\qquad\square$

For $s = 3$, Shen [12] proved another approximate version of the conjecture: if the minimum outdegree is at least $\mu \cdot n$, then there is a directed triangle, where $\mu > 1/3$ is a "multiplicative" constant (see also [14, Section 2.3]). This result, combined with Lemma 2, yields the following lower bound:

Theorem 5. *For any n, it holds that $B(n,3) \geq 1+n-\mu \cdot n$, where $\mu = 3-\sqrt{7} = 0.3542\cdots$.*

3   The Minimum Competitive Bound

In this section, we turn our attention to "less naive" algorithms which can be obtained with the method described in Section 1.2. In particular, we study the bound in (1), again when $\mathcal{F}$ consists of *all* s-*subsets* of the n processors (for the sake of readability, we omit the subscript '$\mathcal{F}$'):

Definition 2. *The* minimum competitive bound *is*

$$C(n, s) := \min_{M \in Feas(n,s)} \{c(M)\},$$

where $Feas(n, s)$ *consists of all functions* $M(\cdot)$ *mapping every* s-*subset of* n *processors into a non-empty subset* $M(S) \subseteq S$.

Notice that we have a trivial upper bound $C(n, s) \leq n/s$ (see Equation 2). We prove that $C(n, s) = n/s$, unless we disprove Conjecture 1. That is, the trivial upper bound is likely the best possible. We will first prove lower bounds for some special cases (these results do not require Conjecture 1).

Lemma 3. *Let* $M \in Feas(n, s)$ *such that* $|M(S)| \geq 2$, *for all* s-*subset* S. *Then, there exists an* s-*subset* T *for which* $|\mathrm{Adversary}(T, M)| = n$.

Proof. Without loss of generality, we can assume that $|M(S)| = 2$, for every S. Indeed, if we shrink all $M(S)$ into a two-set $M'(S) \subseteq M(S)$, we obtain a function $M' \in Feas(n, s)$ satisfying $\mathrm{Adversary}(S, M') \subseteq \mathrm{Adversary}(S, M)$.

We use $\mathrm{Adversary}(S)$ as a shorthand for $\mathrm{Adversary}(S, M)$ and assume, by way of contradiction, that $|\mathrm{Adversary}(S)| < n$, for all S of size s. Using this fact, we give an iterative way to define a suitable sequence $B^1 \subset B^2 \subset \cdots \subset B^k$ as follows. We start from an arbitrary s-subset S^1 and let $B^1 := M(S^1)$. At each iteration i, we "expand" the current B^i into a new set $B^{i+1} := B^i \cup \{b_i\} \cup M(S^{i+1})$, where b_i and S^{i+1} are defined as follows. Each S^i is an s-subset and thus the hypothesis $|\mathrm{Adversary}(S^i)| < n$ implies that we can chose $b_i \notin \mathrm{Adversary}(S^i)$. We then define S^{i+1} as an s-subset such that $b_i \in M(S^{i+1})$, if such a set exists; otherwise, S^{i+1} is an arbitrarily chosen s-subset containing B^i and b_i. Below we will show that the set B^{i+1} adds 2 or 3 elements to the set B^i, thus implying that we can stop when $s - 2 \leq |B^k| \leq s$.

Claim (1). $M(S)$ cannot intersect B^i if S is an s-subset containing $B^i \cup \{b_i\}$.

Proof of Claim (1). We proceed by induction on i. For $i = 1$, if $M(S)$ intersects $B^1 = M(S^1)$, then $\mathrm{Adversary}(S^1)$ contains S. Since $b_1 \in S$, this contradicts the definition of b_1. Now assume the claim holds for $i - 1$ and let S be an s-subset containing $B^i \cup \{b_i\}$. Since $B^i = B^{i-1} \cup \{b_{i-1}\} \cup M(S^i)$, S contains $B^{i-1} \cup \{b_{i-1}\}$, and the inductive hypothesis implies that $M(S)$ cannot intersect B^{i-1}. If $M(S)$ intersects $M(S^i)$ then, since $b_{i-1} \in S$, we have the contradiction $b_i \in \mathrm{Adversary}(S^i)$. If $M(S)$ contains b_{i-1}, the definition of S^i implies that $b_{i-1} \in M(S^i)$. (Recall that $b_{i-1} \notin M(S)$ only in the case there is no s-subset S with $b_{i-1} \in M(S)$.) But then $M(S)$ would again intersect $M(S^i)$, which leads to the same contradiction as above. The inductive step thus follows from $B^i = B^{i-1} \cup \{b_{i-1}\} \cup M(S^i)$. The claim thus follows. $\square$

Since $|M(S)| = 2$, Claim (1) implies that B^{i+1} is obtained from B^i by adding at least two (and at most three) new elements not in B^i. We can thus define k as the first integer such that $s - 2 \leq |B^k| \leq s$. We next show that in each of the three cases a contradiction arises:

1. For $|B^k| = s - 2$, we consider *any* s-subset $S(x) := B^k \cup \{b_k\} \cup \{x\}$, with $x \notin B^k \cup \{b_k\}$. Claim (1) implies $M(S(x)) = \{b_k, x\}$, and thus Adversary$(S(x))$ contains also all elements not in $B^k \cup \{b_k\}$, that is, $|\text{Adversary}(S(x))| = n$.
2. For $|B^k| = s - 1$, we simply observe that for $S := B^{k-1} \cup \{b_{k-1}\}$ Claim (1) yields $M(S) = \{b_{k-1}\}$, contradicting the hypothesis $|M(S)| \geq 2$ for all s-subsets S.
3. For $|B^k| = s$, $B^{k-1} \cup \{b_{k-1}\}$ must have size $s - 2$, since $|B^{k-1}| < s - 2$. For every s-subset $S(x, y) := B^{k-1} \cup \{b_{k-1}, x, y\}$ Claim (1) implies $M(S(x, y)) = \{x, y\}$. If we keep x fixed and consider all y not in this set, we obtain the contradiction $|\text{Adversary}(S(x, y))| = n$.

This concludes the proof of the lemma. □

Observe that the above result says that, if $C(n, s) < n/s$, then the corresponding M must be such that $|M(S)| = 1$ for at least one S. In order to prove the lower bound $C(n, s) = n/s$, we will make use of the following result showing that, without loss of generality, we can restrict ourselves to optimal modifications M having a "canonical" structure (the result applies to any family $\mathcal{F}$ of s-subsets):

Lemma 4. *For any* $M \in Feas(\mathcal{F})$, *there exists an* $M_c \in Feas(\mathcal{F})$ *such that* $c_\mathcal{F}(M_c) \leq c_\mathcal{F}(M)$ *and* M_c *is canonical, that is,* $M_c(S) \not\subset M_c(T)$ *for all* $S, T \in \mathcal{F}$.

Proof. Consider two s-subsets S and T such that $M(S) \subset M(T)$. (Otherwise the lemma holds.) If we shrink $M(T)$ to $M(S)$, what we obtain is a new $M' \in Feas(\mathcal{F})$ such that $M'(T) = M(S)$ and $M'(U) = M(U)$ for $U \neq T$. This implies Adversary$_\mathcal{F}(U, M') \subseteq$ Adversary$_\mathcal{F}(U, M)$ for all s-subsets U, and that

$$\frac{|\text{Adversary}_\mathcal{F}(T, M')|}{|M'(T)|} \leq \frac{|\text{Adversary}_\mathcal{F}(S, M)|}{|M(S)|};$$

$$\max_{U \neq T} \frac{|\text{Adversary}_\mathcal{F}(U, M')|}{|M'(U)|} \leq \max_{U \neq T} \frac{|\text{Adversary}_\mathcal{F}(U, M)|}{|M(U)|}.$$

This yields $c_\mathcal{F}(M') \leq c_\mathcal{F}(M)$. To obtain the final family M_c it suffices to iterate the above transformation at most $|\mathcal{F}|$ times. (At every iteration we let M being the family obtained in the previous iteration and pick S and T as above with $M(S)$ not containing another $M(U)$.) The lemma thus follows. □

We first give a tight bound for some special cases for which we do not need the Caccetta-Häggkvist conjecture:

Theorem 6. *For every* n, *if* $s \geq \sqrt{n}$ *or* $s = 2$, *then it holds that* $C(n, s) = n/s$.

Proof. Let M be such that $c(M) = C(n, s)$. We first consider $s \geq \sqrt{n}$. If there exists one S with $|M(S)| = 1$, then $c(M) \geq |\text{Adversary}(S, M)| \geq s \geq n/s$, where the two inequalities follow from $S \in$ Adversary(S, M) and from $s \geq \sqrt{n}$, respectively. Otherwise, in the case $|M(S)| \neq 1$ for every S, Lemma 3 implies that $c(M) \geq n/s$. (Recall that $|M(S)| \leq |S| \leq s$.)

Let us now consider the case $s = 2$. From Lemma 4 we can assume that M is canonical. This implies that the n processors are partitioned into two sets N_1 and N_2 such that the following holds. For every two-subset $S \subseteq N_i$, it holds that $|M(S)| = i$, with $i = 1, 2$. Let $n_1 := |N_1|$ and $n_2 := |N_2|$, and let $\mathcal{F}_i$ denote the family of all two-subsets of N_i, for $i = 1, 2$. Let M′ be the function M restricted $\mathcal{F}_1$ and observe that $M' \in Blind(n_1, 2)$. Hence, there is one $S \subseteq N_1$ for which $|\text{Adversary}_{\mathcal{F}_1}(S, M')| \geq B(n_1, 2) = n_1/2$ (see Remark 1). That is, $\text{Adversary}(S, M)$ contains at least $n_1/2$ elements from N_1. We next show that it must also contain *all* elements x in N_2. Indeed, for every two-subset $S(x)$ consisting of $x \in N_2$ and $M(S)$, Lemma 4 implies that $M(S(x)) = M(S)$, and thus $x \in \text{Adversary}(S, M)$. Hence, $|\text{Adversary}(S, M)| \geq n_1/2 + n_2 = n_1/2 + (n - n_1) = n - n_1/2 \geq n/2$, where the last inequality follows from $n_1 \leq n$.   □

Finally, from Lemma 3 we obtain the main result of this section:

Theorem 7. *For every n and $2 < s < \sqrt{n}$, it holds that $C(n, s) = n/s$. The lower bound holds unless Conjecture 1 is false. Hence, the trivial upper bound $C(n, s) \leq n/s$ is the best possible one.*

Proof. Let $M \in Feas(n, s)$ with $c(M) = C(n, s)$. From Lemma 4, we can assume M being canonical. Because of Lemma 3, the theorem holds if $|M(S)| \geq 2$ for all s-subsets S. Otherwise, we consider the subset N_1 of those processors x such that $\{x\} = M(S)$, for some S. Let N_2 be the complement of N_1, that is, the subset of processors not in N_1. From the hypothesis, we have $3 \leq s < n/s$. We consider the following two cases for $n_2 := |N_2|$:

1. $n_2 < n/s$. In this case $n_1 := |N_1| = n - n_2 > n - n/s$. Since M is canonical, for every s-subset T contained in N_1, it must be the case $|M(T)| = 1$. Let $\mathcal{F}_1$ denote the family of all s-subsets of N_1 and let M′ be the function obtained by restricting M to $\mathcal{F}_1$. Observe that M′ is a function in $Blind(n_1, s)$. Hence, there exists $S \in N_1$ such that $C(n, s) \geq |\text{Adversary}(S, M')| \geq B(n_1, s)$. From the proof of Theorem 3, if Conjecture 1 holds, then $B(n_1, s) \geq 1 + n_1 - \max\{s, \lceil n_1/s \rceil\} > 1 + n - n/s - \max\{s, \lceil n_1/s \rceil\} \geq 1 + n - n/s - \max\{s, \lceil n/s \rceil\} = 1 + n - n/s - \lceil n/s \rceil > n - 2n/s$, where the last inequality follows from $\lceil n/s \rceil < 1 + n/s$. Since $s \geq 3$, we have $n - 2n/s \geq n/s$, thus implying $C(n, s) \geq B(n_1, s) > n/s$.

2. $n_2 \geq n/s$. By definition of N_1 and N_2, every s-subset S contained in N_2 must satisfy $|M(S)| \geq 2$. Since $s < \sqrt{n}$, we have $n_2 \geq n/s > s$ and thus N_2 contains some s-subset. Let us consider the function M′ obtained by restricting M to the s-subsets of N_2. Observe that $M' \in Feas(n_2, s)$. Lemma 3 implies that there exists $S \subseteq N_2$ with $\text{Adversary}(S, M)$ containing all the elements in N_2. If the set $\text{Adversary}(S, M)$ contains also N_1, then we clearly have $C(n, s) \geq n/s$. Otherwise, we consider an $x \in N_1$ with $x \notin \text{Adversary}(S, M)$. For $\{x\} = M(T)$, if $\text{Adversary}(T, M)$ contains N_2, then $C(n, s) \geq n_2 > n/s$ (recall that $|M(T)| = 1$) and the theorem holds. Otherwise, we can pick a

$y \in N_2$ with $y \notin \text{Adversary}(T, \text{M})$. Observe that $\text{M}(S)$ must contain at least $s - 1$ elements, unless $C(n, s) \geq n/s$. We can thus construct an s-subset $S' := \{x\} \cup R'$ with R' containing y and other $s - 2$ elements from $\text{M}(S)$. Observe that $\text{M}(S')$ cannot contain x since otherwise $\text{M}(S') \cap \text{M}(T) = \{x\} \neq \emptyset$ and thus $y \in S' \subseteq \text{Adversary}(T, \text{M})$, contradicting the definition of y. Similarly, $\text{M}(S')$ cannot contain any of the elements in $R' \setminus \{y\}$ since otherwise $\text{M}(S') \cap \text{M}(S) \neq \emptyset$ and thus $x \in S' \subseteq \text{M}(S)$, contradicting the definition of x. Hence, it must be the case $\text{M}(S') = \{y\}$, contradicting $y \notin N_1$ (since $y \in N_2$). □

4   Conclusions and Open Questions

We have applied the approach by Crescenzi *et al* [8] to a natural class of s-uniform instances, which model the problem version in which the only available information is that each task is assignable to s out of the n processors, for some known constant s. We have shown that this approach is unlikely to lead to any "satisfactory" upper bound. Namely, the minimum competitive bound $C(n, s)$ is equal to the trivial n/s upper bound, unless we find a counterexample to the Caccetta-Häggkvist conjecture [14]. Even for rather limited algorithms, for which the analysis in [8] is tight, an exact answer is "equivalent" to the conjecture above. That is, the competitive ratio $B(n, s)$ of the best algorithm in this class can be determined for all s and n if and only if we resolve the conjecture.

We consider the study of these algorithms interesting by itself since they only require *local* information. Indeed, the online load balancing problem considered here arises in many practical situations (e.g., when connecting mobile devices requiring different bandwidth to one of the "geographically close" base stations). The natural greedy algorithm can have a rather unsatisfactory competitive ratio in several cases [3,5], which motivated the development of more sophisticated "ad-hoc" algorithms [4,5]. The latter are *not* local, though their competitive ratio is significantly better than greedy one. To the best of our knowledge, there is no prior study of *local* online algorithms for this problem version (apart from the tight bound $\Theta(n^{2/3})$ on the greedy [4]). Online local algorithms for a different task allocation problem have been studied by Kuhn *et al* [9]. In their problem, the goal is to maintain (roughly) the same number of tasks on each processor, and tasks can be moved only "locally", i.e., between adjacent processors.

We conclude observing that our results might be used to write a computer program to check the Caccetta-Häggkvist conjecture. Observe that, if we believe the conjecture is true, then a program which verifies it for a fixed n and d, will have to go through *all* possible digraphs on n nodes and minimum outdegree d. This is because we have to show that there is no way to avoid a directed cycle with $\lceil n/d \rceil$ nodes. Theorem 3 gives an alternative that is to come up with an (efficient) algorithm to compute $B(n, s)$. Obviously, this algorithm should not rely on the Caccetta-Häggkvist conjecture, that is, it should be possible to prove its correctness *independently* from the conjecture (e.g., the algorithm returns an optimal modification M for any given $\mathcal{F}$ containing only s-subsets). Notice that, once again, the case $s = 3$ seems to be the "first" difficult one. Indeed, for $s = 2$ the optimal modification M for any family $\mathcal{F}$ reduces to the problem of orienting the edges of an indirected graph in order to minimize the maximum indegree (see Aichholzer *et al* [1] and Nash-Williams [11]). Such optimal orientation can be computed with standard flow techniques, thus yielding an optimal algorithm for $s = 2$ (see the full version of this work [10]). Unfortunately, the results do not apply to $s = 3$, which remains an interesting open question.

Acknowledgments. We wish to thank Noga Alon for a very useful discussion and for pointing us to the Caccetta-Häggkvist conjecture via the proxy of Janos Körner. We also thank Pilu Crescenzi and Carmine Ventre for several comments on a previous version of this work.

References

1. Aichholzer, O., Aurenhammer, F., Rote, G.: Optimal graph orientation with storage applications. SFB Report Series 51, TU-Graz (1995)
2. Azar, Y.: On-line Load Balancing. In: Fiat, A., Wöginger, G.J. (eds.) Online Algorithms. LNCS, vol. 1442, Springer, Heidelberg (1998)
3. Azar, Y., Broder, A., Karlin, A.: On-line load balancing. Theoretical Computer Science 130(1), 73–84 (1994)
4. Azar, Y., Kalyanasundaram, B., Plotkin, S., Pruhs, K., Waarts, O.: On-line load balancing of temporary tasks. Journal of Algorithms 22, 93–110 (1997)
5. Bar-Noy, A., Freund, A., Naor, J.: On-line load balancing in a hierarchical server topology. SIAM Journal on Computing 31(2), 527–549 (2001)
6. Behzad, M., Chartrand, G., Wall, C.: On minimal regular digraph with given girth. Fundamenta Mathematicae 69, 227–231 (1970)
7. Caccetta, L., Häggkvist, R.: On minimal regular digraph with given girth. In: Proc. of 9th S-E Conf. Combinatorics, Graph Theory and Computing, pp. 181–187 (1978)
8. Crescenzi, P., Gambosi, G., Nicosia, G., Penna, P., Unger, W.: On-line load balancing made simple: Greedy strikes back. Journal of Discrete Algorithms 5(1), 162–175 (2007)
9. Kuhn, F., Schmid, S., Wattenhofer, R.: A self-repairing peer-to-peer system resilient to dynamical adversarial churn. In: Castro, M., van Renesse, R. (eds.) IPTPS 2005. LNCS, vol. 3640, pp. 13–23. Springer, Heidelberg (2005)
10. Monti, A., Penna, P., Silvestri, R.: An Equivalent Version of the Caccetta-Häggkvist Conjecture in an Online Load Balancing Problem (2007), Technical report available at http://www.dia.unisa.it/~penna
11. Nash-Williams, C.: Edge-disjoint spanning trees of finite graphs. J. London Math. Soc. 36, 445–450 (1961)
12. Shen, J.: Directed triangles in digraphs. Journal of Combinatorial Theory 74, 405–407 (1998)
13. Shen, J.: On the Caccetta-Häggkvist conjecture. Graphs and Combinatorics 18(3), 645–654 (2003)
14. Sullivan, B.D.: A Summary of Results and Problems Related to the Caccetta-Häggkvist Conjecture. In: ARCC Workshop: The Caccetta-Haggkvist conjecture (2006), available at http://www.aimath.org/pastworkshops/caccetta.html

Mixing 3-Colourings in Bipartite Graphs

Luis Cereceda[1], Jan van den Heuvel[1], and Matthew Johnson[2,*]

[1] Centre for Discrete and Applicable Mathematics, Department of Mathematics,
London School of Economics, Houghton Street, London WC2A 2AE, U.K.
{jan,luis}@maths.lse.ac.uk
[2] Department of Computer Science, Durham University,
Science Laboratories, South Road, Durham DH1 3LE, U.K.
matthew.johnson2@dur.ac.uk

Abstract. For a 3-colourable graph G, the 3-colour graph of G, denoted $\mathcal{C}_3(G)$, is the graph with node set the proper vertex 3-colourings of G, and two nodes adjacent whenever the corresponding colourings differ on precisely one vertex of G. We consider the following question: given G, how easily can we decide whether or not $\mathcal{C}_3(G)$ is connected? We show that the 3-colour graph of a 3-chromatic graph is never connected, and characterise the bipartite graphs for which $\mathcal{C}_3(G)$ is connected. We also show that the problem of deciding the connectedness of the 3-colour graph of a bipartite graph is coNP-complete, but that restricted to planar bipartite graphs, the question is answerable in polynomial time.

1 Introduction

Throughout this paper a graph $G = (V, E)$ is simple, loopless and finite. We always regard a k-vertex-colouring of a graph G as proper; that is, as a function $\alpha : V \to \{1, 2, \ldots, k\}$ such that $\alpha(u) \neq \alpha(v)$ for any $uv \in E$. For a positive integer k and a graph G, we define the *k-colour graph of G*, denoted $\mathcal{C}_k(G)$, as the graph that has the k-colourings of G as its node set, with two k-colourings joined by an edge in $\mathcal{C}_k(G)$ if they differ in colour on just one vertex of G. We say that G is *k-mixing* if $\mathcal{C}_k(G)$ is connected.

Continuing a theme begun in an earlier paper [2], we investigate the connectedness of $\mathcal{C}_k(G)$ for a given G. The connectedness of the k-colour graph is an issue of interest when trying to obtain efficient algorithms for almost uniform sampling of k-colourings of a given graph. In particular, $\mathcal{C}_k(G)$ needs to be connected for the single-site Glauber dynamics of G (a Markov chain defined on the k-colour graph of G) to be rapidly mixing. For further details, see, for example, [5,6] and references therein.

In [2] it was shown that if G has chromatic number k for $k = 2, 3$, then G is not k-mixing, but that, on the other hand, for $k \geq 4$, there are k-chromatic graphs that are k-mixing and k-chromatic graphs that are not k-mixing. In this

---

* Research partially supported by Nuffield grant no. NAL/32772.

A. Brandstädt, D. Kratsch, and H. Müller (Eds.): WG 2007, LNCS 4769, pp. 166–177, 2007.

paper, we look further at the case $k = 3$: we know 3-chromatic graphs are not 3-mixing, but what about bipartite graphs? Examples of 3-mixing bipartite graphs include trees and C_4, the cycle on 4 vertices. On the other hand, all cycles except C_4 are not 3-mixing — see [2] for details. In Theorem 1, we distinguish between 3-mixing and non-3-mixing bipartite graphs in terms of their structure and the possible 3-colourings they may have. As G is k-mixing if and only if every connected component of G is k-mixing, we will take our "argument graph" G to be connected.

Some terminology is required to state the result. If v and w are vertices of a bipartite graph G at distance two, then a *pinch* on v and w is the identification of v and w (and the removal of any double edges produced). And G is *pinchable* to a graph H if there exists a sequence of pinches that transforms G into H.

Given a 3-colouring α, the *weight* of an edge $e = uv$ oriented from u to v is

$$w(\overrightarrow{uv}, \alpha) = \begin{cases} +1, \text{ if } \alpha(u)\alpha(v) \in \{12, 23, 31\}; \\ -1, \text{ if } \alpha(u)\alpha(v) \in \{21, 32, 13\}. \end{cases} \tag{1}$$

To *orient* a cycle means to orient each edge on the cycle so that a directed cycle is obtained. If C is a cycle, then by $\overrightarrow{C}$ we denote the cycle with one of the two possible orientations. The *weight* $W(\overrightarrow{C}, \alpha)$ of an oriented cycle $\overrightarrow{C}$ is the sum of the weights of its oriented edges.

Theorem 1. *Let G be a connected bipartite graph. The following are equivalent:*

(i) The graph G is not 3-mixing.
(ii) There exists a cycle C in G and a 3-colouring α of G with $W(\overrightarrow{C}, \alpha) \neq 0$.
(iii) The graph G is pinchable to the 6-cycle C_6.

We also determine the computational complexity of the following decision problem.

3-MIXING
Instance: A connected bipartite graph G.
Question: Is G 3-mixing?

Theorem 2. *The decision problem* 3-MIXING *is* coNP*-complete.*

We also prove, however, that there is a polynomial algorithm for the restriction of 3-MIXING to planar graphs. We remark that this difference in complexity contrasts with many other well-known graph colouring problems where the planar case is no easier to solve.

Theorem 3. *Restricted to* planar *bipartite graphs, the decision problem* 3-MIXING *is in the complexity class* P.

Organization of the paper: we prove Theorems 1, 2 and 3 in Sections 2, 3 and 4 respectively.

2   Characterising 3-Mixing Bipartite Graphs

To prove Theorem 1, we need some definitions, terminology and lemmas.

For the rest of this section, let $G = (V, E)$ denote a connected bipartite graph with vertex bipartition X, Y. We use $\alpha, \beta, \ldots$ to denote specific colourings, and, having defined the colourings as nodes of $C_3(G)$, the meaning of, for example, the path between two colourings should be clear. We denote the cycle on n vertices by C_n, and will often describe a colouring of C_n by just listing the colours as they appear on consecutive vertices.

Given a 3-colouring α of G, we define a *height function for α with base X* as a function $h : V \to \mathbb{Z}$ satisfying the following conditions. (See [1,4] for other, similar height functions.)

H1 For all $v \in X$, $h(v) \equiv 0 \pmod 2$; for all $v \in Y$, $h(v) \equiv 1 \pmod 2$.

H2 For all $uv \in E$, $h(v) - h(u) = w(\overrightarrow{uv}, \alpha)$ ($\in \{-1, +1\}$).

H3 For all $v \in V$, $h(v) \equiv \alpha(v) \pmod 3$.

If $h : V \to \mathbb{Z}$ satisfies conditions H2, H3 and also

H1′ For all $v \in X$, $h(v) \equiv 1 \pmod 2$; while for $v \in Y$, $h(v) \equiv 0 \pmod 2$.

then h is said to be a height function for α with base Y.

Observe that for a particular colouring of a given G, a height function might not exist. An example of this is the 6-cycle C_6 coloured 1-2-3-1-2-3.

Conversely, however, a function $h : V \to \mathbb{Z}$ satisfying conditions H1 and H2 induces a 3-colouring of G: the unique $\alpha : V \to \{1, 2, 3\}$ satisfying condition H3, and h is in fact a height function for this α. Observe also that if h is a height function for α with base X, then so are $h + 6$ and $h - 6$; while $h + 3$ and $h - 3$ are height functions for α with base Y. Because we will be concerned solely with the question of *existence* of height functions, we assume henceforth that for a given G, all height functions have base X. Thus we let $\mathcal{H}_X(G)$ be the set of height functions with base X corresponding to some 3-colouring of G, and define a metric m on $\mathcal{H}_X(G)$ by setting

$$m(h_1, h_2) = \sum_{v \in V} |h_1(v) - h_2(v)|,$$

for $h_1, h_2 \in \mathcal{H}_X(G)$. Note that condition H1 above implies that $m(h_1, h_2)$ is always even.

For a given height function h, $h(v)$ is said to be a *local maximum* (respectively, *local minimum*) if $h(v)$ is larger than (respectively, smaller than) $h(u)$ for all neighbours u of v. Following [4], we define the following *height transformations* on h.

– An *increasing height transformation* takes a local minimum $h(v)$ of h and transforms h into the height function h' given by $h'(x) = \begin{cases} h(x) + 2, & \text{if } x = v; \\ h(x), & \text{if } x \neq v. \end{cases}$

– A *decreasing height transformation* takes a local maximum $h(v)$ of h and transforms h into the height function h' given by $h'(x) = \begin{cases} h(x) - 2, & \text{if } x = v; \\ h(x), & \text{if } x \neq v. \end{cases}$

Notice that these height transformations give rise to transformations between the corresponding colourings. Specifically, if we let α' be the 3-colouring corresponding to h', an increasing transformation yields $\alpha'(v) = \alpha(v) - 1$, while a decreasing transformation yields $\alpha'(v) = \alpha(v) + 1$, where addition is modulo 3.

The following lemma shows that colourings with height functions are connected in $C_3(G)$. It is a simple extension of the range of applicability of a similar lemma appearing in [4].

Lemma 1 ([4]). *Let α, β be two 3-colourings of G with corresponding height functions h_α, h_β. Then there is a path between α and β in $C_3(G)$.*

Proof. We use induction on $m(h_\alpha, h_\beta)$. The lemma is trivially true when $m(h_\alpha, h_\beta) = 0$, since in this case α and β are identical.

Suppose therefore that $m(h_\alpha, h_\beta) > 0$. We show that there is a height transformation transforming h_α into some height function h with $m(h, h_\beta) = m(h_\alpha, h_\beta) - 2$, from which the lemma follows.

Without loss of generality, let us assume that there is some vertex $v \in V$ with $h_\alpha(v) > h_\beta(v)$, and let us choose v with $h_\alpha(v)$ as large as possible. We show that such a v must be a local maximum of h_α. Let u be any neighbour of v. If $h_\alpha(u) > h_\beta(u)$, then it follows that $h_\alpha(v) > h_\alpha(u)$, since v was chosen with $h_\alpha(v)$ maximum, and $|h_\alpha(v) - h_\alpha(u)| = 1$. If, on the other hand, $h_\alpha(u) \leq h_\beta(u)$, we have $h_\alpha(v) \geq h_\beta(v) + 1 \geq h_\beta(u) \geq h_\alpha(u)$, which in fact means $h_\alpha(v) > h_\alpha(u)$.

Thus $h_\alpha(v) > h_\alpha(u)$ for all neighbours u of v, and we can apply a decreasing height transformation to h_α at v to obtain h. Clearly $m(h, h_\beta) = m(h_\alpha, h_\beta) - 2$. $\square$

The next lemma tells us that for a given 3-colouring, non-zero weight cycles are, in some sense, the obstructing configurations forbidding the existence of a corresponding height function.

Lemma 2. *Let α be a 3-colouring of G with no corresponding height function. Then G contains a cycle C for which $W(\overrightarrow{C}, \alpha) \neq 0$.*

Proof. For a path P in G, let $\overrightarrow{P}$ denote one of the two possible directed paths obtainable from P, and let

$$W(\overrightarrow{P}, \alpha) = \sum_{e \in E(\overrightarrow{P})} w(e, \alpha),$$

where $w(e, \alpha)$ takes values as defined in (1).

Notice that if a colouring does have a height function, it is possible to construct one by fixing a vertex $x \in X$, giving x an appropriate height (satisfying properties H1–H3) and then assigning heights to all vertices in V by following a breadth-first ordering from x.

Whenever we attempt to construct a height function h for α in such a fashion, we must come to a stage in the ordering where we attempt to give some vertex v a height $h(v)$ and find ourselves unable to because v has a neighbour u

with a previously assigned height $h(u)$ and $|h(u) - h(v)| > 1$. Letting P be a path between u and v formed by vertices that have been assigned a height, and choosing the appropriate orientation of P, we have $w(\overrightarrow{P}, \alpha) = |h(u) - h(v)|$. The lemma now follows by letting C be the cycle formed by P and the edge uv. $\square$

The following lemma is obvious.

Lemma 3. *Let u and v be vertices on a cycle C in a graph G, and suppose there is a path P between u and v in G internally disjoint from C. Let α be a 3-colouring of G. Let C' and C'' be the two cycles formed from P and edges of C, and let $\overrightarrow{C'}, \overrightarrow{C''}$ be the orientations of C', C'' induced by an orientation $\overrightarrow{C}$ of C (so the edges of P have opposite orientations in $\overrightarrow{C'}$ and $\overrightarrow{C''}$). Then $W(\overrightarrow{C}, \alpha) = W(\overrightarrow{C'}, \alpha) + W(\overrightarrow{C''}, \alpha)$.*

Note this tells us that $W(\overrightarrow{C}, \alpha) \neq 0$ implies $W(\overrightarrow{C'}, \alpha) \neq 0$ or $W(\overrightarrow{C''}, \alpha) \neq 0$.

Proof of Theorem 1.   Let G be a connected bipartite graph.

(i) $\implies$ (ii). Suppose $\mathcal{C}_3(G)$ is not connected. Take two 3-colourings of G, α and β, in different components of $\mathcal{C}_3(G)$. By Lemma 1 we know at least one of them, say α, has no corresponding height function, and, by Lemma 2, there is a cycle C in G with $W(\overrightarrow{C}, \alpha) \neq 0$.

(ii) $\implies$ (iii). Let G contain a cycle C with $W(\overrightarrow{C}, \alpha) \neq 0$ for some 3-colouring α of G. Because $W(\overrightarrow{C_4}, \beta) = 0$ for any 3-colouring β of C_4, it follows that $C = C_n$ for some even $n \geq 6$. If $G = C$, then it is easy to find a sequence of pinches that will yield C_6. If G is C plus some chords, then, by Lemma 3, there is a smaller cycle C' with $W(\overrightarrow{C'}, \alpha) \neq 0$. Thus if $G \neq C$, we can assume that $V(G) \neq V(C)$, and we describe how to pinch a pair of vertices so that (ii) remains satisfied (for a specified cycle with G replaced by the graph created by the pinch and α replaced by its restriction to that graph; also denoted α); by repetition, we can obtain a graph that is a cycle and, by the previous observations, the implication is proved.

We shall choose vertices coloured alike to pinch so that the restriction of α to the graph obtained is well-defined and proper. If C has three consecutive vertices u, v, w with $\alpha(u) = \alpha(w)$, pinching u and w yields a graph containing a cycle $C' = C_{n-2}$ with $W(\overrightarrow{C'}, \alpha) = W(\overrightarrow{C}, \alpha)$. Otherwise C is coloured 1-2-3-$\cdots$-1-2-3. We can choose u, v, w to be three consecutive vertices of C, such that there is a vertex $x \notin V(C)$ adjacent to v. Suppose, without loss of generality, that $\alpha(x) = \alpha(u)$, and pinch x and u to obtain a graph in which $W(\overrightarrow{C}, \alpha)$ is unchanged.

(iii) $\implies$ (i). Suppose G is pinchable to C_6. Take two 3-colourings of C_6 not connected by a path in $\mathcal{C}_3(C_6)$ — 1-2-3-1-2-3 and 1-2-1-2-1-2, for example. Considering the appropriate orientation of C_6, note that the first colouring has weight 6 and the second has weight 0. We construct two 3-colourings of G not connected by a path in $\mathcal{C}_3(G)$ as follows. Consider the reverse sequence of pinches that gives G from C_6. Following this sequence, for each colouring of C_6, give every pair of new vertices introduced by an "unpinching" the same colour as the vertex from which they originated. In this manner we obtain two 3-colourings of G, α and β, say. Observe that every unpinching maintains a cycle in G which has weight 6 with respect to the colouring induced by the first colouring of C_6 and weight 0 with respect to the second induced colouring. This means G will contain a cycle C for which $W(\overrightarrow{C}, \alpha) = 6$ and $W(\overrightarrow{C}, \beta) = 0$, showing that α and β cannot possibly be in the same connected component of $C_3(G)$.

This completes the proof of the theorem. $\qquad\qquad\qquad\qquad\qquad\qquad$ □

3   The Complexity of 3-Mixing for Bipartite Graphs

Observing that Theorem 1 gives us two polynomial-time verifiable certificates for when G is *not* 3-mixing, we immediately obtain that 3-MIXING is in the complexity class coNP. By the same theorem, the following decision problem is the complement of 3-MIXING.

PINCHABLE-TO-C_6
Instance: A connected bipartite graph G.
Question: Is G pinchable to C_6?

Our proof will in fact show that PINCHABLE-TO-C_6 is NP-complete. We will obtain a reduction from the following decision problem.

RETRACTABLE-TO-C_6
Instance: A connected bipartite graph G with an induced 6-cycle S.
Question: Is G *retractable* to S? That is, does there exist a homomorphism $r : V(G) \to V(S)$ such that $r(v) = v$ for all $v \in V(S)$?

In [7] it is mentioned, without references, that Tomás Feder and Gary MacGillivray have independently proved the following result: for completeness, we give a sketch of a proof.

Theorem 4 (Feder, MacGillivray, see [7]). RETRACTABLE-TO-C_6 *is NP-complete.*

Sketch of proof of Theorem 4. It is clear that RETRACTABLE-TO-C_6 is in NP.

Given a graph G, construct a new graph G' as follows: subdivide every edge uv of G by inserting a vertex y_{uv} between u and v. Also add new vertices a, b, c, d, e together with edges za, ab, bc, cd, de, ez, where z is a particular vertex of G (any one will do). The graph G' is clearly connected and bipartite, and the vertices z, a, b, c, d, e induce a 6-cycle S. We will prove that G is 3-colourable if and only if G' retracts to the induced 6-cycle S.

Assume that G is 3-colourable and take a 3-colouring τ of G with $\tau(z) = 1$. From τ we construct a 6-colouring σ of G'. For this, first set $\sigma(x) = \tau(x)$, if $x \in V(G)$. For the new vertices y_{uv} set $\sigma(y_{uv}) = \begin{cases} 4, \text{ if } \tau(u) = 1 \text{ and } \tau(v) = 2, \\ 5, \text{ if } \tau(u) = 2 \text{ and } \tau(v) = 3, \\ 6, \text{ if } \tau(u) = 3 \text{ and } \tau(v) = 1. \end{cases}$

And for the cycle S we take $\sigma(a) = 4$, $\sigma(b) = 2$, $\sigma(c) = 5$, $\sigma(d) = 3$ and $\sigma(e) = 6$. Now define $r : V(G') \to V(S)$ by setting $r(x) = z$, if $\sigma(x) = 1$; $r(x) = a$, if $\sigma(x) = 4$; $r(x) = b$, if $\sigma(x) = 2$; $r(x) = c$, if $\sigma(x) = 5$; $r(x) = d$, if $\sigma(x) = 3$; and $r(x) = e$, if $\sigma(x) = 6$. It is easy to check that r is a retraction of G' to S.

Conversely, suppose G' retracts to S. We can use this retraction to define a 6-colouring of G' in a similar way to that in which we defined r from σ in the preceeding paragraph. The restriction of this 6-colouring to G yields a 3-colouring of G, completing the proof. □

Proof of Theorem 2. We have established that it is sufficient to describe a polynomial reduction from RETRACTABLE-TO-C_6 to PINCHABLE-TO-C_6. We shall describe the reduction but leave the remainder of the proof — which is a simple matter of checking a number of cases and, though straightforward, is lengthy — to the reader.

The reduction we use follows that used in [7] to prove the NP-completeness of the following problem:

COMPACTABLE-TO-C_6
Instance: A connected bipartite graph G.
Question: Is G compactable to C_6? That is, does there exist an edge-surjective homomorphism $c : V(G) \to V(C_6)$?

Consider an instance of RETRACTABLE-TO-C_6: a connected bipartite graph G and an induced 6-cycle S. From G we construct, in time polynomial in the size of G, an instance G' of PINCHABLE-TO-C_6 such that

$$G \text{ retracts to } S \text{ if and only if } G' \text{ is pinchable to } C_6. \qquad (*)$$

Assume G has vertex bipartition (G_A, G_B). Let $V(S) = S_A \cup S_B$, where $S_A = \{h_0, h_2, h_4\}$ and $S_B = \{h_1, h_3, h_5\}$, and assume $E(S) = \{ h_0 h_1, \ldots, h_4 h_5, h_5 h_0 \}$. The construction of G' is as follows.

- For every vertex $a \in G_A \backslash S_A$, add to G new vertices $u_1^a, u_2^a, w_1^a, y_1^a, y_2^a$, together with edges $u_1^a h_0, a u_2^a, w_1^a h_3, a w_1^a, u_1^a w_1^a, y_1^a h_5, y_2^a h_2, u_1^a y_1^a, w_1^a y_2^a, u_1^a u_2^a, y_1^a y_2^a$.

- For every vertex $b \in G_B \backslash S_B$, add to G new vertices $u_1^b, w_1^b, w_2^b, y_1^b, y_2^b$, together with edges $u_1^b h_0, b u_1^b, w_1^b h_3, b w_2^b, u_1^b w_1^b, y_1^b h_5, y_2^b h_2, u_1^b y_1^b, w_1^b y_2^b, w_1^b w_2^b, y_1^b y_2^b$.

- For every edge $ab \in E(G) \backslash E(S)$, with $a \in G_A \backslash S_A$ and $b \in G_B \backslash S_B$, add two new vertices: x_a^{ab} adjacent to a and u_1^a; and x_b^{ab} adjacent to b, w_1^b and x_a^{ab}.

It is clear that G' is connected and bipartite and that G' contains G as an induced subgraph. Note also that the subgraphs constructed around a vertex $a \in G_A \setminus S_A$ and a vertex $b \in G_B \setminus S_B$ are isomorphic; these subgraphs are depicted below in Fig. 1 and Fig. 2.

It is now easy to prove $(*)$ by considering a number of cases. The details are omitted.

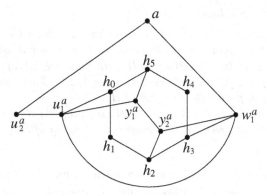

Fig. 1. The subgraph of G' added around a vertex $a \in G_A \setminus S_A$, together with the 6-cycle S

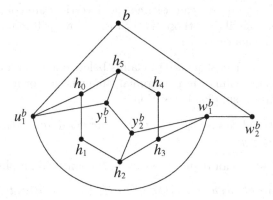

Fig. 2. The subgraph of G' added around a vertex $b \in G_B \setminus S_B$, together with the 6-cycle S

4   A Polynomial-Time Algorithm for Planar Bipartite Graphs

Now let G denote a bipartite *planar* graph. To prove Theorem 3 we need some technical results.

Lemma 4. *Let P be a shortest path between distinct vertices u and v in a bipartite graph H. Then H is pinchable to P.*

Proof. Let P have vertices $u = v_0, v_1, \ldots, v_{k-1}, v_k = v$, and let T be a breadth-first spanning tree of H rooted at u that contains P (we can choose T so that it contains P since P is a shortest path). Now, working in T, pinch all vertices at distance one from u to v_1. Next pinch all vertices at distance two from u to v_2. Continue until all vertices at distance k from u are pinched to $v_k = v$. If necessary, arbitrary pinches on the vertices at distance at least $k+1$ from u will yield P.   □

Lemma 5. *Let H be a bipartite graph.*

(i) Let u and v be two vertices in H properly pre-coloured with colours from 1, 2, 3. Then this colouring can be extended to a proper 3-colouring of H.

(ii) Let u, v and w be three vertices in H with uv, vw ∈ E(H). Suppose u, v, w are properly pre-coloured with colours from 1, 2, 3. Then this colouring can be extended to a proper 3-colouring of H.

(iii)Suppose the vertices of a 4-cycle in H are properly 3-coloured. Then this 3-colouring can be extended to a proper 3-colouring of H.

Proof. (i) is trivial.

(ii) Without loss of generality we can assume that the colouring of u, v, w is 1-2-1 or 1-2-3. In the first instance, since H is bipartite, we can extend the colouring of u, v, w to a colouring of H using colours 1 and 2 only. For the second case, we can use the same 1,2-colouring, except leaving w with colour 3.

(iii) Since any 3-colouring of a C_4 has two vertices with the same colour, without loss of generality we can assume the 4 vertices are coloured 1-2-1-2 or 1-2-1-3. Colourings similar to those used in (ii) above will immediately lead to the appropriate 3-colourings of H.   □

Proof of Theorem 3. The sequence of claims below outlines an algorithm that, given G as input, determines in polynomial time whether or not G is 3-mixing.

The first claim is a simple observation.

Claim 1. *If G is not connected, then G is 3-mixing if and only if every component of G is 3-mixing.*

We next show how we can reduce the case to 2-connected graphs.

Claim 2. *Suppose G has a cut-vertex v. Let H_1 be a component of $G-v$. Denote by G_1 the subgraph of G induced by $V(H_1) \cup \{v\}$, and let G_2 be the subgraph induced by $V(G) \setminus V(H_1)$. Then G is 3-mixing if and only if both G_1 and G_2 are 3-mixing.*

Proof. If G is 3-mixing, then clearly so are G_1 and G_2. Conversely, if G is not 3-mixing, we know by Theorem 1 that there must exist a 3-colouring α of G and a cycle C in G such that $W(\overrightarrow{C}, \alpha) \neq 0$. But because C must lie completely in G_1 or G_2, we have that G_1 or G_2 is not 3-mixing.   □

Now we can assume that G is 2-connected. In the next claim we will show that we can actually assume G to be 3-connected.

Claim 3. *Suppose G has a 2-vertex-cut $\{u, v\}$. Let H_1 be a component of $G - \{u, v\}$. Denote by G_1 the subgraph of G induced by $V(H_1) \cup \{u, v\}$, and let G_2 be the subgraph induced by $V(G) \setminus V(H_1)$. For $i = 1, 2$, let ℓ_i be the distance between u and v in G_i.*

Then only the following cases can occur:

(i) We have $\ell_1 = \ell_2 = 1$. Then G is 3-mixing if and only if both G_1 and G_2 are 3-mixing.

(ii) We have $\ell_1 = \ell_2 = 2$. *(So for $i = 1, 2$, there is a vertex $w_i \in V(G_i)$ so that $uw_i, vw_i \in E(G_i)$.)* Let G_1^* be the subgraph of G induced by $V(G_1) \cup \{w_2\}$ and let G_2^* be the subgraph induced by $V(G_2) \cup \{w_1\}$. Then G is 3-mixing if and only if both G_1^* and G_2^* are 3-mixing.

(iii) We have $\ell_1 + \ell_2 \geq 6$. Then G is not 3-mixing.

Proof. Because G is bipartite, ℓ_1 and ℓ_2 must have the same parity. If $\ell_1 = 1$ or $\ell_2 = 1$, then there is an edge uv in G, and this same edge must appear in both G_1 and G_2. This guarantees that both $\ell_1 = \ell_2 = 1$, and shows that we always have one of the three cases.

(i) In this case we have an edge uv in all of G, G_1, G_2. If one of G_1 and G_2 is not 3-mixing, say G_1, we must have a 3-colouring α of G_1 and a cycle C in G_1 for which $W(\overrightarrow{C}, \alpha) \neq 0$. By Lemma 5 (i) we can easily extend α to the whole of G, showing that G is not 3-mixing. On the other hand, if G is not 3-mixing, we know we must have a 3-colouring β of G and a cycle D in G for which $W(\overrightarrow{D}, \beta) \neq 0$. If D is contained entirely in one of G_1 or G_2, we are done. If not, D must pass through u and v. For $i = 1, 2$, consider the cycle D^i formed from the part of D that is in G_i together with the edge uv. From Lemma 3 it follows that one of D^1 and D^2 has non-zero weight under β, showing that G_1 or G_2 is not 3-mixing.

(ii) If one of G_1^* and G_2^* is not 3-mixing, we can use a similar argument as in (i) (now using Lemma 5 (ii)) to conclude that G is not 3-mixing. For the converse we assume G is not 3-mixing. So there is a 3-colouring α of G and a cycle C in G for which $W(\overrightarrow{C}, \alpha) \neq 0$. If C is contained entirely in one of G_1^* or G_2^*, we are done. If not, C must pass through u and v. If C does not contain w_1, then for $i = 1, 2$, consider the cycle C^i formed by the part of C that is in G_i^* plus the path uw_1v. From Lemma 3 it follows that one of C^1, C^2 has non-zero weight under α, showing that G_1^* or G_2^* is not 3-mixing. If w_1 is contained in C, then we can use the same argument but now using the edge uw_1 or vw_1 as the path (at least one of these edges is not on C since C is not contained entirely in G_2^*).

(iii) For $i = 1, 2$, let P_i be a shortest path between u and v in G_i, so P_i has length ℓ_i. Then, by Lemma 4, we can see that G is pinchable to $C_{\ell_1 + \ell_2}$ (follow, in G, the sequence of pinches that transforms G_1 into P_1 and G_2 into P_2). Since $\ell_1 + \ell_2 \geq 6$, $C_{\ell_1 + \ell_2}$ is of course pinchable to C_6, and hence G is not 3-mixing. $\square$

From now on we consider G to be 3-connected, and can therefore use the following result of Whitney — for details, see, for example, [3] pp. 78–80.

Theorem 5 (Whitney). *Any two planar embeddings of a 3-connected graph are equivalent.*

Henceforth, we identify G with its (essentially unique) planar embedding. For a cycle D in G, denote by $\mathrm{Int}(D)$ and $\mathrm{Ext}(D)$ the set of vertices inside and outside of D, respectively. If both $\mathrm{Int}(D)$ and $\mathrm{Ext}(D)$ are non-empty, D is *separating* and we define $G_{\mathrm{Int}}(D) = G - \mathrm{Ext}(D)$ and $G_{\mathrm{Ext}}(D) = G - \mathrm{Int}(D)$.

We next consider the case that G has a separating 4-cycle.

Claim 4. *Suppose G has a separating 4-cycle D. Then G is 3-mixing if and only if $G_{\mathrm{Int}}(D)$ and $G_{\mathrm{Ext}}(D)$ are both 3-mixing.*

Proof. To prove necessity, we show that if one of $G_{\text{Int}}(D)$ or $G_{\text{Ext}}(D)$ is not 3-mixing, then G is not 3-mixing. Without loss of generality, suppose that $G_{\text{Int}}(D)$ is not 3-mixing, so there exists a 3-colouring α of $G_{\text{Int}}(D)$ and a cycle C in $G_{\text{Int}}(D)$ with $W(\overrightarrow{C}, \alpha) \neq 0$. The 3-colouring of the vertices of the 4-cycle D can be extended to a 3-colouring of $G_{\text{Ext}}(D)$ (use Lemma 5 (iii)). The combination of the 3-colourings of $G_{\text{Int}}(D)$ and $G_{\text{Ext}}(D)$ gives a 3-colouring of G with a non-zero weight cycle, showing G is not 3-mixing.

To prove sufficiency, we show that if G is not 3-mixing, then at least one of $G_{\text{Int}}(D)$ and $G_{\text{Ext}}(D)$ must fail to be 3-mixing. Suppose that α is a 3-colouring of G for which there is a cycle C with $W(\overrightarrow{C}, \alpha) \neq 0$. If C is contained entirely within $G_{\text{Int}}(D)$ or $G_{\text{Ext}}(D)$ we are done; so let us assume that C has some vertices in $\text{Int}(D)$ and some in $\text{Ext}(D)$. Then applying Lemma 3 (repeatedly, if necessary) we can find a cycle C' contained entirely in $G_{\text{Int}}(D)$ or $G_{\text{Ext}}(D)$ for which $W(\overrightarrow{C'}, \alpha) \neq 0$, completing the proof.   □

We call a face of G with k edges in its boundary a *k-face*, and a face with at least k edges in its boundary a *$\geq k$-face*. The number of ≥ 6-faces in G — which now we can assume is a 3-connected bipartite planar graph with no separating 4-cycle — will lead to our final claim.

Claim 5. *Let G be a 3-connected bipartite planar graph with no separating 4-cycle. Then G is 3-mixing if and only if it has at most one ≥ 6-face.*

Proof. We first prove sufficiency. Suppose G has no ≥ 6-faces, so has only 4-faces. Let α be any 3-colouring of G and let C be any cycle in G. We show $W(\overrightarrow{C}, \alpha) = 0$ by induction on the number of faces inside C. If there is just one face inside C, C is in fact a facial 4-cycle and $W(\overrightarrow{C}, \alpha) = 0$. For the inductive step, let C be a cycle with $r \geq 2$ faces in its interior. If, for two consecutive vertices u, v of C, we have vertices $a, b \in \text{Int}(C)$ together with edges ua, ab, bv in G, let C' be the cycle formed from C by the removal of the edge uv and the addition of edges ua, ab, bv. If not, check whether for three consecutive vertices u, v, w of C, there is a vertex $a \in \text{Int}(C)$ with edges ua, aw in G. If so, let C' be the cycle formed from C by the removal of the vertex v and the addition of the edges ua, aw. If neither of the previous two cases apply, we must have, for u, v, w, x four consecutive vertices of C, an edge ux inside C. In such a case, let C' be the cycle formed from C by the removal of vertices v, w and the addition of the edge ux. In all cases we have that C' has $r - 1$ faces in its interior, so, by induction, we can assume $W(\overrightarrow{C'}, \alpha) = 0$. From Lemma 3 we then obtain $W(\overrightarrow{C}, \alpha) = 0$.

Suppose now that G contains exactly one ≥ 6-face. Without loss of generality we can assume that this face is the outside face, and hence the argument above will work exactly the same to show that G is 3-mixing.

To prove necessity we show that if G contains at least two ≥ 6-faces, then G is pinchable to C_6. For f a ≥ 6-face in G, a separating cycle D is said to be *f-separating* if f lies inside D. Let f and f_o be two ≥ 6-faces in G, where we can assume f_o is the outer face of G, and let C be the cycle bounding f. Our claim is that we can successively pinch vertices into a cycle of length at least 6 without ever introducing an f-separating 4-cycle — we will initially do this around C.

Let x, y, z be any three consecutive vertices of C with y having degree at least 3 — if there is no such vertex y, then G is simply a cycle of length at least 6 and we are done. Let a be a neighbour of y distinct from x and z, such that the edges ya and yz form part of the boundary of a face adjacent to f. If the result of pinching a and z introduces no f-separating 4-cycle, then pinch a and z and repeat the process. If pinching a and z *does* result in the creation of an f-separating 4-cycle, this must be because the path ay, yz forms part of an f-separating 6-cycle D. We now show how we can find alternative pinches which do not introduce an f-separating 4-cycle. The fact that D is f-separating means there is a path $P \subseteq D$ of length 4 between a and z. Note that P cannot contain y, for this would contradict the fact that G has no separating 4-cycle. Consider the graph $G' = G_{\text{Int}}(D) - \{yz\}$. We claim that the path $P' = P \cup \{ay\}$ is a shortest path between y and z in G'. To see this, remember that G is bipartite, so any path between y and z in G has to have odd length. We cannot have another edge $yz \in E(G')$ since G is simple. Finally, any path between y and z in G' would, together with the edge yz, form an f-separating cycle in G. Hence a path of length 3 between y and z would contradict the fact that G has no separating 4-cycle. By Lemma 4, we see G' is pinchable to P'. Using the same sequence of pinches in G will pinch $G_{\text{Int}}(D)$ into D. Note this introduces no separating 4-cycle into the resulting graph. If necessary, we can repeat the process by pinching vertices into D, which now bounds a 6-face. This completes the proof.   □

The sequence of Claims 1 – 5 can easily be used to obtain a polynomial-time algorithm to check if a given planar bipartite graph G is 3-mixing. This completes the proof of Theorem 3.   □

Acknowledgements. We are indebted to Gary MacGillivray for helpful discussions and for bringing reference [7] to our attention.

References

1. Baxter, R.J.: Exactly Solved Models in Statistical Mechanics. Academic Press, New York (1982)
2. Cereceda, L., van den Heuvel, J., Johnson, M.: Connectedness of the graph of vertex-colourings. Discrete Math. (to appear)
3. Diestel, R.: Graph Theory, 2nd edn. Springer, Heidelberg (2000)
4. Goldberg, L.A., Martin, R., Paterson, M.: Random sampling of 3-colorings in $\mathbb{Z}^2$. Random Structures Algorithms 24, 279–302 (2004)
5. Jerrum, M.: A very simple algorithm for estimating the number of k-colourings of a low degree graph. Random Structures Algorithms 7, 157–165 (1995)
6. Jerrum, M.: Counting, Sampling and Integrating: Algorithms and Complexity. Birkhäuser Verlag, Basel (2003)
7. Vikas, N.: Computational complexity of compaction to irreflexive cycles. J. Comput. Syst. Sci. 68, 473–496 (2004)

Minimum-Weight Cycle Covers and Their Approximability*

Bodo Manthey**

Yale University, Department of Computer Science
P.O. Box 208285, New Haven, CT 06520-8285, USA
manthey@cs.yale.edu

Abstract. A cycle cover of a graph is a set of cycles such that every vertex is part of exactly one cycle. An L-cycle cover is a cycle cover in which the length of every cycle is in the set $L \subseteq \mathbb{N}$.

We investigate how well L-cycle covers of minimum weight can be approximated. For undirected graphs, we devise a polynomial-time approximation algorithm that achieves a constant approximation ratio for all sets L. On the other hand, we prove that the problem cannot be approximated within a factor of $2 - \varepsilon$ for certain sets L.

For directed graphs, we present a polynomial-time approximation algorithm that achieves an approximation ratio of $O(n)$, where n is the number of vertices. This is asymptotically optimal: We show that the problem cannot be approximated within a factor of $o(n)$.

To contrast the results for cycle covers of minimum weight, we show that the problem of computing L-cycle covers of maximum weight can, at least in principle, be approximated arbitrarily well.

1 Introduction

A cycle cover of a graph is a spanning subgraph that consists solely of cycles such that every vertex is part of exactly one cycle. Cycle covers are an important tool for the design of approximation algorithms for different variants of the traveling salesman problem [2,4,8,9,10,16], for the shortest common superstring problem from computational biology [7,24], and for vehicle routing problems [13].

In contrast to Hamiltonian cycles, which are special cases of cycle covers, cycle covers of minimum weight can be computed efficiently. This is exploited in the above mentioned algorithms, which in general start by computing a cycle cover and then join cycles to obtain a Hamiltonian cycle. Short cycles limit the approximation ratios achieved by such algorithms. Roughly speaking, the longer the cycles in the initial cover, the better the approximation ratio. Thus, we are interested in computing cycle covers without short cycles. Moreover, there are algorithms that perform particularly well if the cycle covers computed do

---

* A full version of this work is available at http://arxiv.org/abs/cs/0609103.

** Supported by the Postdoc-Program of the German Academic Exchange Service (DAAD). On leave from Saarland University, Department of Computer Science, P.O. Box 151150, 66041 Saarbrücken, Germany.

A. Brandstädt, D. Kratsch, and H. Müller (Eds.): WG 2007, LNCS 4769, pp. 178–189, 2007.

not contain cycles of odd length [4]. Finally, some vehicle routing problems [13] require covering vertices with cycles of bounded length. Therefore, we consider *restricted cycle covers*, where cycles of certain lengths are ruled out a priori: For a set $L \subseteq \mathbb{N}$, an *L-cycle cover* is a cycle cover in which the length of each cycle is in L. Unfortunately, computing L-cycle covers is hard for almost all sets L [15,20]. Thus, in order to fathom the possibility of designing approximation algorithms based on computing cycle covers, our aim is to find out how well L-cycle covers can be approximated.

Beyond being a basic tool for approximation algorithms, cycle covers are interesting in their own right. Matching theory and graph factorization are important topics in graph theory. The classical matching problem is the problem of finding one-factors, i. e., spanning subgraphs in which every vertex is incident to exactly one edge. Cycle covers of undirected graphs are also called two-factors since every vertex is incident to exactly two edges in a cycle cover. Both structural properties of graph factors and the complexity of finding graph factors have been the topic of a considerable amount of research (cf. Lovász and Plummer [17] and Schrijver [23]).

1.1 Preliminaries

Let $G = (V, E)$ be a graph. If G is undirected, then a **cycle cover** of G is a subset $C \subseteq E$ of the edges of G such that all vertices in V are incident to exactly two edges in C. If G is a directed graph, then a cycle cover of G is a subset $C \subseteq E$ such that all vertices are incident to exactly one incoming and one outgoing edge in C. Thus, the graph (V, C) consists solely of vertex-disjoint cycles. The length of a cycle is the number of edges it consists of. We are concerned with simple graphs, i. e., the graphs do not contain multiple edges or loops. Thus, the shortest cycles of undirected and directed graphs are of length three and two, respectively. We call a cycle of length λ a **λ-cycle** for short.

An **L-cycle cover** of an undirected graph is a cycle cover in which the length of every cycle is in the set $L \subseteq \mathcal{U} = \{3, 4, 5, \ldots\}$. An L-cycle cover of a directed graph is analogously defined except that $L \subseteq \mathcal{D} = \{2, 3, 4, \ldots\}$. A special case of L-cycle covers are **k-cycle covers**, which are $\{k, k+1, \ldots\}$-cycle covers. Let $\overline{L} = \mathcal{U} \setminus L$ in the case of undirected graphs, and let $\overline{L} = \mathcal{D} \setminus L$ in the case of directed graphs (whether we consider undirected or directed cycle covers will be clear from the context).

Given edge weights $w : E \to \mathbb{N}$, the **weight $w(C)$** of a subset $C \subseteq E$ of the edges of G is $w(C) = \sum_{e \in C} w(e)$. In particular, this defines the weight of a cycle cover since we view cycle covers as sets of edges.

Min-L-UCC is the following optimization problem: Given an undirected complete graph with non-negative edge weights that satisfy the triangle inequality ($w(\{u, v\}) \leq w(\{u, x\}) + w(\{x, v\})$ for all $u, x, v \in V$) find an L-cycle cover of minimum weight. **Min-k-UCC** is defined for $k \in \mathcal{U}$ like Min-L-UCC except that k-cycle covers rather than L-cycle covers are sought. The triangle inequality is not only a natural restriction, it is also necessary: If finding L-cycle covers in graphs is NP-hard, then Min-L-UCC without the triangle inequality does not allow for any approximation at all.

Min-L-DCC and **Min-k-DCC** are defined for directed graphs like Min-L-UCC and Min-k-UCC for undirected graphs except that $L \subseteq \mathcal{D}$ and $k \in \mathcal{D}$ and the triangle inequality is of the form $w(u, v) \leq w(u, x) + w(x, v)$.

Finally, **Max-L-UCC**, **Max-k-UCC**, **Max-L-DCC**, and **Max-k-DCC** are analogously defined except that cycle covers of maximum weight are sought and that the edge weights do not have to fulfill the triangle inequality.

1.2   Previous Results

Min-$\mathcal{U}$-UCC, i.e., the undirected cycle cover problem without any restrictions, can be solved in polynomial time via Tutte's reduction to the classical perfect matching problem [17]. By a modification of an algorithm of Hartvigsen [12], also 4-cycle covers of minimum weight in graphs with edge weights one and two can be computed efficiently. For Min-k-UCC restricted to graphs with edge weights one and two, there exists a factor $7/6$ approximation algorithm for all k [6]. Hassin and Rubinstein [14] presented a randomized approximation algorithm for Max-$\{3\}$-UCC that achieves an approximation ratio of $83/43 + \epsilon$. Max-L-UCC admits a factor 2 approximation algorithm for arbitrary sets L [18,20]. Goemans and Williamson [11] showed that Min-k-UCC and Min-$\{k\}$-UCC can be approximated within a factor of 4. Min-L-UCC is NP-hard and APX-hard if $\overline{L} \not\subseteq \{3\}$, i.e., for all but a finite number of sets L [15,19,20,25].

Min-$\mathcal{D}$-DCC, which is also known as the *assignment problem*, can be solved in polynomial time by a reduction to the minimum weight perfect matching problem in bipartite graphs [1]. The only other L for which Min-L-DCC can be solved in polynomial time is $L = \{2\}$. For all $L \subseteq \mathcal{D}$ with $L \neq \{2\}$ and $L \neq \mathcal{D}$, Min-L-DCC and Max-L-DCC are APX-hard and NP-hard [19,20].

There is a $4/3$ approximation for Max-3-DCC [5] as well as for Min-k-DCC for $k \geq 3$ with the restriction that the only edge weights allowed are one and two [3]. Max-L-DCC can be approximated within a factor of $8/3$ for all L [20].

If Min-L-UCC or Min-L-DCC is NP-hard, then the triangle inequality is necessary for efficient approximations of this problem; without the triangle inequality, the problems cannot be approximated at all.

1.3   New Results

While L-cycle covers of *maximum* weight allow for constant factor approximations, only little is known so far about the approximability of computing L-cycle covers of *minimum* weight. Our aim is to close this gap.

We present an approximation algorithm for Min-L-UCC that works for all sets $L \subseteq \mathcal{U}$ and achieves a constant approximation ratio (Section 2.1). Its running-time is $O(n^2 \log n)$. On the other hand, we show that the problem cannot be approximated within a factor of $2 - \varepsilon$ for general L (Section 2.2).

Our approximation algorithm for Min-L-DCC achieves a ratio of $O(n)$, where n is the number of vertices (Section 3.1). This is asymptotically optimal: There exist sets L for which no algorithm can approximate Min-L-DCC within a factor of $o(n)$ (Section 3.2). Furthermore, we argue that Min-L-DCC is harder to approximate than the other three variants even for more "natural" sets L than the sets used to show the inapproximability (Section 3.3).

Finally, to contrast our results for Min-L-UCC and Min-L-DCC, we show that Max-L-UCC and Max-L-DCC can be approximated arbitrarily well at least in principle (Section 4).

2 Approximability of Min-L-UCC

2.1 An Approximation Algorithm for Min-L-UCC

The aim of this section is to devise an approximation algorithm for Min-L-UCC that works for all sets $L \subseteq \mathcal{U}$. The catch is that for most L it is impossible to decide whether some cycle length is in L since there are uncountably many sets L: If, for instance, L is not a recursive set, then deciding whether a cycle cover is an L-cycle cover is impossible. One option would be to restrict ourselves to sets L such that the unary language $\{1^\lambda \mid \lambda \in L\}$ is in P. For such L, Min-L-UCC and Min-L-DCC are NP optimization problems. Another possibility for circumventing the problem would be to include the permitted cycle lengths in the input. While such restrictions are mandatory if we want to compute optimum solutions, they are not needed for our approximation algorithms.

A complete n-vertex graph contains an L-cycle cover as a spanning subgraph if and only if there exist (not necessarily distinct) lengths $\lambda_1, \ldots, \lambda_k \in L$ for some $k \in \mathbb{N}$ with $\sum_{i=1}^k \lambda_i = n$. We call such an n **L-admissible** and define $\langle L \rangle = \{n \mid n \text{ is } L\text{-admissible}\}$. Although L can be arbitrarily complicated, $\langle L \rangle$ always allows efficient membership testing: For all $L \subseteq \mathbb{N}$, there exists a finite set $L' \subseteq L$ with $\langle L' \rangle = \langle L \rangle$ [20].

Let g_L be the greatest common divisor of all numbers in L. Then $\langle L \rangle$ is a subset of the set of natural numbers divisible by g_L. There exists a minimum $p_L \in \mathbb{N}$ such that $\eta g_L \in \langle L \rangle$ for all $\eta > p_L$. The number p_L is the Frobenius number [22] of $\{\lambda \mid g_L \lambda \in L\}$, which is L scaled down by g_L.

In the following, it suffices to know such a finite set $L' \subseteq L$. The L-cycle covers computed by our algorithm will in fact be L'-cycle covers. In order to estimate the approximation ratio, this cycle cover will be compared to an optimal $\langle L' \rangle$-cycle cover. Since $L' \subseteq L \subseteq \langle L' \rangle$, every L'- or L-cycle cover is also a $\langle L' \rangle$-cycle cover. Thus, the weight of an optimal $\langle L' \rangle$-cycle cover provides a lower bound for the weight of both an optimal L'- and an optimal L-cycle cover. For simplicity, we do not mention L' in the following. Instead, we assume that already L is a finite set, and we compare the weight of the L-cycle cover computed to the weight of an optimal $\langle L \rangle$-cycle cover to bound the approximation ratio.

Goemans and Williamson have presented a technique for approximating constrained forest problems [11], which we will exploit. Let $G = (V, E)$ be an undirected graph, and let $w : E \to \mathbb{N}$ be non-negative edge weights. Let 2^V denote the power set of V. A function $f : 2^V \to \{0, 1\}$ is called a **proper function** if it satisfies

– $f(S) = f(V \setminus S)$ for all $S \subseteq V$ (symmetry),
– if A and B are disjoint, then $f(A) = f(B) = 0$ implies $f(A \cup B) = 0$ (disjointness), and
– $f(V) = 0$.

The aim is to find a set F of edges such that there is at least one edge connecting S to $V \setminus S$ for all $S \subseteq V$ with $f(S) = 1$. (The name "constrained forest problems" comes from the fact that it suffices to consider forests as solutions; cycles only increase the weight of a solution.) Goemans and Williamson have presented an approximation algorithm [11, Fig. 1] for constrained forest problems that are characterized by proper functions. We will refer to their algorithm as GOEWILL.

Theorem 1 (Goemans and Williamson [11]). *Let ℓ be the number of vertices v with $f(\{v\}) = 1$. Then GOEWILL is a $(2 - \frac{2}{\ell})$-approximation for the constrained forest problem defined by a proper function f.*

In particular, the function f_L given by

$$f_L(S) = \begin{cases} 1 \text{ if } |S| \not\equiv 0 \pmod{g_L} \text{ and} \\ 0 \text{ if } |S| \equiv 0 \pmod{g_L} \end{cases}$$

is proper if $|V| = n$ is divisible by g_L. (If n is not divisible by g_L, then G does not contain an L-cycle cover at all.) Given this function, a solution is a forest $H = (V, F)$ such that the size of every connected component of H is a multiple of g_L. In particular, if $g_L = 1$, then $f_L(S) = 0$ for all S, and an optimum solution are n isolated vertices.

If the size of all components of the solution obtained are in $\langle L \rangle$, we are done: By duplicating all edges, we obtain Eulerian components. Then we construct an $\langle L \rangle$-cycle cover by traversing the Eulerian components and taking shortcuts whenever we come to a vertex that we have already visited. Finally, we divide each λ-cycle into paths of lengths $\lambda_1 - 1, \dots, \lambda_k - 1$ for some k such that $\lambda_1 + \dots + \lambda_k = \lambda$ and $\lambda_i \in L$ for all i. By connecting the respective endpoints of each path, we obtain cycles of lengths $\lambda_1, \dots, \lambda_k$. We perform this for all components to get an L-cycle cover. A careful analysis shows that the ratio achieved is 4. The details for the special case of $L = \{k\}$ are spelled out by Goemans and Williamson [11].

However, this procedure does not work for general sets L since the sizes of some components may not be in $\langle L \rangle$. This can happen if $p_L > 0$ (for $L = \{k\}$, for which the algorithm works, we have $p_L = 0$).

In the following, our aim is to add edges to the forest $H = (V, E)$ output by GOEWILL such that the size of each component is in $\langle L \rangle$. This will lead to an approximation algorithm for Min-L-UCC with a ratio of $4 \cdot (p_L + 4)$, which is constant for each L. Let F^* denote the set of edges of a minimum-weight forest such that the size of each component is in $\langle L \rangle$. The set F^* is a solution to G, w, and f_L, but not necessarily an optimum solution.

By Theorem 1, we have $w(F) \leq 2 \cdot w(F^*)$ since $w(F^*)$ is at least the weight of an optimum solution to G, w, and f_L. Let $C = (V', F')$ be any connected component of F with $|V'| \notin \langle L \rangle$. The optimum solution F^* must contain an edge that connects V' to $V \setminus V'$. The weight of this edge is at least the weight of the minimum-weight edge connecting V' to $V \setminus V'$.

We will add edges until the sizes of all components is in $\langle L \rangle$. Our algorithm acts in phases as follows: Let $H = (V, F)$ be the graph at the beginning of the current phase, and let $C_1, \ldots, C_a$ be its connected components, where V_i is the vertex set of C_i. We will construct a new graph $\tilde{H} = (V, \tilde{F})$ with $\tilde{F} \supseteq F$. Let $C_1, \ldots, C_b$ be the connected components with $|V_i| \notin \langle L \rangle$. We call these components **illegal**. For $i \in \{1, \ldots, b\}$, let e_i be the cheapest edge connecting V_i to $V \setminus V_i$. (Note that $e_i = e_j$ for $i \neq j$ is allowed.) We add all these edges to F to obtain $\tilde{F} = F \cup \{e_1, \ldots, e_b\}$. Since e_i is the cheapest edge connecting V_i to $V \setminus V_i$, the graph $\tilde{H} = (V, \tilde{F})$ is a forest. (If some e_i are not uniquely determined, cycles may occur. We can avoid these cycles by discarding some of the e_i to break the cycles. For the sake of simplicity, we ignore this case in the following analysis.) If $\tilde{H}$ still contains illegal components, we set H to be $\tilde{H}$ and iterate the procedure.

Now we have $w(\tilde{F}) \leq w(F) + 2 \cdot w(F^*)$, i.e., in the overall weight increases by at most $2 \cdot w(F^*)$ in every phase. Furthermore, after at most $\lfloor p_L/2 \rfloor + 1$ phases, $\tilde{H}$ does not contain any illegal components.

Eventually, we obtain a forest that consists solely of components whose sizes are in $\langle L \rangle$. We call this forest $\tilde{H} = (V, \tilde{F})$. Then we proceed as already described above: We duplicate each edge, thus obtaining Eulerian components. After that, we take shortcuts to obtain an $\langle L \rangle$-cycle cover. Finally, we break edges and connect the endpoints of each path to obtain an L-cycle cover. The weight of this L-cycle cover is at most $4 \cdot w(\tilde{F})$.

Overall, we obtain APXUNDIR (Algorithm 1) and the following theorem.

Theorem 2. *For every $L \subseteq \mathcal{U}$, APXUNDIR is a factor $(4 \cdot (p_L + 4))$ approximation algorithm for Min-L-UCC. Its running-time is $O(n^2 \log n)$.*

Proof. Let C^* be a minimum-weight $\langle L \rangle$-cycle cover. The weight of $\tilde{F}$ is bounded from above by $w(\tilde{F}) \leq \left(\lfloor \frac{p_L}{2} \rfloor + 1 \right) \cdot 2 \cdot w(F^*) + 2 \cdot w(F^*) \leq (p_L + 4) \cdot w(C^*)$. Combining this with $w(C^{\mathrm{apx}}) \leq 4 \cdot w(\tilde{F})$ yields the approximation ratio.

Executing GOEWILL takes time $O(n^2 \log n)$. All other operations can be implemented to run in time $O(n^2)$. $\square$

We conclude the analysis of this algorithm by mentioning that the approximation ratio of the algorithm depends indeed linearly on p_L.

2.2 Unconditional Inapproximability of Min-L-UCC

In this section, we provide a lower bound for the approximability of Min-L-UCC as a counterpart to the approximation algorithm of the previous section. We show that the problem cannot be approximated within a factor of $2 - \varepsilon$. This inapproximability result is unconditional, i.e., it does not rely on complexity theoretic assumptions like $\mathsf{P} \neq \mathsf{NP}$.

The key to the inapproximability of Min-L-UCC are **immune sets** [21]: An infinite set $L \subseteq \mathbb{N}$ is called an immune set if L does not contain an infinite recursively enumerable subset. Our result limits the possibility of designing general

---

Algorithm 1. APXUNDIR

Input. undirected complete graph $G = (V, E)$, $|V| = n$; edge weights $w : E \to \mathbb{N}$
   satisfying the triangle inequality
Output. an L-cycle cover C^{apx} of G if n is L-admissible, $\perp$ otherwise
1: **if** $n \notin \langle L \rangle$ **then**
2:     return $\perp$
3: run GOEWILL using the function f_L described in the text to obtain $H = (V, F)$
4: **while** the size of some connected components of H is not in $\langle L \rangle$ **do**
5:     let $C_1, \ldots, C_a$ be the connected components of H, where V_i is the vertex set of
         C_i; let $C_1, \ldots, C_b$ be its illegal components
6:     let e_i be the lightest edge connecting V_i to $V \setminus V_i$
7:     add $e_1, \ldots, e_b$ to F
8:     **while** H contains cycles **do**
9:         remove one e_i to break a cycle
10: duplicate each edge to obtain a multi-graph consisting of Eulerian components
11: **for all** components of the multi-graph **do**
12:     walk along an Eulerian cycle
13:     take shortcuts to obtain a Hamiltonian cycle
14:     discard edges to obtain a collection of paths, the number of vertices of each of
         which is in L
15:     connect the two endpoints of every path in order to obtain cycles
16: the union of all cycles constructed forms C^{apx}; return C^{apx}

---

approximation algorithms for L-cycle covers. To obtain algorithms with a ratio
better than 2, we have to design algorithms tailored to specific sets L.

Finite variations of immune sets are again immune sets. Thus for every $k \in \mathbb{N}$,
there exist immune sets L containing no number smaller than k.

Theorem 3. *Let $\varepsilon > 0$ be arbitrarily small. Let $k > 2/\varepsilon$, and let $L \subseteq \{k, k + 1, \ldots\}$ be an immune set. Then Min-L-UCC cannot be approximated within a factor of $2 - \varepsilon$.*

Theorem 3 is tight since L-cycle covers can be approximated within a factor of 2
by L'-cycle covers for every set $L' \subseteq L$ with $\langle L' \rangle = \langle L \rangle$. For finite sets L', all L'-
cycle cover problems are NP optimization problems. This means that in principle
optimum solutions can be found, although this may take exponential time. The
following Theorem 4 holds in particular for finite sets L'. In order to actually
get an approximation algorithm for Min-L-UCC out of it, we have to solve Min-
L'-UCC finite L', which is NP-hard and APX-hard. But the proof of Theorem 4
shows also that any approximation algorithm for Min-L'-UCC for finite sets L'
that achieves an approximation ratio of r can be turned into an approximation
algorithm for the general problem with a ratio of $2r$. Let $\min_L(G, w)$ denote the
weight of a minimum-weight L-cycle cover of G with edge weights w.

Theorem 4. *Let $L \subseteq \mathcal{U}$ be a non-empty set, and let $L' \subseteq L$ with $\langle L' \rangle = \langle L \rangle$. Then $\min_{L'}(G, w) \leq 2 \cdot \min_L(G, w)$ for all undirected graphs G with edge weights w satisfying the triangle inequality.*

Algorithm 2. APXDIR

Input. directed complete graph $G = (V, E)$, $|V| = n$; edge weights $w : E \to \mathbb{N}$ satisfying the triangle inequality

Output. an L-cycle cover C^{apx} of G if n is L-admissible, $\perp$ otherwise

1: **if** $n \notin \langle L \rangle$ **then**
2:    return $\perp$
3: construct an undirected complete graph $G_U = (V, E_U)$ with edge weights $w_U(\{u, v\}) = w(u, v) + w(v, u)$
4: run APXUNDIR on G_U and w_U to obtain C_U^{apx}
5: **for all** cycles c_U of C_U^{apx} **do**
6:    c_U corresponds to a cycle of G that can be oriented in two ways; put the orientation c that yields less weight into C^{apx}
7: return C^{apx}

3 Approximability of Min-L-DCC

3.1 An Approximation Algorithm for Min-L-DCC

In this section, we present an approximation algorithm for Min-L-DCC. The algorithm exploits APXUNDIR to achieve an approximation ratio of $O(n)$. The hidden factor depends on p_L again. This result matches asymptotically the lower bound of Section 3.2 and shows that Min-L-DCC can be approximated at least to some extent. (For instance, without the triangle inequality, no polynomial-time algorithm achieves a ratio of $O(\exp(n))$ for an NP-hard L-cycle cover problem unless P = NP.)

In order to approximate Min-L-DCC, we reduce the problem to a variant of Min-L-UCC, where also 2-cycles are allowed: We obtain a 2-cycle of an undirected graph by taking an edge $\{u, v\}$ twice. Let $G = (V, E)$ be a directed complete graph with n vertices and edge weights $w : E \to \mathbb{N}$ that fulfill the triangle inequality. The corresponding undirected complete graph $G_U = (V, E_U)$ has weights $w_U : E_U \to \mathbb{N}$ with $w_U(\{u, v\}) = w(u, v) + w(v, u)$.

Let C be any cycle cover of G. The corresponding cycle cover C_U of G_U is given by $C_U = \{\{u, v\} \mid (u, v) \in C\}$. Note that we consider C_U as a multiset: If both (u, v) and (v, u) are in C, i.e., u and v form a 2-cycle, then $\{u, v\}$ occurs twice in C_U. We can bound the weight of C_U in terms of the weight of C: For every cycle cover C of G, we have $w_U(C_U) \leq n \cdot w(C)$.

Our algorithm computes an L'-cycle cover for some finite $L' \subseteq L$ with $\langle L' \rangle = \langle L \rangle$. As in Section 2.1, the weight of the cycle cover computed is compared to an optimum $\langle L \rangle$-cycle cover rather than an optimum L-cycle cover. Thus, we can again assume that already L is a finite set.

The algorithm APXUNDIR, which was designed for undirected graphs, remains to be an $O(1)$ approximation if we allow $2 \in L$. The numbers p_L and g_L are defined in the same way as in Section 2.1.

Let C_U^{apx} be the L-cycle cover output by APXUNDIR on G_U. We transfer C_U^{apx} into an L-cycle cover C^{apx} of G. For every cycle c_U of C_U^{apx}, we can orient the corresponding directed cycle c in two directions. We take the orientation that yields less weight, thus $w(C^{\text{apx}}) \leq w_U(C_U^{\text{apx}})/2$. Overall, we obtain APXDIR (Algorithm 2), which achieves an approximation ratio of $O(n)$ for every L.

Theorem 5. *For every $L \subseteq \mathcal{D}$, APXDIR is a factor $(2n \cdot (p_L + 4))$ approximation algorithm for Min-L-DCC. Its running-time is $O(n^2 \log n)$.*

Proof. Theorem 2 yields $w_U(C_U^{\text{apx}}) \leq 4 \cdot (p_L + 4) \cdot w_U(C_U^*)$, where C_U^* is an optimal $\langle L \rangle$-cycle cover of G_U. Now consider an optimum $\langle L \rangle$-cycle cover C^* of G. Overall, we obtain $w(C^{\text{apx}}) \leq \frac{1}{2} \cdot w_U(C_U^{\text{apx}}) \leq 2 \cdot (p_L + 4) \cdot w_U(C_U^*) \leq 2n \cdot (p_L + 4) \cdot w(C^*)$.

The running-time is dominated by the time needed to execute GOEWILL in APXUNDIR, which is $O(n^2 \log n)$. $\square$

3.2 Unconditional Inapproximability of Min-L-DCC

For undirected graphs, both Max-L-UCC and Min-L-UCC can be approximated efficiently to within constant factors. Surprisingly, in case of directed graphs, this holds only for the maximization variant of the directed L-cycle cover problem. Min-L-DCC cannot be approximated within a factor of $o(n)$ for certain sets L, where n is the number of vertices of the input graph. In particular, APXDIR achieves asymptotically optimal approximation ratios for Min-L-DCC.

Similar to the case of Min-L-UCC, this result shows that to find approximation algorithms, specific properties of the sets L have to be exploited. Moreover, as we will discuss in Section 3.3, Min-L-DCC seems to be much harder a problem than the other three variants, even for more practical sets L.

Theorem 6. *Let $L \subseteq \mathcal{U}$ be an immune set. Then no approximation algorithm for Min-L-DCC achieves an approximation ratio of $o(n)$, where n is the number of vertices of the input graph.*

Min-L'-DCC for a finite set L' is an NP optimization problem. Thus, it can be solved, although this may take exponential time. Therefore, the following result shows that Min-L-DCC can be approximated for all L within a ratio of n/s for arbitrarily large constants s, although this may also take exponential time. In this sense, Theorem 6 is tight.

Theorem 7. *For every L and every $s > 1$, there exists a finite set $L' \subseteq L$ with $\langle L' \rangle = \langle L \rangle$ such that $\min_{L'}(G, w) \leq \frac{n}{s} \cdot \min_L(G, w)$ for all directed graphs G with edge weights w satisfying the triangle inequality.*

3.3 Remarks on the Approximability of Min-L-DCC

It might seem surprising that Min-L-DCC is much harder to approximate than Min-L-UCC or the maximization problems Max-L-UCC and Max-L-DCC. In the following, we give some reasons why Min-L-DCC is more difficult than the other three L-cycle cover problems. In particular, even for "easy" sets L, for which membership testing can be done in polynomial time, it seems that Min-L-DCC is much harder to approximate than the other three variants.

Why is minimization harder than maximization? To get a good approximation ratio in the case of maximization problems, it suffices to detect a few "good", i. e., heavy edges. If we have a decent fraction of the heaviest edges, their total weight is already within a constant factor of the weight of an optimal L-cycle cover. In order to form an L-cycle cover, we have to connect the heavy edges using other edges. These other edges might be of little weight, but they do not decrease the weight that we have already obtained from the heavy edges. For approximating cycle covers of minimum weight, it does not suffice to detect a couple of "good", i. e., light edges: Once we have selected a couple of good edges, we might have to connect them with heavy-weight edges. These heavy-weight edges can worsen the approximation ratio dramatically.

Why is Min-L-DCC harder than Min-L-UCC? If we have a cycle in an undirected graph whose length is in $\langle L \rangle$ but not in L (or not in L' but we do not know if it is in L), then we can decompose it into smaller cycles all lengths of which are in L. This can be done such that the weight at most doubles. However, by decomposing a long cycle of a directed graph into smaller ones, the weight can increase tremendously.

Finally, a question that arises naturally is if we can do better if all allowed cycle lengths are known a priori. This can be achieved by restricting ourselves to sets L that allow efficient membership testing. Another option is to include the allowed cycle lengths in the input, i. e., in addition to an n-vertex graph and edge weights, we are given a subset of $\{2, 3, \ldots, n\}$ of allowed cycle lengths. A constant factor approximation for either variant would, however, yield an approximation algorithm for the asymmetric traveling salesman problem (ATSP) with dramatically improved approximation ratio.

4   Properties of Maximum-Weight Cycle Covers

To contrast our results for Min-L-UCC and Min-L-DCC, we show that their maximization counterparts Max-L-UCC and Max-L-DCC can, at least in principle, be approximated arbitrarily well; their inapproximability is solely due to their APX-hardness and not to the difficulties arising from undecidable sets L.

Let $\max_L(G, w)$ be the weight of a maximum-weight L-cycle cover of G with edge weights w. The edge weights w do not have to fulfill the triangle inequality. We will show that $\max_L(G, w)$ can be approximated arbitrarily well by $\max_{L'}(G, w)$ for finite sets $L' \subseteq L$ with $\langle L' \rangle = \langle L \rangle$. Thus, any approximation algorithm for Max-L'-UCC or Max-L'-DCC for finite sets L' immediately yields an approximation algorithm for general sets L with an only negligibly worse approximation ratio.

The following theorem for directed cycle covers contains the case of undirected graphs as a special case.

Theorem 8. *Let $L \subseteq \mathcal{D}$ be any non-empty set, and let $\varepsilon > 0$. Then there exists a finite subset $L' \subseteq L$ with $\langle L' \rangle = \langle L \rangle$ such that $\max_{L'}(G, w) \geq (1 - \varepsilon) \cdot \max_L(G, w)$ for all graphs G with edge weights w.*

5 Concluding Remarks

First of all, we would like to know if there is a general upper bound for the approximability of Min-L-UCC: Does there exist an r (independent of L) such that Min-L-UCC can be approximated within a factor of r? We conjecture that such an algorithm exists. If such an algorithm works also for the slightly more general problem Min-L-UCC with $2 \in L$ (see Section 3.1), then we would obtain a factor $rn/2$ approximation for Min-L-DCC as well.

While the problem of computing L-cycle cover of minimum weight can be approximated efficiently in the case of undirected graphs, the directed variant seems to be much harder. We are interested in developing approximation algorithms for Min-L-DCC for particular sets L or for certain classes of sets L. For instance, how well can Min-L-DCC be approximated if L is a finite set? Are there non-constant lower bounds for the approximability of Min-L-DCC, for instance bounds depending on $\max(L)$?

References

1. Ahuja, R.K., Magnanti, T.L., Orlin, J.B.: Network Flows: Theory, Algorithms, and Applications. Prentice-Hall, Englewood Cliffs (1993)
2. Bläser, M.: A 3/4-approximation algorithm for maximum ATSP with weights zero and one. In: Jansen, K., Khanna, S., Rolim, J.D.P., Ron, D. (eds.) RANDOM 2004 and APPROX 2004. LNCS, vol. 3122, pp. 61–71. Springer, Heidelberg (2004)
3. Bläser, M., Manthey, B.: Approximating maximum weight cycle covers in directed graphs with weights zero and one. Algorithmica 42(2), 121–139 (2005)
4. Bläser, M., Manthey, B., Sgall, J.: An improved approximation algorithm for the asymmetric TSP with strengthened triangle inequality. Journal of Discrete Algorithms 4(4), 623–632 (2006)
5. Bläser, M., Shankar Ram, L., Sviridenko, M.I.: Improved approximation algorithms for metric maximum ATSP and maximum 3-cycle cover problems. In: Dehne, F., López-Ortiz, A., Sack, J.-R. (eds.) WADS 2005. LNCS, vol. 3608, pp. 350–359. Springer, Heidelberg (2005)
6. Bläser, M., Siebert, B.: Computing cycle covers without short cycles. In: Meyer auf der Heide, F. (ed.) ESA 2001. LNCS, vol. 2161, pp. 368–379. Springer, Heidelberg (2001)
7. Blum, A.L., Jiang, T., Li, M., Tromp, J., Yannakakis, M.: Linear approximation of shortest superstrings. Journal of the ACM 41(4), 630–647 (1994)
8. Böckenhauer, H.-J., Hromkovič, J., Klasing, R., Seibert, S., Unger, W.: Approximation algorithms for the TSP with sharpened triangle inequality. Information Processing Letters 75(3), 133–138 (2000)
9. Sunil Chandran, L., Shankar Ram, L.: On the relationship between ATSP and the cycle cover problem. Theoretical Computer Science 370(1-3), 218–228 (2007)

10. Chen, Z.-Z., Okamoto, Y., Wang, L.: Improved deterministic approximation algorithms for Max TSP. Information Processing Letters 95(2), 333–342 (2005)
11. Goemans, M.X., Williamson, D.P.: A general approximation technique for constrained forest problems. SIAM Journal on Computing 24(2), 296–317 (1995)
12. Hartvigsen, D.: An Extension of Matching Theory. PhD thesis, Department of Mathematics, Carnegie Mellon University, Pittsburgh, Pennsylvania, USA (September 1984)
13. Hassin, R., Rubinstein, S.: On the complexity of the k-customer vehicle routing problem. Operations Research Letters 33(1), 71–76 (2005)
14. Hassin, R., Rubinstein, S.: Erratum to "An approximation algorithm for maximum triangle packing" [Discrete Applied Mathematics 154, 971–979 (2006)]. Discrete Applied Mathematics 154(18), 2620 (2006)
15. Hell, P., Kirkpatrick, D.G., Kratochvíl, J., Kríz, I.: On restricted two-factors. SIAM Journal on Discrete Mathematics 1(4), 472–484 (1988)
16. Kaplan, H., Lewenstein, M., Shafrir, N., Sviridenko, M.I.: Approximation algorithms for asymmetric TSP by decomposing directed regular multigraphs. Journal of the ACM 52(4), 602–626 (2005)
17. Lovász, L., Plummer, M.D.: Matching Theory. North-Holland Mathematics Studies, vol. 121. Elsevier, Amsterdam (1986)
18. Manthey, B.: Approximation algorithms for restricted cycle covers based on cycle decompositions. In: Fomin, F.V. (ed.) WG 2006. LNCS, vol. 4271, pp. 336–347. Springer, Heidelberg (2006)
19. Manthey, B.: On approximating restricted cycle covers. In: Erlebach, T., Persinao, G. (eds.) WAOA 2005. LNCS, vol. 3879, pp. 282–295. Springer, Heidelberg (2006)
20. Manthey, B.: On approximating restricted cycle covers. Report 07-011, Electronic Colloquium on Computational Complexity (ECCC) (2007)
21. Odifreddi, P.: Classical Recursion Theory. Studies in Logic and The Foundations of Mathematics, vol. 125. Elsevier, Amsterdam (1989)
22. Ramírez Alfonsín, J.L.: The Diophantine Frobenius Problem. Oxford Lecture Series in Mathematics and Its Applications, vol. 30. Oxford University Press, Oxford (2005)
23. Schrijver, A.: Combinatorial Optimization: Polyhedra and Efficiency. Algorithms and Combinatorics, vol. 24. Springer, Heidelberg (2003)
24. Sweedyk, Z.: A $2\frac{1}{2}$-approximation algorithm for shortest superstring. SIAM Journal on Computing 29(3), 954–986 (1999)
25. Vornberger, O.: Easy and hard cycle covers. Technical report, Universität/ Gesamthochschule Paderborn (1980)

On the Number of α-Orientations

Stefan Felsner and Florian Zickfeld

Technische Universität Berlin, Fachbereich Mathematik
Straße des 17. Juni 136, 10623 Berlin, Germany
{felsner,zickfeld}@math.tu-berlin.de

Abstract. We deal with the asymptotic enumeration of combinatorial structures on planar maps. Prominent instances of such problems are the enumeration of spanning trees, bipartite perfect matchings, and ice models. The notion of an α-orientation unifies many different combinatorial structures, including the afore mentioned. We ask for the number of α-orientations and also for special instances thereof, such as Schnyder woods and bipolar orientations. The main focus of this paper are bounds for the maximum number of such structures that a planar map with n vertices can have. We give examples of triangulations with 2.37^n Schnyder woods, 3-connected planar maps with 3.209^n Schnyder woods and inner triangulations with 2.91^n bipolar orientations. These lower bounds are accompanied by upper bounds of 3.56^n, 8^n, and 3.97^n, respectively. We also show that for any planar map M and any α the number of α-orientations is bounded from above by 3.73^n and we present a family of maps which have at least 2.598^n α-orientations for n big enough.

1 Introduction

A *planar map* is a planar graph together with a plane drawing. Many different structures on planar maps have attracted the attention of researchers. Among them are spanning trees, bipartite perfect matchings (or more generally bipartite f-factors), Eulerian orientations, Schnyder woods, bipolar orientations and 2-orientations of quadrangulations. The concept of α-orientations is a quite general one. Remarkably, all the above structures can be encoded by certain α-orientations. Let a planar map M with vertex set V and a function $\alpha : V \to \mathbb{N}$ be given. An orientation X of the edges of M is an α-orientation if every vertex v has out-degree $\alpha(v)$.

For some of the structures mentioned above it is not obvious how to encode them as α-orientations. For Schnyder woods on triangulations the encoding by 3-orientations goes back to de Fraysseix and de Mendez [6]. For bipolar orientations an encoding was proposed by Woods and independently by Tamassia and Tollis [22]. Bipolar orientations of M are one of the structures which cannot be encoded as α-orientations on M, an auxiliary map M' (the angle graph of M) has to be used instead. For Schnyder woods on 3-connected planar maps as well as bipartite f-factors and spanning trees Felsner [10] describes encodings as α-orientations. He also proves that the set of α-orientations of a planar map

A. Brandstädt, D. Kratsch, and H. Müller (Eds.): WG 2007, LNCS 4769, pp. 190–201, 2007.
© Springer-Verlag Berlin Heidelberg 2007

M is a distributive lattice. This structure on the set of α-orientations found applications in drawing algorithms for example in [3], and for enumeration and random sampling of graphs in [12].

Given the existence of a combinatorial structure on a class $\mathcal{M}_n$ of planar maps with n vertices, one of the questions of interest is how many such structures there are for a given map $M \in \mathcal{M}_n$. Especially, one is interested in the minimum and maximum that this number attains on the maps from $\mathcal{M}_n$. This question has been treated quite successfully for spanning trees and bipartite perfect matchings. For spanning trees the Kirchhoff Matrix Tree Theorem comes into the game and allows to bound the maximum number of spanning trees of a planar graph with n vertices between 5.02^n and 5.34^n, see [19]. Pfaffian orientations can be used to efficiently calculate the number of bipartite perfect matchings in the planar case, see for example [16]. Kasteleyn has shown, that the $k \times \ell$ square grid has about $e^{0.29 \cdot k\ell} \approx 1.34^{k\ell}$ perfect matchings. The number of Eulerian orientations is studied in statistical physics under the name of ice models. In particular Lieb [15] has shown that the square grid on the torus has $(8\sqrt{3}/9)^{k\ell} \approx 1.53^{k\ell}$ Eulerian orientations and Baxter [1] has worked out the asymptotics for the triangular grid on the torus as $(3\sqrt{3}/2)^{k\ell} \approx 2.598^{k\ell}$.

In many cases it is relatively easy to see which maps in a class $\mathcal{M}_n$ carry a unique object of a certain type, while the question about the maximum number is rather intricate. Therefore, we focus on finding the asymptotics or lower and upper bounds for the maximum number of α-orientations that a map from $\mathcal{M}_n$ can carry. The next table gives an overview of the results of this paper for different instances of $\mathcal{M}_n$ and α. The entry c in the "Upper Bound" column is to be read as $O(c^n)$, in the "Lower Bound" column as $\Omega(c^n)$ and for the "$\approx c$" entries the asymptotics are known.

The paper is organized as follows. In Section 2 we treat the most general case, where $\mathcal{M}_n$ is the class of all planar maps with n vertices and α can be any integer valued function. Apart from giving lower and upper bounds we briefly discuss the complexity of counting α-orientations and a reduction to counting perfect

Graph class and orientation type	Lower bound	Upper bound
α-orientations on planar maps	2.59	3.73
Eulerian orientations on planar maps	2.59	3.73
Schnyder woods on triangulations	2.37	3.56
Schnyder woods on the square grid	≈ 3.209	
Schnyder woods on 3-connected planar maps	3.209	8
2-orientations on quadrangulations	1.53	1.91
bipolar orientations on stacked triangulations	≈ 2	
bipolar orientations on outerplanar maps	≈ 1.618	
bipolar orientations on the square grid	2.18	2.62
bipolar orientations on planar maps	2.91	3.97

matchings of bipartite graphs. In Section 3 we consider Schnyder woods on plane triangulations and the more general case of Schnyder woods on 3-connected planar maps. We split the treatment of Schnyder woods because the more direct encoding of Schnyder woods on triangulations as α-orientations yields stronger bounds. We also discuss the asymptotic number of Schnyder woods on the square grid. In Section 4, we study bipolar orientations on the square grid and outerplanar maps as well as general planar maps. The upper bound for planar maps relies on a new encoding of bipolar orientations of inner triangulations. We conclude with some open problems.

In Sections 2 and 3 we include some proofs since many of the other proofs use these results or similar techniques. Proofs omitted due to space constraints are available in a full version of this paper on the authors' homepage.

2   Counting α-Orientations

A *planar map* M is a simple planar graph G together with a fixed crossing-free embedding of G in the Euclidean plane. In particular M has a designated outer (unbounded) face. We denote the sets of vertices, edges and faces of a given planar map by V, E, and $\mathcal{F}$, and their respective cardinalities by n, m and f. The degree of a vertex v will be denoted by $d(v)$.

Let M be a planar map and $\alpha : V \to \mathbb{N}$. An edge orientation X of M is an *α-orientation* if every $v \in V$ has exactly $\alpha(v)$ edges directed away from it in X.

Let X be an α-orientation of G and let C be a directed cycle in X. Define X^C as the orientation obtained from X by reversing all edges of C. Since the reversal of a directed cycle does not affect out-degrees the orientation X^C is also an α-orientation of M. The plane embedding of M allows us to classify a directed simple cycle as clockwise if the interior is to the right of C or as counterclockwise otherwise. If C is a ccw-cycle of X then we say that X^C is *left of* X. The set of α-orientations of M endowed with the transitive closure of the 'left of' relation is a distributive lattice [10].

The following observation is easy but very useful. Let M and $\alpha : V \to \mathbb{N}$ be given, $W \subset V$ and E_W the edges of M with one endpoint in W and one in $V \backslash W$. Suppose all edges of E_W are directed away from W in some α-orientation X_0 of M. Then, the demand of W for $\sum_{w \in W} \alpha(w)$ outgoing edges forces all edges in E_W to be directed away from W in every α-orientation of M. Edges with the same direction in every α-orientation are called *rigid*.

We denote the number of α-orientations of M by $r_\alpha(M)$. Most of this paper is concerned with lower and upper bounds for $\max_{M \in \mathcal{M}} r_{\alpha_M}(M)$ for some class $\mathcal{M}$ of planar maps. A planar map has at most 2^m α-orientations as every edge can be directed in at most two ways.

Lemma 1. *Let M be a planar map and $A \subset E$ be a cycle free subset of edges of M. Then, there are at most $2^{m-|A|}$ α-orientations of M. This holds for every function $\alpha : V \to \mathbb{N}$. Furthermore, M has less than 4^n α-orientations.*

Proof. Let X be an arbitrary but fixed orientation out of the $2^{m-|A|}$ orientations of the edges of $E \setminus A$. It suffices to show that X can be extended to an

α-orientation of M in at most one way. We proceed by induction over $|A|$. The base case $|A| = 0$ is trivial. If $|A| > 0$, then, as A is cycle free, there is a vertex v which is incident to exactly one edge e of A. If v has out-degree $\alpha(v)$ respectively $\alpha(v) - 1$ in X, then e must be directed towards respectively away from v. In either case the direction of e is determined by X, and by induction there is at most one way to extend the resulting orientation of $E \setminus (A - e)$ to an α-orientation of M. If v does not have $\alpha(v)$ or $\alpha(v) - 1$ outgoing edges, then there is no extension of X to an α-orientation of M. The bound $2^{m-n+1} < 4^n$ follows by choosing A to be a spanning forest and applying Euler's formula. $\square$

To improve on this bound we state Lemma 2, but the proof is omitted here.

Lemma 2. *Let M be a planar map with n vertices that has an independent set of n_2 vertices, which have degree 2 in M. Then, M has at most $(3n-6)-(n_2-1)$ edges.*

Proposition 1. *Let M be a planar map, $\alpha : V \rightarrow \mathbb{N}$, and $I = I_1 \cup I_2$ an independent set of M, where I_2 is the set of vertices in I, which have degree 2 in M. Then, M has at most*

$$2^{2n-4-|I_2|} \cdot \prod_{v \in I_1} \left(\frac{1}{2^{d(v)-1}} \binom{d(v)}{\alpha(v)} \right) \tag{1}$$

α-orientations.

Proof. We may assume that M is connected. Let M_i, for $i = 1, \ldots c$, be the components of $M - I$. We claim that M has at most $(3n-6)-(c-1)-(|I_2|-1)$ edges. Note, that every component C of $M - I$ must be connected to some other component C' via a vertex $v \in I$ such that the edges vw and vw' with $w \in C$ and $w' \in C'$ form an angle at v. As w and w' are in different connected components the edge ww' is not in M and we can add it without destroying planarity. We can add at least $c - 1$ edges not incident to I in this fashion. Thus, by Lemma 2 we have that $m + (c - 1) \leq 3n - 6 - (I_2 - 1)$.

Let S' be a spanning forest of $M - I$, and let S be obtained from S' by adding one edge incident to every $v \in I$, S has $n - c$ edges. By Lemma 1 M has at most $2^{m-|S|}$ α-orientations and we note that $m - |S| \leq (3n - 6) - (c - 1) - (|I_2| - 1) - (n - c) = 2n - 4 - |I_2|$.

For every vertex $v \in I_1$ there are $2^{d(v)-1}$ possible orientations of the edges of $M - S$ at v. Only the orientations with $\alpha(v)$ or $\alpha(v) - 1$ outgoing edges at v can potentially be completed to an α-orientation of M. Since I_1 is an independent set it follows that M has at most

$$2^{m-|S|} \cdot \prod_{v \in I_1} \frac{\binom{d(v)-1}{\alpha(v)} + \binom{d(v)-1}{\alpha(v)-1}}{2^{d(v)-1}} \leq 2^{2n-4-|I_2|} \cdot \prod_{v \in I_1} \frac{\binom{d(v)}{\alpha(v)}}{2^{d(v)-1}} \tag{2}$$

α-orientations. $\square$

Proposition 2. *For every α and M we have that $r_\alpha(M) \leq 3.73^n$. There are infinitely many graphs with more than 2.59^n Eulerian orientations.*

Proof. By the Four Color Theorem every planar map allows for an independent set I of size $n/4$. Edges incident to degree 1 vertices are always rigid, so we assume $d(v) \geq 3$ for $v \in I_1$. We use Proposition 1 and upper bound the right hand side of (2) to conclude that $r_\alpha(M) \leq 2^{2n-4-|I_2|} \cdot \left(\frac{3}{4}\right)^{|I_1|} \leq 3.73^n$. The lower bound uses a planarization of the triangular grid on the torus, which was mentioned in the introduction and is shown on the right in Figure 2.   □

Given a planar map M and some $\alpha : V \to \mathbb{N}$, what is the complexity of computing the number of α-orientations of M? In some instances this number can be computed efficiently, e.g. for perfect matchings and spanning trees of general planar maps, as mentioned in the introduction.

Recently, Creed [4] has shown that counting Eulerian orientations is #P-complete even for planar maps. In the full paper we use Creed's method and a reduction from perfect matchings of k-regular bipartite graphs, see [5], to show the following. Counting α-orientations is #P-complete for 4-regular planar maps with $\alpha : V \to \{1, 2, 3\}$ as well as for planar maps with vertex degrees in $\{3, 4, 5\}$ and $\alpha(v) = 2$ for all $v \in V$.

In general, computing the number of α-orientations can be reduced to counting f-factors in bipartite planar graphs and thus to counting perfect matchings in bipartite graphs [23]. This reduction is useful because bipartite perfect matchings have been the subject of extensive research (for example [16,13,18]). We mention two useful facts that follow from this relation. First, it can be tested in polynomial time if the α-orientations of a given map can be counted using Pfaffian orientations. If this is not the case, there is a fully polynomial randomized approximation scheme for approximating this number.

3   Counting Schnyder Woods

Schnyder woods for triangulations have been introduced as a tool for graph drawing and graph dimension theory in [20,21] and for 3-connected planar maps in [8]. Here we review the definitions and encodings as α-orientations, for a comprehensive introduction, see e.g. [9].

Let M be a planar map and a_1, a_2, a_3 be three vertices occurring in clockwise order on the outer face of M. A suspension M^σ of M is obtained by attaching a half-edge that reaches into the outer face to each of these *special vertices*. A *Schnyder wood* rooted at a_1, a_2, a_3 is an orientation and coloring of the edges of M^σ with the colors $1, 2, 3$ satisfying the following rules, see the left part of Figure **??**. Every edge e is oriented in one direction or in two opposite directions. The directions of edges are colored such that if e is bidirected the two directions have distinct colors. The half-edge at a_i is directed outwards and colored i. Every vertex v has out-degree one in each color. The edges e_1, e_2, e_3 leaving v in colors $1, 2, 3$ occur in clockwise order. Each edge entering v in color i enters v in the clockwise sector from e_{i+1} to e_{i-1}. There is no interior face the boundary of which is a monochromatic directed cycle. Note that for triangulations only the three outer edges are bidirected. The next theorem is from [6].

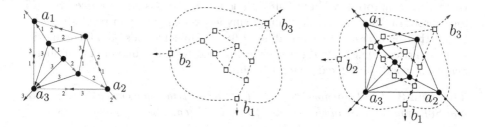

Fig. 1. Schnyder wood on a map M^σ, the suspension dual M^{σ^*}, the completion $\widetilde{M}$

Theorem 1. *Let T be a plane triangulation, let $\alpha_T(v) := 3$ if v is an internal vertex and $\alpha_T(v) := 0$ if v lies on the outer face. Then, there is a bijection between the Schnyder woods of T and the α_T-orientations of the interior edges of T.*

In the sequel we refer to an α_T-orientation simply as a 3-orientation. We now explain how Schnyder woods of non-triangular maps are encoded as α-orientations.

Let M^σ be a 3-connected planar map plus three rays emanating from three outer vertices a_1, a_2, a_3 into the unbounded face. The *suspension dual* M^{σ^*} of M^σ is obtained from the dual M^* of M as follows, see also Figure **??**. Replace the vertex v_∞^*, which represents the unbounded face of M in M^*, by a triangle on three new vertices b_1, b_2, b_3. Let P_i be the path from a_{i-1} to a_{i+1} on the outer face of M which avoids a_i. In M^{σ^*} the edges dual to those on P_i are incident to b_i instead of v_∞^*. Adding a ray to each of the b_i yields M^{σ^*}. The completion $\widetilde{M}$ of M^σ and M^{σ^*} is obtained by superimposing the two graphs such that exactly the primal dual pairs of edges cross. In the completion $\widetilde{M}$ the common subdivision of each crossing pair of edges is replaced by a new edge-vertex. Note that the rays emanating from the three special vertices of M^σ cross the three edges of the triangle induced by b_1, b_2, b_3 and thus produce edge vertices. The six rays emanating into the unbounded face of the completion end at a new vertex v_∞ placed in this unbounded face. Let a function α_M be defined on M^{σ^*} by $\alpha_M(v) = 3$ for every primal or dual vertex v, $\alpha_M(v_e) = 1$ for every edge vertex v_e, and $\alpha_M(v_\infty) = 0$ for the special closure vertex v_∞. For a proof of the next theorem see [10].

Theorem 2. *The Schnyder woods of a suspended planar map M^σ are in bijection with the α_M-orientations of $\widetilde{M}$.*

The full paper contains a constructive characterization of all maps with a unique Schnyder wood, thus generalizing the known characterization for triangulations.

Bonichon [2] found a bijection between Schnyder woods on triangulations with n vertices and pairs of non-crossing Dyck-paths, which implies that there are $C_{n+2}C_n - C_{n+1}^2$ Schnyder woods on triangulations with n vertices, where C_n denotes the nth Catalan number. Thus, there are asymptotically about 16^n Schnyder woods on triangulations with n vertices. Tutte's classic result says, that there are asymptotically about 9.48^n plane triangulations on n vertices. See [17] for a proof of Tutte's formula using Schnyder woods. The two results together imply, that on average there are about 1.68^n Schnyder woods on a triangulation with n vertices. The next theorem is concerned with the maximum number of Schnyder woods on a fixed triangulation.

Theorem 3. *Let $\mathcal{T}_n$ denote the set of all plane triangulations with n vertices and $\mathcal{S}(T)$ the set of Schnyder woods of $T \in \mathcal{T}_n$. Then, $2.37^n \le \max_{T \in \mathcal{T}_n} |\mathcal{S}(T)| \le 3.56^n$.*

Proof. The upper bound follows from Proposition 1 by using that $\binom{d(v)}{3} \cdot 2^{1-d(v)} \le \frac{5}{8}$ for $d(v) \ge 3$. For the proof of the lower bound we introduce the *triangular grid*, which is derived from the *square grid* $G_{k,\ell}$. The graph $G_{k,\ell}$ has vertex set $\{(i,j) \mid 1 \le i \le k,\ 1 \le j \le \ell\}$ and all edges of the form $\{(i,j),(i,j+1)\}$ and $\{(i,j),(i+1,j)\}$. The triangular grid $T_{k,\ell}$ is obtained by adding the edges of the form $\{(i,j),(i-1,j+1)\}$. The grid triangulation $T^*_{k,\ell}$ is derived from $T_{k,\ell}$ by augmenting it with a triangle as shown in Figure 2.

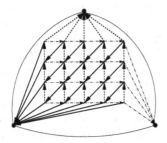 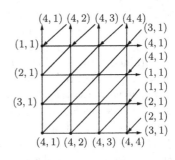

Fig. 2. The graphs $T^*_{4,5}$ with a canonical Schnyder wood and $T_{4,4}$ with the additional edges simulating Baxter's boundary conditions

Intuitively, $T^*_{k,\ell}$ promises to be a good candidate for a lower bound because the canonical orientation shown in Figure 2 on the left has many directed cycles. We formalize this now by showing that $T_{k,k}$ has at least $2^{5/4(k-1)^2}$ Schnyder woods, which yields the claimed bound for k big enough. Instead of working with the 3-orientations of $T^*_{k,\ell}$ we use α^*-orientations of $T_{k,\ell}$ where $\alpha^*(i,j) = 3$ if $2 \le i \le k-1$ and $2 \le i \le \ell - 1$ $\alpha^*(i,j) = 1$ if $(i,j) \in \{(1,1),(1,\ell),(k,\ell)\}$ and $\alpha^*(i,j) = 2$ otherwise. For simplicity, we refer to α^*-orientations of $T_{k,\ell}$ as 3-orientations.

The boundaries of the triangles of $T_{k,k}$ can be partitioned into two classes $\mathcal{C}$ and $\mathcal{C}'$ of directed cycles of cardinality $(k-1)^2$ each. No two cycles from the same class share an edge and $C \in \mathcal{C}'$ shares an edge with three cycles from $\mathcal{C}$ if it does not include a boundary edge.

For any subset D of $\mathcal{C}$ reversing all the cycles in D yields a 3-orientation of $T_{k,k}$, and we can encode this orientation as a 0-1-sequence of length $(k-1)^2$. After performing the flips of a given 0-1-sequence a a cycle $C' \in \mathcal{C}'$ is directed if and only if either all or none of the three cycles sharing an edge with C' have been reversed. Thus the number of different cycle flip sequences on $\mathcal{C} \cup C'$ is bounded from below by $\sum_{a \in \{0,1\}^{(k-1)^2}} 2^{\sum_{C' \in \mathcal{C}'} X_{C'}(a)}$. Here $X_{C'}(a)$ is an indicator function, which takes value 1 if C' is directed after performing the flips of a and 0 otherwise.

We now assume that every $a \in \{0,1\}^{(k-1)^2}$ is chosen uniformly at random. The expected value of the above function is then

$$\mathbb{E}[2^{\sum X_{C'}}] = \frac{1}{2^{(k-1)^2}} \sum_{a \in \{0,1\}^{(k-1)^2}} 2^{\sum_{C' \in \mathcal{C}'} X_{C'}(a)}.$$

Jensen's inequality $\mathbb{E}[\varphi(X)] \geq \varphi(\mathbb{E}[X])$ holds for a random variable X and a convex function φ. We derive that $\mathbb{E}[2^{\sum X_{C'}}] \geq 2^{\mathbb{E}[\sum X_{C'}]} = 2^{\sum \mathbb{P}[C' \; flippable]}$. The probability that C' is flippable is at least $1/4$. For C' which does not include a boundary edge the probability depends only on the three cycles from $\mathcal{C}$ that share an edge with C' and two out of the eight flip vectors for these three cycles make C' flippable. A similar reasoning applies for C' including a boundary edge. Altogether this yields that $\sum_{a \in \{0,1\}^{(k-1)^2}} 2^{\sum_{C' \in \mathcal{C}'} X_{C'}(a)} \geq 2^{(k-1)^2} \cdot 2^{1/4 \cdot (k-1)^2}$. Different cycle flip sequences yield different Schnyder woods. The orientation of an edge is determined by whether the two cycles on which it lies are both flipped or not. We can tell a flip sequence apart from its complement by looking at the boundary edges. $\square$

Remark. We relate Baxter's result [1] for Eulerian orientations of the triangular grid on the torus to Schnyder woods on $T^*_{k,\ell}$. Every 3-orientation of $T_{k,\ell}$ plus the wrap-around edges oriented as shown in Figure 2 on the right yields a Eulerian orientation on the torus. We deduce that $T^*_{k,\ell}$ has at most 2.599^n Schnyder woods. There are only $2^{2(k+\ell)-1}$ different orientations of these wrap-around edges. By the pigeon hole principle there is an orientation $\alpha_{k,\ell}$ of these edges which can be extended to a Eulerian orientation in asymptotically $2.598^{k\ell}$ ways. Thus, there is an $\alpha_{k,\ell}$ for $T^*_{k,\ell}$, which deviates from a 3-orientation only on the special vertices and the border of the grid such that there are asymptotically 2.598^n $\alpha_{k,\ell}$-orientations. Note, however, that directing all the wrap-around edges away from the vertex to which they are attached in Figure 2 induces a unique Eulerian orientation of $T^*_{k,\ell}$. We have not been able to show that $T_{k,\ell}$ has $2.598^{k\ell}$ 3-orientations, i.e. to verify that Baxter's result also gives a lower bound for the number of 3-orientations.

Now we discuss bounds on the number of Schnyder woods on 3-connected planar maps. The lower bound comes from the grid. The upper bound for this case is much larger than the one for triangulations. This is due to the encoding of Schnyder woods by 3-orientations on the primal dual completion graph, which has more vertices.

Theorem 4. *Let $\mathcal{M}_n$ be the set of 3-connected planar maps with n vertices and $\mathcal{S}(M)$ denote the set of Schnyder woods of $M \in \mathcal{M}_n$. Then, $3.209^n \leq |\mathcal{S}(M)| \leq 8^n$.*

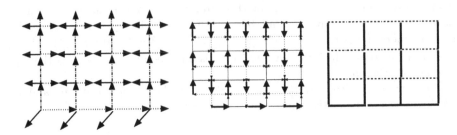

Fig. 3. A Schnyder wood on the map $G^*_{4,4}$, the reduced primal dual completion $G_{7,7} -$ $(7,1)$ with the corresponding orientation and the associated spanning tree

The example used for the lower bound is the square grid graph $G_{k,\ell}$. Enumeration and counting of different combinatorial structures on the grid graph has received a lot of attention in the literature, see e.g. [15].

As Schnyder woods are defined on 3-connected graphs we augment $G_{k,\ell}$ by an outer triangle $\{a_1, a_2, a_3\}$ and edges from the boundary vertices of the grid to the vertices of this triangle. Figure 2 shows an augmented triangular grid; removing the diagonals in the squares yields $G^*_{4,5}$.

Theorem 5. *For k, ℓ big enough the number of Schnyder woods of the augmented grid $G^*_{k,\ell}$, is $|S(G^*_{k,\ell})| \approx 3.209^{k\ell}$.*

The proof uses a bijection between Schnyder woods on $G^*_{k,\ell}$ and perfect matchings of $G_{2k-1,2\ell-1}$ minus on corner. This bijection also shows that Schnyder woods of $G^*_{k,\ell}$ are in bijection with spanning trees of $G_{k,\ell}$ by a result of Temperley, see [14] for more on the topic.

The proof of the upper bound stated in Theorem 4 uses the upper bound for Schnyder woods on plane triangulations. A triangulation T_M is derived from a planar map M by taking barycentric subdivisions of all non-triangular bounded faces. The number of Schnyder woods of T_M is shown to be an upper bound for the number of Schnyder woods of M and a specialization of the techniques from Proposition 1 yields the claimed bound.

Felsner et al. [11] present a theory of 2-orientations of plane quadrangulations, which shows many similarities with the theory of Schnyder woods for triangulations. We have studied the number of 2-orientations of quadrangulations and obtained the following results.

Theorem 6. *Let Q_n denote the set of all plane quadrangulations with n vertices and $Z(Q)$ the set of 2-orientations of $Q \in Q_n$. Then, for n big enough $1.53^n \leq \max_{Q \in Q_n} |Z(Q)| \leq 1.91^n$.*

4   Counting Bipolar Orientations

A good starting point for reading about bipolar orientations is [7]. Let G be a connected graph and $e = st$ a distinguished edge of G. An orientation X of the edges of G is an *e-bipolar orientation* of G if it is acyclic, s is the only vertex without incoming edges and t is the only vertex without outgoing edges. We call s and t the source respectively sink of X. There are many equivalent definitions of bipolar orientations, c.f. [7]. For our considerations any choice of two vertices s, t on the outer face will do, they need not be adjacent. We will simply refer to bipolar orientations instead of e-bipolar orientations. At this point we restrict ourselves to giving the encoding of bipolar orientations as α-orientations, which is introduced in [7].

Theorem 7. *Let M be a planar map and $\widehat{M}$ its angle graph. Let $\hat{\alpha} : V \cup \mathcal{F} \to \mathbb{N}$ be such that $\hat{\alpha}(F) = 2 = \hat{\alpha}(v)$ for $F \in \mathcal{F}$ and $v \in V \setminus \{s,t\}$. The source s and sink t have $\hat{\alpha}(s) = \hat{\alpha}(t) = 0$. Then, the bipolar orientations of M are in bijection with the $\hat{\alpha}$-orientations of $\widehat{M}$.*

Below, in Theorem 10, we give another encoding of bipolar orientations, which will turn out to be useful to upper bound the number of bipolar orientations.

Theorem 8. *Let $\mathcal{B}(G_{k,\ell})$ denote the set of bipolar orientations of $G_{k,\ell}$ with source $(1,1)$ and sink (k,ℓ) respectively $(k,1)$. For k, ℓ big enough the number of bipolar orientations of the grid $G_{k,\ell}$ is bounded by $2.18^{k\ell} \leq |\mathcal{B}(G_{k,\ell})| \leq 2.619^{k\ell}$.*

The lower bound uses reorientations of a canonical 2-orientation of the angle graph of $G_{k,\ell}$, which looks like a tilted grid. For the upper bound we use Lieb's bijection [15] between 2-orientations of the grid and 3-colorings of the squares of the grid. We encode such a 3-coloring by a sparse sequence, that is a 0-1-sequence without consecutive 1s.

Theorem 9. *Let $\mathcal{M}_n$ denote the set of all planar maps with n vertices and $\mathcal{B}(M)$ the set of all bipolar orientations of $M \in \mathcal{M}_n$. Then, for n big enough $2.91^n \leq \max_{M \in \mathcal{M}_n} |\mathcal{B}(M)| \leq 3.97^n$.*

Since it is not hard to prove that adding edges to non-triangular faces of a planar map M can only increase the number of bipolar orientations we restrict our considerations to plane inner triangulations. In the full paper it is shown that the set of all outerplanar maps with n vertices has $\max_{M \in \mathcal{O}_n} |\mathcal{B}(M)| = F_{n-1}$, where F_n is the nth Fibonacci number, and that $|\mathcal{B}(T_{2,\ell})|$ attains this value. The proof of the lower bound makes use of this by glueing together $k-1$ copies of $T_{2,\ell}$, which yields again a triangular grid.

The following relation is useful to upper bound the number of bipolar orientations for general plane inner triangulations. Let $\mathcal{F}_b$ be the set of bounded faces of M and $\mathcal{B}$ the set of bipolar orientations of M. Fix a bipolar orientation B. The boundary of every triangle $\Delta \in \mathcal{F}_b$ consists of a path of length 2 and an edge from the source to the sink of Δ. We say that Δ is a $+$ triangle of B if looking along the direct source-sink edge the triangle is on the left. Otherwise, if the triangle is on the right of the edge we speak of a $-$ triangle. We use this notation to define a mapping $G_B : \mathcal{F}_b \to \{-, +\}$.

The next result immediately yields an upper bound of 4^n for the number of bipolar orientations. The improvement can be made using the observation that every vertex is incident to faces of both types.

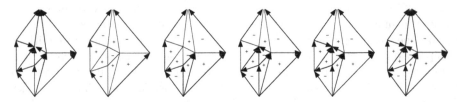

Fig. 4. A bipolar orientation, the corresponding $+/-$ encoding and an illustration of the decoding algorithm

Theorem 10. *Let M be a plane inner triangulation and B a bipolar orientation of M. Given G_B, i.e., the signs of bounded faces, it is possible to recover B. In other words the function $B \to G_B$ is injective from $\mathcal{B}(M) \to \{-,+\}^{|\mathcal{F}_b|}$.*

5   Conclusions

In this paper we have studied the maximum number of α-orientations for different classes of planar maps and different α. In most cases we have exponential upper and lower bounds c_L^n and c_U^n for this number. The obvious problem is to improve on the constants c_L and c_U for the different instances. We think, that in particular improving the upper bound of 8^n for the number of Schnyder woods on 3-connected planar maps is worth further efforts.

Results by Lieb [15] and Baxter [1] yield the exact asymptotic behavior of the number of Eulerian orientations for the square and triangular grid on the torus. This yields upper bounds for the number of 2-orientations on the square grid and the number of Schnyder woods on triangular grids of specific dimensions. We could not yet utilize these results for improving the lower bounds for the number of 2-orientations respectively Schnyder woods.

We mentioned several #P-completeness results in Section 2. This contrasts with spanning trees and planar bipartite perfect matchings for which polynomial algorithms are available. It remains open to determine the complexity of counting Schnyder woods and bipolar orientations on planar maps.

Acknowledgments. We would like to thank Graham Brightwell for interesting discussions and valuable hints in connection with Lieb's 3-coloring of the square grid. We thank Christian Krattenthaler for directing us to reference [14], Mark Jerrum for bringing Páidí Creed's work to our attention and Páidí Creed for sending us a preliminary version of his proof. Florian Zickfeld was supported by the Studienstiftung des deutschen Volkes.

References

1. Baxter, R.J.: F model on a triangular lattice. J. Math. Physics 10, 1211–1216 (1969)
2. Bonichon, N.: A bijection between realizers of maximal plane graphs and pairs of non-crossing dyck paths. Discrete Mathematics, FPSAC 2002 Special Issue 298, 104–114 (2005)

3. Bonichon, N., Felsner, S., Mosbah, M.: Convex drawings of 3-connected planar graphs. In: Pach, J. (ed.) GD 2004. LNCS, vol. 3383, pp. 60–70. Springer, Heidelberg (2005)

4. Creed, P.: Counting Eulerian orientations is planar graphs is #P-complete. Personal Communication (2007)

5. Dagum, P., Luby, M.: Approximating the permanent of graphs with large factors. Theoretical Computer Science 102, 283–305 (1992)

6. de Fraysseix, H., de Mendez, P.O.: On topological aspects of orientation. Discrete Math. 229, 57–72 (2001)

7. de Fraysseix, H., de Mendez, P.O., Rosenstiehl, P.: Bipolar orientations revisited. Discrete Appl. Math. 56, 157–179 (1995)

8. Felsner, S.: Convex drawings of planar graphs and the order dimension of 3-polytopes. Order 18, 19–37 (2001)

9. Felsner, S.: Geometric Graphs and Arrangements. Vieweg Verlag (2004)

10. Felsner, S.: Lattice structures from planar graphs. Elec. J. Comb. R15 (2004)

11. Felsner, S., Huemer, C., Kappes, S., Orden, D.: Binary labelings for plane quadrangulations and their relatives (preprint, 2007)

12. Fusy, É., Poulalhon, D., Schaeffer, G.: Dissections and trees, with applications to optimal mesh encoding and to random sampling. In: SODA, pp. 690–699 (2005)

13. Jerrum, M., Sinclair, A., Vigoda, E.: A polynomial-time approximation algorithm for the permanent of a matrix with non-negative entries. In: ACM STOC, pp. 712–721 (2001)

14. Kenyon, R.W., Propp, J.G., Wilson, D.B.: Trees and matchings. Elec. J. Comb. 7 (2000)

15. Lieb, E.H.: The residual entropy of square ice. Physical Review 162, 162–172 (1967)

16. Lovász, L., Plummer, M.D.: Matching Theory. Annals of Discrete Mathematics, vol. 29. North-Holland, Amsterdam (1986)

17. Poulalhon, D., Schaeffer, G.: Optimal coding and sampling of triangulations. In: Baeten, J.C.M., Lenstra, J.K., Parrow, J., Woeginger, G.J. (eds.) ICALP 2003. LNCS, vol. 2719, pp. 1080–1094. Springer, Heidelberg (2003)

18. Robertson, N., Seymour, P.D., Thomas, R.: Permanents, Pfaffian orientations, and even directed circuits. Ann. of Math. 150, 929–975 (1999)

19. Rote, G.: The number of spanning trees in a planar graph. In: Oberwolfach Reports. EMS, vol. 2, pp. 969–973 (2005),
http://page.mi.fu-berlin.de/rote/about_me/publications.html

20. Schnyder, W.: Planar graphs and poset dimension. Order 5, 323–343 (1989)

21. Schnyder, W.: Embedding planar graphs on the grid. Proc. 1st ACM-SIAM Sympos. Discrete Algorithms 5, 138–148 (1990)

22. Tamassia, R., Tollis, I.G.: A unified approach to visibility representations of planar graphs. Discrete Comput. Geom. 1, 321–341 (1986)

23. Tutte, W.: A short proof of the factor theorem for finite graphs. Canadian Journal of Mathematics 6, 347–352 (1954)

Complexity and Approximation Results for the Connected Vertex Cover Problem

Bruno Escoffier[1], Laurent Gourvès[1], and Jérôme Monnot[1]

CNRS-LAMSADE, Université Paris-Dauphine, Place du Maréchal De Lattre de Tassigny, F-75775 Paris Cedex 16, France
{escoffier,laurent.gourves,monnot}@lamsade.dauphine.fr

Abstract. We study a variation of the vertex cover problem where it is required that the graph induced by the vertex cover is connected. We prove that this problem is polynomial in chordal graphs, has a PTAS in planar graphs, is APX-hard in bipartite graphs and is 5/3-approximable in any class of graphs where the vertex cover problem is polynomial (in particular in bipartite graphs).

Keywords: Connected vertex cover, chordal graphs, bipartite graphs, planar graphs, **APX**-complete, approximation algorithm.

1 Introduction

In this paper, we study a variation of the vertex cover problem where the subgraph induced by any feasible solution must be connected. Formally, a vertex cover of a simple graph $G = (V, E)$ is a subset of vertices $S \subseteq V$ which covers all edges, *i.e.* which satisfies: $\forall e = \{x, y\} \in E$, $x \in S$ or $y \in S$. The vertex cover problem (MINVC in short) consists in finding a vertex cover of minimum size. MINVC is known to be **APX**-complete in cubic graphs [1] and **NP**-hard in planar graphs, [13]. MINVC is 2-approximable in general graphs, [3] and admits a polynomial approximation scheme in planar graphs, [5]. On the other hand, MINVC is polynomial for several classes of graphs such as bipartite graphs, chordal graphs, graphs with bounded treewidth, etc. [7].

The connected vertex cover problem, denoted by MINCVC, is the variation of the vertex cover problem where, given a connected graph $G = (V, E)$, we seek a vertex cover S^* of minimum size such that the subgraph induced by S^* is connected. This problem has been introduced by Garey and Johnson, [12] where it is proved to be **NP**-hard in planar graphs of maximum degree 4. As indicated in [19], this problem has some applications in the domain of wireless network design. In such a model, the vertices of the network are connected by transmission links. We want to place a minimum number of relay stations on vertices such that any pair of relay stations are connected (by a path which uses only relay stations) and every transmission link is incident to a relay station. This is exactly the connected vertex cover problem.

A. Brandstädt, D. Kratsch, and H. Müller (Eds.): WG 2007, LNCS 4769, pp. 202–213, 2007.

1.1 Previous Related Works

The main complexity and approximability results known on this problem are the following: in [21], it is shown that MINCVC is polynomially solvable when the maximum degree of the input graph is at most 3. However, it is **NP**-hard in planar bipartite graphs of maximum degree 4, [10], as well as in 3-connected graphs, [22]. Concerning the positive and negative results of the approximability of this problem, MINCVC is 2-approximable in general graphs, [20,2] but it is **NP**-hard to approximate within ratio $10\sqrt{5} - 21$, [10]. Finally, recently the fixed-parameter tractability of MINCVC with respect to the vertex cover size or to the treewidth of the input graph has been studied in [10,14,17,18,19]. More precisely, in [10] a parameterized algorithm for MINCVC with complexity $O^*(2.9316^k)$ is presented improving the previous algorithm with complexity $O^*(6^k)$ given in [14] where k is the size of an optimal connected vertex cover. Independently, the authors of [17,18] have also obtained FPT algorithms for MINCVC and they obtain in [18] an algorithm with complexity $O^*(2.7606^k)$. In [19], the author gives a parameterized algorithm for MINCVC with complexity $O^*(2^t \cdot t^{3t+2}n)$ where t is the treewidth of the graph and n the number of vertices.

MINCVC is related to the unweighted version of tree cover. The tree cover problem has been introduced in [2] and consists, given a connected graph $G = (V, E)$ with non-negative weights w on the edges, in finding a tree $T = (S, E')$ of G with $S \subseteq V$ and $E' \subseteq E$ which spans all edges of G and such that $w(T) = \sum_{e \in E'} w(e)$ is minimum. In [2], the authors prove that the tree cover problem is approximable within factor 3.55 and the unweighted version is 2-approximable. Recently, (weighted) tree cover has been shown to be approximable within a factor of 3 in [16], and a 2-approximation algorithm is proposed in [11]. Clearly, the unweighted version of tree cover is (asymptotically) equivalent to the connected version since S is a connected vertex cover of G iff there exists a tree cover $T' = (S, E')$ for some subset E' of edges. Since in this latter case, the weight of T' is $|S| - 1$, the result follows.

1.2 Our Contribution

In this article, we mainly deal with complexity and approximability issues for MINCVC in particular classes of graphs. More precisely, we first present some structural properties on connected vertex covers (Section 2). Using these properties, we show that MINCVC is polynomial in chordal graphs (Section 3). Then, in Section 4, we prove that MINCVC is **APX**-complete in bipartite graphs of maximum degree 4, even if each vertex of one block of the bipartition has a degree at most 3. On the other hand, if each vertex of block part of the bipartition has a degree at most 2 and the vertices of the other block have an arbitrary degree, then MINCVC is polynomial. Section 5 deals with the approximability of MINCVC. We first show that MINCVC is 5/3-approximable in any class of graphs where MINVC is polynomial (in particular in bipartite graphs, or more generally in perfect graphs). Then, we present a polynomial approximation scheme for MINCVC in planar graphs.

Notation. All graphs considered are undirected, simple and without loops. Unless otherwise stated, n and m will denote the number of vertices and edges, respectively, of the graph $G = (V, E)$ considered. $N_G(v)$ denotes the *neighborhood* of v in G, ie., $N_G(v) = \{u \in V : \{u, v\} \in E\}$ and $d_G(v)$ its *degree* that is $d_G(v) = |N_G(v)|$. Finally, $G[S]$ denotes the subgraph of G induced by S.

2   Structural Properties

We present in this subsection some properties on vertex covers or connected vertex covers. These properties will be useful in the rest of the article to devise polynomial algorithms that solve MINCVC either optimally (chordal graphs) or approximately (bipartite graphs,...).

2.1   Vertex Cover and Graph Contraction

For a subset $A \subseteq V$ of a graph $G = (V, E)$, the *contraction* of G with respect to A is the simple graph $G_A = (V', E')$ where we replace A in V by a new vertex v_A (so, $V' = (V \setminus A) \cup \{v_A\}$) and $\{x, y\} \in E'$ iff either $x, y \notin A$ and $\{x, y\} \in E$ or $x = v_A$, $y \neq v_A$ and there exists $v \in A$ such that $\{v, y\} \in E$. The *connected contraction* of G following $V' \subseteq V$ is the graph $G^c_{V'}$ corresponding to the iterated contractions of G with respect to the connected components of V' (note that contraction is associative and commutative). Formally, $G^c_{V'}$ is constructed in the following way: let $A_1, \cdots, A_q$ be the connected components of the subgraph induced by V'. Then, we inductively apply the contraction with respect to A_i for $i = 1, \cdots, q$. Thus, $G^c_{V'} = G_{A_1 \circ \cdots \circ A_q}$. Finally, let $New(G^c_{V'}) = \{v_{A_1}, \cdots, v_{A_q}\}$ be the new vertices of $G^c_{V'}$ (those resulting from the contraction). The following Lemma concerns contraction properties that will, in particular, be the basis of the approximation algorithm presented in Subsection 5.1.

Lemma 1. *Let $G = (V, E)$ be a connected graph and let $S \subseteq V$ be a vertex cover of G. Let $G_0 = (V_0, E_0) = G^c_S$ be the connected contraction of G following S where $A_1, \cdots, A_q$ are the connected components of the subgraph induced by S. The following assertions hold:*

(i) G_0 is connected and bipartite.

(ii) If $S = S^*$ is an optimal vertex cover of G, then $New(G_0)$ is an optimal vertex cover of G_0.

(iii) If $S = S^*$ is an optimal vertex cover of G and $v \in V \setminus S^*$ with $d_{G^c_{S^*}}(v) \geq 2$, then $New(G_0)$ is an optimal vertex cover of $G_0 = G^c_{S^* \cup \{v\}}$.

2.2   Connected Vertex Covers and Biconnectivity

Now, we deal with connected vertex covers. It is easy to see that if the removal of a vertex v disconnects the input graph (v is called a *cut-vertex*, or an *articulation point*), then v has to be in any connected vertex cover. In this section we show that, informally, solving MINCVC in a graph is equivalent to solve it on the biconnected components of the graph, under the constraint of including all cut vertices.

Formally, a connected graph $G = (V, E)$ with $|V| \geq 3$ is *biconnected* if for any two vertices x, y there exists a simple cycle in G containing both x and y. A *biconnected component* (also called *block*) $G_i = (V_i, E_i)$ is a maximal connected subgraph of G that is biconnected. For a connected graph $G = (V, E)$, V_c denotes the set of cut-vertices of G and $V_{i,c}$ its restriction to V_i.

Lemma 2. *Let $G = (V, E)$ be a connected graph. $S \subseteq V$ is a connected vertex cover of G iff for each biconnected component $G_i = (V_i, E_i)$, $i = 1, \cdots, p$, $S_i = S \cap V_i$ is a connected vertex cover of G_i containing $V_{i,c}$.*

Lemma 2 allows us to characterize the optimal connected vertex covers of G.

Corollary 1. *Let $G = (V, E)$ be a connected graph. $S^* \subseteq V$ is an optimal connected vertex cover of G iff for each biconnected component $G_i = (V_i, E_i)$, $i = 1, \cdots, p$, $S_i^* = S^* \cap V_i$ is an optimal connected vertex cover of G_i among the connected vertex covers of G_i containing $V_{i,c}$.*

For instance, using Corollary 1, we deduce that for the class of *trees* or *split graphs* MINCVC is polynomial. More generally, we will see in Section 3 that this result holds in chordal graphs. If we denote by MINPREXTCVC (by analogy with the well known PreExtension Coloring problem) the variation of MINCVC where given $G = (V, E)$ and $V_0 \subseteq V$, we seek a connected vertex cover S of G containing V_0 and of minimal size, we obtain the following result:

Lemma 3. *Let $\mathcal{G}$ be a class of connected graphs defined by a hereditary property. Solving MINCVC in $\mathcal{G}$ polynomially reduces to solve MINPREXTCVC in the biconnected graphs of $\mathcal{G}$. Moreover, if $\mathcal{G}$ is closed by pendant addition (ie., is closed under addition of a new vertex v and a new edge $\{u, v\}$ where $u \in V$), then they are polynomially equivalent.*

3   Chordal Graphs

The class of *chordal graphs* is a very well known class of graphs which arises in many practical situations. A graph G is chordal if any cycle of G with a size at least 4 has a chord (*i.e.*, an edge linking two non-consecutive vertices of the cycle). There are many characterizations of chordal graphs, see for instance [7].

In this section, we devise a polynomial time algorithm to compute an optimal CVC in chordal graphs. To achieve this, we need the following lemma.

Lemma 4. *Let $G = (V, E)$ be a connected chordal graph and let S be a vertex cover of G. The following properties hold:*

(i) The connected contraction $G_0 = (V_0, E_0) = G_S^c$ of G following S is a tree.
(ii) If G is biconnected, then S is a connected vertex cover of G.

Proof. Let S be a vertex cover of G.

For (i): from Lemma 1, we know that $G_0 = (V_0, E_0) = G_S^c$ is bipartite and connected. Assume that G_0 is not a tree, and let Γ be a cycle of G_0 with a minimal size. By construction, Γ is chordless, has a size at least 4 and alternates vertices of $New(G_0)$ and vertices of $V \setminus S$. From Γ, we can build a cycle Γ' of G using the following rule: if $\{x, v_{A_i}\} \in \Gamma$ and $\{v_{A_i}, y\} \in \Gamma$ where $x, y \notin S$ and $v_{A_i} \in New(G_0)$ (where we recall that A_i is some connected component of $G[S]$), then we replace these two edges by a shortest path $\mu_{x,y}$ from x to y in G among the paths from x to y in G which only use vertices of A_i (such a path exists since A_i is connected and is linked to x and y); by repeating this operation, we obtain a cycle Γ' of G with $|\Gamma'| \geq |\Gamma| \geq 4$. Let us prove that Γ' is chordless which will lead to a contradiction since G is assumed to be chordal. Let v_1, v_2 be two non consecutive vertices of Γ'. If $v_1 \notin S$ and $v_2 \notin S$, then $\{v_1, v_2\} \notin E$ since otherwise Γ would have a chord in G_0. So, we can assume that $v_1 \in (\mu_{x,y} \setminus \{x, y\})$ and $v_2 \in \mu_{x,y}$ (since there is no edge linking two vertices of disjoint paths $\mu_{x,y}$ and $\mu_{x',y'}$); in this case, using edge $\{v_1, v_2\}$, we could obtain a path which uses strictly less edges than $\mu_{x,y}$.

For (ii): Suppose that S is not connected. Then, from (i) we deduce that G_0 is not a star and thus, there are two edges $\{v_{A_i}, x\}$ and $\{x, v_{A_j}\}$ in G_0 where A_i and A_j are two connected components of S. We deduce that x would be a cut-vertex of G, contradiction since G is assumed to be biconnected.

In particular, using (ii) of Lemma 4, we deduce that any optimal vertex cover S^* of a biconnected chordal graph G is also an optimal connected vertex cover. Now, we give a simple linear algorithm for computing an optimal connected vertex cover of a chordal graph.

Theorem 1. MINCVC *is polynomial in chordal graphs. Moreover, an optimal solution can be found in linear time.*

Proof. Following Lemma 3, solving MINCVC in a chordal graph $G = (V, E)$ can be done by solving MINPREXTCVC in each of the biconnected components $G_i = (V_i, E_i)$ of G. Since G_i is both biconnected and chordal, by Lemma 4, MINPREXTCVC is the same problem as MINPREXTVC (in G_i). But, by adding a pendant edge to vertices required to be taken in the vertex cover, we can easily reduce MINPREXTVC to MINVC (note that the graph remains chordal). Since computing the biconnected components and solving MINVC in a chordal graph can be done in linear time (see [7]), the result follows.

4   Bipartite Graphs

A bipartite graph $G = (V, E)$ is a graph where the vertex set is partitioned into two independent sets L and R. Using the result of [10], we already know that MINCVC is **NP**-hard in planar bipartite graphs of maximum 4. Using Lemma 3, we can strengthen this result:

Lemma 5. MINCVC *is* **NP**-*hard in biconnected planar bipartite graphs of maximum degree 4.*

Now, one can show that MINCVC has no PTAS in bipartite graphs of maximum degree 4.

Theorem 2. MINCVC *is not* 1.001031-*approximable in connected bipartite graphs* $G = (L, R; E)$ *where* $\forall l \in L$, $d_G(l) \leq 4$ *and* $\forall r \in R$, $d_G(r) \leq 3$, *unless* **P=NP**.

In Theorem 2, we proved in particular that MINCVC is **NP**-hard when all the vertices of one part of the bipartition have a degree at most 3. It turns out that if all the vertices of one part of this bipartition have a degree at most 2, the problem becomes easy. This property will be very useful to devise our approximation algorithm in Subsection 5.1.

Lemma 6. MINCVC *is polynomial in bipartite graphs* $G = (L, R; E)$ *such that* $\forall r \in R$, $d_G(r) \leq 2$. *Moreover, if* $L_2 = \{l \in L : d_G(l) \geq 2\}$, *then* $opt(G) = |L| + |L_2| - 1$.

5 Approximation Results

MINCVC is trivially **APX**-complete in k-connected graphs for any $k \geq 2$ since starting from graph $G = (V, E)$, instance of MINVC, we can add a clique K_k of size k and link each vertex of G to each vertex of K_k. This new graph G' is obviously k-connected and S is a vertex cover of G iff S union the k vertices of K_k (we can always assume that $S \neq V$) is a connected vertex cover of G'. Thus, using the negative result of [15] it is quite improbable that one can improve the approximation ratio of 2 for MINCVC, even in k-connected graphs. Thus, in this subsection we deal with the approximability of MINCVC in particular classes of graphs.

In Subsection 5.1, we devise a 5/3-approximation algorithm for any class of graphs where the classical vertex cover problem is polynomial. In Subsection 5.2, we show that MINCVC admits a PTAS in planar graphs.

5.1 When MinVC Is Polynomial

Let $\mathcal{G}$ be a class of connected graphs where MINVC is polynomial (for instance, the connected bipartite graphs). The underlying idea of the algorithm is simple: we first compute an optimal vertex cover, and then try to connect it by adding vertices (either using high degree vertices or Lemma 6). The analysis leading to the ratio 5/3 is based on Lemma 1 which deals with graph contraction.

Now, let us formally describe the algorithm. Recall that given a vertex set V', $G^c_{V'}$ denotes the connected contraction of V following V', and $New(G^c_{V'})$ denotes the set of new vertices (one for each connected component of $G[V']$).

---

algo$_{CVC}$ input: A graph $G = (V, E)$ of $\mathcal{G}$ with at least 3 vertices.

1 Find an optimal vertex cover S^* of G such that in G^c_{S*}, $\forall v \in New(G^c_{S*})$, $d_{G^c_{S*}}(v) \geq 2$;
2 Set $G_1 = G^c_{S*}$, $N_1 = New(G^c_{S*})$, $S = S^*$ and $i = 1$;
3 While $|N_i| \geq 2$ and there exists $v \notin N_i$ such that v is linked in G_i to at least 3 vertices of N_i do

    3.1 Set $S := S \cup \{v\}$ and $i := i + 1$;
    3.2 Set $G_i := G^c_S$ and $N_i = New(G^c_S)$;

4 If $|N_i| \geq 2$, apply the polynomial algorithm of Lemma 6 on G_i (let S' be the produced solution) and set $S := S \cup (V \cap S')$;
5 Output S;

Now, we show that algo$_{CVC}$ outputs a connected vertex cover of G in polynomial time. First of all, given an optimal vertex cover S^* of a graph G (assumed here to be computable in polynomial time), we can always transform it in such a way that $\forall v \in New(G^c_{S*})$, $d_{G^c_{S*}}(v) \geq 2$. Indeed, if a vertex of G^c_{S*} corresponding to a connected component of S^* has only one neighbor in G^c_{S*}, then we can take this neighbor in S^* and remove one vertex on this connected component (and the number of such 'leaf' connected components decreases, as soon as G^c_{S*} has at least 3 vertices). Now, using (ii) of Lemma 1, we know that $New(G^c_{S*})$ is an optimal vertex cover of G^c_{S*}. Then, from $New(G^c_{S*})$, we can find such a solution within polynomial time.

Moreover, using (i) of Lemma 1 with S^*, we deduce that the graph G_i is bipartite, for each possible value of i. Assume that $G_i = (N_i; R_i, E_i)$ for iteration i where N_i is the left set corresponding to the contracted vertices and R_i is the right set corresponding to the remaining vertices and let p be the last iteration. Clearly, if $|N_p| = 1$, the the output solution S is connected. Otherwise, the algorithm uses step 4; we know that G_p is bipartite and by construction $\forall r \in R_p$, $d_{G_p}(r) \leq 2$. Thus, we can apply Lemma 6 on G_p. Moreover, a simple proof also gives that $\forall l \in N_p$, $d_{G_p}(l) \geq 2$. Indeed, otherwise there exists $l \in N_p$ such that l has a unique neighbor $r_0 \in R_p$. Let $\{x_1, \cdots, x_j\} \subseteq N_{p-1}$ with $j \geq 3$ and r_1 be the vertices contracted in G_{p-1} in order to obtain G_p. We conclude that the neighborhood of $\{x_1, \cdots, x_j\}$ is $\{r_0, r_1\}$ in G_{p-1} which is impossible since on the one hand, N_{p-1} is an optimal vertex cover of G_{p-1} (using (iii) of Lemma 1), and on the other hand, by flipping $\{x_1, \cdots, x_j\}$ with $\{r_0, r_1\}$, we obtain another vertex cover of G_{p-1} with smaller size than N_{p-1}! Finally, using Lemma 6, an optimal connected vertex cover of G_p consists of taking N_p and $|N_p| - 1$ of R_p. In conclusion, S is a connected vertex cover of G.

We now prove that this algorithm improves the ratio 2.

Theorem 3. *Let $\mathcal{G}$ be a class of connected graphs where* MinVC *is polynomial. Then,* algo$_{CVC}$ *is a 5/3-approximation for* MinCVC *in $\mathcal{G}$.*

Proof. Let $G = (V, E) \in \mathcal{G}$. Let S be the approximate solution produced by algo_{CVC} on G. Using the previous notations and Lemma 6, the solution S has a value $apx(G)$ satisfying:

$$apx(G) = |S^*| + p - 1 + |N_p| - 1 \qquad (1)$$

where p is the number of iterations of step 3. Obviously, we have:

$$opt(G) \geq |S^*| \qquad (2)$$

Now let us prove that for any $i = 1, \cdots, p - 1$, we also have $opt(G_i) \geq opt(G_{i+1}) + 1$. Let S_i^* be an optimal connected vertex cover of G_i. Let $r_i \in R_i$ be the vertex added to S during iteration i and let $N_{G_i}(r_i)$ be the neighbors of r_i in G_i. The graph G_{i+1} is obtained from the contraction of G_i with respect to the subset $S_i = \{r_i\} \cup N_{G_i}(r_i)$. Thus, if v_{S_i} denotes the new vertex resulting from the contraction of S_i, then $(S_i^* \setminus S_i) \cup \{v_{S_i}\}$ is a connected vertex cover of G_{i+1}. Moreover, $|S_i^* \cap S_i| \geq 2$ since either $r_i \in S_i^*$ and at least one of these neighbors must belong to S_i^* (S_i^* is connected and $i < p$) or $N_{G_i}(r_i) \subseteq S_i^*$ since S_i^* is a vertex cover. Thus $opt(G_{i+1}) \leq |S_i^* \setminus S_i| + 1 = opt(G_i) - |S_i^* \cap S_i| + 1 \leq opt(G_i) - 1$. Summing up these inequalities for $i = 1$ to $p - 1$, and using that $opt(G) \geq opt(G_1)$, we obtain:

$$opt(G) \geq opt(G_p) + p - 1 \qquad (3)$$

Moreover, thanks to Lemma 6, we know that $opt(G_p) = 2|N_p| - 1$. Together with equation (3), we get:

$$opt(G) \geq 2|N_p| - 1 + p - 1 \qquad (4)$$

Finally, since each vertex chosen in step 3 has degree at least 3, we get $|N_{i+1}| \leq |N_i| - 2$. This immediately leads to $|N_1| \geq |N_p| + 2(p - 1)$. Since $|S^*| \geq |N_1|$, we get:

$$|S^*| \geq |N_p| + 2(p - 1) \qquad (5)$$

Combination of equations (2), (4) and (5) with coefficients 4, 1 and 1 (respectively) gives:

$$5opt(G) \geq 3|S^*| + 3|N_p| - 1 + 3(p - 1) \qquad (6)$$

Then, equation (1) allows to conclude.

5.2   Planar Graphs

Given a planar embedding of a planar graph $G = (V, E)$, the level of a vertex is defined as follows (see for instance [4]): the vertices on the exterior face are at level 1. Given vertices at level i, let f be an interior face of the subgraph induced by vertices at level i. If G_f denotes the subgraph induced by vertices included

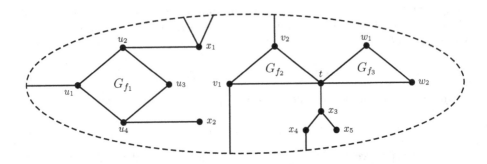

Fig. 1. Level of a planar graph

in f, then the vertices on the exterior face of G_f are at level $i + 1$. The set of vertices at level i is called the layer L_i.

This is illustrated on Figure 1. The dashed ellipse represents an interior face on level $i - 1$. Depicted vertices are at level i. There are 3 interior faces (constituted respectively by the u_i's, by $\{v_1, v_2, t\}$ and $\{t, w_1, w_2\}$).

Baker gave in [4] a polynomial time approximation scheme for several problems including vertex cover in planar graphs. The underlying idea is to consider k-outerplanar subgraphs of G constituted by k consecutive layers. The polynomiality of vertex cover in k-outerplanar graphs (for a fixed k) allows to achieve a $(k + 1)/k$ approximation ratio.

We adapt this technique in order to achieve an approximation scheme for MINCVC (MINCVC is **NP**-hard in planar graphs, see [12]). First of all, note that k-outerplanar graphs have treewidth bounded above by $3k - 1$, [6]. Since MINCVC is polynomially solvable for graphs with bounded treewidth, [19], MINCVC is polynomial for k-outerplanar graphs.

Theorem 4. MINCVC *admits an approximation scheme in planar graphs.*

Proof. Given an embedding of a planar (connected) graph G, we define, as previously, the layers $L_1, \cdots, L_q$ of G. For each layer L_i, we define F_i as the set of vertices of L_i that are in an interior face of L_i. For instance, in Figure 1, all vertices but the x_i's are in F_i.

Following the principle of the approximation scheme for vertex cover, we define an algorithm for any integer $k > 0$. Let $V_i = F_i \cup L_{i+1} \cup L_{i+2} \cup \ldots \cup L_{i+k}$, and G_i be the graph induced by V_i. Note that G_i is not necessarily connected since for example there can be several disjoint faces in F_i (there are two connected components in Figure 1).

Let S^* be an optimum connected vertex cover on G with value $opt(G)$, and $S_i^* = S^* \cap V_i$. Then of course S_i^* is a vertex cover of G_i. However, even restricted to a connected component of G_i, it is not necessarily connected. Indeed, S^* is connected but the path(s) connecting two vertices of S^* in a connected component of G_i may use vertices out of this connected component. To overcome this problem, notice that only vertices in F_i or in F_{i+k} connect V_i to $V \setminus V_i$. Hence,

$S_i^* \cup F_i \cup F_{i+k}$ can be partitioned into a set of connected vertex covers on each of the connected components of G_i (since F_i and F_{i+k} are made of cycles). Now, take an optimum connected vertex cover on each of these connected components, and define S_i as the union of these optimum solutions. Then, we have :

$$|S_i^* \cup F_i \cup F_{i+k}| \geq |S_i| \qquad (7)$$

Now, let $p \in \{1, \ldots, k\}$. Let $V_0 = L_1 \cup L_2 \cup \ldots \cup L_p$, G_0 be the subgraph of G induced by V_0, $S_0^* = S^* \cap V_0$, and S_0 be an optimum vertex cover on G_0. With similar arguments as previously, we have:

$$|S_0^* \cup F_p| \geq |S_0| \qquad (8)$$

We build a solution S^p on the whole graph G as follows. S^p is the union of S_0 and of all S_i's for $i = p \mod k$. Of course, S^p is a vertex cover of G, since any edge of G appears in at least one G_i (or G_0). Moreover, it is connected since:

- S_0 is connected, and each S_i is made of connected vertex covers on the connected components of G_i;
- each of these connected vertex covers in S_i is connected to S_{i-k} (or to S_0 if $i = p$): this is due to the fact that F_i belongs to V_i and to V_{i-k} (or V_0). Hence, a level i interior face f is common to S_{i-k} (or S_0) and to the connected vertex cover of S_i we are dealing with. Both partial solutions cover all the edges of this face f. Since f is a cycle, the two solutions are necessarily connected. In other words, each connected component of S_i is connected to S_{i-k} (or S_0) and, by recurrence, to S_0. Consequently, the whole solution S^p is connected.

Summing up equation (7) for each $i = p \mod k$ and equation (8), we get:

$$|S_0^* \cup F_p| + \sum_{i=p \mod k} |S_i^* \cup F_i \cup F_{i+k}| \geq |S_0| + \sum_{i=p \mod k} |S_i| \qquad (9)$$

By definition of S^p, we have $|S^p| \leq |S_0| + \sum_{i=p \mod k} |S_i|$. On the other hand, since only vertices in F_i ($i = p \mod k$) appear in two different V_i's ($i = 0$ or $i = p \mod k$), we get that $|S_0^* \cup F_p| + \sum_{i=p \mod k} |S_i^* \cup F_i \cup F_{i+k}| \leq |S^*| + 2\sum_{i=p \mod k} |F_i|$. This leads to:

$$opt(G) + 2 \sum_{i=p \mod k} |F_i| \geq |S^p| \qquad (10)$$

If we consider the best solution S with value $apx(G)$ among the S^p's ($p \in \{1, \ldots, k\}$), we get :

$$opt(G) + \frac{2}{k} \sum_{i=1}^{q} |F_i| \geq apx(G) \qquad (11)$$

To conclude, we observe that the following property holds:

Property 1. S^* takes at least one fourth of the vertices of each F_i.

To see this property of $S^* \cap F_i$, consider F_i and the set E_i of edges of G that belong to one and only one interior face of F_i (for example, in Figure 1, if there were edges $\{u_2, u_4\}$ and $\{u_3, v_1\}$, they would not be in E_i). Let n_i be the number of vertices in F_i, and m_i the number of edges in E_i. This graph is a collection of edge-disjoint (but not vertex-disjoint, as one can see in Figure 1) interior faces (cycles). Of course, $S^* \cap F_i$ is a vertex cover of this graph. Since this graph is a collection of interior faces (cycles), on each of these faces f $S^* \cap F_i$ cannot reject more than $|f|/2$ vertices. In all,

$$|S^* \cap F_i| \geq n_i - \sum_{f \in F_i} \frac{|f|}{2} \tag{12}$$

But since faces are edge-disjoint, $\sum_{f \in F_i} |f| = m_i$. On the other hand, if N_f denotes the number of interior faces in F_i, since each face contains at least 3 vertices, $m_i = \sum_{f \in F_i} |f| \geq 3N_f$. Since the graph is planar, using Euler formula we get $1 + m_i = n_i + N_f \leq n_i + m_i/3$. Hence, $m_i \leq 3n_i/2$. Finally, $|S^* \cap F_i| \geq n_i - m_i/2 \geq n_i/4$.

Based on this property, we get:

$$opt(G) \left(1 + \frac{8}{k}\right) \geq apx(G) \tag{13}$$

Taking k sufficiently large leads to a $1 + \varepsilon$ approximation. The polynomiality of this algorithm follows from the fact that each subgraph we deal with is (at most) $k + 1$-outerplanar, hence for a fixed k we can find an optimum solution in polynomial time.

References

1. Alimonti, P., Kann, V.: Some APX-completeness results for cubic graphs. Theor. Comput. Sci. 237(1-2), 123–134 (2000)
2. Arkin, E.M., Halldórsson, M.M., Hassin, R.: Approximating the tree and tour covers of a graph. Inf. Process. Lett. 47(6), 275–282 (1993)
3. Ausiello, G., Crescenzi, P., Gambosi, G., Kann, V., Marchetti-Spaccamela, A., Protasi, M.: Complexity and Approximation (Combinatorial Optimization Problems and Their Approximability Properties). Springer, Berlin (1999)
4. Baker, B.S.: Approximation algorithms for NP-complete problems on planar graphs. J. ACM 41(1), 153–180 (1994)
5. Bar-Yehuda, R., Even, S.: On approximating a vertex cover for planar graphs. In: STOC, pp. 303–309 (1982)
6. Bodlaender, H.L.: A partial -arboretum of graphs with bounded treewidth. Theor. Comput. Sci. 209(1-2), 1–45 (1998)
7. Brandstadt, A., Le, V.B., Spinrad, J.: Graph classes: a survey. Society for Industrial and Applied Mathematic, Philadelphia (1999)
8. Dawar, A., Grohe, M., Kreutzer, S., Schweikardt, N.: Approximation schemes for first-order definable optimisation problems. In: LICS, pp. 411–420 (2006)
9. Demaine, E.D., Hajiaghayi, M.T.: Bidimensionality: new connections between fpt algorithms and ptass. In: SODA, pp. 590–601 (2005)

10. Fernau, H., Manlove, D.: Vertex and edge covers with clustering properties: Complexity and algorithms. In: Algorithms and Complexity in Durham, pp. 69–84 (2006)

11. Fujito, T.: How to trim an mst: A 2-approximation algorithm for minimum cost tree cover. In: Bugliesi, M., Preneel, B., Sassone, V., Wegener, I. (eds.) ICALP 2006. LNCS, vol. 4051, pp. 431–442. Springer, Heidelberg (2006)

12. Garey, M.R., Johnson, D.S.: The rectilinear steiner tree problem in NP complete. SIAM Journal of Applied Mathematics 32, 826–834 (1977)

13. Garey, M.R., Johnson, D.S.: Computer and Intractability: A Guide to the Theory of NP-Completeness. W. H. Freeman, New York (1979)

14. Guo, J., Niedermeier, R., Wernicke, S.: Parameterized complexity of generalized vertex cover problems. In: Dehne, F., López-Ortiz, A., Sack, J.-R. (eds.) WADS 2005. LNCS, vol. 3608, pp. 36–48. Springer, Heidelberg (2005)

15. Khot, S., Regev, O.: Vertex cover might be hard to approximate to within $2 - \varepsilon$. In: IEEE Conference on Computational Complexity, pp. 379–386. IEEE Computer Society Press, Los Alamitos (2003)

16. Könemann, J., Konjevod, G., Parekh, O., Sinha, A.: Improved approximations for tour and tree covers. Algorithmica 38(3), 441–449 (2003)

17. Mölle, D., Richter, S., Rossmanith, P.: Enumerate and expand: Improved algorithms for connected vertex cover and tree cover. In: Grigoriev, D., Harrison, J., Hirsch, E.A. (eds.) CSR 2006. LNCS, vol. 3967, pp. 270–280. Springer, Heidelberg (2006)

18. Mölle, D., Richter, S., Rossmanith, P.: Enumerate and expand: New runtime bounds for vertex cover variants. In: Chen, D.Z., Lee, D.T. (eds.) COCOON 2006. LNCS, vol. 4112, pp. 265–273. Springer, Heidelberg (2006)

19. Moser, H.: Exact algorithms for generalizations of vertex cover. Master's thesis, Institut für Informatik, Friedrich-Schiller-Universität Jena (2005)

20. Savage, C.D.: Depth-first search and the vertex cover problem. Inf. Process. Lett. 14(5), 233–237 (1982)

21. Ueno, S., Kajitani, Y., Gotoh, S.: On the nonseparating independent set problem and feedback set problem for graphs with no vertex degree exceeding three. Discrete Mathematics 72, 355–360 (1988)

22. Wanatabe, T., Kajita, S., Onaga, K.: Vertex covers and connected vertex covers in 3-connected graphs. In: IEEE International Sympoisum on Circuits and Systems, pp. 1017–1020. IEEE Computer Society Press, Los Alamitos (1991)

Segmenting Strings Homogeneously Via Trees

Peter Damaschke

Department of Computer Science and Engineering
Chalmers University, 41296 Göteborg, Sweden
ptr@cs.chalmers.se

Abstract. We divide a string into k segments, each with only one sort of symbols, so as to minimize the total number of exceptions. Motivations come from machine learning and data mining. For binary strings we develop a linear-time algorithm for any k. Key to efficiency is a special-purpose data structure, called W-tree, which reflects relations between repetition lengths of symbols. Existence of algorithms faster than obvious dynamic programming remains open for non-binary strings. Our problem is also equivalent to finding weighted independent sets of prescribed size in paths. We show that this problem in bounded-degree graphs is FPT.

1 Introduction

Segmentation problems appear in image and language processing, classification, and machine learning. Application examples include the automatic segmentation of text into words in Asian languages, or into parts dealing with one main subject. The putative quality of segmentations is measured by some objective function based on plausibility of segments. One type of measure is similarity within the segments [8]. Segmentation of more abstract (e.g., numerical) linearly ordered data is a related topic. Efficient algorithms for cutting time series into somehow homogeneous segments are given in [10], along with interesting motivations for their use. A similar problem with other demands on the segments is studied in [9]. Segmentations of labeled point sets in higher dimensions by linear discriminators according to various homogeneity criteria lead to nontrivial computational problems which are relevant in neural network and decision tree learning. We refer to the series of papers [1,2,3,4].

Dynamic programming is the obvious design technique for string segmentation (see [11, Chapter 6, Exercise 5]), but the precise complexity depends on the objective. In the present work we consider a specific objective. Let S be a string of n symbols from an alphabet of size b. The *segment* $S[i..j]$ is the substring of S from position i to position j. A *segmentation* splits S into segments and assigns a *designated symbol* to each. An *exception* in a segment is any occurence of a symbol distinct from the designated one. Our objective is:

SEGMENTATION WITH MINIMIZED EXCEPTIONS (SME)
Given a string S of n symbols from an alphabet of size b, and some $k < n$, compute a segmentation of S with at most k segments that minimizes the total number of exceptions. By SME-Binary we denote SME over alphabet $\{0, 1\}$.

A. Brandstädt, D. Kratsch, and H. Müller (Eds.): WG 2007, LNCS 4769, pp. 214–225, 2007.

Segmentation in general received much interest, and the number of exceptions is one of the natural quality measures for string segmentation, but in spite of that, SME has not been studied in the algorithms literature, to our best knowledge. To give a motivation of SME-Binary from machine learning, suppose that some binary feature f of objects depends randomly on a real-valued attribute x, in such a way that, for value x, we would observe $f = 1$ with probability $p(x)$. Function p is unknown but supposed to be "smooth", i.e., p oscillates only a few times on the x-axis. When we have sampled many values of f along the x-axis, these data sorted by x are naturally described as a string S of symbols 0 and 1. In order to predict f of yet unseen objects with other values x, we would split S in homogeneous segments, as good as possible, and predict always the majority symbol in the segment containing the probed x. Then the rate of prediction errors is roughly the total number of exceptions divided by $|S|$. Optimal segmentation can also serve as a subroutine in more complex machine learning tasks where real-valued attributes must be discretized, as in decision tree learning. A similar discussion applies to SME over larger alphabets. As for the choice of k, there is a trade-off between homogeneity of segments and reliability: A larger k results in fewer exceptions in the string S of training data, but on the other hand, predictions in smaller segments are less reliable due to overfitting. Thus we study SME with an arbitrary, user-defined parameter k.

Our contributions: SME is easy to solve by dynamic programming, but for SME-Binary we will develop less obvious, faster algorithms. In Section 2 we observe that we never need to split the *runs* of S, that is, the maximal substrings of only one sort of symbols. Let r denote the number of runs of S. Runs and their lengths are trivially computed in an $O(n)$ time preprocessing, thus we give all further time bounds as functions of b, k, r, where we count arithmetic operations with integers as $O(1)$ time operations. The time used for the actual segmentation is significant especially if S comes from noisy data with many short runs. Dynamic programming solves SME in $O(br \cdot \min\{k, r - k\})$ time. Then, we focus on SME-Binary. In Section 3 we show that SME-Binary is equivalent to computing a minimum weight independent set of exactly m vertices in a path of (essentially) r vertices, where $m \approx (r - k)/2$. A greedy algorithm solves SME-Binary in $O(r + m \log m)$ time, which beats dynamic programmming if $k = \omega(\log r)$. In the more relevant case $k \ll r$ this would become $O(r \log r)$. In Section 4 we use the greedy algorithm just in order to show correctness of yet another algorithm that works in $O(r)$ time for any k. This result is achieved by a special data structure that we call the W-tree. From this tree we can read off an optimal solution for any desired number of segments. It seems that we cannot get $O(r)$ time much simpler: In Section 5 we briefly discuss another dynamic programming approach that penalizes every new segment. The catch is that one has to find the appropriate penalty. On the other hand, our method for SME-Binary breaks down if $b > 2$, and dynamic programming is still the best method that we currently have in the non-binary case. In Section 6 we show that computing weighted independent sets of exactly m vertices in general graphs is fixed-parameter tractable (FPT), with m and the maximum degree d as combined parameters. Section 7 concludes the paper with some open problems.

2   Runs and Exceptions

We collect a few simple properties of an optimal solution to SME. We call a segmentation *alternating* if the designated symbols of any two neighbored segments differ. A *run* in string S is a maximal contiguous substring with only one sort of symbol. We call a segmentation *regular* if every segment is a concatenation of runs. Throughout the paper, r is the number of runs in S.

Lemma 1. *There exists an optimal segmentation which is alternating and regular, and where any symbol at an end of a segment is different from the designated symbol of the adjacent segment.*

Proof. In a non-alternating segmentation we could merge two neighbored segments having the same designated symbol, without creating new exceptions.

Next consider, in an alternating segmentation, any two neighbored segments $S[i..j]$ and $S[j+1..m]$., w.l.o.g. with designated symbol 0 and 1, respectively. Assume, e.g., that $S[j] = 1$, the designated symbol of the segment to the right. Then we could move the border and replace the two segments with $S[i..j-1]$ and $S[j..m]$, thereby reducing the number of exceptions. Hence we can always suppose $S[j] \neq 1$, and similarly $S[j+1] \neq 0$, in some optimal solution. We remark that, if $i = j$, the left segment becomes empty, in which case the number of segments decreases as well, and perhaps further segments with identical designated symbols can be merged.

Finally, if a segmentation with the above properties cuts through some run, then the symbol in the run is different from the designated symbols of two neighbored segments. Hence every position in the run is an exception in both segments, and we can move borders without changing the number of exceptions, until the run is entirely in one segment, or a segment disappears.   □

Theorem 1. *SME can be solved in $O(br \cdot \min\{k, r-k\})$ time.*

Proof. We apply dynamic programming. First consider the case $k < r - k$. Let $E(i, j, s)$, with $i \leq k$, $j \leq r$, and symbol s, indicate the optimal number of exceptions in an alternating, regular segmentation of the first j runs into exactly i segments, with s as designated symbol of the jth run. (By Lemma 1 we need to consider such segmentations.) For $i > j$ let $E(i, j, s) = \infty$. It is easy to initialize $E(i, 1, s)$ for any i, s. For the $O(bkr)$ bound, we show for each $j > 1$ how to compute all $E(i, j, s)$ from all $E(i, j-1, s)$ in $O(bk)$ time. Let the jth run consist of l_j copies of symbol s_j. The following recursion is obvious:

$$E(i, j, s) = \min\{E(i, j-1, s), \min_{t \neq s}\{E(i-1, j-1, t)\}\} + q_{j,s}l_j,$$

where $q_{j,s} = 0$ if $s = s_j$, and 1 otherwise. For any fixed i we apply it simultaneously to all symbols s as follows. Find $\min_t\{E(i-1, j-1, t)\}$ in $O(b)$ time. The minimum is attained by some t_0, ties are broken arbitrarily. For each $s \neq t_0$ we have $\min_{t \neq s}\{E(i-1, j-1, t)\} = E(i-1, j-1, t_0)$, and $\min_{t \neq t_0}\{E(i-1, j-1, t)\}$ is also found in $O(b)$ time. Hence we can compute the second term in the recursion in $O(b)$ time for all s together, for any fixed i, and then we can evaluate the recursion in $O(bk)$ time for all i, s.

Case $k \geq r - k$ is similar, but now we let i count the runs that continue the previous interval, and adjust the recursive formula for E accordingly.   □

3   Greedy Algorithm for Binary Strings

In a regular segmentation, we call a run x an *exception run* if the symbol in x differs from the designated symbol of the segment x belongs to. By Lemma 1, every instance (S, k) of SME-Binary has an optimal segmentation which is alternating and regular, and where no segment begins or ends with an exception run. (However, the first or last run of S may be an exception run.) Thus, no two exception runs are adjacent. Conversely, any set X of pairwise non-adjacent runs describes a segmentation with X as the set of exception runs. Thus we can henceforth characterize a segmentation simply by the set X.

Adding to X a new exception run x (not adjacent to any member of X) decreases the number of segments by 2, unless x is the first or last run in S, in which case we get only one segment less. We decide in the beginning which of the two outer runs be in X. Only two options exist for any r, k: If $r - k$ is even, either no or both outer runs are in X. If $r - k$ is odd, exactly one outer run must be in X. For given (S, k) we consider the two options separately. That is, it remains to decide which inner runs be in X. An outer run put in X is removed from the instance, so that its neighbor becomes an outer run, and it cannot be added to X anymore. Since now *every* new exception is an inner run and decreases the number of segments by 2, our instance of SME-Binary is reduced to two instances of the following problem: *Given a binary string and a number m, find m pairwise non-adjacent inner runs of minimum total length.*

After trivial preprocessing, S is already given as the sequence of lengths of runs. Note that $m \approx (r - k)/2$, up to a small additive constant depending on the cases above. Obviously, the latter problem can be further rephrased:

Lemma 2. *As instance of SME-Binary is linear-time reducible to two instances of the following problem: Given a path P with r weighted vertices and a number $m < r/2$, compute a minimum weight independent set X with exactly m vertices, avoiding the two outer vertices of P.*   □

We still refer to this "graph-theoretic" formulation of the problem as SME-Binary, and denote the number of vertices by r, this will not cause confusion. Since the weights of the outer vertices are irrelevant, we may set them to any huge constant. These two dummy vertices will simplify the presentation of the algorithms. In the following, the *length* of a path is the number of vertices, the weight of vertex v is $l(v)$, and $l(X)$ is the total weight of a set X of vertices.

Greedy Algorithm for SME-Binary:

(1) Do the following m times. Merge an inner vertex v with minimum $l(v)$ and its neigbors u, w into a new vertex z with $l(z) := l(u) + l(w) - l(v)$.
(2) After termination of (1) we have obtained a path of vertices labeled with odd-length subpaths of the original path, in the obvious sense. Output as members of X all original vertices which have even positions in these subpaths.

Clearly, the segments of our segmentation correspond to the vertices of the path produced in (1), and these segments have always odd length. In the following we prove correctness of the greedy algorithm. Trivially, X is an independent set. Induction on m easily shows $|X| = m$. It remains to prove that $l(X)$ is minimized among all independent sets of size m.

Lemma 3. *Let v be some inner vertex with minimum $l(v)$. There exists an optimal solution X that contains either v or both its neighbors u, w.*

Proof. Let X be a solution with $v \notin X$. If also $u, w \notin X$, then replace any vertex in X with v and obtain another solution that is no worse. If $u \in X$ but $w \notin X$, replace u with v (the other case is symmetric). □

Theorem 2. *The greedy algorithm solves SME-Binary correctly and can be implemented to work in $O(r + (r - k) \log(r - k))$ time.*

Proof. We show that the first step of (1) transforms a given instance (P, m) into an equivalent instance $(Q, m - 1)$, where Q is the path obtained from P by merging u, v, w into z. Consider any solution X to (P, m) that enjoys the property of Lemma 3. Let Y be the solution to $(Q, m - 1)$ which contains all vertices of X other than u, v, w, and where $z \in Y$ iff $u, w \in X$. Since X consists of inner vertices only, so does Y. We also have $|Y| = m - 1$ and $l(Y) = l(X) - l(v)$. Conversely, for any solution Y to $(Q, m - 1)$, let X be the solution to (P, m) which contains all vertices of Y other than z, and where $u, w \in X$ if $z \in Y$, and $v \in X$ if $z \notin Y$. Note that X has the property as in Lemma 3, $|X| = m$, $l(X) = l(Y) + l(v)$, and X consists of inner vertices only (since Y does). This correspondence makes the problem instances (P, m) and $(Q, m - 1)$ equivalent: By our definition of $l(z)$, the objective function is shifted in both directions by the same amount $l(v)$. In particular, the transformations turn an optimal X into an optimal Y and vice versa.

By induction, loop (1) eventually yields an equivalent trivial instance $(R, 0)$ with $\emptyset$ as the only solution. In order to get back an optimal solution to (P, m) we may trace back (1), expanding in each step the vertices merged by (1) and applying the above transformation. We show that (2) gives the same result in a simpler way. This is done by induction on the (odd) length of any subpath of P merged into one vertex of R: A subpath of one vertex has no vertices with even positions, and after any expansion of a vertex into three, the solution still contains exactly the vertices with even positions in their respective subpaths. This shows the correctness.

Phase (2) is done in $O(r)$ time. Phase (1) may be implemented with a doubly linked list for the vertices and a priority queue that returns the minimum and supports deletions and insertions in logarithmic time (in the maximum size of the queue) per operation. Note that the priority queue is needed only for the $m \approx (r - k)/2$ smallest weights. □

The greedy algorithm is faster than dynamic programming (Theorem 1) if $k = \omega(\log r)$. In the next section we get rid of the priority queue and the logarithmic factor by more structural preprocessing on the sequence. We build and use, in linear time, a special tree describing monotonicity relations in the sequence of weights. This tree gives us the results of the greedy algorithm (hence correct results) without actually performing its steps.

4   Tree-Based Linear-Time Algorithm for Binary Strings

Remember that, after some preprocessing, an instance of SME-Binary is represented as a path of weighted vertices $(v_1, \ldots, v_r)$, where $l(v_1), l(v_r)$ do not matter, hence we may set them to a huge number, e.g., larger than the sum of all other weights. Then $l(v_1), l(v_r)$ remain always maximal, even after any merge operations. In the following we build our supporting data structure.

W-trees: We multiply the weights of vertices in their given order alternatingly by -1 and $+1$, the start sign is arbitrary. Distinguish carefully between these *signed weights* and their absolute values, simply called *weights* further on. We will sometimes compare disjoint segments by their total (signed) weights. For tie-breaking when equality occurs, we apply some arbitrary priority rule, e.g., the left interval is always considered the "smaller" one in such cases.

Definition 1. *An ordered set of at least three vertices is a W-segment if the weights of all vertices are, alternatingly, local maxima and minima, where the weight of the first and last vertex, respectively, is larger than the weight of its only neighbor in the segment.*

The name W-segment is inspired by the "zigzag" up-and-down pattern of weights. Now we construct a rooted and ordered tree which we call a *W-tree* of the sequence. To avoid confusion, we speak of *nodes* rather than vertices in the tree. The term *ordered tree* means that a left-to-right order is defined among the children of any node. This naturally induces an order on the leaves. A *sibship* is the orderd set of all children of some node.

Definition 2. *A rooted and ordered tree with signed and weighted nodes is called a W-tree of a sequence if it enjoys the following properties.*
(i) The leaf nodes correspond to the vertices in the given sequence, in the given left-to-right order and with the given signed weights.
(ii) The signed weight of any non-leaf is the sum of signed weights of its children.
(iii) The ordered set of children of any node is a W-segment (except perhaps the root which may have exactly two children).
(iv) The weight of a node is no larger than the weight of its parent.

(v) Every node (except the leaves and perhaps the root) has an odd number of children. In every sibship of odd size, we call a node odd/even if it has an odd/even position in the sibship (where the leftmost node has number 1, etc).
(vi) The weight of an even node is never larger than the weights of its (odd) neighbored siblings.

Properties (v) and (vi) follow from (iii), but we stated them for later use.

Lemma 4. *We can construct a W-tree of a weighted path in $O(r)$ time.*

Proof. We describe an algorithm that constructs a W-tree. At any moment we maintain an ordered set of trees and consider the sequence of their roots in their natural left-to-right order. Initally, all these trees consist of only one node, hence the sequence of roots is just the given sequence.

In every step we take an arbitrary W-segment of current roots and connect them to a *new* root, the signed weight of which is defined according to (ii). In the new sequence of roots, the signs of weights still alternate, this is obvious from Definition 1. Thus we can iterate this step. Since the two outermost roots have huge weights, we find some W-segment in every step, i.e., the process stops only when everything is merged into one tree (r odd) or two trees (r even). In the latter case we connect the two huge-weight roots again to a new root.

We obtain in fact a W-tree: (i) holds since the sequence of leaves and their signed weights are never changed. Properties (ii) and (iii) are true by construction. These and Definition 1 yield (iv), and the rest follows from (iii).

Finally we argue that the construction can be done in $O(r)$ time. The current sequence of roots is stored as a doubly-linked list to support fast local changes. Observe that some W-segment is found around an arbitrary local minimum in the sequence of weights. More precisely, a local minimum together with its two neighbors form a W-segment (which might be extended by pairs of further elements in both directions, if their weights satisfy Definition 1). Adding the signed weights and replacing a W-segment with a new root costs $O(1)$ time per node. By storing all local minima separately in a queue, we can take a local minimum in each step in $O(1)$ time. This is correct because each local minimum is preserved until we use the corresponding node in a W-segment: Since, by (iv), a parent's weight is not smaller than the children's weights, the weights of the neighbors of any local minimum can only increase. In summary, every node is processed $O(1)$ times. Since inner nodes of the W-tree have at least three children, the W-tree has size $O(r)$, which gives the $O(r)$ time bound.   □

Tracing: Now we "trace" our greedy algorithm for SME-Binary: For every step we update our W-tree so as to obtain a W-tree of the shorter string.

Recall that the greedy algorithm merges some vertex v with minimum $l(v)$ and its neighbors u, w, and assigns the merged vertex the weight $l(u) + l(w) - l(v)$. Due to (iv) and (vi), we may always take an even leaf as v. Let u' and w' be the siblings of v next to the left and right, respectively, in the W-tree. Either we have $u' = u$, or u' is an ancestor of u, and similarly with w' and w. Moreover, u', w' are odd nodes. We proceed step by step as follows in the resulting cases.

(1) *None of u', w' is a leaf:* Then merge u' and w' to a new odd node z'. The children of z' are, from left to right, the children of u', node v, and the children of w'. The signed weight of z' is the sum of signed weights of u', v, w'.

(2) *Exactly one of u', w' is a leaf, say w' (the other case is symmetric):* Then u' adopts v and w' as two new rightmost children. To the signed weight of u' add the signed weights of v and w'.

(3) *Both u' and w' are leaves:* Then merge u', v, w' to a leaf z'. The signed weight of z' is the sum of signed weights of u', v, w'.

These steps are repeated, with updated u', w', until case (3) appears. Furthermore, whenever a node retains only one child, parent and child have the same signed weight, and we identify them immediately by contracting the edge.

The procedure will always terminate, as v "goes down the tree". Check that, due to the choice of v, every step preserves properties (i)-(vi) of a W-tree. Hence, upon termination we get a W-tree of the updated sequence, as claimed.

The following properties of the tracing procedure are crucial: An even node remains even until it is merged with its neighbored odd siblings into a new odd node. Odd nodes are merged into odd nodes only. Hence an even node never changes its (signed) weight until it disappears. The even node that disappears next is always one with smallest weight. It follows that the even nodes are processed in their fixed order of increasing weights. Hence, after any number of steps of the greedy algorithm, the W-tree obtained by tracing contains exactly those even nodes with weights above some threshold thr, plus certain odd nodes.

Getting the high nodes: Given a threshold thr, we call a node *high/low* if its weight is larger/smaller than thr. The following procedure *Extract(thr)* constructs, from the initial W-tree of the given string, the W-tree obtained by tracing up to weight thr, but without actually performing the tracing.

Extract(thr):
(1) Retain the high even nodes and all their ancestors and siblings. Delete all other nodes.
(2) In every sibship that contains high even nodes, do the following. Between any two high even nodes, and on the left/right side of the leftmost/rightmost high even node in the sibship, merge all siblings into one node and and let the signed weight of the new node be the sum of signed weights of merged siblings.

Lemma 5. *Procedure Extract(thr) computes, in $O(r)$ time, a W-tree of the segmentation obtained by running the greedy algorithm for SME-Binary as long as the weight of the middle vertex in a merging step does not exceed thr.*

Proof. (sketch) Every low even node becomes a leaf during tracing, and then it is, together with its two neighbored siblings, replaced with one new node in the sibship. It follows that all siblings between the high even nodes are eventually merged. Furthermore, invariant (ii) holds, so that adding the signed weights is also correct. The time bound is obvious. □

Once we have the W-tree, it is easy to reconstruct the segmentation in linear time, since its leaves represent the segments.

Number of even nodes and final result: To get out the desired number k of segments, we have to use the appropriate *thr* in *Extract(thr)*.

Lemma 6. *A W-tree with e even nodes has $k = 2e + 1$ leaves if the root has an odd number of children, and $k = 2e + 2$ leaves if the root has two children.*

Proof. It suffices to show the first part, then the second part follows immediately. We proceed by induction. The Lemma is true if the W-tree is just a root with an odd number of leaves as children. If we make some leaf the parent of a new sibship of, say, $2l + 1$ nodes, then e and k increases by l and $2l$, respectively. □

By Lemma 6, all we need is the $(k-1)/2$ or $(k-2)/2$ even nodes with largest weights, to get a W-tree of a segmentation into k segments. Summarizing the process, we can now state the main result. Regarding linear-time selection see, e.g., [6], or common textbooks on algorithm design. *Selection* means to determine a prescribed number of largest elements in an unsorted set of numbers.

Theorem 3. *SME-Binary can be solved in $O(r)$ time for any k.*

Proof. From the given string, build a W-tree as in Lemma 4. By Selection, mark the number of high even nodes specified in Lemma 6. Let *thr* be the smallest weight of the high even nodes. Compute a W-tree of a segmentation by *Extract(thr)* as in Lemma 5, and from the leaves of this W-tree straightforwardly recover the segmentation. □

There remain some practical issues. Selection in worst-case linear time suffers from a large hidden constant. We may use randomized selection with pivot elements and content ourselves with expected linear time (which, however, should be faster in general). For $k \ll r$, we may determine the high even nodes in time $O(r + k \log k)$ (with small hidden constant) using a priority queue. We would consider the even nodes of the W-tree at increasing distance from the root, since due to (iv) we can prune a subtree below a current node x if $l(x)$ is already smaller than the kth largest weight found so far.

The algorithm is easy to apply. We illustrate this for string 001010001110011, where k is even, and the outer runs are not exception runs. The dendrogram-like table shows a W-tree. Note that the huge-weight dummy vertices are not displayed. The even nodes (printed in bold) with largest weights yield the following segment borders for $k = 2, 4, 6$: 00101000|1110011, 00101000|111|00|11, 00|101|000|111|00|11. The other cases (where both outer runs are exception runs, or k is odd) give worse solutions for this string.

-2	$+1$	$-\mathbf{1}$	$+1$	-3	$+\mathbf{3}$	$-\mathbf{2}$	$+2$
		$+1$			$+\mathbf{3}$		
		-4					

5   Dynamic Programming with Penalties

In this section we judge another dynamic programming approach to SME that easily comes to mind. Let us scan the given string S from left to right and fix some penalty $p > 0$ for changing the designated symbol, thus starting a new segment. Define the *score* of a segmentation as the number of exceptions plus the sum of penalties. A segmentation with minimum score is computable in $O(br)$ time: For each symbol s in the alphabet and each prefix of S, maintain a segmentation of this prefix, with s as the last designated symbol, and minimum score. The $O(b)$ calculations needed for extending the prefix by a new run are obvious. The catch is that we must find a penalty p which gives the desired number k of segments. In the following we discuss only case $b = 2$ closer.

If k is even/odd, we force the "penalty algorithm" to output an even/odd number j of segments by appending an empty segment in the end, if j has the wrong parity. That is, we add another p to the score in that case, and get $j := j+1$ segments. This modification is motivated by an observation that might be interesting in itself:

Theorem 4. *For a given binary string, let $x(k)$ be the minimum number of exceptions in a segmentation with k segments, where argument k is restricted to the even/odd positive integers. Then, function $x(k)$ is monotone decreasing and convex.*

Proof. Both assertions follow from correctness of the greedy algorithm for SME-Binary: Each step reduces k by 2, increases the number of exceptions by the weight of the middle vertex of the merged triple, and these weights increase. ☐

Remark: Remember that the greedy optimal solutions for all k with the same parity are "nested", i.e., new borders of segments are inserted when k grows. Interestingly this is in general not true for different parities. An example is $S = 00110001$. The only segmentation with $k = 2$ and 2 exceptions is $0011000|1$, and the only segmentation with $k = 3$ and 1 exception is $00|11|0001$.

Back to dynamic programming: The score of a segmentation with k segments and x exceptions is $x + pk$. Convexity (Theorem 4) and simple geometric reasoning yields that, for any k of the considered parity (even or odd) there exists an integer p so that an optimal segmentation with k segments has also the optimal score with respect to penalty p. That is, in principle we can solve SME-Binary via $O(r)$-time dynamic programming with a suitable penalty p, however it seems that binary search is needed to find this p. We conclude that the penalty approach looks promising at first glance, but it does not give us a simpler and (practically) faster algorithm.

6   A Parameterized Weighted Independent Set Problem

The graph problem specified in Lemma 2 (for general graphs rather than just paths) is interesting in itself. In the following we call it CARDINALITY MINIMUM WEIGHT INDEPENDENT SET (CMWIS). (Here, *minimum* may also be *maximum* without changing the problem, since m is fixed.) By Theorem 3, CMWIS on paths of length r is solable in $O(r)$ time. Slightly more generally, CMWIS is linear for graphs with maximum degree $d = 2$. By way of contrast we have:

Theorem 5. *CMWIS is NP-complete for graphs of any fixed degree $d \geq 3$.*

Proof. We reduce MAXIMUM INDEPENDENT SET for degree $d \geq 3$ to CMWIS: Assign unit weights to every vertex. An independent set of size *at least* m exists iff an independent set of size *exactly* m and weight *at most* m exists. □

Due to this hardness result it is sensible to consider parameterized algorithms (see [7,12] for introductions). An alternative as in Lemma 3 gives:

Theorem 6. *CMWIS on graphs of degree d, with exactly m vertices in the solution, can be solved in $O^*(x^m)$ time, where $x = (\sqrt{2d^2 - 2d + 1} + 1)/2 \approx d/\sqrt{2}$.*

Proof. We give a search tree algorithm. Whenever a new vertex has been added to the independent set, remove this vertex and its neighbors from the graph. Consider any node in the search tree where fewer than m vertices are chosen so far. Let v be a vertex in the rest graph with minimum $l(v)$. In one branch, add v to the solution. In all other branches (where v is not taken) we can add *at least two* of v's neighbors to the solution, because: If only one neighbor u is taken, we can replace u with v in any solution and get an independent set with smaller weight. If no neighbor of v is taken, some other vertex w adjacent to neither v nor previously taken vertices must be in the solution (since size m is not yet exhausted), but then we can replace w with v. Hence we reduce m by 1 in one branch and by 2 in at most $\binom{d}{2}$ branches. Characteristic equation $x^2 = x + d(d-1)/2$ yields the base. □

The time is likely to be improved by more advanced techniques for parameterized algorithms. CMWIS is solved in [5] for the more general class of d-degenerate graphs, however the time as a function of the parameters is $O^*(2^{dm})$ there. Finally we observe that a small problem kernel is very easy to obtain:

Theorem 7. *CMWIS has a kernel of at most $(d + 1)(m - 1) + 1$ vertices.*

Proof. We claim that the set K of $(d + 1)(m - 1) + 1$ vertices with smallest weights contains an optimal solution: Consider any X, $|X| = m$, where $K \cap X$ is independent and a vertex $v \in X \setminus K$ exists. At most $m - 1$ vertices of X are in K, each having at most d neighbors in K. Hence some $u \in K$ is neither in X nor adjacent to any vertex of $K \cap X$. Replacing v with u in X reduces $l(X)$, and $K \cap X$ is still independent. In particular, if we start from an independent X, some iterations give an independent set of smaller weight, contained in K. □

Computing a minimum weight independent set with m vertices is (probably) not FPT with parameter m alone: If all weights are equal, we have to figure out existence of an independent set with m vertices, but this is known to be $W[1]$-complete. However, the same problem with dual parameter $n - m$ is a variation of WEIGHTED VERTEX COVER, thus it is FPT even for unbounded degree.

7   Some Open Problems

Our main result is a linear-time algorithm for partitioning a binary string into segments with one sort of symbols, so as to minimize the total number of exceptions. Our algorithm is based on a specific tree data structure. It remains open whether similar ideas can beat dynamic programming also for alphabet size $b > 2$. We announce that, in the full paper, we give an $O(r(k + \log b))$ time algorithm that still uses dynamic programming, but after a deeper problem analysis extending some observations from Section 2-3. Other interesting directions are SME on data streams, and partioning sequences of real numbers into segments where large and small numbers abound, with varying thresholds.

References

1. Bennett, K.P.: Decision tree construction via linear programming. In: 4th Midwest AI and Cognitive Science Soc. Conf., pp. 97–101 (1992)
2. Bennett, K.P., Mangasarian, O.L.: Robust linear programming discrimination of two linearly inseparable sets. Optimization Methods and Software 1, 23–34 (1992)
3. Bennett, K.P., Mangasarian, O.L.: Bilinear separation of two sets in n-space. Computational Optimization and Applications 2, 207–227 (1993)
4. Bennett, K.P., Mangasarian, O.L.: Multicategory separation via linear programming. Optimization Methods and Software 3, 27–39 (1993)
5. Cai, L., Chan, S.M., Chan, S.O.: Random separation: A new metod for solving fixed-cardinality optimization problems. In: Bodlaender, H.L., Langston, M.A. (eds.) IWPEC 2006. LNCS, vol. 4169, pp. 239–250. Springer, Heidelberg (2006)
6. Dor, D.: Selection algorithms, PhD thesis, Tel-Aviv Univ. (1995)
7. Downey, R.G., Fellows, M.R.: Parameterized Complexity. Springer, Heidelberg (1999)
8. Fragkou, P., Petridis, V., Kehagias, A.: A dynamic programming algorithm for linear text segmentation. J. Intell. Inf. Syst. 23, 179–197 (2004)
9. Haiminen, N., Gionis, A.: Unimodal segmentation of sequences. In: Perner, P. (ed.) ICDM 2004. LNCS (LNAI), vol. 3275, pp. 106–113. Springer, Heidelberg (2004)
10. Himberg, J., Korpiaho, K., Mannila, H., Tikanmäki, J., Toivonen, H.: Time series segmentation for context recognition in mobile devices. In: 1st IEEE Int. Conf. on Data Mining ICDM 2001, pp. 203–210 (2001)
11. Kleinberg, J., Tardos, E.: Algorithm Design. Pearson/Addison-Wesley (2006)
12. Niedermeier, R.: Invitation to Fixed-Parameter Algorithms. Oxford Lecture Series in Mathematics and its Applications. Oxford Univ. Press, Oxford (2006)

Characterisations and Linear-Time Recognition of Probe Cographs

Van Bang Le and H.N. de Ridder

Institut für Informatik, Universität Rostock, 18051 Rostock, Germany
{le,hnridder}@informatik.uni-rostock.de

Abstract. Cographs are those graphs without induced path on four vertices. A graph G is a *probe cograph* if its vertex set can be partitioned into two sets, N (non-probes) and P (probes) where N is independent and G can be extended to a cograph by adding edges between certain non-probes. A *partitioned* probe cograph is a probe cograph with a given partition in N and P.

We characterise probe cographs in several ways. Moreover, we characterise partitioned probe cographs in terms of five forbidden induced subgraphs. Using the forbidden induced subgraph characterisation, we give a linear-time recognition algorithm for probe cographs, improving the recent quadratic-time recognition algorithm by Chandler et al. Our algorithm is a modification of the linear-time recognition algorithm for cographs by Corneil et al.

1 Introduction

In 1994, in the context of genome research, Zhang [26] introduced probe interval graphs. A graph is a probe interval graph if its vertex set can be partitioned into two sets, probes P and non-probes N, such that N is independent and new edges can be added between certain non-probes in such a way that the resulting graph is an interval graph. This definition can of course readily be generalized to some graph class $\mathcal{C}$: A graph is probe $\mathcal{C}$ if its vertex set can be partitioned into two sets, probes P and non-probes N, such that N is independent and new edges can be added between non-probes in such a way that the resulting graph is in $\mathcal{C}$. If a partition into probes and non-probes is given, we talk about a *partitioned* probe class, otherwise about an *unpartitioned* one.

Probe interval graphs have been further discussed in [19,22,23]. In [25] and [24], probe interval graphs which are trees, respectively, 2-trees are considered.

Actually, already in 1989 — before Zhang's article — Hertz [15] defined what he called *slim* graphs, which are in fact probe Meyniel graphs, and proved that these are perfect. Hoàng and Maffray [16] used Hertz' construction to define probe Gallai graphs.

In a series of recent articles [2,3,5,6,7,8,10,13,21] several probe classes are discussed. So far, only probe graphs of perfect graphs have been considered; here, perfect graphs are those graphs in which the chromatic number of each induced subgraph is equal to its clique number, and it is known that perfect graphs can

A. Brandstädt, D. Kratsch, and H. Müller (Eds.): WG 2007, LNCS 4769, pp. 226–237, 2007.

be recognized in polynomial time. The results established lead to the following interesting conjectures.

Probe Perfect Graph Conjecture (PPGC) ([10]). *Partitioned probe perfect graphs are polynomially recognizable.*

Strong Probe Perfect Graph Conjecture (SPPGC) ([10]). *Probe perfect graphs are polynomially recognizable.*

Note that if recognizing a class $\mathcal{C}$ is NP-complete, then recognizing partitioned probe graphs of $\mathcal{C}$ is also NP-complete: consider partitioned graphs with at most one non-probe. We conjecture that the converse holds, too.

Probe Graph Conjecture (PGC). *Partitioned probe graphs of $\mathcal{C}$ are polynomially recognizable whenever $\mathcal{C}$ is polynomially recognizable.*

Strong Probe Graph Conjecture (SPGC). *Probe graphs of $\mathcal{C}$ are polynomially recognizable whenever $\mathcal{C}$ is polynomially recognizable.*

As perfect graphs can be recognized in polynomial time, the truth of each of our conjectures implies the corresponding probe perfect graph conjecture.

Cographs (graphs without induced path on four vertices) form, in a sense, a basic class of perfect graphs and are well-investigated in the literature. In [10], Chang et al. give an $O(n^3)$ recognition for partitioned probe cographs and an $O(n^5)$ recognition for probe cographs (n is the number of vertices of the input graph). Chandler et al. [5] improved this to $O(n^2)$ by showing that the recognition of unpartitioned probe cographs can be reduced to the partitioned case in linear time and reducing recognition of partitioned probe cographs to the recognition of partitioned probe distance hereditary graphs, for which they give an $O(n^2)$ algorithm.

In this article we will characterise probe cographs in several ways. As a by-product, we give several characterisations for partitioned probe cographs, one of which is in terms of five forbidden induced subgraphs. Using this forbidden subgraph characterisation we give a linear recognition algorithm for partitioned probe cographs by modifying the linear-time recognition algorithm for cographs by Corneil et al. [12]. As mentioned above, the recognition of unpartitioned probe cographs can be reduced to the partitioned case in linear time, therefore unpartitioned probe cographs can be recognized in linear time, as well.

All graphs considered are finite, undirected and simple. Thus, the edge set E of a graph $G = (V, E)$ is a subset of $\binom{V}{2}$, the set of all 2-element subsets of the vertex set V. We often write $xy \in E$ for $\{x, y\} \in E$. The complement of G is written $\overline{G}$. For two disjoint sets of vertices X and Y, $X \text{①} Y$ ($X \text{⓪} Y$) (pronounced "join" and "cojoin", respectively) means that every vertex in X is adjacent (nonadjacent) to every vertex in Y. A set of vertices is called *independent* or *stable* if the vertices are pairwise non-adjacent and it is called a *clique* if they are pairwise adjacent. The *neighbourhood* $N(v)$ of a vertex v is the set of its neighbours. For a set of vertices X we write $N(X) = \bigcup_{v \in X} N(x) \setminus X$. If necessary, the graph in which the neighbourhood is considered is written as a subscript: $N_G(x)$. Given a set of vertices $X \subseteq V$, the subgraph induced by X is written $G[X]$ and $G - X$ stands for $G[V \setminus X]$. We often identify a subset of vertices with the subgraph induced by that subset, and vice versa.

For a set X and a singleton $\{x\}$, we usally write $X + x$ for $X \cup \{x\}$ and $X - x$ for $X \setminus \{x\}$. A *module* M in a graph is a set of vertices such that every vertex outside M is either adjacent to all vertices in M, or to none; if $M = \{x, y\}$, then x, y are called *twins*.

P_n denotes the chordless path on n vertices. The vertices of degree one are called the *endpoints* and the others the *midpoints* of the path. C_n denotes the chordless cycle on n vertices.

For convenience, we will use the following notion. For a graph class $\mathcal{C}$ and a graph $G = (V, E)$, a partition $V = P \cup N$ with an independent set N is called a *valid partition* (with respect to $\mathcal{C}$) if there exists $E' \subseteq \binom{N}{2}$ such that $G' = (V, E \cup E')$ is a member of $\mathcal{C}$. Such a graph G' is then called a *valid extension* for G. Thus, G is a probe graph of $\mathcal{C}$ if and only if G has a valid partition, and $G = (P, N, E)$ is a partitioned probe graph of $\mathcal{C}$ if and only if $V(G) = P \cup N$ is a valid partition.

2   Probe Cographs Versus Singular Cograph Contractions

Recall that cographs (or complement-reducible graphs) [11,12] are those graphs that can be constructed from a single vertex by repeated applications of complement and disjoint union. They are precisely the P_4-free graphs and are known under many different names and definition, see for example [18]. Cographs are discussed in more detail in section 4.

Cographs were generalized to cograph contractions as follows. A graph G is called a *cograph contraction* if there exists a cograph H together with some pairwise disjoint independent sets $S_1, \ldots, S_t$ in H such that G is obtained from H by contracting each S_i to a single vertex s_i and then making the vertices $s_1, \ldots, s_t$ to a clique. Cograph contractions have been introduced and investigated in [17] in connection to graph precoloring extensions and perfection, and have been characterised in [20]. In [28] it is shown that the set of minimal forbidden induced subgraphs for cograph contractions is finite.

A special case of cograph contractions where all independent sets $S_1, \ldots, S_t$ are one-element sets has a close relationship to probe cographs. We formulate this situation in a more general setting.

Definition 1. *Let $\mathcal{C}$ be a class of graphs. A graph G is called a singular $\mathcal{C}$ contraction if there exists a graph $H \in \mathcal{C}$ together with a (not necessarily independent) set S of vertices of H such that G is obtained from H by making S to a clique.*

Let co-$\mathcal{C}$ be the class consisting of the complements of all graphs in $\mathcal{C}$. $\mathcal{C}$ is *self-complementary* if $\mathcal{C}$ and co-$\mathcal{C}$ coincide. Singular $\mathcal{C}$ contractions are related to probe graphs of $\mathcal{C}$ as follows.

Proposition 1. *A graph G is a probe graph of C if and only if $\overline{G}$ is a singular co-C contraction. In particular, if C is self-complementary, then probe graphs of C coincide with the complements of singular C contractions.*

Proof. Suppose that $G = (V, E)$ is probe graph of C, and let $V = P \cup N$ be a valid partition with N independent and $E' \subseteq \binom{N}{2}$ such that $H = (V, E \cup E')$ is a member of C. Then $\overline{G}$ is obtained from $\overline{H} \in$ co-C by making N to a clique. That is, $\overline{G}$ is a singular co-C contraction.

Conversely, suppose $\overline{G}$ is a singular co-C contraction. Let $H \in$ co-C together with a subset $S \subseteq V(H)$ such that $\overline{G}$ is obtained from H by making S to a clique. Then S is an independent set of G, and the graph $G' = (V, E \cup E')$ coincides with $\overline{H}$, where E' is the set of edges in $\overline{H}$ with both endvertices in S. Thus, as $G' \in G$, G is a probe graph of C.   □

Corollary 1. *Probe cographs are exactly complements of singular cograph contractions. Also, probe perfect graphs are exactly complements of singular perfect contractions.*

A *bull* consists of five vertices, a, b, c, d and e, and a, b, c form a triangle and d is adjacent exactly to a, e is adjacent exactly to b. Singular cograph contractions, hence (complements of) probe cographs, can be characterised as follows.

Theorem 1. *A graph G is a singular cograph contraction if and only if G has a clique Q such that every P_4 has two vertices in Q, every $\overline{P_5}$ has its triangle in Q, and every bull has a vertex of degree one or a vertex of degree two in Q.*

The proof of Theorem 1 will be given in the full version of this paper.

3   Characterising Probe Cographs

In this section we characterise unpartitioned probe cographs in several ways.

Let N be an independent set in $G = (V, E)$. Following [5] we call two vertices x, y in G *twins with respect to N* if both x, y are in N or outside N and x, y are twins in G, or $x \in V \setminus N$, $y \in N$ and $N(y) - x = N(x) \setminus N$.

In a bull, vertices belonging to the triangle of the bull are called *mid-points* of the bull.

Theorem 2. *The following statements are equivalent for any graph G.*

(i) *G is a probe cograph;*
(ii) *G admits an independent set N meeting every P_4 in two vertices, every P_5 in three vertices, and every bull in a mid-point;*
(iii) *$\overline{G}$ is a singular cograph contraction;*
(iv) *G admits an independent set N such that, for each induced subgraph H of G with at least two vertices, H has two twins with respect to $N \cap V(H)$.*

Proof. By Corollary 1, (i) ⇔ (iii). By Theorem 1, (iii) ⇔ (ii).

(i) ⇔ (iv): A corresponding characterisation for partitioned probe distance-hereditaray graphs has been observed in [5]. The proof of [5, Theorem 1] can be adopted here by noting that any cograph with at least two vertices has always two twins.   □

A graph is called *weakly chordal* ([14]) if it does not contain any C_ℓ or $\overline{C_\ell}$, $\ell \geq 5$, as an induced graph.

Corollary 2. *Probe cographs are P_6-free weakly chordal.*

We will see in the next section that cographs can be even recognized in linear time (hence also singular cograph contractions). At this time we are unable to find a complete list of (minimal) forbidden induced subgraphs for probe cographs. In the partitioned case, however, we have a characterisation in terms of five forbidden induced subgraphs.

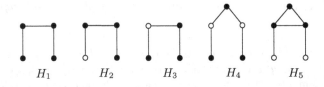

Fig. 1. Forbidden subgraphs for partitioned probe cographs (black vertices probes, white vertices non-probes)

Theorem 3. *Let $G = (P, N, E)$ be a partitioned graph with an independent set N. Then the following statements are equivalent.*

(i) *G is a partitioned probe cograph;*

(ii) *In G, every P_4 has two vertices, every P_5 has three vertices, and every bull has a mid-point in N;*

(iii) *None of the graphs $H_1, \ldots, H_5$ depicted in Fig. 1 is an induced subgraph in G;*

(iv) *$\overline{G}$ is obtained from a cograph on the same vertex set by making N to a clique;*

(v) *Each induced subgraph H of G with at least two vertices has two twins with respect to $N \cap V(H)$.*

Proof. Note that (ii) and (iii) are clearly equivalent. The rest follows imediately from (the proof of) Theorem 2.   □

4   Recognizing Probe Cographs

In this section we give a linear time recognition algorithm for partitioned probe cographs. As mentioned, the unpartitioned case can be reduced to the partitioned case in linear time, probe cographs therefore can be recognized in linear time.

Cographs. Our recognition algorithm for partitioned probe cographs is based on Algorithm 1 below for recognition of cographs due to Corneil at al. [12]. As (the linear time bound of) Algorithm 1 is far from being obvious, we assume that the readers are familiar with this algorithm. Note that we reformulated this algorithm in comparison to [12] in order to make it more readable.

Algorithm 1. COGRAPH-RECOGNITION [12] Given a graph $G = (V, E)$ with $V = \{v_1, \ldots, v_n\}$, determine whether G is a cograph and construct the cotree T for G if it is.

1: Create a new ①-node R
2: **if** $v_1 v_2 \in E$ **then**
3:     add v_1, v_2 as children of R
4: **else**
5:     create a new ⓪-node N with children v_1, v_2; add N as a child of R
6: **for** $x \leftarrow v_3, \ldots, v_n$ **do**
7:     MARK(x)
8:     **if** all nodes of T are marked and unmarked **then**
9:         add x as a child of R
10:        **continue** with the next iteration of the for-loop
11:    **if** all nodes of T are not marked **then**
12:        **if** $deg(R) = 1$ **then**
13:            add x as a child of the only child of R
14:        **else**
15:            create a new ①-node R with one child (a new ⓪-node) and two grand-children: x and the old root
16:        **continue** with the next iteration of the for-loop
17:    $u \leftarrow$ FIND-LOWEST
18:    Let A denote the set of children of u that are marked and unmarked and B the set of children of u that are not marked
19:    **if** $label(u) = 0 \wedge |A| = 1$ **then**
20:        **if** $w \in A$ is a leaf **then**
21:            add a new ①-node in place of w and make w, x children of this node
22:        **else**
23:            add x as a new child of w
24:    **else if** $label(u) = 1 \wedge |B| = 1$ **then**
25:        **if** $w \in B$ is a leaf **then**
26:            add a new ⓪-node in place of w and make w, x children of this node
27:        **else**
28:            add x as a new child of w
29:    **else**
30:        remove all elements of A from u and add them as children of a new node y with $label(y) = label(u)$
31:        **if** u is a ⓪-node **then**
32:            add a new ①-node as a child of u; children of this new ①-node are x, y.
33:        **else**
34:            remove u from its parent and add y in its place; add a new ⓪-node as a child of y; children of this new ⓪-node are x, u.

For a node x in the cotree, $deg(x)$ returns the number of children in the cotree of x, $label(x)$ returns 0 if x is a ⓪-node and 1 if x is a ①-node. For the definition of the procedures MARK and FIND-LOWEST, we refer to [12] (but see below). Note that the algorithm always creates a cotree with a ①-node as the root; if G is disconnected, the root will have just one single child, a ⓪-node. Alg. 1 builds the cotree for G incrementally. It starts with the cotree T for $G[v_1, v_2]$ (line 1–5) and then tries to add the other vertices to T one by one (the for-loop in line 6).

Assume T is the cotree for $G' = G[v_1, \ldots, v_i]$ (thus, G' is a cograph) and $x = v_{i+1}$. First, in line 7, information about the neighbours of x is propagated up T by MARK, using a marking scheme. Every node t in T can be in one of three states: It is either 'not marked', or it is 'marked' or it is 'marked' and subsequently 'unmarked'. If x is adjacent to all or no vertices in T, this is enough to determine that $G' + x$ is a cograph and how x should be added to T (line 8–16). Otherwise FIND-LOWEST is called. For FIND-LOWEST the following properties hold:

Proposition 2. *Let T be the cotree for the cograph G'. If $G' + x$ is not a cograph then* FIND-LOWEST(x) *will return an error. Otherwise a node u in T is returned. Let A be the set of children of u that were marked and B the set of children that were not marked. Then the following properties holds:*

1. *$G' + x$ is a cograph.*
2. *A, B are both non-empty.*
3. *$A \cup B \cup \{x\}$ is a module in $G' + x$.*
4. *x is adjacent to all vertices in the subtrees rooted at a node in A.*
5. *x is non-adjacent to all vertices in the subtrees rooted at a node in B.*

If in the for-loop the current vertex x cannot be added to the tree, the call to FIND-LOWEST will fail and the algorithm will terminate. Otherwise x will be added in one of the following lines:

line 9 x is adjacent to all vertices in the cotree.

line 13,15 x is adjacent to no vertices in the cotree. x gets inserted in line 13 if the cotree represents a disconnected graph and in line 15 if it represents a connected graph.

line 21,26 x is a twin of w; a true twin if u is a ⓪-node, a false twin if u is a ①-node.

line 23,28 Let M be the module under w. Every vertex outside M that is adjacent (non-adjacent) to M is also adjacent (non-adjacent) to x and x is adjacent to all vertices in M if w is a ①-node and non-adjacent to all vertices in M if w is a ⓪-node.

line 30–34 x is adjacent to all vertices in A and non-adjacent to all vertices in B. Other vertices have adjacencies to x as to $A \cup B$ (so $A \cup B \cup x$ is a module).

Note that the adjacencies between the vertices that are already in the cotree are the same before and after adding x.

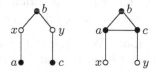

Fig. 2. Labeled H_4 and H_5

Recognizing partitioned probe cographs. We start with some observations concerning probe cographs and Algorithm 1: For a partitioned graph $G = (P, N, E)$ to be a probe cograph, it is necessary that $G[P]$ and for all $x \in N$: $G[P + x]$ are cographs. This is exemplified by the forbidden subgraphs H_1, H_2, and H_3 in Fig. 1. Assume that $G[P+x]$ is a cograph for all $x \in N$ and, moreover, that we have constructed the cotree for $G[P]$. Label the vertices of H_4 and H_5 as in Fig. 2.

If G contains an H_4 as an induced subgraph, then we can add either x or y to the cotree, but not both; see Fig. 3. x, y would be added in line 26–30 of Algorithm 1 with $A = \{a, b\}, B = \{c\}$ for x and $A = \{b, c\}, B = \{a\}$ for y. The problem here is that both adding x and adding y requires the introduction of a new ⓪-node in the cotree, but that the sets of children of these new vertices are neither disjoint, nor is one contained in the other. Note that if a, b, c do not have a common parent in the cotree then there is a probe p such that $\{a, b, c, p\}$ together with x or y contain an induced H_2 or H_3 in G. The case for H_5 is symmetrical, with the roles of ①- and ⓪-nodes in the cotree exchanged. Fortunately, all of those cases can be handled by some modifications of COGRAPH-RECOGNITION (Algorithm 1). Our recognition algorithm for probe cographs is described by Algorithm 2. (The sets UNIVERSAL, ISOLATED, $A(x), B(x)$ for each non-probe x, TWIN(w) for each probe w, and MODULE(w) for each internal node w of the cotree T of $G[P]$ are initially empty. For the procedure MARK and function FIND-LOWEST see [12].)

Given $G = (P, N, E)$, Algorithm 2 first builds a cotree T for $G[P]$. This succeeds if and only if G does not contain an induced H_1. Then it tests (via FIND-LOWEST) if, for each non-probe $x \in N$, $G[P]+x$ is a cograph. This succeeds if and only if G does not contain an induced H_2 or H_3. In case $G[P] + x$ is a cograph, it determines how x should be added to T, without actually doing so (that is, T is the cotree of $G[P]$ throughout lines 5-26). Instead of adding

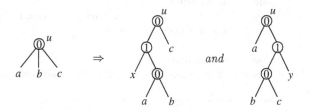

Fig. 3. Attempting to create a cotree for H_4.

x to the cotree, we save the following information (compare the discussion of Algorithm 1; the line numbers refer to Algorithm 1):

line 9 x is adjacent to all probes: We add x to the set UNIVERSAL.

line 13,15 x is adjacent to no probe: We add x to the set ISOLATED.

line 21,26 x is a twin of w: We add x to the set TWIN(w).

line 23,28 x is going to be a member of the module under w: We add x to the set MODULE(w).

Algorithm 2. PROBE-COGRAPH-RECOGNITION Decide whether $G = (P, N, E)$ is a probe cograph and create the cotree T of a valid extension of G if it is.

1: Call COGRAPH-RECOGNITION($G[P]$)
2: **if** $G[P]$ is not a cograph **then**
3:    return false and terminate
4: Let T be the cotree of $G[P]$ returned by COGRAPH-RECOGNITION($G[P]$)
5: **for all** $x \in N$ **do**
6:    MARK(x)
7:    **if** all nodes of T are marked and unmarked **then**
8:       add x to UNIVERSAL
9:       **continue** with the next iteration of the for-loop
10:    **if** all nodes of T are not marked **then**
11:       add x to ISOLATED
12:       **continue** with the next iteration of the for-loop
13:    $u \leftarrow$ FIND-LOWEST
14:    Let A denote the set of children of u that are marked and unmarked and B the set of children of u that are not marked
15:    **if** $label(u) = 0 \wedge |A| = 1$ **then**
16:       **if** $w \in A$ is a leaf **then**
17:          add x to TWIN(w)
18:       **else**
19:          add x to MODULE(w)
20:    **else if** $label(u) = 1 \wedge |B| = 1$ **then**
21:       **if** $w \in B$ is a leaf **then**
22:          add x to TWIN(w)
23:       **else**
24:          add x to MODULE(w)
25:    **else**
26:       $u(x) \leftarrow u$; $A(x) \leftarrow A$
27: **for all** nodes t in T **do**
28:    **if** t is a ⓪-node (①-node) **then**
29:       **if** the sets $A(x)$ ($B(x)$) with $u(x) = t$ are nested or disjoint **then**
30:          add non-probes x to the cotree (see text)
31:       **else**
32:          return false and terminate
33: Add all other non-probes in UNIVERSAL, ISOLATED, TWIN and MODULE to the cotree (see text).

---

line 30–34 We save u, A as $u(x), A(x)$. B needs not be saved as $B(x)$ equals the children of $u(x)$ minus those in $A(x)$. We are going to use this information in the next step.

In line 30 of PROBE-COGRAPH-RECOGNITION, we add the non-probes for line 30–34 of COGRAPH-RECOGNITION appropriately. Let t be an internal $\textcircled{0}$-node ($\textcircled{1}$-node). If i, j are non-probes with $u(i) = u(j) = t$ then $A(i)$ and $A(j)$ ($B(i)$ and $B(j)$) are either disjoint or one is contained in the other:

Assume not, and let $t_a \in A(i) \setminus A(j)$, $t_b \in A(j) \setminus A(i)$ and $t_c \in A(i) \cap A(j)$ for t a $\textcircled{0}$-node. Let a be a vertex in the subtree rooted at t_a, b a vertex in the subtree rooted at t_b and c a vertex in the subtree rooted at t_c. Then a, b, c, i, j induce an H_4. When t is a $\textcircled{1}$-node, symmetrically an H_5 is induced.

Thus, we can order the vertices x with $u(x) = t$ into a rooted tree F such that x_i is a descendant of x_j if and only if $A(x_i) \subseteq A(x_j)$ ($B(x_i) \subseteq B(x_j)$). Then we use F top-down to add the non-probes to T. If x_i is a child of x_j in F with $A(x_i) \subset A(x_j)$ we follow line 30–34 of Algorithm 1, but for t a $\textcircled{0}$-node use $u(x_i) := y(x_j)$; see Fig. 4. If t is a $\textcircled{1}$-node, then TWIN$(u(x_i))$ and MODULE$(u(x_i))$ must be moved to $y(x_i)$. If $A(x_i) = A(x_j)$ ($B(x_i) = B(x_j)$), then x_i, x_j are added as children of the same $\textcircled{1}$-node ($\textcircled{0}$-node). We see that the type ($\textcircled{0}/\textcircled{1}$) of the least common ancestor of vertices already in the cotree does not get changed and that the non-probes x get added so that the least common ancestor of x and a probe p is a $\textcircled{0}$-vertex if $xp \notin E$ and a $\textcircled{1}$-vertex if $xp \in E$.

In line 33 of PROBE-COGRAPH-RECOGNITION, we add the other non-probes to the cotree as well: Let w be a probe with TWIN(w) non-empty. If the parent of w is a $\textcircled{0}$-node ($\textcircled{1}$-node), then we replace w by a new $\textcircled{1}$-node ($\textcircled{0}$-node) with children w and the vertices in TWIN(w). Let w be an internal node of the cotree with MODULE(w) non-empty. We add the vertices in MODULE(w) as children to w. Vertices in UNIVERSAL are added as children of the root R. Finally, if ISOLATED is non-empty, a new $\textcircled{0}$-node R' is created with children R and the vertices in ISOLATED; R' is the new root of the cotree.

It is clear that adding non-probes in this way does not change the adjacencies between vertices already in the cotree and that the non-probes then have the correct adjacencies (compare Proposition 2).

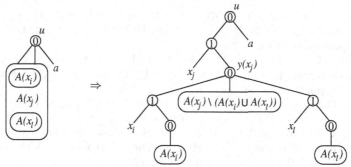

Fig. 4. Illustrating line 30 of Algorithm 2. Note that $|A(x_i)| > 1, |A(x_j)| > 1$ and $|A(x_l)| > 1$.

Thus, if $G = (P, N, E)$ is $(H_1, \ldots, H_5)$-free, we created a cotree for a cograph $G' = (P \cup N, E')$ such that for all probes $p \in P$ and $v \in P \cup N$, $v \neq p$, $pv \in E'$ if and only if $pv \in E$. That is, G' is a valid extension of G.

Theorem 4. *Partitioned and unpartitioned probe cographs can be recognized in linear time. Moreover, a valid extension of a given probe graph can be determined in linear time, too.*

Proof. By the previous discussion and by Theorem 3 (i) $\Leftrightarrow$ (iii), Algorithm 2 correctly recognizes partitioned probe cographs. Concerning the time bound, we will explain in the full version that line 30 is performed in time $O(\sum_{u(x)=t} |A(x)|)$. Then the for-loop at line 27 takes in total $\sum_t O(\sum_{u(x)=t} |A(x)|) \leq O(\sum_{x \in N} |A(x)|) \leq O(\sum_{x \in N} |N_G(x)|) = O(|E(G)|)$ time. The other steps can be done in linear time by linearity of Algorithm 1 and it follows that Algorithm 2 runs in linear time.

By [5, Theorem 5], or [1, Satz 4.17], stating that recognition of unpartitioned probe cographs can be reduced to the recognition of partitioned probe cographs in linear time, recognition of unpartitioned probe cographs is then also linear.

$\square$

References

1. Bayer, D.: Über probe-trivially-perfect und probe-Cographen, Diplomarbeit, Universität Rostock, Institut für Informatik (2006)
2. Bayer, D., Van Bang Le, de Ridder, H.N.: Probe trivially perfect graphs and probe threshold graphs, Manuscript (2006)
3. Berry, A., Golumbic, M.C., Lipshteyn, M.: Recognizing chordal probe graphs and cycle-bicolorable graphs. SIAM J. Discrete Math. 21, 573–591 (2007)
4. Brandstädt, A., Van Bang Le, Spinrad, J.P.: Graph Classes: A Survey, Philadelphia. SIAM Monographs on Discrete Math. Appl., vol. 3 (1999)
5. Chandler, D.B., Chang, M.-S., Kloks, A.J.J., Liu, J., Peng, S.-L.: Recognition of probe cographs and partitioned probe distance hereditary graphs. In: Cheng, S.-W., Poon, C.K. (eds.) AAIM 2006. LNCS, vol. 4041, pp. 267–278. Springer, Heidelberg (2006)
6. Chandler, D.B., Chang, M.-S., Kloks, A.J.J., Liu, J., Peng, S.-L.: Partitioned probe comparability graphs. In: Fomin, F.V. (ed.) WG 2006. LNCS, vol. 4271, pp. 179–190. Springer, Heidelberg (2006)
7. Chandler, D.B., Chang, M.-S., Kloks, A.J.J., Liu, J., Peng, S.-L.: On probe permutation graphs. In: Cai, J.-Y., Cooper, S.B., Li, A. (eds.) TAMC 2006. LNCS, vol. 3959, pp. 494–504. Springer, Heidelberg (2006)
8. Chang, G.J., Kloks, A.J.J., Peng, S.-L.: Probe interval bigraphs. In: 2nd Brazilian Symposium on Graphs, Algorithms and Combinatorics (2005)
9. Chang, G.J., Kloks, A.J.J., Liu, J., Peng, S.-L.: The PIGs full monty - A floor show of minimal separators. In: Diekert, V., Durand, B. (eds.) STACS 2005. LNCS, vol. 3404, pp. 521–532. Springer, Heidelberg (2005)
10. Chang, M.-S., Kloks, A.J.J., Kratsch, D., Liu, J., Peng, S.-L.: On the recognition of probe graphs of some self-complementary classes of perfect graphs. In: Wang, L. (ed.) COCOON 2005. LNCS, vol. 3595, pp. 808–817. Springer, Heidelberg (2005)

11. Corneil, D.G., Perl, Y., Stewart, L.: Cographs: recognition, application and algorithms. Congr. Numerantium 43, 249–258 (1984)
12. Corneil, D.G., Perl, Y., Stewart, L.: A linear recognition algorithm for cographs. SIAM J. Computing 14, 926–934 (1985)
13. Golumbic, M.C., Lipshteyn, M.: Chordal probe graphs. Discrete Appl. Math. 143, 221–237 (2004)
14. Hayward, R.B.: Weakly triangulated graphs. J. Combin. Theory (B) 39, 200–209 (1985)
15. Hertz, A.: Slim graphs. Graphs Comb. 5, 149–157 (1989)
16. Hoàng, C.T., Maffray, F.: On slim graphs, even pairs, and star-cutsets. Discrete Math. 105, 93–102 (1992)
17. Hujter, M., Tuza, Z.: Precoloring extensions III: Classes of perfect graphs. Comb. Prob. Comp. 5, 35–56 (1996)
18. Information System on Graph Classes and their Inclusions (ISGCI), http://wwwteo.informatik.uni-rostock.de/isgci
19. Johnson, J.L., Spinrad, J.P.: A polynomial time recognition algorithm for probe interval graphs. In: Proceedings ACM-SIAM symposium on discrete algorithms, pp. 477–486 (2001)
20. Van Bang Le: A good characterization of cograph contractions. J. Graph Theory 30, 309–318 (1999)
21. Van Bang Le, de Ridder, H.N.: Probe split graphs. Discrete Mathematics and Theoretical Computer Science (to appear)
22. McMorris, F.R., Wang, C., Zhang, P.: On probe interval graphs. Discrete Appl. Math. 88, 315–324 (1998)
23. McConnell, R.M., Spinrad, J.P.: Construction of probe interval models. In: Proceedings ACM-SIAM symposium on discrete algorithms, pp. 866–875 (2002)
24. Przulj, N., Corneil, D.G.: 2-tree probe interval graphs have a large obstruction set. Discrete Appl. Math. 150, 216–231 (2005)
25. Sheng, L.: Cycle free probe interval graphs. Congr. Numerantium 140, 33–42 (1999)
26. Zhang, P.: Probe interval graphs and its application to physical mapping of DNA (unpublished, 1994)
27. Zhang, P., Schon, E.A., Fischer, S.G., Cayanis, E., Weiss, J., Kistler, S., Bourne, P.E.: An algorithm based on graph theory for the assembly of contigs in physical mapping of DNA. Computer Applications in the Biosciences 10, 309–317 (1994)
28. Zverovich, I.E., Zverovich, I.I.: Forbidden induced subgraph characterization of cograph contractions. J. Graph Theory 46, 217–226 (2004)

Recognition of Polygon-Circle Graphs and Graphs of Interval Filaments Is NP-Complete

Martin Pergel

Department of Applied Mathematics, Faculty of Mathematics and Physics,
Charles University, Malostranské náměstí 25, 118 00 Praha 1, Czech Republic
perm@kam.mff.cuni.cz

Abstract. Polygon-circle graphs (PC-graphs) are defined as intersection graphs of polygons inscribed into a circle, graphs of interval filaments (IFA-graphs) are intersection graphs of curves with both endpoints on prescribed line (filament), filaments above two disjoint intervals must not intersect each other. Recognition of these classes has been a long outstanding open problem. We prove that it is NP-complete to recognize both classes.

1 Introduction

Throughout the whole article any graph will be a simple undirected finite graph, that is, an ordered pair of vertices and edges $G = (V, E)$, $E \subseteq \binom{V}{2}$. The complement of a graph means the edge-complement, i.e., co-$G = (V, \binom{V}{2} \setminus E)$. For a class of graphs $\mathcal{C}$ we define co-$\mathcal{C} = \{\text{co-}G | G \in \mathcal{C}\}$. Intersection graphs are graphs with an intersection representation by a set-system: Each vertex is represented by a set, two vertices are adjacent if and only if corresponding sets have non-empty intersection. Any graph has some intersection representation [16], but non-trivial and interesting results are established when only sets of special properties are considered. The class of graphs representable by arc-connected sets in the (Euclidean) plane is the well-known class of STRING graphs [2,11,17]). Many intersection-defined subclasses of STRING-graphs have been considered, e.g., interval graphs (shortly INT-graphs, intersection graphs of intervals on a line), circle graphs (intersection graphs of segments inscribed into a circle), circular-arc graphs (intersection graphs of arcs of a circle) [19,16]. Intersection graphs have applications, for example, in VLSI-circuit design, biology and even in archeology. Some generally NP-hard problems can be solved efficiently when the input graph is given with a representation. Therefore it is important to ask, how difficult it is to find such a representation, or at least to decide, whether it exists. This is what we call *the recognition problem*. For example, for STRING graphs the recognition problem is known to be NP-complete ([17] and [11]) and the membership in NP is quite non-trivial. Several classes (e.g., interval graphs, circle graphs and circular-arc graphs) can be recognized in polynomial time. In this article we solve the recognition problem for two classes for which it has long been open.

A. Brandstädt, D. Kratsch, and H. Müller (Eds.): WG 2007, LNCS 4769, pp. 238–247, 2007.

Polygon-circle graphs, or shortly *PC-graphs,* are intersection graphs of convex polygons inscribed into a circle. Polygon-circle graphs are interesting because they form a common generalization of several other intersection classes, e.g., circle graphs, circular-arc graphs, chordal graphs (all polynomially recognizable [20,3,15]). Maximum clique, maximum independent set and coloring by fixed number of colors is polynomially solvable for circular-arc graphs, coloring is NP-hard for circular-arc graphs and circle graphs, while maximum clique and independent set remain polynomially solvable even for IFA-graphs (even in weighted case [5]).

Koebe in [14] presented many claims that he believed to imply polynomial recognition algorithm for PC-graphs, but this algorithm was never published. This is not surprising as now we show polynomial reduction showing NP-completeness of this problem.

Interval-filament graphs, shortly *IFA-graphs,* are defined as intersection graphs of filaments above a given line in a plane. Filaments are curves with endpoints on the given line. Each filament defines between endpoints on the line an interval, filaments belonging to disjoint intervals *must not intersect each other.* IFA-graphs were defined by Gavril [5] and characterized by mixed property (see below) as complements of co-INT-mixed graphs and they were equivalently described as caterpilar-overlap graphs [8]. Despite of the structural characterization the recognition has been open since. IFA-graphs contain PC-graphs and therefore Gavril's polynomial algorithm for the problem of the maximum weighted clique and maximum weighted independent set [5] immediately yields polynomial algorithm for PC-graphs.

If we consider intervals on the line (which are special cases of filaments), we obtain very well known class of interval graphs. A partial ordering $\leq$ on a set $\{a_1, a_2, \ldots, a_n\}$ is represented by a graph $G = (\{a_1, a_2, \ldots, a_n\}, \{a_i a_j | i \neq j \wedge (a_i \leq a_j) \vee (a_j \leq a_i)\})$. A graph is called a *comparability graph* if it represents some partial ordering on some set. For a class $\mathcal{C}$ of graphs a graph G is called $\mathcal{C}$-mixed if there exist $H = (V, E) \in \mathcal{C}$ and a comparability graph $I = (V, F)$ such that $G = (V, E \cup F)$. For any triple of vertices a, b, c in such a graph, the following must hold. If $ab \in E$ and $b \leq c$, then $ac \in E$. Gavril [5] showed that co-IFA-graphs are co-interval-mixed graphs (i.e., for any IFA-graph, there exists an interval-completion formed by the edges of comparability graph) which yields NP-membership for the recognition problem.

2 PC-Graphs

2.1 Preliminaries

First we show that the recognition problem for PC-graphs is in NP (as for several classes, like segment-graphs or CONV-graphs it is not known whether respective recognition problem is in NP [10]). We use the notion of *alternating sequence,* which was first introduced by Bouchet [1] to recognize circle-graphs:

Definition 1. *A sequence $\mathcal{S} = S_1, \ldots, S_m$ of symbols $\{v_1, \ldots v_n\}$ is called an alternating sequence representing graph $G = (\{v_1, \ldots v_n\}, E)$ if $\forall_{uv \in \{v_1, \ldots v_n\}}$ in $\mathcal{S}$*

symbols u and v alternate (i.e., S contains either a subsequence u...v...u...v or v...u...v...u) if and only if $uv \in E$.

It is a well-known fact that G is a PC-graph if and only if it has an alternating representation [9]. Moreover, for a vertex of degree $d \geq 2$, at most d occurences are sufficient, for vertices of degree at most one two occurences are sufficient [13].

For a vertex v we use P_v to denote the polygon representing v in a PC-representation.

Lemma 1. *The recognition problem for PC-graphs is in NP.*

Proof. We guess an alternating sequence (which is polynomially large with respect to the size of the graph) and verify that exactly those pairs forming an edge alternate.

Definition 2. *The corners of any polygon R divides the bounding circle into circular arcs.*

- *The regions bounded by those arcs and the chords between their endpoints are called R-segments.*
- *If P, Q, R are disjoint polygons (with the same bounding circle) and Q lies in a different P-segment than R, we say that Q is* blocked *from R by P.*
- *For disjoint polygons P and Q, the union of the P-segments not containing Q is called the* place under P *with respect to Q.*
- *A set A of polygons is said to lie around the circle if for no triple of disjoint polygons $P, Q, R \in A$, the polygon Q is blocked by R from P.*
- *We say that a polygon P is* visible *from a point x if there exists a ray starting in x that intersects P in a point y such that no point of the segment xy is contained in any other polygon of the representation.*

3-SAT is a well-known NP-complete problem. We use a special case, which is still NP-complete [18], as it is equivalent to bicolorability of 3-uniform hypergraph [4]:

Definition 3. NAE-SAT *(or E3-PURE-NAE-SAT) is the problem of whether a formula ϕ in conjunctive normal form (i.e., conjunction of disjunctions) having in each clause exactly 3 literals and having all literals in positive form (i.e., without negation) has an evaluation of variables such that in each clause at least one literal is true and at least one in false.*

2.2 The Reduction

Now we reduce the problem NAE-SAT to the recognition of PC-graphs. Given formula ϕ, we define a graph G as follows: We start by two non-neighboring vertices e_1 and e_2. Then we order (and number) variables $v_1, v_2, ..., v_n$ of ϕ. Each variable v contributes to the graph by two important vertices v and $\bar{v}$. Technical details of the construction force us to represent each v_i and $\bar{v}_i$ with polygons which lie in the place only under P_{e_1} with respect to P_{e_2} or under P_{e_2}

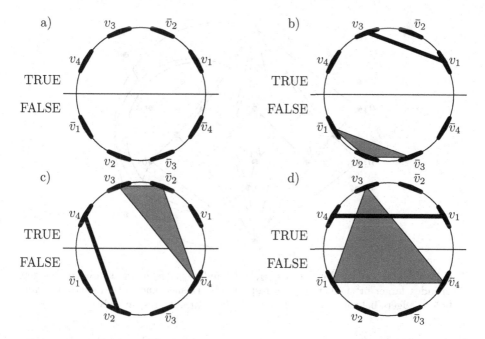

Fig. 1. Picture a) shows how vertices representing literals should be distributed around the circle, b) and c) describe a representation of satisfied clauses, d) is an unsatisfied clause – note that in case d) polygons representing the clause must intersect each other. All other possibilities are equivalent to one of those depicted.

with respect to P_{e_1} (i.e., for no i neither $P_{e_1}, P_{e_2}, P_{v_i}$ nor $P_{e_1}, P_{e_2}, P_{\bar{v}_i}$ lie around the circle). When v gets represented in the place under P_{e_1} and $\bar{v}$ under P_{e_2}, v is assigned TRUE, if v is represented under P_{e_2} and $\bar{v}$ under P_{e_1}, v is FALSE, all other possible placements get obstructed (by not yet described constructions). Under P_{e_1} and P_{e_2} (with respect to each other) vertices v_i and $\bar{v}_i$ get represented, in the ordering of increasing index as depicted in the Figure 1 a) where instead of P_{e_1} and P_{e_2} we show a line separating them. We represent each clause (v_1, v_2, v_3) by two mutually non-adjacent vertices q_1 and q_2, where q_1 is adjacent to $\bar{v}_1, v_2$ and $\bar{v}_3$, and q_2 to v_1 and v_3.

Note that under assumptions of last paragraph the "clause"-vertices (q_1, q_2) can be properly represented if and only if neither all three of v_1, v_2, v_3 are assigned TRUE nor (all three) FALSE. These ideas are demonstrated in the Figure 1.

Now we describe technical the part of the reduction. We give a description of what should be done for three consecutive variables (with respect to the indices of variables) A, B and C and for a clause Q. Variable B contributes 16 vertices: $a_1, \ldots a_5, b_1, \ldots b_5, c_1, \ldots c_4, r_1, r_2$. These vertices behave as in the Figure 2. Variable A contributes 16 vertices with the same labels as for B, but with a prime. These two variables share four vertices in the following way: $a_1' = a_4, a_2' = a_5, b_4' = b_1$ and $b_5' = b_2$. In the same way C contributes 16 vertices with double-primes,

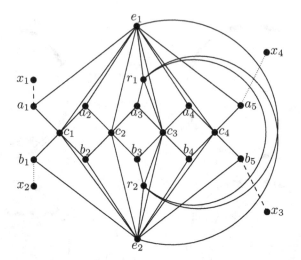

Fig. 2. Variable gadget. Vertices a_1, a_2, b_4, b_5 and a_5, a_4, b_2, b_1 are shared by gadgets with index larger or smaller by one, respectively. Dashed and dotted edges are valid only for gadgets belonging to the first or the last variable, respectively.

and so: $b_1'' = b_4, b_2'' = b_5, a_4'' = a_1$ and $a_5'' = a_2$. All c- and r- vertices of distinct variables are adjacent. We add a 6-cycle $e_1, x_1, x_2, e_2, x_3, x_4$. We make the vertex e_1 adjacent to all c-, r- and a-vertices except all a_3's and the vertex e_2 adjacent to all c-, r- and b-vertices except b_3's. When representing variable v_1, we connect its copy of a_1 to x_1 and b_5 to x_4, similarly for the last variable we connect its b_1 to x_2 and a_5 to x_3.

For each clause Q containing literals l_1, l_2, l_3 we add two nonadjacent vertices q_1 and q_2 make both adjacent to all c- and r-vertices, to e_1, e_2 and to all vertices representing the other clauses. Then we make q_1 adjacent with the b_3-vertices of the gadgets representing variables l_1 and l_3 and to the a_3-vertex of the gadget representing l_2 (note that ϕ has only positive occurences). Finally we connect q_2 with the a_3-vertices of the gadgets representing l_1 and l_3 (not for l_2, it is okay to do so, but more convenient not to).

Note that the a_3- and b_3-vertices of respective variables are exactly the vertices v and $\bar{v}$ introduced in the idea of the reduction. When we prove that a_3's and b_3's appear around the circle in desired ordering (i.e., such that there are two disjoint half-circles in each of which one of a_3 and b_3 corresponding to each variable v_i occur, and do so in order of increasing index of the corresponding v_i), we are done. This gets proven by following lemma with easy consequence that the variable-gadget must be represented as depicted in the Figure 3.

Lemma 2. *When representing one fixed variable gadget, polygons representing its $a_1, a_2, a_4, a_5, b_5, b_4, b_2$ and b_1 appear around the circle and, moreover, in this ordering.*

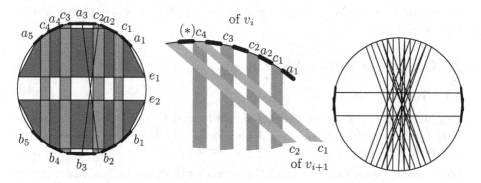

Fig. 3. Picture on the left describes the unique possibility how to represent the variable-gadget (up to switching a_3 and b_3), picture in the middle describes how representants of neighboring variables are represented and share vertices, polygon (*) represents a_5 of v_i and simultaneously a_2 of v_{i+1}, picture on the right sketches representation of c-vertices belonging to three consecutive variables.

Note that sharing vertices with "neighboring" gadgets propagates this orientation to all other variable-gadgets. Proof in full version.

Lemma 3. *The constructed graph without clause-vertices can be represented only in such a way that polygons representing the a_3's and b_3's appear around the circle, and consecutive literals (around the circle) have increasing indices (mod n).*

Proof. Corollary of Lemma 2.

Now we know that unsatisfiable formula cannot be represented (because polygons representing corresponding vertices must intersect each other although they represent non-adjacent vertices). Thus we have to show:

Lemma 4. *For a satisfiable formula ϕ, it is possible to represent the graph obtained from our construction.*

Proof. After representing the variable-gadgets as depicted in the Figure 3 for each clause we add two vertices q_1 (having three neighbors among the a_3's and b_3's) and q_2 (having two such neighbors) in such a way to intersect relevant P_{a_3}'s and P_{b_3}'s and extend them to correctly intersect all auxiliary polygons. We choose one variable v_i of l_1 and l_3 which is the "endvariable" for P_{q_1} (i.e., which is not blocked by P_{q_1} from P_{a_3}'s of l_1 and l_3)[1]. We extend P_{q_1} and P_{q_2} to cover almost half of the circle (we proceed for P_{q_1}, P_{q_2} is similar): P_{q_1} has three corners intersecting relevant P_{a_3} and P_{b_3}'s. We add a corner intersecting P_{c_2} of v_{i-1}. Now except P_{c_3} and P_{c_4} of v_{i-1} and v_i we are intersecting exactly what we should. So we add a corner under P_{c_4} of v_{i-1} with respect to P_{x_1} and P_{x_3} and simultaneously P_{c_1} of v_i and we are done.

Lemma 4 finishes the proof of the theorem:

---

[1] Formally when exploring this blocking we consider P_{q_1} restricted not to intersect P_{b_3} of l_1 and l_3 because blocking is defined only for disjoint polygons.

Theorem 1. *It is NP-complete to recognize whether a graph G has a PC-representation.*

3 Graphs of Interval Filaments

3.1 Technical Notions

Definition 4. *Let G be an interval-filament graph G and R be its representation. For each vertex $v \in V(G)$ we denote by F_v the filament corresponding to v and by I_v the underlying interval (of the filament).*

We are interested in the "topological" position for three mutually nonintersecting filaments. Note that two principal positions are that either from each filament either both other are visible or neither are. This visibility can be defined in the following technical way:

Definition 5. *For a triple of non-intersecting filaments F_a, F_b and F_c we say that F_a is covered by F_b against F_c if either $I_a \subset I_b \subset I_c$ (or the reverse) or $I_a \subset I_b$ and $I_b \cap I_c = \emptyset$ (or the reverse). For a set $\mathcal{F}$ of filaments we say this set lies consecutively on the base-line if F_a is not covered by F_b against F_c for any triple $F_a, F_b, F_c \in \mathcal{F}$.*

Theorem 2. *It is NP-complete to decide whether a given graph is an IFA-graph.*

The theorem is proven in two parts. First it is necessary to show that the recognition problem is in NP. This was already proven in [6]:

Theorem 3. *G is IFA-graph if and only if it is co-(co-INT-mixed)-graph.*

Corollary 1. *The recognition problem of IFA-graphs is in NP.*

Proof. We guess an appropriate INT-graph H (with an interval representation), an appropriate partial-ordering P of vertices and check correctness.

3.2 The Reduction

For the reduction we use almost the same graph as for PC-graphs and the same basic ideas. We just subdivide each a- and b-vertex into two new vertices for a_i named f_i, g_i and for b_i we name them h_i and j_i (to obtain 6-cycles instead of 4-cycles in the variable-gadget). New variable-gadget is depicted in the Figure 4. The e_1 is adjacent to all f_i's and g_i's again, except f_3's and g_3's, and analogously the e_2 is adjacent to all h- and j-vertices. The clause gadget is created analogously as for PC-graphs, both new polygons P_{f_3} and P_{g_3} (or P_{h_3} and P_{j_3}) are intersected instead of P_{a_3}'s (or P_{b_3}'s, respectively). Note that for a satisfiable formula this graph is a PC-graph. Now we show that for an unsatisfiable formula such a graph does not even have an IFA-representat. The argumentation is analogous to the case of PC-graphs, there are just more cases to analyze.

Observation: Filaments representing f- and j- (and similarly g- and h-) vertices lie consecutively on the base-line.

We generalize the notion of consecutive occurences to *circularly consecutive*:

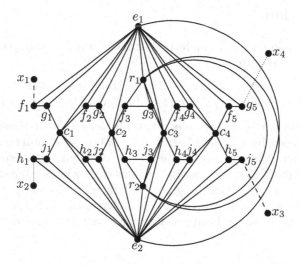

Fig. 4. The variable-gadget for graphs of interval filaments is similar to the gadget for PC-graphs, only C_4's are subdivided into C_6's. Again dashed lines apply only to the first variable while the dotted to the last.

Definition 6. *For a set of disjoint filaments $F_0, ... F_n$ lying consecutively on the base-line we say that this set is* circularly consecutive *if there exists $i \in \{0, ..., n\}$ such that for all $j \neq i$ interval I_j preceeds interval I_{j+1} (i.e., its right endpoint is to the left of the left endpoint of I_{j+1}) where $j + 1$ is considered modulo $n + 1$ and I_i either contains all other intervals or I_i preceeds I_{i+1}.*

Lemma 5. *To represent a variable-gadget up to symmetry the c-vertices must occur as depicted in the Figure 5.*

Proof. By simple case analysis.

We prove following Lemma similar to Lemma 2:

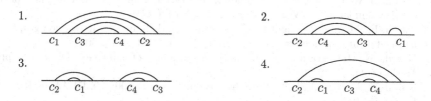

Fig. 5. These are (up to symmetries) only possibilities how to represent vertices c_1, c_2, c_3 and c_4 of the variable gadget

Lemma 6. *Filaments representing $f_1, f_2, f_4, f_5, j_5, j_4, j_2, j_1$ are circularly consecutive.*

Proof. By case-analysis according to the Figure 5.

This Lemma finishes the proof of NP-hardness of IFA-graphs.

4   Conclusion

As the construction for IFA-graphs produces either a PC-graph, or a graph that is not an IFA-graph, we proved:

Theorem 4. *No polynomially recognizable class can be sandwiched between PC-graphs and IFA-graphs (i.e., C such that $PC \subseteq C \subseteq IFA$), unless $P=NP$.*

As open problems remain the recognition problem of other classes of filament graphs introduced by Gavril [5,7] (circular-, subtree-, 2dim-interval-...). These classes are interesting, e.g., due to existence of polynomial algorithm for the maximum weighted clique. As well it remains open whether IFA-graphs can be recognized in polynomial time for graphs of large girth. Recently we have proven that PC-graphs with girth at least 5 can be polynomially, while, e.g., for segment graphs the recognition problem remains hard even with arbitrarily large girth [12].

Acknowledgement

The author would like to thank to Jan Kratochvíl and Jessica Enright for discussions, and to Jan Kratochvíl for the inspiration to revisit these problems. This work is supported by the grant GAUK 154907.

References

1. Bouchet, A.: Circle graph obstructions. Journal of Combinatorial Theory, Series B 60(1), 107–144 (1994)
2. Ehrlich, G., Even, S., Tarjan, R.E.: Intersection graphs of curves in the plane. Journal of Combinatorial Theory, Series B 21, 8–20 (1976)
3. Eschen, E.M., Spinrad, J.: An O(n) algorithm for circular-arc graph recognition. In: SODA, pp. 128–137 (1993)
4. Garey, M.R., Johnson, D.S.: Computers and Intractability: A Guide to the Theory of NP-Completeness. W. H. Freeman, New York (1979)
5. Gavril, F.: Maximum weight independent sets and cliques in intersection graphs of filaments. Information Processing Letters 73(5-6), 181–188 (2000)
6. Gavril, F.: Intersection graphs of helly families of subtrees. Discrete Appl. Math. 66(1), 45–56 (1996)
7. Gavril, F.: k-interval-filament graphs, Technical report 2004-30, DIMACS, Rutgers University (2004)
8. Chalopin, J., Goncalves, D., Ochem, P.: On graph classes defined by overlap and intersection models. In: 6th Czech-Slovak International Symposium on Combinatorics. Graph Theory, Algorithms and Applications (2006)
9. Kostochka, A., Kratochvíl, J.: Covering and coloring polygon-circl e graphs. Discrete Math. 163, 299–305 (1997)
10. Kratochvíl, J., Matoušek, J.: Intersection graphs of segments. Journal of Combinatorial Theory, Series B 62(2), 289–315 (1994)
11. Kratochvíl, J.: String graphs II. Recognizing string graphs is NP-hard. Journal of Combinatorial Theory, Series B 52, 67–78 (1991)

12. Kratochvíl, J., Pergel, M.: Geometric intersection graphs: Do short cycles help? In: Lin, G.(ed) COCOON 2007. LNCS, vol. 4598, pp. 118–128. Springer, Heidelberg (2007)
13. Kratochvíl, J., Pergel, M.: Two Results on Intersection Graphs of Polygons. In: Liotta, G. (ed.) GD 2003. LNCS, vol. 2912, pp. 59–70. Springer, Heidelberg (2004)
14. Koebe, M.: On a new class of intersection graphs. In: Proceedings of the Fourth Czechoslovak Symposium on Combinatorics Prachatice, pp. 141–143 (1990)
15. Lueker, G., Rose, D., Tarjan, R.: Algorithmic aspects of vertex elimination on graphs. SIAM J. Comput. 5, 266–283 (1976)
16. McKee, T.A., McMorris, F.R.: Topics on Intersection Graphs. SIAM (1999)
17. Schaefer, M., Sedgewick, E., Štefankovič, D.: Recognizing string graphs in NP. J. Comput. Syst. Sci. 67(2), 365–380 (2003)
18. Schaefer, T.J.: The complexity of satisfiability problems. In: Proceedings of the tenth annual ACM symposium on Theory of computing, pp. 216–226 (1978)
19. Spinrad, J.: Efficient Graph Representations, American Mathematical Society (2003)
20. Spinrad, J.: Recognition of circle graphs. J. Algorithms 16(2), 264–282 (1994)

Proper Helly Circular-Arc Graphs

Min Chih Lin[1,*], Francisco J. Soulignac[1,**], and Jayme L. Szwarcfiter[2,***]

[1] Universidad de Buenos Aires, Facultad de Ciencias Exactas y Naturales,
Departamento de Computación, Buenos Aires, Argentina
[2] Universidade Federal do Rio de Janeiro, Inst. de Matemática, NCE and COPPE,
Caixa Postal 2324, 20001-970 Rio de Janeiro, RJ, Brasil
{oscarlin,fsoulign}@dc.uba.ar, jayme@nce.ufrj.br

Abstract. A circular-arc model $\mathcal{M} = (C, \mathcal{A})$ is a circle C together with a collection $\mathcal{A}$ of arcs of C. If no arc is contained in any other then $\mathcal{M}$ is a proper circular-arc model, if every arc has the same length then $\mathcal{M}$ is a unit circular-arc model and if $\mathcal{A}$ satisfies the Helly Property then $\mathcal{M}$ is a Helly circular-arc model. A (proper) (unit) (Helly) circular-arc graph is the intersection graph of the arcs of a (proper) (Helly) circular-arc model. Circular-arc graphs and their subclasses have been the object of a great deal of attention in the literature. Linear time recognition algorithms have been described both for the general class and for some of its subclasses. In this article we study the circular-arc graphs which admit a model which is simultaneously proper and Helly. We describe characterizations for this class, including one by forbidden induced subgraphs. These characterizations lead to linear time certifying algorithms for recognizing such graphs. Furthermore, we extend the results to graphs which admit a model which is simultaneously unit and Helly, also leading to characterizations and a linear time certifying algorithm.

Keywords: algorithms, forbidden subgraphs, Helly circular-arc graphs, proper circular-arc graphs, unit circular-arc graphs.

1 Introduction

Circular-arc (CA) graphs and its subclasses are interesting families of graphs that have been receiving much of attention recently. The most common subclasses of circular-arc graphs [1,3,12] are proper circular-arc (PCA) graphs, unit circular-arc (UCA) graphs and Helly circular-arc (HCA) graphs. The classes of PCA and UCA graphs have been characterized by families of (infinite) forbidden subgraphs [14], and for HCA graphs there is a characterization by forbidden CA graphs [4]. Each of the graphs of the mentioned classes is the intersection graph of a collection of arcs of a circle. The circle and the collection of arcs form a

---

* Partially supported by UBACyT Grants X184 and X212, CNPq under PROSUL project Proc. 490333/2004-4.
** Partially supported by UBACyT Grant X184 and CNPq under PROSUL project Proc. 490333/2004-4.
*** Partially supported by CNPq and FAPERJ, Brasil.

A. Brandstädt, D. Kratsch, and H. Müller (Eds.): WG 2007, LNCS 4769, pp. 248–257, 2007.
© Springer-Verlag Berlin Heidelberg 2007

model for the corresponding graph. The model is called CA,PCA,UCA or HCA, according to whether the corresponding graph is CA,PCA,UCA or HCA. For each of the four classes of graphs (CA, PCA, UCA, HCA) there is a linear time recognition algorithm (see [6,9,10] for CA, [2,5] for PCA, [5,7] for UCA and [4] for HCA).

In this paper, we study the class of circular-arc graphs which admit a model which is simultaneously PCA and HCA. We call these graphs (models) as proper Helly circular-arc graphs (models). Clearly, such a class of graphs is contained in the intersection of the classes of PCA and HCA graphs. However, the containment is proper. Similarly, we define UHCA graphs as those admitting a model which is simultaneously UCA and HCA. We describe characterizations for the classes of PHCA and UHCA graphs, including a characterization by forbidden subgraphs. The characterizations lead to linear time algorithms for recognizing PHCA and UHCA graphs. Besides, we also describe how to obtain certificates, both positive and negative, and authenticate them in linear time.

As a motivation for this work, besides its theoretical interest, we mention that PCA and HCA are two classes of graphs, which behave differently in various aspects. For instance, PCA graphs can be efficiently colored, while problems involving cliques tend to be easier for HCA graphs. For instance, interval graphs are exactly those admitting a clique matrix having the consecutive ones property on the columns. By extending the consecutive ones property to the circular consecutive ones property leads to a broader class, which is exactly the HCA graphs. In this context, arises the problem of characterizing the clique graphs of HCA graphs. The latter class is related to the PHCA graphs [8].

Let G be a graph, $V(G)$ and $E(G)$ its sets of vertices and edges, respectively, $|V(G)| = n$ and $|E(G)| = m$. For $v \in V(G)$, denote by $N(v)$ the set of vertices adjacent to v, and $N[v] = N(v) \cup \{v\}$. A vertex v of G is **universal** if $N[v] = V(G)$. Two vertices v and w are **twins** in G if $N[v] = N[w]$. A **clique** is a maximal subset of pairwise adjacent vertices.

A **circular-arc** (CA) model $\mathcal{M}$ is a pair $(C, \mathcal{A})$, where C is a circle and $\mathcal{A}$ is a collection of arcs of C. When traversing the circle C, we will always choose the clockwise direction, unless explicitly stated. If s, t are points of C, write (s, t) to mean the arc of C defined by traversing the circle from s to t. Call s, t the **extremes** of (s, t), while s is the **start point** and t the **end point** of the arc. For $A_i \in \mathcal{A}$, write $A_i = (s_i, t_i)$ and $\overline{A_i} = (t_i, s_i)$. The **extremes** of $\mathcal{A}$ are those of all arcs $A_i \in \mathcal{A}$. An **s-sequence** (**t-sequence**) is a maximal sequence of start points (end points) of $\mathcal{A}$, in a traversal of C. Without loss of generality, all arcs of C are considered as open arcs, no two extremes of distinct arcs of $\mathcal{A}$ coincide and no single arc entirely covers C. We will say that $\epsilon > 0$ is **small enough** if ϵ is smaller than the minimum arc distance between two consecutive extremes of $\mathcal{A}$.

When no arc of $\mathcal{A}$ contains any other, $(C, \mathcal{A})$ is a **proper** circular-arc (PCA) model, while when every arc has the same length it is called **unit** circular-arc (UCA) model. When every set of pairwise intersecting arcs share a common point, $(C, \mathcal{A})$ is called a **Helly** circular-arc (HCA) model. If no two arcs of $\mathcal{A}$

cover C then the model is called **normal**. A PHCA (UHCA) model is one which is both HCA and PCA (UCA). Finally, a (proper) interval model is a (proper) CA model where $\bigcup_{A \in \mathcal{A}} A \neq C$.

A CA (PCA) (UCA) (HCA) (PHCA) (UHCA) graph is the intersection graph of a CA (PCA) (UCA) (HCA) (PHCA) (UHCA) model. We may use the same terminology used for vertices when talking about arcs. For example, we say an arc in a CA model is **universal** when its corresponding vertex in the intersection graph is universal. Similarly, a **connected** model is one whose intersection graph is connected. Say that two CA models are **equivalent** when they have the same intersection graph. If $\mathcal{M}$ is a CA model, denote by $\mathcal{M}_1$ a model obtained from $\mathcal{M}$ by removing all its universal arcs, if existing, except precisely one. Also, $\mathcal{M}_0$ represents the model where all universal arcs have been removed. Clearly, if $\mathcal{M}$ has no universal arcs then $\mathcal{M} = \mathcal{M}_1 = \mathcal{M}_0$, while if $\mathcal{M}$ contains exactly one universal arc $\mathcal{M} = \mathcal{M}_1$. In an HCA graph, each clique $Q \subseteq V(G)$ can be represented by a **clique-point** $q \in C$, which is a point of the circle common to all those arcs of $\mathcal{A}$, which correspond to the vertices of Q.

2 Preliminaries

In this section we present some simple observations and basic propositions, that we employ throughout the article.

Lemma 1. *Let v, w be two twin vertices in a graph G. Then G is a PCA (UCA) (HCA) (proper interval) graph if and only if $G \setminus \{v\}$ is a PCA (UCA) (HCA) (proper interval) graph. Moreover G has a normal PCA (UCA) (HCA) model if and only if $G \setminus \{v\}$ also has a normal PCA (UCA) (HCA) model.*

Proof. $\Longleftarrow$) Let $\mathcal{M} = (C, \mathcal{A})$ be a PCA (UCA) (HCA) (proper interval) model of $H \setminus \{v\}$ and $A_i \in \mathcal{A}$ be the arc corresponding to w. It is easy to see that $\mathcal{M}' = (C, \mathcal{A} \cup \{(s_i + \epsilon, t_i + \epsilon)\})$ is a PCA (UCA) (HCA) (proper interval) model of H, for every small enough ϵ. Moreover, if $\mathcal{M}$ is normal then we obtain that $\mathcal{M}'$ is normal. The converse is clear. $\square$

Lemma 2. *If G is a PCA graph with at most one universal vertex then every PCA model of it is normal.*

Proof. If two arcs cover the circle of a PCA model then they must be both universal, thus the graph has more than one universal vertex. $\square$

Theorem 1. *[4] A circular-arc model $(C, \mathcal{A})$ is HCA if and only if*

  (i) if three arcs of $\mathcal{A}$ cover C then two of them also cover it, and
  (ii) the intersection graph of $\overline{\mathcal{A}}$ is chordal.

Corollary 1. *If a normal PCA model $(C, \mathcal{A})$ is not HCA then three arcs of $\mathcal{A}$ cover C.*

Proof. On the contrary, suppose that no three arcs of $\mathcal{A}$ cover C. Then, by Theorem 1, the intersection graph of $\overline{\mathcal{A}} = \{\overline{A} : A \in \mathcal{A}\}$) has a hole $v_1, \ldots, v_k$ for some $k \geq 4$. Thus, the arcs $\overline{A_1}, \overline{A_3} \in \overline{\mathcal{A}}$ corresponding to vertices v_1 and v_3 do not intersect and therefore A_1, A_3 are arcs of $\mathcal{A}$ that cover C, contradicting the fact that $(C, \mathcal{A})$ is normal. $\qquad\square$

3   The Characterizations

In this section, we describe characterizations for PHCA graphs including a characterization by forbidden subgraphs. The forbidden subgraphs for a PCA graph to be a PHCA graph are the 4-wheel, denoted by W_4 and depicted in Figure 1 and the 3-sun, which appears in Figure 2.

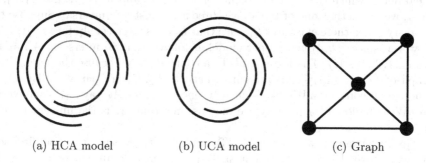

(a) HCA model   (b) UCA model   (c) Graph

Fig. 1. HCA and PCA models of W_4

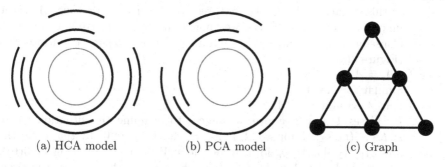

(a) HCA model   (b) PCA model   (c) Graph

Fig. 2. HCA and PCA models of the 3-sun

Theorem 2. *Let G be a PCA graph and $\mathcal{M}$ be a PCA normal model of it. The following affirmatives are equivalent:*

(i) $\mathcal{M}$ is equivalent to a PHCA model.
(ii) $\mathcal{M}$ contains no 4-wheels nor 3-suns, as submodels.
(iii) $\mathcal{M}_1$ is HCA or $\mathcal{M}_0$ is an interval model.

Proof. $(i) \implies (ii)$: By hypothesis, $\mathcal{M}$ is equivalent to a PHCA model of G. We know that a PCA model and a HCA model both exist for W_4 (Figure 1). However, we show that no PHCA model exists for W_4. Every HCA model for W_4 contains four clique points, such that the universal arc in the submodel corresponding to W_4 covers these four points, while each of the other arcs of the submodel covers exactly two of them. Consequently, no HCA model can be normal. By Theorem 1, the latter implies that whenever $\mathcal{M}$ is equivalent to a PHCA model, $\mathcal{M}$ does not contain a submodel of W_4. The proof for the 3-sun is similar.

$(ii) \implies (iii)$: Let $\mathcal{M}$ be a normal PCA model, containing no submodels of W_4s nor 3-suns. Suppose $\mathcal{M}_1$ is not an HCA model. Since $\mathcal{M}_1$ is also a PCA normal model, Corollary 1 implies that $\mathcal{M}_1$ contains three arcs A_1, A_2, A_3 covering C. By Lemma 2 no two arcs cover C, thus we may assume that in a traversal of C the order in which the extremes points of these arcs appear is $s_1, t_3, s_2, t_1, s_3, t_2$. First, we prove that one of the above three arcs must be universal. Suppose the contrary. Then there exist arcs B_i, such that B_i does intersect A_i, for $i \in \{1, 2, 3\}$. However, since $(C, \mathcal{A})$ is a proper model, it follows that B_i intersects A_j, A_k for $j, k \in \{1, 2, 3\} \setminus \{i\}$. The latter leads to a contradiction because the intersection graph of $\{A_i, B_i\}_{i \in \{1,2,3\}}$ is a 3-sun, when B_1, B_2, B_3 are pairwise disjoint, or otherwise it contains a W_4. Consequently, one of A_1, A_2, A_3, say A_1, is a universal arc. Without loss of generality, we may assume that A_1 is the universal arc of $\mathcal{M}_1$.

Next, we examine arc A_1 in $\mathcal{M}_1$. Traverse (s_1, t_1) in the clockwise direction. We will prove that no start point $s_l \in A_1$ can precede an end point $t_r \in A_1$. To obtain a contradiction for this fact, assume the contrary and discuss the following alternatives.

Case 1 : $A_r = A_3$

    In this situation, A_l, A_2, A_3 are three arcs covering C. Because A_1 is the unique universal arc of $\mathcal{M}_1$, we know that A_l, A_2, A_3 are not universal. Consequently, as above, $\mathcal{M}_1$ contains a W_4 or a 3-sun, a contradiction (Figure 3(a))

Case 2 : $A_l = A_2$

    Similar to Case 1.

Case 3 : $A_r \neq A_3$ and $A_l \neq A_2$

    By Cases 1 and 2, above, it suffices to examine the situation where $s_l, t_r \in (t_3, s_2)$. Suppose $A_l \cap A_3 = A_r \cap A_2 = \emptyset$. In this case, the arcs A_1, A_2, A_3, A_l, A_r form a forbidden W_4, impossible (Figure 3(b)). Alternatively, let $A_l \cap A_3 \neq \emptyset$. Then the arcs A_3, A_l, A_r cover the circle and none of them is the universal arc A_1, an impossibility (Figure 3(c)). The situation $A_r \cap A_2 \neq \emptyset$ is similar.

By the above cases, we conclude that all end points must precede the start points in (s_1, t_1). Let t_{last} and s_{first} be the last end point and the first start point inside (s_1, t_1), respectively. Taking into account that A_1 is universal, we conclude that any point of the arc $(t_{\text{last}}, s_{\text{first}}) \subset A_1$ of C can not be contained in any arc of $\mathcal{A}$ except A_1. Hence $\mathcal{M}_0 = \mathcal{M}_1 - \{A_1\}$ is an interval model.

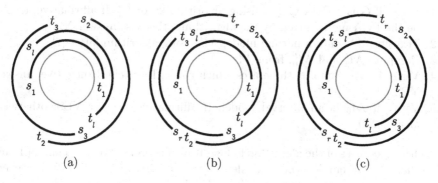

Fig. 3. Theorem 2

$(iii) \implies (i)$: Suppose $\mathcal{M}_1$ is a HCA model. By Lemma 1, we can include in the model all the universal arcs that have been possibly removed from it, obtaining $\mathcal{M}$ as both a PCA and HCA model. Then $\mathcal{M}$ is a PHCA model and G a PHCA graph.

Next, suppose $\mathcal{M}_0$ is an interval model. Since $\mathcal{M}_0$ is a PCA model, it is in fact a proper interval model. The extreme points of $\mathcal{M}_0$ form a linear ordering. If $\mathcal{M}_0 = \mathcal{M}$ there is nothing to prove. Otherwise, examine the alternatives.

Case 1 : $\mathcal{M}_0$ is connected

Let t_l be the first end point and s_r the last start point in the ordering of the extreme points of $\mathcal{M}_0$. Then insert the arc $(t_l - \epsilon, s_r + \epsilon)$, for small enough ϵ. It follows that such an arc must be universal and the model still proper, because $\mathcal{M}$ is so. In addition, include the possibly remaining universal arcs, as in Lemma 1. Hence $\mathcal{M}$ is a PHCA model.

Case 2 : $\mathcal{M}_0$ has two connected components

Traversing C in the clockwise direction, let t_l be the first end point in the first of the connected components, and s_r the last start point in the second of the components. Then include the arc $(s_r + \epsilon, t_l - \epsilon)$, for small enough ϵ. The new model is equivalent to $\mathcal{M}_1$. By Lemma 1, include the remaining universal arcs. Therefore $\mathcal{M}$ is equivalent to a PHCA model, i.e. G is a PHCA graph.

Case 3 : $\mathcal{M}_0$ has more than two connected components

This case can not occur, as $\mathcal{M}$ would not be a PCA model.

The proof is complete.   □

4   The Algorithms

The characterizations described in the last section lead directly to an algorithm for recognizing PHCA graphs. Let G be a given graph. The algorithm answers YES, if G is a PHCA graph, and NO otherwise. The formulation is as follows.

1. Verify if G is a PCA graph, using algorithm [2] or [5]. If affirmative, let $\mathcal{M}$ be the PCA model given by the algorithm; otherwise answer NO.
2. Transform $\mathcal{M}$ into a normal model $\mathcal{M}'$, using algorithm [5] or [7].
3. Compute $\mathcal{M}'_1$ and $\mathcal{M}'_0$, from $\mathcal{M}$.
4. Verify if $\mathcal{M}'_1$ contains three arcs which cover the circle. If negative, answer YES.
5. Verify if $\mathcal{M}_0$ is an interval model. If affirmative, answer YES; otherwise answer NO.

The correctness of the algorithm follows from Theorems 1 and 2, Lemma 1 and Corollary 1. We employ the equivalence $(i) \Longleftrightarrow (iii)$ of Theorem 2. However, the check of whether the model $\mathcal{M}'_1$ is HCA is replaced by the simpler check of whether $\mathcal{M}'_1$ contains three arcs covering the circle, according to Corollary 1.

Next, we discuss the complexity of the algorithm. Step 1 requires $O(n + m)$ time [2], [5]. All the remaining steps can be implemented in $O(n)$ time, as follows. The construction of normal models can be done in $O(n)$ time, using algorithms [5] or [7]. In order to implement Step 3, we need to find the set of universal arcs of $\mathcal{M}'$. Such a construction can be done easily, by observing that in a proper normal model, any arc A_i is universal precisely when it contains exactly one of the extremes of each of the other arcs of $\mathcal{A}$. Consequently, to check whether A_i is universal, it suffices to check whether the n-th extreme point after s_i, in the ordering, is the end point t_i. This can be done in constant time, for each A_i, hence in overall $O(n)$ time. Step 4 can be implemented in time $O(n)$, also in a simple manner, as follows. For any arc $A_i \in \mathcal{A}$, denote by $PREV(A_i)$ the arc of $\mathcal{A}$ whose start point is closest to the end point of A_i, in the counterclockwise direction. Let $t_1, \ldots, t_k$ be a t-sequence of a set of arcs $A_1, \ldots, A_k$. Clearly, $PREV(A_i) = PREV(A_1)$ for every $1 \leq i \leq k$. Therefore, by a simple examination of the t-sequences, we can find $PREV(A_i)$ for every arc A_i in $O(n)$ time. If A_1 covers C with two other arcs, then $A_i, PREV(A_i), PREV(PREV(A_i))$ must cover C because the model is proper. Thus, it is enough to check if $A_i, PREV(A_i), PREV(PREV(A_i))$ cover the circle for every $1 \leq i \leq n$, which can be done in $O(n)$ time. Finally, for Step 5, we need to verify whether $\mathcal{M}'_0$ is an interval model, which can be easily checked in $O(n)$ time.

The above algorithm produces certificates, as a by-product. When the algorithm answers YES, we exhibit a PHCA model for the graph, while in the NO answer, we show either a forbidden subgraph or a corresponding argument for the graph not to be PHCA.

The YES answers appear in Steps 4 and 5. The certificate corresponding to Step 4 is just the model $\mathcal{M}'$, constructed in Step 2. For obtaining the YES certificate relative to Step 5, we refer to the proof of Theorem 2, in particular, the Case 2 of the implication $(iii) \Longrightarrow (i)$. In this situation, $\mathcal{M}'_1$ contains three arcs covering the circle, but $\mathcal{M}'_0$ is a disconnected interval model, formed by two connected components. To obtain the required model, we include in $\mathcal{M}'_0$ the arc $(s_r + \epsilon, t_l - \epsilon)$, where s_r is the last start point in the second connected component of $\mathcal{M}'_0$, while t_l is the first end point in the first component. Such a model contains exactly one universal arc. Finally, using Lemma 1, include the possibly remaining universal arcs that have been removed in Step 3. The model so obtained is a PHCA model for G. The complexity for producing such certificates is $O(n)$.

Next, we discuss the authentication of the YES certificates. Denote by $\mathcal{M}''$ the certificate obtained by the algorithm, which we are required to authenticate, as a PHCA model of the input graph G. That is, we ought to verify that $\mathcal{M}''$ is a model for G and that $\mathcal{M}''$ is proper, normal and Helly. The first task is simple, just compare the adjacencies of the vertices of G with the intersections of the arcs of $\mathcal{M}''$. This requires $O(n + m)$ time. The remaining authentications involve only operations on the arcs of $\mathcal{M}''$ and can be done in $O(n + m)$ time, as below described.

We employ an observation that a model $\mathcal{M}''$ is simultaneously proper and normal if and only if no arc A_i of it contains both extremes of some other arc $A_j \neq A_i$. This observation leads to an algorithm where we traverse each arc A_i of $\mathcal{M}''$ in the order $i = 1, \ldots, n$, and if we detect both extremes of a same arc during the traversal of A_i, the algorithm reports the certificate to be false and stops. After traversing the last arc A_n it authenticates $\mathcal{M}''$ as being proper and normal. Consequently, the authentication for the model to be proper and normal can be easily implemented in $O(n + m)$ time. Finally, for checking whether $\mathcal{M}''$ is Helly, we apply Corollary 1 and just confirm that $\mathcal{M}''$ does not contain three arcs covering the circle. With this purpose, we run Step 4 of the Algorithm, terminating within $O(n)$ time.

The NO answers are in Steps 1 and 5. The negative answer in Step 1 occurs when G is not a PCA graph. A certificate of this fact is given by algorithm [5]. The task is divided into two cases. If G is co-bipartite then [5] employs the characterization proved in [13] and exhibits a forbidden subgraph for a co-bipartite graph to be a PCA graph. When G is not co-bipartite, the certificate obtained is that for a matrix not to have the consecutive ones property in its columns. Such a certificate is given by the algorithm [11]. In any of the cases, the certificates can be obtained in $O(n + m)$ time. Finally, the NO answer in Step 5 corresponds to a certificate where we exhibit a forbidden subgraph for a PCA graph to be PHCA. According to Theorem 2, such a subgraph is either a W_4 or a 3-sun, and can be constructed as follows. Let A_1, A_2, A_3 be the three arcs which cover the circle, obtained in Step 4. If none of these arcs is universal then by Theorem 2, we know that there are three arcs B_1, B_2, B_3, such that B_i intersects A_j and not A_i, for all $1 \leq i \neq j \leq 3$. Then the arcs $A_1, A_2, A_3, B_1, B_2, B_3$ either form a 3-sun or contain a W_4. Finally, if one among A_1, A_2, A_3, say A_1, is a universal arc then there are arcs A_l, A_r, such that s_l precedes t_r in A_1. In this situation, A_1, A_2, A_3, A_l, A_r form a forbidden W_4. There is no difficulty to obtain such certificates in $O(n)$ time.

Finally, we discuss the authentication of the NO certificates. The one obtained in Step 1 is given by algorithm [5] and can be authenticated in $O(n)$ time [5]. The NO certificate constructed in Step 5 is one of the forbidden subgraphs W_4 or 3-sun. They can be easily authenticated in $O(n)$ time, as being an induced subgraph of G.

5   Unit Helly Circular-Arc Graphs

The characterization of PHCA graphs, described in Section 3, can be extended
so as to characterize UHCA graphs. The formulation is below presented. The
proof is similar to that of Theorem 2.

Theorem 3. *Let G be a UCA graph and $\mathcal{M}$ be a UCA normal model of it. The
following affirmatives are equivalent:*

(i) $\mathcal{M}$ is equivalent to a UHCA model.
(ii) $\mathcal{M}$ contains no 4-wheels, as submodels.
(iii) $\mathcal{M}_1$ is HCA or $\mathcal{M}_0$ is an interval model.

The above characterization, together with Lemma 1 and Corollary 1, lead to a
linear time algorithm for recognizing UHCA graphs and exhibiting certificates.
The algorithm is similar to that for PHCA graphs, except that we employ algorithms [5] or [7] in Step 1. The positive certificate is a UHCA model and can be
obtained by algorithm [7], whereas the negative certificate of Step 1 is obtained
by [5].

6   Conclusions

We have described characterizations and recognition algorithms for PHCA and
UHCA graphs. The characterizations imply a complete family of forbidden subgraphs for these classes. That is, adding W_4 and the 3-sun to Tucker's forbidden
subgraphs for PCA graphs, we obtain the complete list of forbidden subgraphs
for PHCA graphs. Similarly, adding W_4 to Tucker's forbidden family for UCA
graphs, we obtain all the subgraphs forbidden for UHCA graphs.

The algorithms have complexity $O(n + m)$. Moreover, if the input consists of
a circularly ordered PCA model, the complexity drops to $O(n)$.

References

1. Brandstädt, A., Le, V., Spinrad, J.: Graph Classes: A Survey. SIAM, Philadelphia
   (1999)
2. Deng, X., Hell, P., Huang, J.: Linear time representation algorithms for proper
   circular-arc graphs and proper interval graphs. SIAM Journal on Computing 25,
   390–403 (1996)
3. Golumbic, M.: Algorithmic Graph Theory and Perfect Graphs. Academic Press,
   London (1980)
4. Joeris, B., Lin, M.C., McConnell, R.M., Spinrad, J., Szwarcfiter, J.L.: Linear time
   recognition of Helly circular-arc models and graphs. (manuscript, 2007) (Presented
   at COCOON 2006 and SIAM DM 2006 Confs.)
5. Kaplan, H., Nussbaum, Y.: Certifying algorithms for recognizing proper circular-
   arc graphs and unit circular-arc graphs. In: Fomin, F.V. (ed.) WG 2006. LNCS,
   vol. 4271, pp. 289–300. Springer, Heidelberg (2006)

6. Kaplan, H., Nussbaum, Y.: A simpler linear-time recognition of circular-arc graphs. In: Arge, L., Freivalds, R. (eds.) SWAT 2006. LNCS, vol. 4059, pp. 41–52. Springer, Heidelberg (2006)
7. Lin, M.C., Szwarcfiter, J.L.: Efficient construction of unit circular-arc models. In: SODA 2006. Proceedings of the 17th ACM-SIAM Symposium on Discrete Algorithms, pp. 309–315 (2006)
8. Lin, M.C., McConnell, R.M., Soulignac, F.J., Szwarcfiter, J.L.: On cliques of Helly circular-arc graphs (in preparation)
9. McConnell, R.M.: Linear-time recognition of circular-arc graphs. In: FOCS 2001. Proceedings of the 42nd Annual IEEE Symposium on Foundations of Computer Science, pp. 386–394 (2001)
10. McConnell, R.M.: Linear-time recognition of circular-arc graphs. Algorithmica 37(2), 93–147 (2003)
11. McConnell, R.M.: A certifying algorithm for the consecutive ones property. In: SODA 2004. Proceedings of the 15th Annual ACM-SIAM Symposium on Discrete Algorithms, pp. 716–770 (2004)
12. Spinrad, J.: Efficient Graph Representations. American Mathematical Society, Providence, Rhode Island (2003)
13. Tucker, A.: Matrix characterizations of circular-arc graphs. Pacific Journal of Mathematics 38, 535–545 (1971)
14. Tucker, A.: Structure theorems for some circular-arc graphs. Discrete Mathematics 7, 167–195 (1974)

Pathwidth of Circular-Arc Graphs

Karol Suchan[2,3] and Ioan Todinca[1]

[1] LIFO, Université d'Orléans, 45067 Orléans Cedex 2, France
Ioan.Todinca@univ-orleans.fr
[2] Departamento de Ingeniería Matemática, Universidad de Chile, Santiago, Chile
[3] Department of Discrete Mathematics, Faculty of Applied Mathematics, AGH,
Cracow, Poland
karol@suchan.info

Abstract. The pathwidth of a graph G is the minimum clique number of H minus one, over all interval supergraphs H of G. Although pathwidth is a well-known and well-studied graph parameter, there are extremely few graph classes for which pathwidh is known to be tractable in polynomial time. We give in this paper an $\mathcal{O}(n^2)$-time algorithm computing the pathwidth of circular-arc graphs.

1 Introduction

A graph is an *interval graph* if it is the intersection graph of a finite set of intervals on a line. The *pathwidth* of an arbitrary graph G is the minimum clique number of H, over all interval supergraphs H of G, on the same vertex set. Pathwidth has been introduced in the first article of Robertson and Seymour's Graph Minor series [13]. It is a well-known and well studied graph parameter, and not surprisingly NP-hard to compute. Note that the pathwidth has been redefined and studied in different contexts under different names. The parameter is also equal to the vertex separation number, and to the interval thickness or node search number of the graph, minus one (see [1] for a survey).

The pathwidth problem is fixed parameter tractable. Indeed, since the class of graphs of pathwidth at most k is minor-closed, using Roberston and Seymour's results on graph minors there exists an $\mathcal{O}(n^2)$ algorithm for recognizing graphs of pathwidth at most k, for any constant k; unfortunately this technique is not constructive. Bodlaender and Kloks [3] give a constructive linear-time algorithm deciding, for any fixed k, if the pathwidth of an input graph is at most k. Their algorithm first computes a tree decomposition of the input graph of width at most k, using the fact that the treewidth of a graph is at most its pathwidth (see Section 3 for definitions). Then this tree decomposition is used to decide if the pathwidth is at most k. The best approximation algorithms for pathwidth are also based on approximation algorithms for treewidth, combined with the fact that $\mathrm{pwd}(G) \leq \mathrm{twd}(G) \cdot \log n$ (see [1]).

By [3], the pathwidth is polynomially tractable for all classes of graphs of bounded treewidth. For trees and forests there exist several algorithms solving

A. Brandstädt, D. Kratsch, and H. Müller (Eds.): WG 2007, LNCS 4769, pp. 258–269, 2007.

the problem in $\mathcal{O}(n \log n)$ time [12,5]; only recently, Skodinis [14] gave a linear time algorithm. Even the unicyclic graphs, obtained from a tree by adding one edge, require more complicated algorithms in order to obtain an $\mathcal{O}(n \log n)$ time bound [4]. There exist some other graph classes for which the pathwidth problem is polynomial, e.g. permutation graphs, but this is mainly because for these classes the pathwidth equals the treewidth. Roughly speaking, almost everything that we know about computing pathwidth comes from its relationship with treewidth. More surprisingly, even for classes of graphs of very small treewidth, like (biconnected) outerplanar or Halin graphs, there are interesting approximation algorithms for pathwidth [2,6]. Although in these cases the parameter is polynomially tractable by the algorithm of [3], the running time is huge and the dynamic programming technique of [3] could not be translated into simple, combinatorial algorithms.

In this paper we give the first polynomial time algorithm computing the pathwidth of circular-arc graphs. A graph is a *circular-arc* graph if it is the intersection graph of a finite set of arcs on a circle. See [7] for classical results on circular-arc graphs. The pathwidth of these graphs can be easily approximated within a factor of 2. Nevertheless, for circular-arc graphs the pathwidth is not necessarily equal to the treewidth, and clearly this class is not of bounded treewidth, therefore we cannot use the classical techniques for computing pathwidth. Our algorithm is based on a study of interval completions of circular-arc graphs. An interval completion of a graph G is an interval supergraph H, on the same vertex set. If no interval completion H' of G is a strict subgraph of H, we say that H is a minimal interval completion of G. We study the minimal interval completions of circular-arc graphs and we characterize a subclass of these minimal interval completions, containing optimal solutions to the pathwidth problem. Based on this combinatorial result, we give an $O(n^2)$ algorithm computing the pathwidth of circular-arc graphs.

2   Definitions and Basic Results

Let $G = (V, E)$ be a finite, undirected and simple graph. Moreover, we only consider connected graphs — in the non connected case each connected component can be treated separately. Denote $n = |V|$, $m = |E|$. If $G = (V, E)$ is a subgraph of $G' = (V', E')$ (i.e. $V \subseteq V'$ and $E \subseteq E'$) we write $G \subseteq G'$. The *neighborhood* of a vertex v in G is $N_G(v) = \{u \mid \{u, v\} \in E\}$. Similarly, for a set $A \subseteq V$, $N_G(A) = \bigcup_{v \in A} N_G(v) \setminus A$. As usual, the subscript is sometimes omitted. For a set of vertices $A \subset V$, $G[A]$ is the *subgraph* of G *induced* by A, that is the graph (A, E_A), where $E_A = \{\{x, y\} \mid \{x, y\} \subseteq A$ and $\{x, y\} \in E\}$.

The *intersection graph* of a family V of n sets is the graph $G = (V, E)$, where the vertices are the sets and the edges are the pairs of sets that intersect. Every graph is the intersection graph of some family of sets [15]. A graph is an *interval graph* if it is the intersection graph of a finite set of intervals on a line. A graph is a *circular-arc* graph if it is the intersection graph of a finite set of arcs on a circle. A *model* of an interval graph or a circular-arc graph G is a set of intervals or circular arcs that represent G in this way. Without loss of generality we may assume that no two circular arcs of the model share a common endpoint.

Given a graph G, there exists a linear-time algorithm recognizing whether G is a circular-arc graph [11]. The algorithm produces a circular-arc model if such a model exists. Therefore, from now on we assume that our input is a circular arc-graph together with a model M_G.

Given a model M_G of a circular-arc $G = (V, E)$, we introduce some vocabulary concerning the arcs. We identify a vertex $v \in V$ with the corresponding arc in M_G. We call the clockwise endpoint of an arc v the *left* endpoint, denoted by $l(v)$, and the counterclockwise endpoint the *right* endpoint, denoted by $r(v)$. Note that an interval graph is a special case of circular-arc graphs - it is a circular-arc graph that can be modeled with arcs that do not cover a point A on the circle. Just cut the circle at A and straighten it into a line - this yields an interval model. If G is a circular-arc graph with a model M_G, and X is a subset of the vertices, then $M_G[X]$ is the *restriction* of M_G to X, namely, the result of removing from M_G the arcs corresponding to vertices not in M_G. Clearly, $M_G[X]$ is a model of $G[X]$.

To each point p on the circle in M_G we assign the set of arcs $V(p)$ that intersect it. Clearly, $V(p)$ is a clique in G. In particular, we are interested in the cliques maximal with respect to inclusion. To analyze them, we need not consider all points. In fact, it is enough to take the set of right endpoints. Using only the right endpoints r that have the corresponding set of arcs $V(r)$ maximal with respect to inclusion, we define the following structure describing G.

Definition 1. *Given a circular-arc model M_G of $G = (V, E)$, a clique-cycle is the cycle $CC_{G,M} = (\mathcal{X}, C)$. The vertices are the right endpoints in M_G with the corresponding sets of circular-arcs maximal by inclusion among all endpoints in M_G. The edges form a cycle, with every vertex $X \in \mathcal{X}$ adjacent to $X_l, X_r \in \mathcal{X}$, where X_l (X_r) is the element of $\mathcal{X}$ clockwise (counterclockwise) closest to X in M_G.*

3 Connected Decompositions

Definition 2. *Let $\mathcal{X} = \{X_1, ..., X_k\}$ be a set of subsets of V such that X_i, $1 \leq i \leq k$, is a clique in $G = (V, E)$. If, for every $\{v_i, v_j\} \in E$, there is some X_p such that $\{v_i, v_j\} \subseteq X_p$, then $\mathcal{X}$ is called an edge clique cover of G.*

Definition 3. *A connected decomposition of an arbitrary graph $G = (V, E)$ is a graph $D = (\mathcal{X}, A)$, where $\mathcal{X}$ is a family of subsets of V called bags and A is a set of edges on $\mathcal{X}$, such that the following three conditions are satisfied.*

1. *Each vertex $v \in V$ appears in some bag.*
2. *For every edge $\{v_i, v_j\} \in E$ there is a bag containing both v_i and v_j.*
3. *For every vertex $v \in V$, the bags containing v induce a connected subgraph of D.*

$D = (\mathcal{X}, A)$, a connected decomposition of G, is a clique connected decomposition of G if $\mathcal{X}$ is an edge clique cover of G.

A path decomposition *(tree decomposition) is a connected decomposition with D being a path (tree). The width of a decomposition is the size of a largest bag, minus one. The* pathwidth *(treewidth) of a graph G, denoted by* $\mathrm{pwd}(G)$ *($\mathrm{twd}(G)$) is the minimum width over all path decompositions (tree decompositions) of G.*

Lemma 1 (see, e.g., [7]). *A graph G is an interval graph if and only of it has a clique path decomposition.*

Similarly, we can characterize circular-arc graphs through clique cycle decompositions. A *cycle decomposition* is a connected decomposition with D being a cycle, and a *clique cycle decomposition* is a cycle decomposition with all bags being cliques.

By Definition 1, we have:

Lemma 2. *Given a circular-arc model M_G of $G = (V, E)$, $CC_{G,M} = (\mathcal{X}, C)$ is a clique cycle decomposition of G.*

We easily deduce:

Theorem 1. *A graph G is a circular-arc graph if and only if it has a clique cycle decomposition.*

Given a clique path decomposition of an interval graph, the intersection of two consecutive cliques form a separator. Given a circular model M_G of $G = (V, E)$ and a clique cycle decomposition $CC_{G,M} = (\mathcal{X}, C)$, we say that the intersection of two consecutive cliques of the cycle is a *semi-separator*.

4   Folding

Given a path decomposition $\mathcal{P}$ of G, let `PathFill`$(G, \mathcal{P})$ be the graph obtained by adding edges to G so that each bag of $\mathcal{P}$ becomes a clique. It is straight forward to verify that `PathFill`$(G, \mathcal{P})$ is an interval supergraph of G, for every path decomposition P. Moreover P is a clique path decomposition of `PathFill`$(G, \mathcal{P})$.

Definition 4 (see also [8]). *Let $\mathcal{X}$ be an edge clique cover of an arbitrary graph G and let $\mathcal{Q} = (Q_1, ..., Q_k)$ be a permutation of $\mathcal{X}$. We say that $(G, \mathcal{Q})$ is a folding of G by $\mathcal{Q}$.*

To any folding of G by an ordered edge clique cover $\mathcal{Q}$ we can naturally associate, by Algorithm `FillFolding` of Figure 1, an interval supergraph $H = $ `FillFolding`$(G, \mathcal{Q})$ of G. The algorithm also constructs a clique path decomposition of H.

Lemma 3. *Given a folding $(G, \mathcal{Q})$ of G, the graph $H = FillFolding(G, \mathcal{Q})$ is an interval completion of G.*

Proof. Observe that after the **for** loops, $\mathcal{P}$ is a path decomposition of H, since every edge is contained in some bag, and for every vertex the bags containing it induce a subpath of $\mathcal{P}$. Hence, since $H = $ `PathFill`$(G, \mathcal{P})$, it is an interval completion of G.   □

Algorithm `FillFolding`
Input: Graph $G = (V, E)$ and $\mathcal{Q} = (Q_1, ..., Q_k)$, a sequence of subsets of V;
Output: A supergraph H of G;

$\mathcal{P} = \mathcal{Q}$;
for each vertex v of G **do**
    $s = \min\{i \mid x \in Q_i\}$;
    $t = \max\{i \mid x \in Q_i\}$;
    for $j = s + 1$ to $t - 1$ **do**
        $P_j = P_j \cup \{v\}$;
end-for
$H = $ `PathFill`$(G, \mathcal{P})$;

Fig. 1. The `FillFolding` algorithm

We shall also say that the graph $H = $ `FillFolding`$(G, \mathcal{Q})$ is *defined* by the folding $(G, \mathcal{Q})$. The graph defined by a folding is not necessarily a minimal interval completion of G. Nevertheless, we prove in Theorem 2 that every minimal interval completion of G is defined by some folding.

Theorem 2. *Let H be a minimal interval completion of a graph G with an edge clique cover $\mathcal{X}$. Then there exists a folding $(G, \mathcal{Q})$, where $\mathcal{Q}$ is a permutation of $\mathcal{X}$, such that $H = FillFolding(G, \mathcal{Q})$.*

Proof. Let $\mathcal{X} = \{X_i \mid 1 \le i \le p\}$ and $\mathcal{K} = \{P_i \mid 1 \le i \le k\}$ denote an enumeration of $\mathcal{X}$ and the set of maximal cliques of H, respectively. Let $\mathcal{P} = (P_1, \ldots, P_k)$ be a clique-path of H. It defines a linear order on the set $\mathcal{K}$. Let us use it to construct a linear order on $\mathcal{X}$.

In a natural way, $\mathcal{P}$ defines a linear pre-order on $\mathcal{X}$ by

$$X_a \le X_b \text{ if } \exists i, j \text{ such that } X_a \subseteq P_i, X_b \subseteq P_j, \text{ where } 1 \le i \le j \le k, 1 \le a, b \le p,$$

where for a clique X_i that is contained in several maximal cliques of H, consider just the first occurrence. Transform it into a linear order (sequence) $\mathcal{Q}$ by fixing any permutation inside the equivalence classes.

Let us define $H' = $ `FillFolding`$(G, \mathcal{Q})$, and prove that $H' = H$. By Lemma 3, H' is an interval completion of G. Moreover, $E(H') \subseteq E(H)$, since $xy \in E(H')$ only if the interval between the first and the last element in $\mathcal{Q}$ that contains x intersects the one corresponding to y. In this case, the corresponding intervals in $\mathcal{P}$ intersect as well, so there is $xy \in E(H)$. By minimality of H, there is $H = H'$.   □

It is well-known (see also Lemma 1) that the pathwidth of G is the minimum, over all interval completions H of G, of the cliquesize of H minus one. Clearly, we can restrict to minimal interval completions. Theorem 2 tells us that an optimal interval completion for the pathwidth problem is defined by some folding of the graph.

5   Folding of Circular-Arc Graphs

Let $(\mathcal{X}, C)$ be a clique cycle of the circular-arc graph $G = (V, E)$, obtained like in Definition 1. Consider a permutation $\mathcal{Q}$ of the set of bags $\mathcal{X}$. In the case of circular-arc graphs, we study the permutation $\mathcal{Q}$ with respect to the circular ordering of $\mathcal{X}$ on the cycle $(\mathcal{X}, C)$. Therefore it is more convenient to think of a folding as a *triple* $(\mathcal{X}, C, \mathcal{Q})$.

Remark 1. Let $(\mathcal{X}, C, \mathcal{Q})$ be a folding of the circular-arc graph G. Consider the clique path decomposition $\mathcal{P}$ produced by the algorithm FillFolding$(G, \mathcal{Q})$. Let us look at the Algorithm FillFolding(see Figure 1). Observe that each bag P of $\mathcal{P}$ is the union of a clique $Q \in \mathcal{Q}$ (which corresponds to the bag P at the initialization step) and some semi-separators of type $Q' \cap Q''$, where Q', Q'' are two cliques consecutive on the cycle, but separated by Q in $\mathcal{Q}$. We say that the clique Q and the semi-separators have been *merged* by the folding. A folding $(\mathcal{X}, C, \mathcal{Q})$ naturally defines an upper part and a lower part of the cycle $(\mathcal{X}, C)$. Let Q_L, Q_R be the leftmost and rightmost element of the permutation $\mathcal{Q}$. Let $\mathcal{X}^{down}$ $(\mathcal{X}^{up})$ denote the cliques counterclockwise (clockwise) between Q_L and Q_R on the cycle. Let $\mathcal{Q}^{down} = (Q_L = Q_{l_1}, Q_{l_2}, \ldots, Q_{l_r} = Q_R)$ denote the restriction of $\mathcal{Q}$ to $\mathcal{X}^{down}$. Similarly let $\mathcal{Q}^{up} = (Q_L = Q_{u_1}, Q_{u_2}, \ldots, Q_{u_t} = Q_R)$ denote the restriction of $\mathcal{Q}$ to $\mathcal{X}^{up}$.

Definition 5. *Given a clique cycle decomposition $(\mathcal{X}, C)$ of G and a permutation $\mathcal{Q}$ of $\mathcal{X}$, we say that a clique $X \in \mathcal{X}$ is a pivot of the folding $(\mathcal{X}, C, \mathcal{Q})$ if its neighbors on the cycle $(\mathcal{X}, C)$ appear on the same side of X in $\mathcal{Q}$. We extend this definition to any subset $\mathcal{X}'$ of $\mathcal{X}$: $X \in \mathcal{X}'$ is a pivot w.r.t. $\mathcal{X}'$ if $X_L, X_R \in \mathcal{X}'$, its closest neighbors on the cycle among the elements of $\mathcal{X}'$, are on the same side of X in $\mathcal{Q}$.*

Definition 6. *Let $(\mathcal{X}, C)$ be a clique cycle decomposition of $G = (V, E)$. A permutation $\mathcal{Q}$ of a subset $\mathcal{X}' \subseteq \mathcal{X}$, is k-monotone if it contains at most k pivots. The monotonicity of $\mathcal{Q}$ is the minimum k such that $\mathcal{Q}$ is k-monotone. The monotonicity of a folding $(\mathcal{X}, C, \mathcal{Q})$ is the monotonicity of $\mathcal{Q}$.*

The main combinatorial result of the paper consists in proving that there exists a 2-monotone folding $(\mathcal{X}, C, \mathcal{Q})$ such that $H = $ FillFolding$(G, \mathcal{Q})$ is an interval completion of G satisfying pwd$(H) = $ pwd(G). Therefore, an optimum interval completion for the pathwidth problem can be found among the completions defined by 2-monotone foldings. In a 2-monotone folding, the only pivots are the first and last element of $\mathcal{Q}$. Moreover, $\mathcal{Q}^{up}$ $(\mathcal{Q}^{down})$ is clockwise (counterclockwise) consecutive on the cycle $(\mathcal{X}, C)$.

The following lemma is straightforward (see also Remark 1).

Lemma 4. *Let $(\mathcal{X}, C, \mathcal{Q})$ be a 2-monotone folding and let $\mathcal{P}$ be the clique path decomposition produced by FillFolding$(G, \mathcal{Q})$. Every bag of $\mathcal{P}$ is the union of a clique $Q \in \mathcal{Q}$ and the unique semi-separator corresponding to the edge $\{Q', Q''\}$ of the cycle, such that Q separates Q' and Q'' in the permutation $\mathcal{Q}$.*

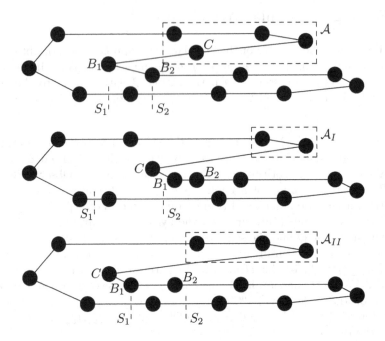

Fig. 2. Reduction of $\mathcal{A}$ (top) to $\mathcal{A}'$: one way (middle) or the other (bottom)

Definition 7. *Let* $(\mathcal{X}, C)$ *be a clique cycle decomposition of* $G = (V, E)$ *and let* $(\mathcal{X}, C, \mathcal{Q})$ *be a 4-monotone folding. Let* $\mathcal{Q}_L, \mathcal{Q}_R$ *be the end-cliques (pivots) of* $\mathcal{Q}$. *Let* B_1, P *be the other pivots, ordered as in* $\mathcal{Q}$. *Assume w.l.o.g. that* B_1, P *belong to* $\mathcal{Q}^{up}$. *The consecutive part of the cycle that appears counterclockwise, starting right after* B_1, *passing through* P, *continuing as long as it stays after* B_1 *in* $\mathcal{Q}$ *is called the* anomaly *(see the top part of Figure 2).*

Notice that for a 4-monotone folding $\mathcal{Q}$, the restriction of $\mathcal{Q}$ to $\mathcal{X} \setminus \mathcal{A}$ is 2-monotone.

One of our main tools (Theorem 3) shows that if $(\mathcal{X}, C, \mathcal{Q})$ is a 4-monotone folding which defines $H = \texttt{FillFolding}(G, \mathcal{Q})$, then there exists a 2-monotone folding $(\mathcal{X}, C, \mathcal{Q}')$ defining an interval graph $H' = \texttt{FillFolding}(G, \mathcal{Q}')$ of pathwidth smaller or equal the pathwidth of H. Here is an informal sketch of the main idea. Consider an anomaly $\mathcal{A}$ of the 4-monotone folding $(\mathcal{X}, C, \mathcal{Q})$. Suppose that the anomaly is in the upper part of the cycle, as on Figure 2.

Notice that when we restrict $(\mathcal{X}, C)$ to the cliques that are not in the anomaly (just remove every $X \in \mathcal{A}$, making the former neighbors of X adjacent), then we obtain a clique cycle of $\overline{G} = G[\bigcup(X \setminus \mathcal{A})]$, an induced subgraph of G. Moreover, we obtain $(\overline{\mathcal{X}}, \overline{C}, \overline{\mathcal{Q}})$, a 2-monotone folding of $\overline{G}$, where $\overline{\mathcal{Q}}$ is the restriction of sequence $\mathcal{Q}$ to the cliques not in the anomaly $\overline{\mathcal{X}} = \mathcal{X} \setminus \mathcal{A}$. It defines an interval completion $\overline{H} = \texttt{FillFolding}(\overline{G}, \overline{\mathcal{Q}})$. Notice that the pathwidth of $\overline{H}$ is at most the pathwidth of H.

Definition 8. *Let* $(\mathcal{X}, C)$ *be a clique cycle decomposition of* $G = (V, E)$ *and let* $(\mathcal{X}, C, \mathcal{Q})$ *be a 4-monotone folding with the anomaly* $\mathcal{A}$. *The* $\mathcal{A}$*-width of* $(\mathcal{X}, C, \mathcal{Q})$ *is the pathwidth of* $\overline{H} = \mathtt{FillFolding}(\overline{G}, \overline{\mathcal{Q}})$, *where* $\overline{G}$ *is the circular-arc graph defined by the clique cycle* $(\mathcal{X}, C)$ *restricted to* $\overline{\mathcal{X}} = \mathcal{X} \setminus \mathcal{A}$ *and* $\overline{\mathcal{Q}}$ *is the sequence* $\mathcal{Q}$ *restricted to* $\overline{\mathcal{X}}$.

One step of the procedure is to slightly modify the folding $(\mathcal{X}, C, \mathcal{Q})$ to obtain $(\mathcal{X}, C, \mathcal{Q}')$ with a strictly smaller anomaly $\mathcal{A}'$, ensuring that the $\mathcal{A}'$-width of $(\mathcal{X}, C, \mathcal{Q}')$ is not bigger than the pathwidth of H. Continue until the anomaly is empty. Eventually, this yields a folding $(\mathcal{X}, C, \mathcal{Q}'')$ which is 2-monotone. Its anomaly $\mathcal{A}''$ being empty, the $\mathcal{A}''$-width of this folding is equal to the pathwidth of $H'' = \mathtt{FillFolding}(G, \mathcal{Q}'')$, and it is not bigger than the pathwidth of H.

Let us give in more detail the construction of $(\mathcal{X}, C, \mathcal{Q}')$ based on $(\mathcal{X}, C, \mathcal{Q})$. Consider the pivots of $(\mathcal{X}, C, \mathcal{Q})$ that are not end-cliques of $\mathcal{Q}$. Let P be the one that belongs to the anomaly $\mathcal{A}$, and let B_1 be the other one. Let B_{k+1} be the neighbor of B_1 on the cycle that belongs to the anomaly (labeled Y in Figure 2). Let $B_2, \ldots, B_k$ be the cliques, which do not belong to the anomaly, that follow B_1 clockwise on the cycle and appear before B_{k+1} in $\mathcal{Q}$. Let $S_1, \ldots, S_{k+1}$ be the semi-separators on the lower part of the cycle, such that S_i is merged with the corresponding B_i in $\mathtt{FillFolding}(G, \mathcal{Q})$. In this setting, we choose a semi-separator S_l and permute $\mathcal{Q}$ in order to put all $B_{k+1}, B_1, \ldots, B_{l-1}$ (in this order) between Q and Q', where Q, Q' are the consecutive cliques in the lower part of the cycle such that $S_l = Q \cap Q'$. We choose S_l such that the new folding $(\mathcal{X}, C, \mathcal{Q}')$ has the desired property. We say that in such a situation we *put* $B_{k+1}, B_1, \ldots, B_{l-1}$ *on* the semi-separator S_l. This construction is illustrated in Figure 2. Informally, the bags $B_{k+1}, B_1, \ldots, B_{l-1}$ slide (without jumping) along the cycle, in the clockwise sense, and they stop above the edge of the cycle corresponding to S_l.

Theorem 3. *Let* $(\mathcal{X}, C)$ *be a clique cycle decomposition of* $G = (V, E)$. *Let* $(\mathcal{X}, C, \mathcal{Q})$ *be a 4-monotone folding and* $H = \mathtt{FillFolding}(G, \mathcal{Q})$. *Then there is a 2-monotone folding* $(\mathcal{X}, C, \mathcal{Q}')$ *such that* $H' = \mathtt{FillFolding}(G, \mathcal{Q}')$ *has pathwidth at most* $\mathrm{pwd}(H)$. *Moreover, we can assume that* $(\mathcal{X}, C, \mathcal{Q}')$ *is such that* $\mathcal{X}'^{up} = \mathcal{X}^{up}$ *and* $\mathcal{X}'^{down} = \mathcal{X}^{down}$.

Proof. Let $(\mathcal{X}, C, \mathcal{Q}')$ be a 4-monotone folding with the anomaly $\mathcal{A}'$, such that $\mathcal{X}'^{up} = \mathcal{X}^{up}$ and $\mathcal{X}'^{down} = \mathcal{X}^{down}$, which satisfies the following Properties:

1. $\mathcal{A}' \subseteq \mathcal{A}$;
2. for any clique Q of $\mathcal{A}'$, the semi-separator S of the lower part of the cycle merged with Q in $\mathtt{FillFolding}(G, \mathcal{Q}')$ is the same as in $\mathtt{FillFolding}(G, \mathcal{Q})$;
3. the $\mathcal{A}'$-width of $(\mathcal{X}, C, \mathcal{Q}')$ is not bigger than the pathwidth of H,
4. the anomaly $\mathcal{A}'$ is inclusion minimal among all such foldings.

Let us show that $\mathcal{A}'$ is in fact empty, thus the pathwidth of H' is not bigger than the pathwidth of H. Suppose $\mathcal{A}'$ is not empty.

We use the notations introduced in the informal description above. Let us use Y and S_Y as shorthands for B_{k+1} and S_{k+1}. By Property 2 and Remark 1, we have:

$$|Y \cup S_Y| \leq \text{pwd}(H) + 1. \tag{1}$$

The semi-separators S_i, $1 \leq i \leq k+1$, can be partitioned as follows:

$$S_i = N_i^j \cup B_i^j \cup Y_i^j \cup I_i^j \text{ , for any } 1 \leq j \leq k \text{ , where}$$
$$N_i^j = S_i \setminus (B_j \cup Y), B_i^j = S_i \cap B_j \setminus Y, Y_i^j = S_i \cap Y \setminus B_j, I_i^j = S_i \cap B_j \cap Y. \tag{2}$$

Claim. For any $1 \leq i \leq k$, $1 \leq p \leq q \leq k+1$, one of the following holds:

$$|B_i \cup S_p| \geq |B_i \cup S_q| \text{ or } |Y \cup S_q| \geq |Y \cup S_p| \tag{3}$$

Proof. Suppose it is not true. By Equation 2, we have:

$$|B_i \cup N_p^i \cup Y_p^i| < |B_i \cup N_q^i \cup Y_q^i| \text{ and } |Y \cup N_q^i \cup B_q^i| < |Y \cup N_p^i \cup B_p^i|,$$

which yields a contradiction: $|N_p^i| < |N_q^i|$ and $|N_q^i| < |N_p^i|$, since for any j, p, q, $1 \leq j \leq k$, $1 \leq p \leq q \leq k+1$ there is $Y_q^j \subseteq Y_p^j$ and $B_p^j \subseteq B_q^j$, by properties of the clique cycle. □

Claim. Let l be the biggest integer such that $|Y \cup S_Y| < |Y \cup S_i|$, for $1 \leq i \leq l-1$. Then

$$|Y \cup S_Y| \geq |Y \cup S_l|, \tag{4}$$

$$|B_i \cup S_i| \geq |B_i \cup S_l|, \text{ for any } 1 \leq i \leq l, \tag{5}$$

Proof. The first equation is clear from the construction. Since $|Y \cup S_l| \leq |Y \cup S_Y|$ and $|Y \cup S_Y| < |Y \cup S_i|$, for any $1 \leq i \leq l-1$, there is $|Y \cup S_l| < |Y \cup S_i|$, for any $1 \leq i \leq l-1$. Now, by Equation 3, for any $1 \leq i \leq l-1$, we get $|B_i \cup S_i| \geq |B_i \cup S_l|$, for any $1 \leq i \leq l-1$. □

Therefore, by putting Y and all B_i, for $1 \leq i \leq l-1$, on S_l, we create a new folding $(\mathcal{X}, C, \mathcal{Q}'')$ with a strictly smaller anomaly $\mathcal{A}''$. Indeed, there is $Y \in \mathcal{A}' \setminus \mathcal{A}''$. Notice there may be other cliques in $\mathcal{A}' \setminus \mathcal{A}''$ as well.

Let us check that the $\mathcal{A}''$-width of the folding $(\mathcal{X}, C, \mathcal{Q}'')$ is at most $\text{pwd}(H)$. For each clique X of $\mathcal{Q}'' \setminus \mathcal{A}''$, let S_X be the (unique) semi-separator of the lower part of the cycle to which X is merged in FillFolding$(G, \mathcal{Q}'')$. The $\mathcal{A}''$-width of $(\mathcal{X}, C, \mathcal{Q}'')$ is the maximum, over all cliques X, of $|X \cup S_X| - 1$. We check that $|X \cup S_X| - 1$ is at most the $\text{pwd}(H)$, for every clique X. If X is also in $\mathcal{Q}' \setminus \mathcal{A}'$, this quantity is upper bounded by the $\mathcal{A}'$-width of $(\mathcal{X}, C, \mathcal{Q}')$ and the conclusion follows by Property 3. If $X = Y$, then $S_X = S_l$ and the conclusion follows from Equations 4 and 1. If X is one of the B_i's, $1 \leq i \leq l-1$, again $S_X = S_l$ and the conclusion follows from Equation 5 and Property 3. Finally, if X is one of the cliques of $\mathcal{A}' \setminus \mathcal{A}''$, different from Y, the conclusion follows from Property 2.

The new folding $(\mathcal{X}, C, \mathcal{Q}'')$ also respects Property 2, since in the permutation $\mathcal{Q}''$ the cliques of $\mathcal{A}''$ have the same position w.r.t. the lower part of the cycle as before.

The construction of $\mathcal{Q}''$ contradicts Property 4 of $\mathcal{Q}'$. So $\mathcal{A}'$ must be empty.

$\square$

The following theorem, which is the main combinatorial result of this paper, shows that not only a 4-monotone but any folding can be reduced to a 2-monotone one without augmenting the pathwidth of the resulting completion.

Theorem 4. *Let $(\mathcal{X}, C)$ be a clique cycle decomposition of $G = (V, E)$. There is a folding $(\mathcal{X}, C, \mathcal{Q})$, with $\mathcal{Q}$ being a permutation of $\mathcal{X}$, such that $\mathcal{Q}$ is 2-monotone and $H = \text{FillFolding}(G, \mathcal{Q})$ is an interval completion of G of pathwidth equal the pathwidth of G.*

Proof. Let $(\mathcal{X}, C, \mathcal{Q})$ be a folding of minimum monotonicity such that the pathwidth of $H = \text{FillFolding}(G, \mathcal{Q})$ is not bigger than the pathwidth of G. We will prove that it is 2-monotone.

Suppose it is not. Assume w.l.o.g. that $\mathcal{X}^{up}$ contains some pivots other than Q_L, Q_R. Let B_1 be the leftmost pivot in $\mathcal{Q}^{up}$ different from Q_L. Let P be the rightmost in $\mathcal{Q}^{up}$ among the pivots which are between Q_L and B_1 clockwise on the cycle $(\mathcal{X}, C)$. Let $\mathcal{Q}_L^{up}$ denote the subsequence of $\mathcal{Q}^{up}$ induced by cliques clockwise between Q_L and P (included) on the cycle. Let $\mathcal{Q}_C^{up}$ denote the subsequence of $\mathcal{Q}^{up}$ induced by cliques between P and B_1 (included), and $\mathcal{Q}_R^{up}$ denote the subsequence of $\mathcal{Q}^{up}$ induced by cliques between B_1 and Q_R (included). Let G_L^{up} be the graph defined by the folding $\mathcal{Q}_L^{up}$, restricted to the corresponding set of bags: $G_L^{up} = \text{FillFolding}(G[\bigcup \mathcal{Q}_L^{up}], \mathcal{Q}_L^{up})$. We denote by P_L^{up} the clique path decomposition produced by the folding algorithm. Similarly, we define G_C^{up}, G_R^{up} and G^{down}, with the corresponding clique path decompositions. Let $\tilde{G}$ be the union of these four graphs. Note that $\tilde{G}$ is a circular-arc graph. A clique cycle decomposition $(\tilde{\mathcal{X}}, \tilde{C})$ of $\tilde{G}$ is obtained by gluing into a cycle the paths P_L^{up}, the reverse of P_C^{up}, P_R^{up}, and to the reverse of P^{down}. The gluing is performed by identifying the bags P, then B_1, Q_R and finally Q_L.

Moreover, this procedure yields a folding $(\tilde{\mathcal{X}}, \tilde{C}, \tilde{\mathcal{Q}})$ of $\tilde{G}$. The bags of $\tilde{\mathcal{X}}$ are in one-to-one correspondence to the bags of $\mathcal{X}$, so the permutation $\mathcal{Q}$ of $\mathcal{X}$ is translated into a permutation $\tilde{\mathcal{Q}}$ of $\tilde{\mathcal{X}}$. Notice that $(\tilde{\mathcal{X}}, \tilde{C}, \tilde{\mathcal{Q}})$ is a 4-monotone folding, since Q_L, B_1, P, Q_R are the only pivots left. Also $\text{FillFolding}(\tilde{G}, \tilde{\mathcal{Q}}) = H = \text{FillFolding}(G, \mathcal{Q})$.

By Theorem 3 on $(\tilde{\mathcal{X}}, \tilde{C}, \tilde{\mathcal{Q}})$ there is a 2-monotone folding $(\tilde{\mathcal{X}}, \tilde{C}, \tilde{\mathcal{Q}}')$ such that the pathwidth of $H' = \text{FillFolding}(\tilde{G}, \tilde{\mathcal{Q}}')$ is not bigger than the pathwidth of H, thus not bigger than the pathwidth of G.

Since $\tilde{\mathcal{Q}}'$ is 2-monotone and $\tilde{\mathcal{X}}'^{up} = \tilde{\mathcal{X}}^{up}$, the only pivots of $\tilde{\mathcal{Q}}'$ are Q_L and Q_R. Notice that there is : $\tilde{\mathcal{Q}}'^{up}$ equals P_L^{up} glued to the reverse of P_C^{up} glued to P_R^{up}, and $\tilde{\mathcal{Q}}'^{down}$ equals P^{down}.

Because of the one-to-one correspondence between the elements of $\mathcal{X}$ and $\tilde{\mathcal{X}}$, we construct the folding $(\mathcal{X}, C, \mathcal{Q}')$ directly from $(\tilde{\mathcal{X}}, \tilde{C}, \tilde{\mathcal{Q}}')$, by just replacing the elements of $\tilde{\mathcal{X}}$ with the corresponding elements of $\mathcal{X}$. Clearly, B_1 and P are not pivots of Q', whereas all the other pivots of Q' are also pivots of Q. Moreover, it is easy to verify that $\mathtt{FillFolding}(G, Q') = \mathtt{FillFolding}(\tilde{G}, \tilde{Q}')$. Therefore, $(\mathcal{X}, C, Q')$ is a folding of strictly smaller monotonicity than $(\mathcal{X}, C, Q')$, which also defines a completion of pathwidth not bigger that the pathwidth of G. A contradiction.   □

6   The Algorithm

Our algorithm computing the pathwidth of circular-arc graphs is very similar to the algorithms computing the treewidth and minimum fill-in for circle and circular-arc graphs [9,10].

Consider a clique cycle $CC_{G,M} = (\mathcal{X}, C)$ of the input graph G, obtained from a circular-arc model like in Definition 1. Subdivide each edge of the cycle by adding a new bag containing the semi-separator corresponding to the edge. We obtain a *clique-semi-separator* cycle alternating original clique bags and semi-separator bags. We should also see this cycle as a (regular) *polygon of scanpoints* $\mathcal{P}_G$, following the terminology of [10]; the scanpoints are the clique and semi-separator bags of the cycle. Therefore we associate to each scanpoint s the set of vertices $V(s)$, corresponding to the clique or semi-separator represented by the scanpoint. For each triangle T formed by three scanpoints s_1, s_2, s_3, define the width $w(T)$ of the triangle as the cardinality of the union $V(s_1) \cup V(s_2) \cup V(s_3)$.

Definition 9. *A* linear (planar) triangulation LP *of the polygon of scanpoints* $\mathcal{P}_G$ *is a planar triangulation such that every triangle contains at most two diagonals. The width $w(LT)$ of the linear triangulation is the maximum width over its faces (triangles).*

Theorem 5. *For any circular-arc graph G, its pathwidth is the minimum, over all linear planar triangulations LT of $\mathcal{P}_G$, of $w(LT) - 1$.*

Proof. Let us only sketch the proof of this theorem. We first show that from any planar linear triangulation LT one can construct a path decomposition of G, of width $w(LT) - 1$. Conversely, given a 2-monotone folding $(\mathcal{X}, C, Q)$ one can construct a planar linear triangulation of width at most $\mathtt{FillFolding}(G, Q)$, plus one. The conclusion follows by Theorem 4.   □

Due to space restrictions, we do not describe the dynamic programming algorithm for computing a linear planar triangulation of $\mathcal{P}_G$ of minimum width. We point out that it is very similar to the one of [9,10], which describe algorithms computing (non necessarily linear) planar triangulations of a polygon playing the same role as $\mathcal{P}_G$.

Theorem 6. *The pathwidth of circular-arc graphs can be computed in $\mathcal{O}(n^2)$ time.*

References

1. Bodlaender, H.: A Partial k-Arboretum of Graphs with Bounded Treewidth. Theoretical Computer Science 209(1-2), 1–45 (1998)
2. Bodlaender, H., Fomin, F.: Approximating the pathwidth of outerplanar graphs. Journal of Algorithms 43(2), 190–200 (2002)
3. Bodlaender, H., Kloks, T.: Efficient and constructive algorithms for the pathwidth and treewidth of graphs. Journal of Algorithms 21(2), 358–402 (1996)
4. Ellis, J.A., Markov, M.: Computing the vertex separation of unicyclic graphs. Information and Computation 192(2), 123–161 (2004)
5. Ellis, J.A., Sudborough, I.H., Turner, J.S.: The Vertex Separation and Search Number of a Graph. Information and Computation 113(1), 50–79 (1994)
6. Fomin, F., Thilikos, D.: A 3-approximation for the pathwidth of Halin graphs. Journal of Discrete Algorithms 4(4), 499–510 (2006)
7. Golumbic, M.C.: Algorithmic Graph Theory and Perfect Graphs. Academic Press, London (1980)
8. Heggernes, P., Suchan, K., Todinca, I., Villanger, Y.: Characterizing minimal interval completions: towards better understanding of profile and pathwidth. In: Thomas, W., Weil, P. (eds.) STACS 2007. LNCS, vol. 4393, pp. 236–247. Springer, Heidelberg (2007)
9. Kloks, T.: Treewidth of circle graphs. Int. J. Found. Comput. Sci. 7(2), 111–120 (1996)
10. Kloks, T., Kratsch, D., Wong, C.K.: Minimum Fill-in on Circle and Circular-Arc Graphs. J. Algorithms 28(2), 272–289 (1998)
11. McConnell, R.M.: Linear-Time Recognition of Circular-Arc Graphs. Algorithmica 37(2), 93–147 (2003)
12. Meggido, N., Hakimi, S.L., Garey, M.R., Johson, D.S., Papadimitriou, C.H.: The complexity of searching a graph. Journal of the ACM 35, 18–44 (1988)
13. Robertson, N., Seymour, P.D.: Graph minors. I. Excluding a forest. Journal of Combinatorial Theory. Series B 35, 39–61 (1983)
14. Skodinis, K.: Construction of linear tree-layouts which are optimal with respect to vertex separation in linear time. J. Algorithms 47(1), 40–59 (2003)
15. Szpilrajn-Marczewski, E.: Sur deux propriétés des classes d'ensembles. Fundamenta Mathematicae 33, 303–307 (1945)

Characterization and Recognition of Digraphs of Bounded Kelly-width

Daniel Meister, Jan Arne Telle, and Martin Vatshelle

Institutt for Informatikk, Universitetet i Bergen, 5020 Bergen, Norway

Abstract. Kelly-width is a parameter of directed graphs recently proposed by Hunter and Kreutzer as a directed analogue of treewidth. We give several alternative characterizations of directed graphs of bounded Kelly-width in support of this analogy. We apply these results to give the first polynomial-time algorithm recognizing directed graphs of Kelly-width 2. For an input directed graph $G = (V, A)$ the algorithm will output a vertex ordering and a directed graph $H = (V, B)$ with $A \subseteq B$ witnessing either that G has Kelly-width at most 2 or that G has Kelly-width at least 3, in time linear in H.

1   Introduction

The tractability of large classes of NP-complete problems when parameterized by the treewidth of the input graph counts among the strongest results in algorithmic graph theory. The algorithms behind this tractability have two stages: first an algorithm computing treewidth, then an algorithm solving the specific problem using the tree-structure discovered in the first stage. See for example [2] for a recent overview of these algorithms. For directed graphs (digraphs) there have been several proposals for a parameter analogous to treewidth: 'directed treewidth' of Johnson, Robertson, Seymour, Thomas [4], 'D-width' of Safari [7], 'DAG-width' of Berwanger, Dawar, Hunter, Kreutzer [1] and independently Obdržálek [6], and 'Kelly-width' of Hunter and Kreutzer [3]. Which of these proposed parameters is the better analogue of treewidth? In this paper we give evidence in support of the Kelly-width parameter.

The success of a model depends on a balance between the modeling power, which measures how general its domain of application is, and the analytical power, which measures how good it is as an analytical tool. The two are typically in conflict. This is also the case for the above proposals for tree-like parameters of digraphs. The better the modeling power, e.g. the larger the class of digraphs that have bounded parameter value, the worse the analytical power, e.g. the smaller the chance of successfully emulating both stages of the algorithmic results for treewidth. We do not go into details of the modeling and analytical powers of each of the proposed digraph parameters, but simply note that from a purely algorithmic point of view there is at least not yet a clear winner. How then to choose the digraph parameter which is the most natural directed analogue of treewidth? Note that while some concepts of undirected graphs have

A. Brandstädt, D. Kratsch, and H. Müller (Eds.): WG 2007, LNCS 4769, pp. 270–279, 2007.

unambiguous natural translations to directed graphs, e.g. from paths to directed paths, there are other concepts, e.g. cliques and separators, for which the translation is less clear. The treewidth parameter is known to have many equivalent characterizations. If we start with a characterization of treewidth that uses only concepts that have unambiguous translations to directed graphs then we should arrive at a directed graph parameter which is a natural analogue of treewidth. This is the approach we take in this paper. In Section 3, we give a new characterization of digraphs of Kelly-width at most k arising from a characterization of treewidth that uses the fairly unambiguous concepts of vertex orderings, paths and neighbours.

We also enhance the algorithmic argument in favour of Kelly-width. Digraphs of Kelly-width 1 are the directed acyclic graphs and recognizable by a simple algorithm. For all larger values of k the only algorithms that were known for recognizing digraphs of Kelly-width k had running time exponential in the size of the input digraph [3]. Using the given characterizations we are able to present a fast algorithm recognizing digraphs of Kelly-width 2 in Section 4. For an input digraph $G = (V, A)$ this algorithm will output a vertex ordering and a digraph $H = (V, B)$ with $A \subseteq B$ witnessing either that G has Kelly-width at most 2 or that G has Kelly-width at least 3, in time linear in H. In the positive case the witness can be used to easily find a decomposition of the digraph into a tree-like structure.

Due to space restrictions proofs are omitted. Complete proofs can be found in the corresponding technical report [5].

2   Graph Preliminaries and Digraphs of Bounded Kelly-width

A *simple finite directed graph* G is a pair of sets, (V, A), where V is finite and A is an irreflexive relation over V. The set V is called the *vertex set* of G, and A is called the *arc set* of G. Since we mostly consider simple finite directed graphs, we shortly call them "digraphs". When we deal with undirected graphs, we will explicitly mention it. For an arbitrary digraph H, $V(H)$ and $A(H)$ denote the vertex and arc set of H, respectively. An arc of graph G is denoted as (u, v) and u is the *start vertex* and v is the *end vertex* of (u, v). Let H be a digraph. We say that G is a *subgraph* of H, if $V \subseteq V(H)$ and $A \subseteq A(H)$. If $V = V(H)$ and G is a subgraph of H then G is a *spanning subgraph* or *partial graph* of H. Further definitions are given when they are needed.

Hunter and Kreutzer introduced the notion of Kelly-width [3]. Kelly-width is a parameter for digraphs, and it is the least width of a so-called *Kelly-decomposition*. We will not define Kelly-decompositions here, since we will not use this notion. The authors gave several alternative characterizations of digraphs of bounded Kelly-width by: elimination process, inductive construction, graph game. We will study graphs of bounded Kelly-width starting from the inductive construction. Let $G = (V, A)$ be a digraph. Let u and v be vertices of G. We call v an *in-neighbour* of u, if (v, u) is an arc of G. The *(open) in-neighbourhood* of u, denoted as $N_G^{\text{in}}(u)$, is the set of in-neighbours of u. The *closed in-neighbourhood* of u, denoted as $N_G^{\text{in}}[u]$, is defined as $N_G^{\text{in}}(u) \cup \{u\}$. Similarly, v is an *out-neighbour* of u, if (u,v) is an arc of G. *Open* and *closed out-neighbourhood* of a vertex are defined respectively. The *out-degree* of a vertex is the number of its out-neighbours. Let X be a set of vertices of G. We define the *common in-neighbourhood* of X, denoted as $\bigcap N_G^{\text{in}}[X]$, recursively:

$$X = \emptyset \qquad : \quad \bigcap N_G^{\text{in}}[X] =_{\text{def}} \quad V$$

$$X \neq \emptyset \text{ and } a \in X : \quad \bigcap N_G^{\text{in}}[X] =_{\text{def}} \quad N_G^{\text{in}}[a] \; \cap \; \bigcap N_G^{\text{in}}[X \setminus \{a\}] .$$

The inductive construction characterization of digraphs of bounded Kelly-width by Hunter and Kreutzer started from a basic class of graphs, and the partial graph relation defines the complete class. The basic graphs are called k-*DAGs*. Since certain of our statements become easier, e.g., Theorem 5, we generalise the definition and define k-*GDAGs*.

Definition 1. *Let* $k \geq 0$. *The class of* generalised k-DAGs, k-*GDAGs, for short, is the class of digraphs inductively defined by the two following construction steps:*

(1) a graph on one vertex is a k-GDAG
(2) let G be a k-GDAG and let u be a vertex that does not appear in G. Let X be a set of at most k vertices of G, called the parent vertices *of u. Then, G' is a k-GDAG where G' emerges from G by adding vertex u and the following arc set:*

$$\Big\{ (u,x) : x \in X \Big\} \cup \Big\{ (y,u) : y \in \bigcap N_G^{\text{in}}[X] \Big\} .$$

With a k-GDAG, we associate a sequence $\langle x_1, \ldots, x_n \rangle$ of vertices, where x_1 is the vertex of the start graph in construction step (1) of Definition 1, and x_i, $i \in \{2, \ldots, n\}$, is added to the graph on the vertices $x_1, \ldots, x_{i-1}$, that has already been constructed, according to construction step (2). Let $k \geq 0$, and let $G = (V, A)$ be a k-GDAG. A vertex sequence $\sigma = \langle x_1, \ldots, x_n \rangle$ for G is a *construction sequence* for G, if G can be obtained according to construction steps (1) and (2) adding vertices according to σ and choosing $N_G^{\text{out}}(x_i) \cap \{x_1, \ldots, x_{i-1}\}$ as the parent vertices set of x_i, $i \in \{1, \ldots, n\}$. Parent vertices are always defined with respect to a vertex sequence. The *child vertices* of a vertex x are those vertices that choose x as a parent vertex.

Definition 2. *Let* $k \geq 0$, *and let G be a digraph. G is a* partial k-GDAG *if and only if G is a partial graph of some k-GDAG.*

Hunter-Kreutzer k-DAGs are defined analogous to k-GDAGs with the following difference: instead of starting with a graph on a single vertex in construction step (1), k-DAGs start with a complete graph on k vertices. This means that every k-DAG contains a complete subgraph on k vertices, which is not true for k-GDAGs in general. *Partial k-DAGs* are partial graphs of k-DAGs. Even though the classes of k-GDAGs and k-DAGs do not coincide, the derived classes of partial graphs do.

Lemma 1. *Let $k \geq 0$, and let G be a digraph. G is a partial k-GDAG if and only if G is a partial k-DAG.*

The Kelly-width of a digraph is a width parameter based on the width of Kelly-decompositions. Kelly-width and Kelly-decomposition were introduced by Hunter and Kreutzer as a decomposition counterpart of tree-decompositions for undirected graphs [3]. The authors showed a strong correspondence between partial k-DAGs and graphs of bounded Kelly-width.

Theorem 1 ([3]). *Let $k \geq 0$, and let G be a digraph. G has Kelly-width at most $k + 1$ if and only if G is a partial k-DAG.*

Corollary 1. *Let $k \geq 0$, and let G be a digraph. G has Kelly-width at most $k + 1$ if and only if G is a partial k-GDAG.*

In the following, we will mostly deal with k-GDAGs and partial k-GDAGs. We will also speak of "graphs of bounded Kelly-width". k-DAGs are mentioned only occasionally.

As we motivated in the introduction, the notion of Kelly-width can be viewed as a suitable digraph analogue of treewidth of undirected graphs. The definition of treewidth can also be based on a class of graphs. These are the chordal graphs. An undirected graph is *chordal* if it does not contain an induced cycle of length larger than 3. The *treewidth* of an undirected graph G is the smallest clique-number of a chordal graph that contains G as a subgraph, minus 1. Treewidth is a very fundamental notion and has many different characterizations.

Our first result shows a correspondence between undirected graphs of bounded treewidth and digraphs of bounded Kelly-width. Let $G = (V, A)$ be a digraph. By bi-dir(G), we denote the undirected graph on vertex set V in which two vertices u and v are adjacent if and only if (u, v) and (v, u) are arcs of G.

Theorem 2. *Let $k \geq 0$. An undirected graph G is a chordal graph of treewidth at most k if and only if there is a k-GDAG H such that $G = $ bi-dir(H).*

Corollary 2. *Let $k \geq 0$. An undirected graph G has treewidth at most k if and only if there is a partial k-GDAG H such that $G = $ bi-dir(H).*

3   Characterizations of Graphs of Bounded Kelly-width

So far, graphs of bounded Kelly-width have four different characterizations: via elimination process, inductive construction, cops-robber game, decomposition. These many characterizations were the start point for us to consider the concepts of Kelly-width and Kelly-decompositions as a good digraph counterpart of the concepts of treewidth and tree-decompostion of undirected graphs. Treewidth seems a very natural concept, since undirected graphs of bounded treewidth can be characterised by a long list of different statements. In this section, we will add two further results to the list of characterizations for graphs of bounded Kelly-width. We will see that graphs of bounded Kelly-width have a vertex-ordering characterization, and we show that partial k-GDAGs are the same as subgraphs of k-GDAGs. We begin by recalling the elimination process characterization by Hunter and Kreutzer. This characterization will be used later.

3.1   Elimination Process Characterization

Undirected graphs of bounded treewidth have a nice characterization using an elimination scheme. Let $G = (V, E)$ be an undirected graph on at least two vertices, and let x be a vertex of G. The operation *reducing G by x* yields graph G' that is obtained from G by deleting vertex x and adding the edge set $\{\{u, v\} : u \neq v \text{ and } u, v \in N_G(x)\}$. In words, G' is obtained from G by deleting x and making its neighbourhood (in G) into a clique.

Theorem 3 (folklore). *Let $k \geq 0$, and let $G = (V, E)$ be an undirected graph. Then, G has treewidth at most k if and only if G can be reduced to a graph on one vertex by repeatedly reducing by a vertex of degree at most k.*

The characterization of undirected graphs of bounded treewidth in Theorem 3 can be translated into the world of digraphs. However, the reduction operation must be adjusted. Let $G = (V, A)$ be a digraph on at least two vertices, and let x be a vertex of G. The operation *reducing G by x* yields graph G' that is obtained from G by deleting vertex x and adding the arc set $\{(u, v) : u \neq v \text{ and } u \in N_G^{\text{in}}(x) \text{ and } v \in N_G^{\text{out}}(x)\}$. This definition of the reduction operation is a natural way to translate the completion from the undirected case to the directed case, although it is not the only possibility. Hunter and Kreutzer did this to obtain the following result for digraphs of bounded Kelly-width.

Theorem 4 ([3]). *Let $k \geq 0$, and let $G = (V, A)$ be a digraph. Then, G has Kelly-width at most $k + 1$ if and only if G can be reduced to a graph on one vertex by repeatedly reducing by a vertex of out-degree at most k.*

The result of Theorem 4 implies an easy algorithm for recognizing graphs of bounded Kelly-width. Unfortunately, this algorithm is not a polynomial-time algorithm. A given graph, partial k-GDAG or not, can have more than one vertex of out-degree at most $k - 1$. There is no a priori argument or criterion deciding which one to choose.

3.2   Vertex-Ordering Characterization

In this subsection, we show that graphs of bounded Kelly-width are the graphs whose vertices can be arranged in a linear order to satisfy special conditions. We start with a characterization of k-GDAGs. This characterization is used in most of our proofs about graphs of bounded Kelly-width.

Let $G = (V, A)$ be a digraph. A *path P* in G is a sequence $(x_0, \ldots, x_l)$ of mutually different vertices of G where (x_i, x_{i+1}) is an arc of G for every $i \in \{0, \ldots, l - 1\}$. Let σ be a vertex ordering for G. Path P is called *σ-monotone-left*, if $x_l \prec_\sigma \cdots \prec_\sigma x_0$ holds. P starts at vertex x_0; so, if P is σ-monotone-left, it is a σ-monotone-left path starting at x_0. For a vertex u and an arc (x, y) of G, we say that (x, y) *spans over u with respect to σ*, if $x \prec_\sigma u \prec_\sigma y$ or $y \prec_\sigma u \prec_\sigma x$. If the ordering σ is uniquely determined, we shortly say that (x, y) *spans over u*. Let u be a vertex of G. We say that a σ-monotone-left path in G has the *spanning-vertex u property*, if the pair (P, u) satisfies the following condition: if P contains an arc that spans over u, then P contains a vertex $w \prec_\sigma u$ such that $w \in N_G^{out}(u)$ and the arc of P that spans over u has end vertex w.

Theorem 5. *Let $k \geq 0$, and let $G = (V, A)$ be a digraph. G is a k-GDAG if and only if there is a vertex ordering $\sigma = \langle x_1, \ldots, x_n \rangle$ for G such that the pair (G, σ) satisfies the following two conditions:*

(1) for every $i \in \{1, \ldots, n\}$, $|N_G^{out}(x_i) \cap \{x_1, \ldots, x_{i-1}\}| \leq k$
(2) for every pair u, v of vertices of G where $u \prec_\sigma v$, (u, v) is an arc of G if and only if every σ-monotone-left path starting at v has the spanning-vertex u property.

If G is a k-GDAG, the vertex orderings σ such that the pair (G, σ) satisfies conditions (1) and (2) are exactly the construction sequences for G.

Also for k-DAGs, a characterization theorem in the flavour of Theorem 5 can be formulated. However, it will have a more complex version of condition (1).

We want to extend the characterization result of Theorem 5 for k-GDAGs to digraphs of bounded Kelly-width. Since partial k-GDAGs are just the partial graphs of k-GDAGs, there must be some relaxation in the conditions of Theorem 5. This relaxation affects condition (2). The following lemma defines a subclass of partial k-GDAGs for which a characterization in the flavour of Theorem 5 exists.

Lemma 2. *Let $k \geq 0$, and let $G = (V, A)$ be a digraph. The following two statements are equivalent:*

(A) there is a vertex ordering $\sigma = \langle x_1, \ldots, x_n \rangle$ for G such that the pair (G, σ) satisfies the following two conditions:
   (1) for every $i \in \{1, \ldots, n\}$, $|N_G^{out}(x_i) \cap \{x_1, \ldots, x_{i-1}\}| \leq k$
   (2) for every pair u, v of vertices of G where $u \prec_\sigma v$, if (u, v) is an arc of G then every σ-monotone-left path starting at v has the spanning-vertex u property
(B) there is a k-GDAG H with construction sequence σ such that the triple (G, H, σ) satisfies the following two conditions:
   (3) G is a partial graph of H
   (4) $A(G) \cap \{(u, v) : v \prec_\sigma u\} = A(H) \cap \{(u, v) : v \prec_\sigma u\}$

The crucial point of the characterization in Lemma 2 is condition (4). Informally, the question is whether every partial k-GDAG G can be embedded into a k-GDAG H_G where H_G can be constructed according to the two construction steps such that every vertex chooses only parent vertices that are out-neighbours in

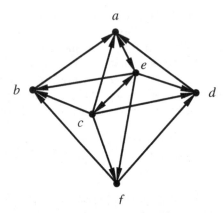

Fig. 1. The depicted graph is a partial 1-GDAG, which can be seen using the construction sequence $\langle e, c, a, f, b, d \rangle$. For a 1-GDAG H and a vertex ordering σ that shall satisfy conditions (3) and (4) of Lemma 2, the last vertex of σ can only be a, b or d; all other vertices have two out-neighbours in G. Distinguishing these cases, it can be shown that part (B) of Lemma 2 cannot be satisfied for G.

G. For partial 0-GDAGs, the question can immediately be answered positively, since 0-GDAGs do not choose any parent vertex. Interestingly, already for partial 1-GDAGs, the answer is negative. The graph depicted in Figure 1 is a partial 1-GDAG for which part (B) of Lemma 2 does not hold. So, for a characterization of partial k-GDAGs, we have to relax the conditions a little more.

Theorem 6. *Let $k \geq 0$, and let $G = (V, A)$ be a digraph. G is a partial k-GDAG if and only if there are a vertex ordering $\sigma = \langle x_1, \ldots, x_n \rangle$ and a set F of arcs such that the triple (G, F, σ) satisfies the following two conditions, where we set $G' =_{\text{def}} G \cup F$:*

(1) for every $i \in \{1, \ldots, n\}$, $|N_{G'}^{\text{out}}(x_i) \cap \{x_1, \ldots, x_{i-1}\}| \leq k$
(2) for every pair u, v of vertices of G where $u \prec_\sigma v$, if (u, v) is an arc of G then every σ-monotone-left path starting at v in G' has the spanning-vertex u property in G'.

Combining the result of Corollary 2 and the characterization of Theorem 6 provides the following characterization of undirected graphs of bounded treewidth. The definitions of σ-monotone-left paths and spanning-vertex u property for undirected graphs are analogous to the definitions for directed graphs.

Corollary 3. *Let $k \geq 0$, and let $G = (V, E)$ be an undirected graph. G has treewidth at most k if and only if there are a vertex ordering $\sigma = \langle x_1, \ldots, x_n \rangle$ for G and a set F of additional edges such that the triple (G, F, σ) satisfies the following two conditions, where we set $G' =_{\text{def}} G \cup F$:*

(1) for every $i \in \{1, \ldots, n\}$, $|N_{G'}(x_i) \cap \{x_1, \ldots, x_{i-1}\}| \leq k$

(2) for every pair u, v of vertices of G where $u \prec_\sigma v$, if uv is an edge of G then every σ-monotone-left path starting at v in G' has the spanning-vertex u property in G'.

The concept of a σ-monotone-left path having spanning-vertex u property is unambiguously translated between undirected graphs and directed graphs. Thus, apart from the binary choice of translating 'neighbours' to either 'in-neighbours' or 'out-neighbours', all undirected graph concepts used in Corollary 3 to characterize treewidth are unambiguously translated to give Theorem 6 characterizing Kelly-width. In our opinion this constitutes a weighty argument that Kelly-width is indeed the natural directed analogue of treewidth.

3.3 Subgraph Characterization

From the characterization result of Theorem 6, we derive yet another characterization of digraphs of bounded Kelly-width. This characterization is not surprising, rather a necessity. It simply says that "partial graph" in the definition of partial k-GDAGs can be replaced by the more natural term "subgraph". This is an analogue to partial k-trees, which are defined as partial graphs of k-trees and can be characterised as subgraphs of k-trees.

Theorem 7. *Let $k \geq 0$, and let $G = (V, A)$ be a digraph. Then, G is a partial k-GDAG if and only if G is a subgraph of a k-GDAG.*

4 A Fast Algorithm for Recognition of Graphs of Kelly-width 2

Theorem 4 gives an algorithm for recognition of digraphs of bounded Kelly-width: a graph has Kelly-width at most $k + 1$ if and only if it can be reduced to a graph on a single vertex by repeatedly reducing by a vertex of out-degree at most k (Theorem 4). A polynomial-time algorithm does not evolve directly from this result, since it is not clear which of the possible vertices to choose. However, in this section we show that it does give a polynomial-time algorithm for Kelly-width 2. In fact, we will show that every choice of a vertex is then a good choice.

For graphs of Kelly-width 2, vertices of out-degree 0 and 1 can be chosen. We treat the two cases separately. The main difference between both cases is that reducing a graph by a vertex of out-degree 0 does not change the remaining graph, whereas reducing by a vertex of out-degree 1 may add new arcs between vertices in the remaining graph. At first, we consider the out-degree 0 case. We can even show a general result: the Kelly-width of a digraph is not influenced by vertices of out-degree 0.

Theorem 8. *Let $k \geq 0$, and let $G = (V, A)$ be a digraph on at least two vertices. Let a be a vertex of G of out-degree 0. Then, G is a partial k-GDAG if and only if the graph obtained from reducing G by a is a partial k-GDAG.*

The proof of the case of reducing by a vertex of out-degree 1 is based on the fact that we can perform an operation on a k-GDAG that corresponds to the reduction on a contained partial graph. This operation is a variant of edge-contraction. Let $G = (V, A)$ be a digraph, and let (a, b) be an arc of G. The graph obtained from *in-contracting arc* (a, b) in G, denoted as $G_{\vartriangleleft_i}(a, b)$, is defined as $(G-a) \cup \{(x, b) : x \neq b \text{ and } x \in N_G^{in}(a)\}$. Informally spoken, arc (a, b) is in-contracted by deleting vertex a and making every in-neighbour of a an in-neighbour of b. If a has out-degree 1, in-contracting arc (a, b) is exactly what we mean by reducing vertex a.

Theorem 9. *Let $G = (V, A)$ be a partial 1-GDAG, and let (a, b) be an arc of G. Then, $G_{\vartriangleleft_i}(a, b)$ is a partial 1-GDAG.*

Corollary 4. *Let $G = (V, A)$ be a digraph, and let a be a vertex of out-degree 1 of G. Then, G is a partial 1-GDAG if and only if the graph obtained from reducing G by a is a partial 1-GDAG.*

The result of Theorem 9 is stronger than the corresponding implication of Corollary 4: reducing a graph by a vertex of out-degree 1 can be simulated by an appropriate in-contraction operation. Equivalence only holds, if the start vertex of the in-contracted arc has out-degree 1. It is a natural question to ask whether Corollary 4 holds for partial k-GDAGs for $k \geq 2$.

Using the two main results about reducing a graph, we obtain a characterization of graphs of Kelly-width 2 that is stronger than Theorem 4.

Theorem 10. *Let $G = (V, A)$ be a digraph. Then, G is a partial 1-GDAG if and only if G can be reduced to a graph on one vertex by repeatedly reducing by an arbitrary vertex of out-degree at most 1.*

The result of Theorem 10 implies a fast algorithm for recognition of graphs of Kelly-width at most 2. The reduction sequence can even be used to construct a witness for being a Kelly-width-2 graph: a 1-GDAG containing the input graph G. In the negative case, our algorithm outputs a graph H' and a vertex sequence $\sigma' = \langle x_r, \ldots, x_n \rangle$ of the following kind: let $G_n =_{\text{def}} G$, and let G_i be obtained from G_{i+1} by reducing by x_{i+1}, $i \in \{r, \ldots, n-1\}$. Then, x_{i+1}, $i \in \{r, \ldots, n-1\}$, has out-degree at most 1 in G_{i+1}, and G_r does not contain any vertex of out-degree at most 1. Due to Theorem 10, G can therefore not be a partial 1-GDAG. The output witness H' is defined as follows: let $H'_r =_{\text{def}} G_r$, and obtain H'_{i+1} from H'_i by adding x_{i+1} according to construction step (2) of Definition 1 choosing as parent vertex the out-neighbour of x_{i+1} in G_{i+1}, if there is one. Then, $H' =_{\text{def}} H'_n$. Using σ' it is easy to verify by the user of the algorithm that H' arises from G and thus the algorithm worked correctly. Moreover, H' is independent of σ' in the sense that on input H' any reduction sequence on vertices of out-degree at most 1 will result in G_r.

Theorem 11. *There is an algorithm that, given a digraph G, decides whether G has Kelly-width at most 2, and if so, outputs a 1-GDAG H that contains G as partial graph. If G has Kelly-width at least 3, the algorithm outputs a witness for* *this case, which is a graph containing G as partial graph and a vertex sequence. The running time and the working space of the algorithm are linear in the size of the output graph.*

For deciding whether a graph has Kelly-width exactly 2, it suffices to run the algorithm only for non-acyclic graphs. Acyclic graphs are exactly the graphs of Kelly-width 1.

5   Final Remarks

Since Kelly-width of a directed graph is a new concept, a lot of problems can still be solved. The most important question, however, affects the status of Kelly-width: does it really capture the notion of treewidth for undirected graphs in the directed setting? We gave good reasons to answer positively: we presented two new characterizations of digraphs of bounded Kelly-width, and we gave an easy algorithm for recognition of digraphs of Kelly-width 2. This recognition algorithm can be considered a directed version of the undirected counterpart: a reduction algorithm for graphs of treewidth at most 1. Recognition algorithms for graphs of bounded Kelly-width are of great interest, since one can expect that they also compute a Kelly-decomposition, which is important for the design of algorithms solving optimization problems on digraphs of bounded Kelly-width.

References

[1] Berwanger, D., Dawar, A., Hunter, P., Kreutzer, S.: DAG-Width and Parity Games. In: Durand, B., Thomas, W. (eds.) STACS 2006. LNCS, vol. 3884, pp. 524–536. Springer, Heidelberg (2006)

[2] Bodlaender, H.L.: Treewidth: Characterizations, Applications, and Computations. In: Fomin, F.V. (ed.) WG 2006. LNCS, vol. 4271, pp. 1–14. Springer, Heidelberg (2006)

[3] Hunter, P., Kreutzer, S.: Digraph Measures: Kelly Decompositions, Games, and Orderings. In: SODA 2007. Proceedings of the Eighteenth Annual ACM-SIAM Symposium on Discrete Algorithms, pp. 637–644. ACM Press, New York (2007)

[4] Johnson, T., Robertson, N., Seymour, P.D., Thomas, R.: Directed Tree-Width. Journal of Combinatorial Theory, Series B 82, 138–154 (2001)

[5] Meister, D., Telle, J.A., Vatshelle, M.: Characterization and recognition of digraphs of bounded Kelly-width. Technical report no. 351, Institutt for Informatikk, Universitetet i Bergen (2007)

[6] Obdržálek, J.: DAG-width – Connectivity Measure for Directed Graphs. In: SODA 2006. Proceedings of the Seventeenth Annual ACM-SIAM Symposium on Discrete Algorithms, pp. 814–821. ACM Press, New York (2006)

[7] Safari, M.A.: D-Width: A More Natural Measure for Directed Tree Width. In: Jedrzejowicz, J., Szepietowski, A. (eds.) MFCS 2005. LNCS, vol. 3618, pp. 745–756. Springer, Heidelberg (2005)

How to Use Planarity Efficiently: New Tree-Decomposition Based Algorithms

Frederic Dorn*

Department of Informatics, University of Bergen, PO Box 7800, 5020 Bergen, Norway
frederic.dorn@ii.uib.no

Abstract. We prove new structural properties for tree-decompositions of planar graphs that we use to improve upon the runtime of tree-decomposition based dynamic programming approaches for several NP-hard planar graph problems. We give for example the fastest algorithm for PLANAR DOMINATING SET of runtime $3^{\text{tw}} \cdot n^{O(1)}$, when we take the treewidth tw as the measure for the exponential worst case behavior. We also introduce a tree-decomposition based approach to solve non-local problems efficiently, such as PLANAR HAMILTONIAN CYCLE in runtime $6^{\text{tw}} \cdot n^{O(1)}$. From any input tree-decomposition, we compute in time $O(nm)$ a tree-decomposition with geometric properties, which decomposes the plane into disks, and where the graph separators form Jordan curves in the plane.

1 Introduction

Many separator results for topological graphs, especially for planar embedded graphs base on the fact that separators have a structure that cuts the surface into two or more pieces onto which the separated subgraphs are embedded on. The celebrated and widely applied (e.g., in many divide-and-conquer approaches) result of Lipton and Tarjan [23] finds in planar graphs a small sized separator. However, their result says nothing about the structure of the separator, it can be any set of discrete points. Applying the idea of Miller for finding small simple cyclic separators [24] in planar triangulations, one can find small separators whose vertices can be connected by a closed curve in the plane intersecting the graph only in vertices, so-called *Jordan curves* (e.g. see [4]). Tree-decompositions have been historically the choice when solving NP-hard optimization and FPT problems with a dynamic programming approach (see for example [6] for an overview). Although much is known about the combinatorial structure of tree-decompositions (a.o, [7,31]), only few results are known to the author relating to the topology of tree-decompositions of planar graphs (e.g., [9]). A branch-decomposition is another tool, that was introduced by Robertson and Seymour in their proof of the Graph Minors Theorem and the parameters of these similar structures, the *treewidth* tw(G) and *branchwidth* bw(G) of the graph G have the relation bw$(G) \leq$ tw$(G) + 1 \leq 1.5$ bw(G) [27]. Recently, branch-decompositions

---

* Supported by the Research Council of Norway.

A. Brandstädt, D. Kratsch, and H. Müller (Eds.): WG 2007, LNCS 4769, pp. 280–291, 2007.

Table 1. Worst-case runtime expressed by treewidth tw and branchwidth bw of the input graph. The PLANAR HAMILTONIAN CYCLE stands representatively for all planar graph problems posted in [16] such as METRIC TSP, whose algorithms we can improve analogously. In [13], only those graph problems are improved upon, which are unweighted or of small integer weights. Therefor, we state the improvements independently for weighted and unweighted graph problems. In some calculations, the fast matrix multiplication constant $\omega < 2.376$ is hidden.

	Previous results	New results
weighted PLANAR DOM SET	$O(n2^{\min\{2\,tw,2.38\,bw\}})$	$O(n2^{1.58\,tw})$
unweighted PLANAR DOM SET	$O(n2^{1.89\,bw})$	$O(n2^{\min\{1.58\,tw,1.89\,bw\}})$
w PLAN INDEPENDENT DOM SET	$O(n2^{\min\{2\,tw,2.28\,bw\}})$	$O(n2^{1.58\,tw})$
uw PLAN INDEPENDENT DOM SET	$O(n2^{1.89\,bw})$	$O(n2^{\min\{1.58\,tw,1.89\,bw\}})$
w PLAN TOTAL DOM SET	$O(n2^{\min\{2.58\,tw,3\,bw\}})$	$O(n2^{2\,tw})$
uw PLAN TOTAL DOM SET	$O(n2^{2.38\,bw})$	$O(n2^{\min\{2\,tw,2.38\,bw\}})$
w PLAN PERF TOTAL DOM SET	$O(n2^{\min\{2.58\,tw,3.16\,bw\}})$	$O(n2^{\min\{2.32\,tw,3.16\,bw\}})$
uw PLAN PERF TOTAL DOM SET	$O(n2^{2.53\,bw})$	$O(n2^{\min\{2.32\,tw,2.53\,bw\}})$
w PLANAR HAM CYCLE	$O(n2^{3.31\,bw})$	$O(n2^{\min\{2.58\,tw,3.31\,bw\}})$
uw PLANAR HAM CYCLE	$O(n2^{2.66\,bw})$	$O(n2^{\min\{2.58\,tw,2.66\,bw\}})$

started to become a more popular tool than tree-decompositions, in particular for problems whose input is a topologically embedded graph [10,19,11,16,15], mainly for two reason: the branchwidth of planar graphs can be computed in polynomial time (yet there is no algorithm known for treewidth) with better constants for the upper bound than treewidth. Secondly, planar branch decompositions have geometrical properties, i.e. they are assigned with separators that form Jordan curves. Thus, one can exploit planarity in the dynamic programming approach in order to get an exponential speedup, as done by [16,13]. We give the first result which employs planarity obtained by the structure of tree-decompositions for getting faster algorithms. This enables us to give the first tree-decomposition based algorithms for planar Hamiltonian-like problems with slight runtime improvements compared to [16]. We emphasize our result in terms of the width parameters tw and bw with the example of DOMINATING SET. The graph problem DOMINATING SET asks for a minimum vertex set S in a graph $G = (V, E)$ such that every vertex in V is either in S or has a neighbor in S. Telle and Proskurowski [30] gave a dynamic programming approach based on tree-decompositions with runtime $9^{tw} \cdot n^{O(1)}$, and that was improved to $4^{tw} \cdot n^{O(1)}$ by Alber et al [1]. Note that in the extended abstract [2], the same authors first stated the runtime wrongly to be $3^{tw} \cdot n^{O(1)}$. Fomin and Thilikos [19] gave a branch-decomposition based approach of runtime $3^{1.5\,bw} \cdot n^{O(1)}$. In [13], the author combined dynamic programming with fast matrix multiplication to get $4^{bw} \cdot n^{O(1)}$ and for PLANAR DOMINATING SET even $3^{\frac{\omega}{2}\,bw} \cdot n^{O(1)}$, where ω is the constant in the exponent of fast matrix multiplication (currently, $\omega \leq 2.376$). Exploiting planarity, we improve further upon the existing bounds and give a $3^{tw} \cdot n^{O(1)}$ algorithm for PLANAR DOMINATING SET, representative for a number of improvements on results of [3,16,17] as shown in Table 1.

Given any tree-decomposition as an input, we show how to compute a geometric tree-decomposition that has the same properties as planar branch decompositions. Employing structural results on minimal graph separators for planar graphs, we create in polynomial time a *parallel* tree-decomposition that is assigned by a set of pairwise parallel separators that form pairwise non-crossing Jordan curves in the plane. In a second step, we show how to obtain a *geometric* tree-decomposition, that has a ternary tree and is assigned Jordan curves that exhaustively decompose the plane into disks (one disk being the infinite disk). In fact, geometric tree-decompositions have all the properties in common with planar branch decompositions, that are algorithmically exploited in [19] and [16].

Organization of the paper: after giving some preliminary results in Section 2, we introduce in Section 3 our algorithm to compute a parallel tree-decomposition. In Section 4, we describe how Jordan curves and separators in plane graphs influence each other and we get some tools for relating Jordan curves and tree-decompositions in Section 5. Finally, we show how to compute geometric tree-decompositions and state in Section 6 their influence on dynamic programming approaches. In Section 7, we argue how our results may lead to faster algorithms when using fast matrix multiplication as in [13].

2 Preliminaries

A *line* is a subset of a surface Σ that is homeomorphic to $[0, 1]$. A closed curve on Σ that is homeomorphic to a cycle is called *Jordan curve*. A planar graph embedded crossing-free onto the sphere $\mathbb{S}_0$ is defined as a *plane* graph, where every vertex is a point of $\mathbb{S}_0$ and each edge a line. In this paper, we consider Jordan curves that intersect with a plane graph only in vertices. For a Jordan curve J, we denote by $V(J)$ the vertices J intersects with.

Given a connected graph $G = (V, E)$, a set of vertices $S \subset V$ is called a *separator* if the subgraph induced by $V \setminus S$ is non-empty and has several components. S is called an u, v-*separator* for two vertices u and v that are in different components of $G[V \setminus S]$. S is a *minimal* u, v-*separator* if no proper subset of S is a u, v-separator. Finally, S is a *minimal separator* of G if there are two vertices u, v such that S is a minimal u, v-separator. For a vertex subset $A \subseteq V$, we *saturate* A by adding edges between every two non-adjacent vertices, and thus, turning A into a clique.

A *chord* in a cycle C of a graph G is an edge joining two non-consecutive vertices of C. A graph H is called *chordal* if every cycle of length > 3 has a chord. A *triangulation* of a graph $G = (V, E)$ is a chordal graph $H = (V, E')$ with $E \subseteq E'$. The edges of $E' \setminus E$ are called *fill edges*. We say, H is a *minimal triangulation* of G if every graph $G' = (V, E'')$ with $E \subseteq E'' \subset E'$ is not chordal. Note that a triangulation of a planar graph may not be planar—not to confuse with the notion of "planar triangulation" that asks for filling the facial cycles with chords. Consider the following algorithm on a graph G that triangulates G, known as the *elimination game* [26]. Repeatedly choose a vertex, saturate its neighborhood, and delete it. Terminate when $V = \emptyset$. The order in which the vertices are deleted is called the *elimination ordering* α, and G_α^+ is the chordal graph obtained by adding all saturating (fill) edges to G. Another way of triangulating a graph G can be obtained by using a tree-decomposition of G.

2.1 Tree-Decompositions

Let G be a graph, T a tree, and let $\mathcal{Z} = (Z_t)_{t \in T}$ be a family of vertex sets $Z_t \subseteq V(G)$, called *bags*, indexed by the nodes of T. The pair $\mathcal{T} = (T, \mathcal{Z})$ is called a *tree-decomposition* of G if it satisfies the following three conditions:

- $V(G) = \cup_{t \in T} Z_t$,
- for every edge $e \in E(G)$ there exists a $t \in T$ such that both ends of e are in Z_t,
- $Z_{t_1} \cap Z_{t_3} \subseteq Z_{t_2}$ whenever t_2 is a vertex of the path connecting t_1 and t_3 in T.

The width $\mathrm{tw}(\mathcal{T})$ of the tree-decomposition $\mathcal{T} = (T, \mathcal{Z})$ is the maximum size over all bags minus one. The *treewidth* of G is the minimum width over all tree-decompositions.

Lemma 1. *[8] Let* $\mathcal{T} = (T, \mathcal{Z}), \mathcal{Z} = (Z_t)_{t \in T}$ *be a tree-decomposition of* $G = (V, E)$, *and let* $K \subseteq V$ *be a clique in* G. *Then there exists a node* $t \in T$ *with* $K \subseteq Z_t$.

As a consequence, we can turn a graph G into another graph H' by saturating the bags of a tree-decomposition, i.e., add an edge in G between any two non-adjacent vertices that appear in a common bag. Automatically, we get that for every clique K in H', there exists a bag Z_t such that $K = Z_t$. Note that the width of the tree-decomposition is not changed by this operation. It is known (e.g. in [31]) that H' is a triangulation of G, actually a so-called *k-tree*. Although there exist triangulations that cannot be computed from G with the elimination game, van Leeuwen [31] describes how to change a tree-decomposition in order to obtain the elimination ordering α and thus $G_\alpha^+ = H'$. For finding a minimal triangulation H that is a super-graph of G and a subgraph of G_α^+, known as the *sandwich* problem, there are efficient $O(nm)$ runtime algorithms (For a nice survey, we refer to [21]).

2.2 Minimal Separators and Triangulations

We want to use triangulations for computing tree-decompositions with "nice" separating properties. By Rose et al [28], we have also the following lemma:

Lemma 2. *Let* H *be a minimal triangulation of* G. *Any minimal separator of* H *is a minimal separator of* G.

Before we give our new tree-decomposition algorithm, we are interested in an additional property of minimal separators. Let $\mathcal{S}_G$ be the set of all minimal separators in G. Let $S_1, S_2 \in \mathcal{S}_G$. We say that S_1 *crosses* S_2, denoted by $S_1 \# S_2$, if there are two connected components $C, D \in G \setminus S_2$, such that S_1 intersects both C and D. Note that $S_1 \# S_2$ implies $S_2 \# S_1$. If S_1 does not cross S_2, we say that S_1 is *parallel* to S_2, denoted by $S_1 \| S_2$. Note that "$\|$" is an equivalence relation on a set of pairwise parallel separators.

Theorem 1. *[25] Let H be a minimal triangulation of G. Then, $\mathcal{S}_H$ is a maximal set of pairwise parallel minimal separators in G.*

3   Algorithm for a New Tree-Decomposition

Before we give the whole algorithm, we need some more definitions. For a graph G, let $\mathcal{K}$ be the set of *maximal cliques*, that is, the cliques that have no superset in $V(G)$ that forms a clique in G. Let $\mathcal{K}_v$ be the set of all maximal cliques of G that contain the vertex $v \in V(G)$. For a chordal graph H we define a *clique tree* as a tree $T = (\mathcal{K}, \mathcal{E})$ whose vertex set is the set of maximal cliques in H, and $T[\mathcal{K}_v]$ forms a connected subtree for each vertex $v \in V(H)$. Vice versa, if a graph H has a clique tree, then H is chordal (see [20]). Even though finding all maximal cliques of a graph is NP-hard in general, there exists a linear time modified algorithm of [29], that exploits the property of chordal graphs having at most $|V(H)|$ maximal cliques. By definition, a clique tree of H is also a tree-decomposition of H (where the opposite is not necessarily true).

Due to [5], a clique tree of a chordal graph H is the maximum weight spanning tree of the intersection graph of maximal cliques of H, and we obtain a linear time algorithm computing the clique tree of a graph H. It follows immediately from Lemma 1 that the treewidth of any chordal graph H equals the size of the largest clique. Let us define an edge (C_i, C_j) in a clique tree T to be equivalent to the set of vertices $C_i \cap C_j$ of the two cliques C_i, C_j in H which correspond to the endpoints of the edge in T. For us, the most interesting property of clique trees is given by [22]:

Theorem 2. *Given a chordal graph H and some clique tree T of H, a set of vertices S is a minimal separator of H if and only if $S = C_i \cap C_j$ for an edge (C_i, C_j) in T.*

We get our lemma following from Theorem 1 and Theorem 2:

Lemma 3. *Given a clique tree $T = (\mathcal{K}, \mathcal{E})$ of a minimal triangulation H of a graph G. Then, T is a tree-decomposition $\mathcal{T}$ of G, where $\mathrm{tw}(\mathcal{T}) = \mathrm{tw}(H)$, and the*

Algorithm TransfTD

Input: Graph G with tree-decomposition $\mathcal{T} = (T, \mathcal{Z}), \mathcal{Z} = (Z_t)_{t \in T}$.

Output: Parallel tree-decomposition $\mathcal{T}'$ of G with $\mathrm{tw}(\mathcal{T}') \leq \mathrm{tw}(\mathcal{T})$.

Triangulation step:
      Saturate every bag $Z_t, t \in T$ to
      obtain the chordal graph H', $E(H') = E(G) \cup F$ with fill edges F.

Minimal triangulation step:
    Compute a minimal triangulation H of G, $E(H) = E(G) \cup F'$, $F' \subseteq F$.

Clique tree step:
      Compute clique tree of H, being simultaneously a tree-decomposition $\mathcal{T}'$ of G.

Fig. 1. Algorithm **TransfTD**

set of all edges (C_i, C_j) in T forms a maximal set of pairwise parallel minimal separators in G.

We call such a tree-decomposition of G *parallel*. We give the algorithm in Figure 1.

The worst case analysis for the runtime of **TransfTD** comes from the Minimal triangulation step, that needs time $O(nm)$ for an input graph G, $(|V(G)| = n$, $|E(G)| = m)$.

4   Plane Graphs and Minimal Separators

In the remainder of the paper, we consider 2-connected plane graphs G. Let $V(J) \subseteq V(G)$ be the set of vertices which are intersected by Jordan curve J. We say that a Jordan curve J is *minimal*, if no proper subset V_A of $V(J)$ with $|V_A| > 2$ forms a Jordan curve. The Jordan curve theorem (e.g. see [12]) states that a Jordan curve J on a sphere $\mathbb{S}_0$ divides the rest of $\mathbb{S}_0$ into two connected parts, namely into two open discs Δ_J and $\Delta_{\bar{J}}$, i.e., $\Delta_J \cup \Delta_{\bar{J}} \cup J = \mathbb{S}_0$. Hence, every Jordan curve J is a separator of a plane graph G if both $\Delta_J \cap G$ and $\Delta_{\bar{J}} \cap G$ are nonempty. Two Jordan curves J, J' then divide $\mathbb{S}_0$ into several regions. We define $V_{J,J'}^{+}$ as the (possibly empty) subset of vertices of $V(J \cap J')$ that are incident to more than two regions. For two Jordan curves J, J', we define $J \Delta J'$ to be the symmetric difference of J and J', and $V(J \Delta J') = V(J \cup J') \setminus V(J \cap J') \cup V_{J,J'}^{+}$. Bouchitté et al [9] use results of [18] to show the following:

Lemma 4. *[9] Every minimal separator S of a 2-connected plane graph G forms the vertices of a Jordan curve.*

That is, in any crossing-free embedding of G in $\mathbb{S}_0$, one can find a Jordan curve only intersecting with G in the vertices of S. Note that a minimal separator S is not necessarily forming a unique Jordan curve. If an induced subgraph G' of G (possibly a single edge) has only two vertices u, v in common with S, and u, v are successive vertices of the Jordan curve J, then G' can be drawn on either side of J. This is the only freedom we have to form a Jordan curve in G, since on both sides of J, there is a connected subgraph of G that is adjacent to all vertices of J. We call two Jordan curves J, J' *equivalent* if they share the same vertex set and intersect the vertices in the same order. Two Jordan curves J, J' *cross* if J and J' are not equivalent and there are vertices $v, w \in V(J')$ such that $v \in V(G) \cap \Delta_J$ and $w \in V(G) \cap \Delta_{\bar{J}}$. The proofs of the following lemmata can be found in [14].

Lemma 5. *Let S_1, S_2 be two minimal separators of a 2-connected plane graph G and each S_i forms a Jordan curve $J_i, i = 1, 2$. If $S_1 \| S_2$, then J_1, J_2 are non-crossing. Vice versa, if two minimal Jordan curves J_1, J_2 in G are non-crossing and $\Delta_{J_i} \cap V(G)$ and $\Delta_{\bar{J_i}} \cap V(G), (i = 1, 2)$ all are non-empty, then the vertex sets $S_i = V(J_i), (i = 1, 2)$ are parallel minimal separators.*

We say that two non-crossing Jordan curves J_1, J_2 *touch* if they intersect in a non-empty vertex set. Note that there may exist two edges $e, f \in E(G) \cap \Delta_{J_1}$ such that $e \in E(G) \cap \Delta_{J_2}$ and $f \in E(G) \cap \Delta_{\overline{J_2}}$.

Lemma 6. *Let two non-crossing Jordan curves J_1, J_2 be formed by two minimal parallel separators S_1, S_2 of a 2-connected plane graph G. If J_1 and J_2 touch, and there exists a Jordan curve $J_3 \subseteq J_1 \Delta J_2$ such that there are vertices of G on both sides of J_3, then the vertices of J_3 form another minimal separator S_3 that is parallel to S_1 and S_2.*

If $J_1 \Delta J_2$ forms exactly one Jordan curve J_3 then we say that J_1 *touches* J_2 *nicely*. Note that if J_1 and J_2 only touch in one vertex, the vertices of $J_1 \Delta J_2$ may not form any Jordan curve. The following lemma gives a property of "nicely touching" that we need later on.

Lemma 7. *If in a 2-connected plane graph G, two non-crossing Jordan curves J_1 and J_2 touch nicely, then $|V_{J_1, J_2}^+| = |V(J_1) \cap V(J_2) \cap V(J_1 \Delta J_2)| \leq 2$.*

5   Jordan Curves and Geometric Tree-Decompositions

We now want to turn a parallel tree-decomposition $\mathcal{T}$ into a *geometric* tree-decomposition $\mathcal{T}' = (T, \mathcal{Z}), \mathcal{Z} = (Z_t)_{t \in T}$ where T is a ternary tree and for every two adjacent edges (Z_r, Z_s) and (Z_r, Z_t) in T, the minimal separators $S_1 = Z_r \cap Z_s$ and $S_2 = Z_r \cap Z_t$ form two Jordan curves J_1, J_2 that touch each other nicely. Unfortunately, we cannot arbitrarily connect two Jordan curves J, J' that we obtain from the parallel tree-decomposition $\mathcal{T}$—even if they touch nicely, since the symmetric difference of J, J' may have more vertices than tw($\mathcal{T}$). With carefully chosen arguments, one can deduce from [9] that for 3-connected planar graphs parallel tree-decompositions are geometric. However, we give a direct proof that enables us to find geometric tree-decompositions for all planar graphs.

For a vertex set $Z \subseteq V(G)$, we define the subset $\partial Z \subseteq Z$ to be the vertices adjacent in G to some vertices in $V(G) \setminus Z$. Let G be planar embedded, Z connected, and ∂Z form a Jordan curve. We define $\overline{\Delta}_Z$ to be the closed disk, onto which Z is embedded and Δ_Z the open disk with the embedding of Z without the vertices of ∂Z. For a non-leaf tree node X with degree d in a parallel tree-decomposition $\mathcal{T}$, let $Y_1, \ldots Y_d$ be its neighbors. Let T_{Y_i} be the subtree including Y_i when removing the edge (Y_i, X) from T. We define $G_{Y_i} \subseteq G$ to be the subgraph induced by the vertices of all bags in T_{Y_i}. For Y_i, choose the Jordan curve J_i formed by the vertex set $\partial G_{Y_i} = Y_i \cap X$ to be the Jordan curve that has all vertices of G_{Y_i} on one side and $V(G) \setminus V(G_{Y_i})$ on the other. For each edge e with both endpoints being consecutive vertices of J_i we choose if $e \in E(G_{Y_i})$ or if $e \in E(G) \setminus E(G_{Y_i})$.

We say that a set $\mathcal{J}$ of non-crossing Jordan curves is *connected* if for every partition of $\mathcal{J}$ into two subsets $\mathcal{J}_1, \mathcal{J}_2$, there is at least one Jordan curve of $\mathcal{J}_1$ that touches a Jordan curve of $\mathcal{J}_2$. A set $\mathcal{J}$ of Jordan curves is *k-connected* if for every partition of $\mathcal{J}$ into two connected sets $\mathcal{J}_1, \mathcal{J}_2$, the Jordan curves of $\mathcal{J}_1$ touch the Jordan curves of $\mathcal{J}_2$ in at least k vertices. Note that if two Jordan curves touch nicely then they intersect in at least two vertices. For a proof of the following lemma, please consult [14].

Lemma 8. *For every inner node X of a parallel tree-decomposition $\mathcal{T}$ of a 2-connected plane graph, the collection $\mathcal{J}_X$ of pairwise non-crossing Jordan curves formed by ∂X is 2-connected.*

Lemma 9. *Every bag X in a parallel tree-decomposition $\mathcal{T}$ can be decomposed into $X_1, \ldots, X_\ell$ such that each vertex set ∂X_i forms a Jordan curve in G and $\bigcup_{i=1}^{\ell} \partial X_i = \partial X$.*

Proof. Let $Y_1, \ldots, Y_d$ be the neighbors of X. By Lemma 8, ∂X forms a 2-connected set of Jordan curves, each bounding a disk inside which one of the subgraphs G_{Y_j} is embedded onto. If we remove the disks Δ_{Y_j} for all $1 \leq j \leq d$ and the set of Jordan curves $\mathcal{J}_X$ from the sphere, we obtain a collection $\mathcal{D}_X$ of ℓ disjoint open disks each bounded by a Jordan curve of $\mathcal{J}_X$. Note that $\ell \leq \max\{d, |X|\}$. Let Z_i be the subgraph in $X \cap \Delta_i$ for such an open disk $\Delta_i \in \mathcal{D}_X$ for $1 \leq i \leq \ell$. Then each Z_i is either empty or consisting only of edges or subgraphs of G and the closed disk $\overline{\Delta}_i$ is bounded by a Jordan curve J_i formed by a subset of ∂X. We set $X_i = Z_i \cup V(J_i)$ with ∂X_i the vertices of J_i.

Lemma 10. *In a decomposition of the sphere $\mathbb{S}_0$ by a 2-connected collection $\mathcal{J}$ of non-crossing Jordan curves, one can repeatedly find two Jordan curves $J_1, J_2 \in \mathcal{J}$ that touch nicely, and substitute J_1 and J_2 by $J_1 \Delta J_2$ in $\mathcal{J}$.*

Proof. Removing $\mathcal{J}$ from $\mathbb{S}_0$ decomposes $\mathbb{S}_0$ into a collection $\mathcal{D}$ of open discs each bounded by a Jordan curve in $\mathcal{J}$. For each $\Delta_1 \in \mathcal{D}$ bounded by $J_1 \in \mathcal{J}$ there is a "neighboring" disk $\Delta_2 \in \mathcal{D}$ bounded by $J_2 \in \mathcal{J}$ such that the intersection $J_1 \cap J_2$ forms a line of $\mathbb{S}_0$. Then, $J_1 \Delta J_2$ bounds $\Delta_1 \cup \Delta_2$. Replace, J_1, J_2 by J_3 in $\mathcal{J}$ and continue until $|\mathcal{J}| = 1$, that is, we are left with one Jordan curve separating $\mathbb{S}_0$ into two open disks.

We get that $X_1, \ldots X_\ell$ and $G_{Y_1}, \ldots, G_{Y_d}$ are embedded inside of closed disks each bounded by a Jordan curve. Thus, the union $\mathcal{D}$ over all these disks together with the Jordan curves $\mathcal{J}_X$ fill the entire sphere $\mathbb{S}_0$ onto which G is embedded. Each subgraph embedded onto $\Delta \cup J$ for a disk $\Delta \in \mathcal{D}$ and a Jordan curve J bounding Δ, forms either a bag X_i or a subgraph G_{Y_j}. Define the collection of bags $\mathcal{Z}^X = \{X_1, \ldots X_\ell, Y_1, \ldots, Y_d\}$. In Figure 2, we give the algorithm **TransfTD II** for creating a geometric tree-decomposition using the idea of Lemma 6.

Since by Lemma 7, $|V(\partial Z_i \cap \partial Z_j \cap \partial Z_{ij})| \leq 2$, we have that at most two vertices in all three bags are contained in any other bag of $\mathcal{Z}^X$. Note that geometric tree-decompositions have a lot in common with *sphere-cut decompositions* (introduced in [16]), namely that both decompositions are assigned with vertex sets that form "sphere-cutting" Jordan curves. For our new dynamic programming algorithm, we use much of the structure results obtained in Subsection [16].

Algorithm **TransfTD II**

Input: Graph G with parallel tree-decomposition $\mathcal{T} = (T, \mathcal{Z}), \mathcal{Z} = (Z_t)_{t \in T}$.

Output: Geometric tree-decomposition $\mathcal{T}'$ of G with $\text{tw}(\mathcal{T}') \leq \text{tw}(\mathcal{T})$.

For each inner bag X with neighbors $Y_1, \ldots, Y_d$ {

Disconnection step: Replace X by $X_1, \ldots X_\ell$ (Lemma 9).

    Set $\mathcal{Z}^X = \{X_1, \ldots X_\ell, Y_1, \ldots, Y_d\}$.

Reconnection step: Until $|\mathcal{Z}^X| = 1$ {

    Find two bags Z_i and Z_j in $\mathcal{Z}^X$ such that Jordan curve $J_i \Delta J_j$

    bounds a disk with $Z_i \cup Z_j$ (Lemma 10);

    Set $Z_{ij} = (Z_i \Delta Z_j) \cup (Z_i \cap Z_j)$ and connect Z_i and Z_j to Z_{ij};

    In $\mathcal{Z}^X$: substitute Z_i and Z_j by Z_{ij}. }}

Fig. 2. Algorithm **TransfTD II**

6 Jordan Curves and Dynamic Programming

The following techniques improve the existing algorithm of Alber et al [1] for weighted PLANAR DOMINATING SET. Their algorithm is based on dynamic programming on *nice tree-decompositions* $\mathcal{T}$ and has the running time $4^{\text{tw}(\mathcal{T})} \cdot n^{O(1)}$. We prove the following theorem by giving an algorithm of similar structure to those in [16] and [19]. Thus, we give here only a sketch of the idea. Namely, to exploit the planar structure of the nicely touching separators to improve upon the runtime.

Theorem 3. *Given a geometric tree-decomposition* $\mathcal{T} = (T, \mathcal{Z}), \mathcal{Z} = (Z_t)_{t \in T}$ *of a planar graph* G. *Weighted* PLANAR DOMINATING SET *on* G *can be solved in time* $3^{\text{tw}(\mathcal{T})} \cdot n^{O(1)}$.

Proof. We root T by arbitrarily choosing a node r as a *root*. Each internal node t of T now has one adjacent node on the path from t to r, called the *parent node*, and two adjacent nodes toward the leaves, called the *children nodes*. To simplify matters, we call them the *left child* and the *right child*.

Let T_t be a subtree of T rooted at node t. G_t is the subgraph of G induced by all bags of T_t. For a subset U of $V(G)$ let $w(U)$ denote the total weight of vertices in U. That is, $w(U) = \sum_{u \in U} w_u$. Define a set of subproblems for each subtree T_t.

Alber et al. [1] introduced the "monotonicity"-property of domination-like problems for their dynamic programming approach that we will use, too. For every node $t \in T$, we use three colors for the vertices of bag Z_t:

black: Represented by 1, meaning the vertex is in the dominating set.

white: Represented by 0, meaning the vertex has a neighbor in G_t that is in the dominating set.

gray: Represented by 2, meaning the vertex has a neighbor in G that is in the dominating set.

For a bag Z_t of cardinality ℓ, we define a *coloring* $c(Z_t)$ to be a mapping of the vertices Z_t to an ℓ-vector over the color-set $\{0, 1, 2\}$ such that each vertex $u \in Z_t$ is assigned a color, i.e., $c(u) \in \{0, 1, 2\}$. We further define the weight $w(c(Z_t))$ to be the minimum weight of the vertices of G_t in the minimum weight dominating set with respect to the coloring $c(Z_t)$. If no such dominating set exists, we set $w(c(Z_t)) = +\infty$. We store all colorings of Z_t, and for two child nodes, we update each two colorings to one of the parent node.

Before we describe the updating process of the bags, let us make the following comments:

We defined the color "gray" according to the monotonicity property: for a vertex u colored gray, we do not have (or store) the information if u is already dominated by a vertex in G_t or if u still has to be dominated in $G \setminus G_t$. Thus, a solution with a vertex v colored white has at least the same the weight as the same solution with v colored gray.

By the definition of bags, for three adjacent nodes r, s, t, the vertices of ∂Z_r have to be in at least on of ∂Z_s and ∂Z_t. The reader may simply recall that the parent bag is formed by the union of the vertices of two nicely touching Jordan curves.

For the sake of a refined analysis, we partition the bags of parent node r and left child s and right child t into four sets L, R, F, I as follows:

- *Intersection* $I := \partial Z_r \cap \partial Z_s \cap \partial Z_t$,
- *Forget* $F := (Z_s \cup Z_t) \setminus \partial Z_r$,
- *Symmetric difference* $L := \partial Z_r \cap \partial Z_s \setminus I$ and $R := \partial Z_r \cap \partial Z_t \setminus I$.

We define F' to be actually those vertices of F that are only in $(\partial Z_s \cup \partial Z_t) \setminus \partial Z_r$. The vertices of $F \setminus F'$ do not exist in Z_r and hence are irrelevant for the continuous update process. We say that a coloring $c(Z_r)$ is *formed* by the colorings $c_1(Z_s)$ and $c_2(Z_t)$ subject to the following rules:

(R1) For every vertex $u \in L \cup R : c(u) = c_1(u)$ and $c(u) = c_2(u)$, respectively.
(R2) For every vertex $u \in F'$ either $c(u) = c_1(u) = c_2(u) = 1$ or $c(u) = 0 \wedge c_1(u), c_2(u) \in \{0, 2\} \wedge c_1(u) \neq c_2(u)$.
(R3) For every vertex $u \in I$ $c(u) \in \{1, 2\} \Rightarrow c(u) = c_1(u) = c_2(u)$ and $c(u) = 0 \Rightarrow c_1(u), c_2(u) \in \{0, 2\} \wedge c_1(u) \neq c_2(u)$.

We define U_c to be the vertices $u \in Z_s \cap Z_t$ for which $c(u) = 1$ and update the weights by: $w(c(Z_r)) = \min\{w(c_1(Z_s)) + w(c_2(Z_t)) - w(U_c) | c_1, c_2 \text{ forms } c\}$.

The number of steps by which $w(c(Z_r))$ is computed for every possible coloring of Z_r is given by the number of ways a color c can be formed by the three rules $(R1), (R2), (R3)$, i.e., $3^{|L|+|R|} \cdot 3^{|F'|} \cdot 4^{|I|}$ steps.

By Lemma 7, $|I| \leq 2$ and since $|L| + |R| + |F| \leq \text{tw}(\mathcal{T})$, we need at most $3^{\text{tw}(\mathcal{T})} \cdot n$ steps to compute all weights $w(c(Z_r))$ that are usually stored in a table assigned to bag Z_r.

In [1], the worst case in the runtime for PLANAR DOMINATING SET is determined by the number of vertices that are in the intersection of three adjacent bags r, s, t. Using the notion of [16] for a geometric tree-decomposition, we partition the vertex sets of three bags Z_r, Z_s, Z_t into sets L, R, F, I, where Z_r is adjacent to Z_s, Z_t. The sets L, R, F represent the vertices that are in exactly two of the bags. Let us consider the *Intersection* set $I := \partial Z_r \cap \partial Z_s \cap \partial Z_t$. By Lemma 7, $|I| \leq 2$. Thus, I is not any more part of the runtime.

7   Conclusion

A natural question to pose, is it possible to solve PLANAR DOMINATING SET in time $2.99^{\text{tw}(\mathcal{T})} \cdot n^{O(1)}$ and equivalently, PLANAR INDEPENDENT SET in $1.99^{\text{tw}(\mathcal{T})} \cdot n^{O(1)}$? Though, we cannot give a positive answer yet, we have a formula that needs "well-balanced" separators in a geometric tree-decomposition $\mathcal{T}$: we assume that the three sets L, R, F are of equal cardinality for every three adjacent bags. Since $|L| + |R| + |F| \leq \text{tw}$, we thus have that $|L|, |R|, |F| \leq \frac{\text{tw}}{3}$. Applying the fast matrix multiplication method from [13] for example to PLANAR INDEPENDENT SET, this leads to a $2^{\frac{\omega}{3}\text{tw}(\mathcal{T})} \cdot n^{O(1)}$ algorithm, where $\omega < 2.376$. Does every planar graph have a geometric tree-decomposition with well-balanced separators?

Acknowledgments. The author thanks Frédéric Mazoit for some enlightening discussion on Theorem 1.

References

1. Alber, J., Bodlaender, H.L., Fernau, H., Kloks, T., Niedermeier, R.: Fixed parameter algorithms for dominating set and related problems on planar graphs. Algorithmica 33, 461–493 (2002)
2. Alber, J., Bodlaender, H.L., Fernau, H., Niedermeier, R.: Fixed parameter algorithms for planar dominating set and related problems. In: Halldórsson, M.M. (ed.) SWAT 2000. LNCS, vol. 1851, pp. 97–110. Springer, Heidelberg (2000)
3. Alber, J., Niedermeier, R.: Improved tree decomposition based algorithms for domination-like problems. In: Rajsbaum, S. (ed.) LATIN 2002. LNCS, vol. 2286, pp. 613–627. Springer, Heidelberg (2002)
4. Arora, S., Grigni, M., Karger, D., Klein, P., Woloszyn, A.: A polynomial-time approximation scheme for weighted planar graph TSP. In: Proceedings of the Ninth Annual ACM-SIAM Symposium on Discrete Algorithms, San Francisco, CA, pp. 33–41. ACM Press, New York (1998)
5. Bernstein, P.A., Goodman, N.: Power of natural semijoins. SIAM Journal on Computing 10, 751–771 (1981)
6. Bodlaender, H.: Treewidth: Algorithmic techniques and results. In: Privara, I., Ružička, P. (eds.) MFCS 1997. LNCS, vol. 1295, pp. 19–36. Springer, Heidelberg (1997)
7. Bodlaender, H.L.: A tourist guide through treewidth. Acta Cybernet. 11, 1–21 (1993)
8. Bodlaender, H.L., Möhring, R.H.: The pathwidth and treewidth of cographs. SIAM Journal on Discrete Mathematics 6, 181–188 (1993)
9. Bouchitté, V., Mazoit, F., Todinca, I.: Chordal embeddings of planar graphs. Discrete Mathematics 273, 85–102 (2003)

10. Cook, W., Seymour, P.: Tour merging via branch-decomposition. INFORMS Journal on Computing 15, 233–248 (2003)
11. Demaine, E.D., Fomin, F.V., Hajiaghayi, M., Thilikos, D.M.: Subexponential parameterized algorithms on graphs of bounded genus and H-minor-free graphs. Journal of the ACM 52, 866–893 (2005)
12. Diestel, R.: Graph theory, 3rd edn. Springer, Heidelberg (2005)
13. Dorn, F.: Dynamic programming and fast matrix multiplication. In: Azar, Y., Erlebach, T. (eds.) ESA 2006. LNCS, vol. 4168, pp. 280–291. Springer, Heidelberg (2006)
14. Dorn, F.: How to use planarity efficiently: new tree-decomposition based algorithms (manuscript, 2007), http://www.ii.uib.no/~frederic/PlanDomSet.pdf
15. Dorn, F., Fomin, F.V., Thilikos, D.M.: Fast subexponential algorithm for non-local problems on graphs of bounded genus. In: Arge, L., Freivalds, R. (eds.) SWAT 2006. LNCS, vol. 4059, pp. 172–183. Springer, Heidelberg (2006)
16. Dorn, F., Penninkx, E., Bodlaender, H.L., Fomin, F.V.: Efficient exact algorithms on planar graphs: Exploiting sphere cut branch decompositions. In: Brodal, G.S., Leonardi, S. (eds.) ESA 2005. LNCS, vol. 3669, pp. 95–106. Springer, Heidelberg (2005)
17. Dorn, F., Telle, J.A.: Two birds with one stone: the best of branchwidth and treewidth with one algorithm. In: Correa, J.R., Hevia, A., Kiwi, M. (eds.) LATIN 2006. LNCS, vol. 3887, pp. 386–397. Springer, Heidelberg (2006)
18. Eppstein, D.: Subgraph isomorphism in planar graphs and related problems. J. Graph Algorithms Appl. 3, 1–27 (1999)
19. Fomin, F.V., Thilikos, D.M.: Dominating sets in planar graphs: branch-width and exponential speed-up. In: SODA 2003. Proceedings of the Fourteenth Annual ACM-SIAM Symposium on Discrete Algorithms, Baltimore, MD, pp. 168–177. ACM Press, New York (2003)
20. Gavril, F.: The intersection graphs of subtrees in trees are exactly the chordal graphs. Journal of Combinatorial Theory Series B 16, 47–56 (1974)
21. Heggernes, P.: Minimal triangulations of graphs: A survey. Discrete Mathematics 306, 297–317 (2006)
22. Ho, C.W., Lee, R.C.T.: Counting clique trees and computing perfect elimination schemes in parallel. Inf. Process. Lett. 31, 61–68 (1989)
23. Lipton, R.J., Tarjan, R.E.: A separator theorem for planar graphs. SIAM J. Appl. Math. 36, 177–189 (1979)
24. Miller, G.L.: Finding small simple cycle separators for 2-connected planar graphs. Journal of Computer and System Science 32, 265–279 (1986)
25. Parra, A., Scheffler, P.: Characterizations and algorithmic applications of chordal graph embeddings. Discrete Applied Mathematics 79, 171–188 (1997)
26. Parter, S.: The use of linear graphs in Gauss elimination. SIAM Review 3, 119–130 (1961)
27. Robertson, N., Seymour, P.D.: Graph minors. X. Obstructions to tree-decomposition. J. Combin. Theory Ser. B 52, 153–190 (1991)
28. Rose, D., Tarjan, R.E., Lueker, G.: Algorithmic aspects of vertex elimination on graphs. SIAM Journal on Computing 5, 146–160 (1976)
29. Tarjan, R.E., Yannakakis, M.: Simple linear-time algorithms to test chordality of graphs, test acyclicity of hypergraphs, and selectively reduce acyclic hypergraphs. SIAM Journal on Computing 13, 566–579 (1984)
30. Telle, J.A., Proskurowski, A.: Algorithms for vertex partitioning problems on partial k-trees. SIAM J. Discrete Math 10, 529–550 (1997)
31. van Leeuwen, J.: Graph algorithms. MIT Press, Cambridge, MA, USA (1990)

Obtaining a Planar Graph by Vertex Deletion

Dániel Marx[1] and Ildikó Schlotter[2]

[1] Institut für Informatik, Humboldt-Universität zu Berlin,
Unter den Linden 6, 10099, Berlin, Germany
dmarx@informatik.hu-berlin.de
[2] Department of Computer Science and Information Theory,
Budapest University of Technology and Economics,
Budapest, H-1521, Hungary
ildi@cs.bme.hu

Abstract. In the PLANAR $+k$ VERTEX problem the task is to find at most k vertices whose deletion makes the given graph planar. The graphs for which there exists a solution form a minor closed class of graphs, hence by the deep results of Robertson and Seymour [19,18], there is an $O(n^3)$ time algorithm for every fixed value of k. However, the proof is extremely complicated and the constants hidden by the big-O notation are huge. Here we give a much simpler algorithm for this problem with quadratic running time, by iteratively reducing the input graph and then applying techniques for graphs of bounded treewidth.

1 Introduction

Planar graphs are subject of wide research interest in graph theory. There are many generally hard problems which can be solved in polynomial time when considering planar graphs, e.g., MAXIMUM CLIQUE, MAXIMUM CUT, and SUBGRAPH ISOMORPHISM [8,12]. For problems that remain NP-hard on planar graphs, we often have efficient approximation algorithms. For example, the problems INDEPENDENT SET, VERTEX COVER, and DOMINATING SET admit an efficient polynomial time approximation scheme (EPTAS) [1,15]. The research for efficient algorithms for problems on planar graphs is still very intensive.

Many results on planar graphs can be extended to almost planar graphs, which can be defined in various ways. For example, we can consider possible embeddings of a graph in a surface other than the plane. The genus of a graph is the minimum number of handles that must be added to the plane to embed the graph without any crossings. Although determining the genus of a graph is NP-hard [20], the graphs with bounded genus are subjects of wide research. A similar property of graphs is their crossing number, i.e., the minimum possible number of crossings with which the graph can be drawn in the plane. Determining crossing number is also NP-hard [10].

In [3] Cai introduced another notation, based on the number of certain elementary modification steps. He defines the distance of a graph G from a graph class $\mathcal{F}$ as the minimum number of modifying steps needed to make G a member

A. Brandstädt, D. Kratsch, and H. Müller (Eds.): WG 2007, LNCS 4769, pp. 292–303, 2007.

of $\mathcal{F}$. Here modification can mean the deletion or addition of edges or vertices. In this paper we consider the following question: given a graph G and an integer k, is there a set of at most k vertices in G, whose deletion makes G planar?

Since this problem was proved to be NP-hard in [14], we cannot hope to find a polynomial-time algorithm for it. Therefore, we study the problem in the framework of parameterized complexity [7]. This approach deals with problems in which besides the input I an integer k is also given. The integer k is referred to as the parameter. In many cases we can solve the problem in time $O(n^{f(k)})$. Clearly, this is also true for the problem we consider. Although this is polynomial time for each fixed k, these algorithms are practically too slow for large inputs, even if k is relatively small. Therefore, the standard goal of parameterized analysis is to take the parameter out of the exponent in the running time. A problem is called *fixed-parameter tractable* (FPT) if it can be solved in time $O(f(k)p(|I|))$, where p is a polynomial not depending on k, and f is an arbitrary function. An algorithm with such a running time is also called FPT.

The standard parameterized version of our problem is the following: given a graph G and a parameter k, the task is to decide whether deleting at most k vertices from G can result in a planar graph. Following Cai [3], we will denote the class of graphs for which the answer is 'yes' by Planar + kv. This family of graphs is closed under taking minors, so thanks to the results of Robertson and Seymour [19,18], we know that there exists an algorithm with running time $O(f(k)n^3)$ which can decide membership for this class. However, this result is inherently non-constructive, and so far there is no direct FPT algorithm known for this problem. In this paper we present an algorithm, which solves the question in $O(f(k)n^2)$ time. The algorithm also returns a solution, i.e., a set of at most k vertices whose deletion from G results in a planar graph.

Our algorithm is strongly based on the ideas used by Grohe in [11] for computing crossing number. Grohe uses the fact that the crossing number of a graph is an upper bound for its genus. Since the genus of a graph in Planar + kv cannot be bounded by a function of k, we need some other ideas. As in [11], we exploit the fact that in a graph with large treewidth we can always find a large grid minor [17]. Examining the structure of the graph with such a grid minor, we can reduce our problem to a smaller instance. Applying this reduction several times, we finally get an instance with bounded treewidth. Then we make use of Courcelle's Theorem [4], which states that every graph property that is expressible in monadic second-order logic can be decided in linear time on graphs of bounded treewidth.

The paper is organized as follows. Section 2 summarizes our notation, Sect. 3 outlines the algorithm, Sect. 4 and 5 describe the two phases of the algorithm.

2 Notation

Graphs in this paper are assumed to be simple, since both loops and multiple edges are irrelevant in the PLANAR + k VERTEX problem. The vertex set and edge set of a graph G are denoted by $V(G)$ and $E(G)$, respectively. The edges of

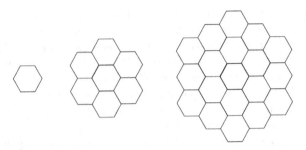

Fig. 1. The hexagonal grids H_1, H_2, and H_3

a graph are unordered pairs of its vertices. If G' is a subgraph of G then $G - G'$ denotes the graph obtained by deleting G' from G. For a set of vertices S in G, we will also use $G - S$ to denote the graph obtained by deleting S from G.

A graph H is a *minor* of a graph G if it can be obtained from a subgraph of G by contracting some of its edges. Here contracting an edge e with endpoints a and b means deleting e, and then identifying vertices a and b.

A graph H is a *subdivision* of a graph G if G can be obtained from H by contracting some of its edges that have at least one endpoint of degree 2. A graph H is a *topological minor* of G if G has a subgraph that is a subdivision of H. G and G' are *topologically isomorphic* if they both are subdivisions of a graph H. If G is a subdivision of H, then an *edge-path* of G with respect to H is a path of G corresponding to exactly one edge of H in the natural way (i.e., a path with inner points only of degree 2, whose vertices are all identified with an endpoint of a certain edge when obtaining H from G as a topological minor).

The $g \times g$ *grid* is the graph $G_{g \times g}$ where $V(G_{g \times g}) = \{v_{ij} \mid 1 \leq i, j \leq g\}$ and $E(G_{g \times g}) = \{v_{ij}v_{i'j'} \mid |i - i'| + |j - j'| = 1\}$.

Instead of giving a formal definition for the *hexagonal grid* of radius r, which we will denote by H_r, we refer to the illustration shown in Fig. 1. A *cell* of a hexagonal grid is one of its cycles of length 6.

A *tree decomposition* of a graph G is a pair $(T, (V_t)_{t \in V(T)})$ where T is a tree, $V_t \subseteq V(G)$ for all $t \in V(T)$, and the following are true:

- for all $v \in V(G)$ there exists a $t \in V(T)$ such that $v \in V_t$,
- for all $xy \in E(G)$ there exists a $t \in V(T)$ such that $x, y \in V_t$,
- if t lies on the path connecting t' and t'' in T, then $V_t \supseteq V_{t'} \cap V_{t''}$.

The *width* of such a tree decomposition is the maximum of $|V_t| - 1$ taken over all $t \in V(T)$. The *treewidth* of a graph G is the smallest possible width of a tree decomposition of G.

3   Problem Definition and Overview of the Algorithm

We are looking for the solution of the following problem:

PLANAR + k VERTEX problem:

Input: A graph $G = (V, E)$ and an integer k.

Task: Find a set X of at most k vertices in V such that $G - X$ is
       planar.

Here we give an algorithm $\mathcal{A}$ which solves this problem in time $O(f(k)n^2)$ for some function f, where n is the number of vertices in the input graph. Algorithm $\mathcal{A}$ works in two phases. In the first phase (Sect. 4) we compress the given graph repeatedly, and finally either conclude that there is no solution for our problem or construct an equivalent problem instance with a graph having bounded treewidth. In the latter case we solve the problem in the second phase of the algorithm (Sect. 5) by applying Courcelle's Theorem concerning the evaluation of MSO-formulae on bounded treewidth graphs.

According to [17,2,16] we know that there is a linear-time algorithm which can solve the following problem, for fixed integers w and r:

Input: A graph G.

Task: If $\mathrm{tw}(G) \leq w$ then find a tree decomposition of width w, or if
      $\mathrm{tw}(G) > w$ then find an $r \times r$ grid minor in G if there is one.

It is a well-known fact that if a graph of maximum degree 3 is a minor of another graph, then it is also contained in it as a topological minor. Hence, it will be convenient to work with hexagonal grids instead of grids. Since a hexagonal grid with radius i is a subgraph of the $(4i - 1) \times (4i - 1)$ grid, we can conclude that for each fixed w and r there is a linear-time algorithm that solves the following modified version of the above problem:

Input: A graph G.

Task: If $\mathrm{tw}(G) \leq w$ then find a tree decomposition of width w, or if
      $\mathrm{tw}(G) > w$ then find a subdivision of H_r in G if there is one.

According to [17] every planar graph with no minor isomorphic to the $r \times r$ grid has treewidth $\leq 6r - 5$. Therefore, it is also true that every planar graph with no minor isomorphic to H_r has treewidth $\leq 6(4r - 1) - 5 = 24r - 11$. But adding k vertices to a graph can increase the treewidth of the graph only by at most k, so if $G \in$ Planar $+ kv$ and H_r is not a minor of G, then $\mathrm{tw}(G) \leq 24r - 11 + k$. We can summarize this in the following simple claim.

Lemma 1. *For arbitrary integers r and k there is a linear-time algorithm $\mathcal{B}$, which can be run with input graph G, and does the following:*

- *it either produces a tree decomposition of G of width $w(r) = 24r - 11 + k$, or*
- *finds a subdivision of H_r in G, or*
- *correctly concludes that $G \notin$ Planar $+ kv$.*

In algorithm $\mathcal{A}$ we will run $\mathcal{B}$ several times. As long as we get a hexagonal grid of radius r as topological minor as a result, we will run Phase I of algorithm

$\mathcal{A}$, which compresses the graph G. If at some step algorithm $\mathcal{B}$ gives us a tree decomposition of width $w(r)$, we run Phase II. (The constant r will be fixed later.) And of course if at some step $\mathcal{B}$ finds out that $G \notin$ Planar $+ kv$, then algorithm $\mathcal{A}$ can stop with the output "No solution."

Clearly, we can assume without loss of generality that the input graph is simple, and it has at least $k + 3$ vertices. So if $G \in$ Planar $+ kv$, then deleting k vertices from G (which means the deletion of at most $k(|V(G)| - 1)$ edges) results in a planar graph, which has at most $3|V(G)| - 6$ edges. Therefore, if $|E(G)| > (k+3)|V(G)|$ then surely $G \notin$ Planar $+ kv$. Since this can be detected in linear time, we can assume that $|E(G)| \le (k+3)|V(G)|$.

4   Phase I of Algorithm $\mathcal{A}$

In Phase I we assume that after running $\mathcal{B}$ on G we get a subgraph H_r' that is a subdivision of H_r. Our goal is to find a set of vertices X such that $G - X$ is planar, and $|X| \le k$. Let PlanarDel(G, k) denote the family of sets of vertices that have these properties, i.e., let PlanarDel$(G, k) = \{X \subseteq V(G) \mid |X| \le k$ and $G - X$ is planar$\}$. Since the case $k = 1$ is very simple we can assume that $k > 1$.

Reduction A: Flat zones. In the following we regard the grid H_r' as a fixed subgraph of G. Let us define z zones in it. Here z is a constant depending only on k, which we will determine later. A *zone* is a subgraph of H_r' which is topologically isomorphic to the hexagonal grid H_{2k+5}. We place such zones next to each other in the well-known radial manner with radius q, i.e., we replace each hexagon of H_q with a subdivision of H_{2k+5}. It is easy to show that in a hexagonal grid with radius $(q-1)(4k+9) + (2k+5)$ we can define this way $3q(q-1)+1$ zones that only intersect in their outer circles. So let $r = (q-1)(4k+9)+(2k+5)$, where we choose q big enough to get at least z zones, i.e., q is the smallest integer such that $3q(q-1)+1 \ge z$. Let the subgraph of these z zones in H_r' be R.

Let us define two types of *grid-components*. An edge which is not contained in R is a grid-component if it connects two vertices of R. A subgraph of G is a grid-component if it is a (maximal) connected component of $G - R$. A grid-component K is *attached* to a vertex v of the grid R if it has a vertex adjacent to v, or (if K is an edge) one of its endpoints is v. The *core* of a zone is the (unique) subgraph of the zone which is topologically isomorphic to H_{2k+3} and lies in the middle of the zone. Let us call a zone Z *open* if there is a vertex in its core that is connected to a vertex v of another zone, $v \notin V(Z)$, through a grid-component. A zone is *closed* if it is not open.

For a subgraph H of R let $T(H)$ denote the subgraph of G spanned by the vertices of H and the vertices of the grid-components which are only attached to H. Let us call a zone Z *flat* if it is closed and $T(Z)$ is planar. Let Z be such a flat zone. A grid-component is an *edge-component* if it is either only attached to one edge-path of Z or only to one vertex of Z. Otherwise, it is a *cell-component* if it is only attached to vertices of one cell. As a consequence of the fact that all embeddings of a 3-connected graph are equivalent (see e.g. [6]), and Z is a subdivision of such a graph, every grid-component attached to some vertex in the core of Z must be one of these two types. Note that we can assume that in an embedding of $T(Z)$ in the plane, all edge-components are embedded in an arbitrarily small neighborhood of the edge-path (or vertex) which they belong to.

Let us define the *ring* R_i ($1 \leq i \leq 2k+4$) as the union of those cells in Z that have common vertices both with the i-th and the $(i+1)$-th concentrical circle of Z. Let R_0 be the cell of Z that lies in its center. The zone Z can be viewed as the union of $2k+5$ concentrical rings, i.e., the union of the subgraphs R_i for $0 \leq i \leq 2k+4$.

Lemma 2. *Let Z be a flat zone in R, and let G' denote the graph $G - T(R_0)$. Then $X \in \mathrm{PlanarDel}(G', k)$ implies $X \in \mathrm{PlanarDel}(G, k)$.*

Proof. Since $G - T(R_0) - X$ is planar, we can fix a planar embedding ϕ of it. If $R_i \cap X = \emptyset$ for some i ($2 \leq i \leq 2k+2$) then let W_i denote the maximal subgraph of $G - T(R_0) - X$ for which $\phi(W_i)$ is in the region determined by $\phi(R_i)$ (including R_i). If $R_i \cap X$ is not empty then let W_i be the empty graph. Note that if $2 \leq i \leq 2k$ then W_i and W_{i+2} are disjoint. Therefore, there exists an index i for which $W_i \cap X = \emptyset$ and W_i is not empty. Let us fix this i.

Let Q_i denote $T(\bigcup_{j=0}^{i} R_j)$. We prove the lemma by giving an embedding for $G - X'$ where $X' = X \setminus V(Q_{i-1})$. The region $\phi(R_i)$ divides the plane in two other regions. We can assume that in the finite region only vertices of Q_{i-1} are embedded, so $G - X' - (Q_{i-1} \cup W_i)$ is entirely embedded in the infinite region. Let U denote those vertices in Q_{i-1} which are adjacent to some vertex in $G - Q_{i-1}$. Observe that the restriction of ϕ to $G - X' - (Q_{i-1} - U)$ has a face whose boundary contains U.

Now let θ be a planar embedding of $T(Z)$, and let us restrict θ to Q_{i-1}. Note that U only contains vertices which are either adjacent to some vertex in R_i or are adjacent to cell-components belonging to a cell of R_i. But θ embeds R_i and its cell-components also, and therefore the restriction of θ to Q_{i-1} results in a face whose boundary contains U. Here we used also that R_i is a subdivision of a 3-connected graph whose embeddings are equivalent.

Now it is easy to see that we can combine θ and ϕ in such a way that we embed $G - X' - (Q_{i-1} - U)$ according to ϕ and, similarly, Q_{i-1} according to θ, and then "connect" them by identifying $\phi(u)$ and $\theta(u)$ for all $u \in U$. This gives the desired embedding of $G - X'$. Finally, we have to observe that $X' \in \mathrm{PlanarDel}(G, k)$ implies $X \in \mathrm{PlanarDel}(G, k)$, since $X' \subseteq X$ and $|X| \leq k$. $\qquad\square$

This lemma has a trivial but crucial consequence: $X \in \mathrm{PlanarDel}(G, k)$ if and only if $X \in \mathrm{PlanarDel}(G - T(R_0), k)$, so deleting $T(R_0)$ reduces our problem to an equivalent instance. Let us denote this deletion as *Reduction A*.

Note that the closedness of a zone Z can be decided by a simple breadth first search, which can also produce the graph $T(Z)$. Planarity can also be tested in linear time [13]. Therefore we can test whether a zone is flat, and if so, we can apply Reduction A on it in linear time.

Later we will see that unless there are some easily recognizable vertices in our graph which must be included in every solution, a flat zone can always be found (Lemma 7). This yields an easy way to handle graphs with large treewidth: compressing our graph by repeatedly applying Reduction A we can reduce the problem to an instance with bounded treewidth.

Reduction B: Well-attached vertices. A subgraph of R is a *block* if it is topologically isomorphic to H_{k+3}. A vertex of a given block is called *inner vertex* if it is not on the outer circle of the block.

Lemma 3. *Let $X \in \mathrm{PlanarDel}(G, k)$. Let x and y be inner vertices of the disjoint blocks B_x and B_y, respectively. If P is an $x - y$ path that (except its endpoints) doesn't contain any vertex from B_x or B_y, then X must contain a vertex from B_x, B_y or P.*

Proof. Let C_x and C_y denote the outer circle of B_x and B_y, respectively. Let us notice that since B_x and B_y are disjoint blocks, there exist at least $k + 3$ vertex disjoint paths between their outer circles, which—apart from their endpoints— do not contain vertices from B_x and B_y. Moreover, it is easy to see that these paths can be defined in a way such that their endpoints that lie on C_x are on the border of different cells of B_x. To see this, note that the number of cells which lie on the border of a given block is $6k + 12$.

At least three of these paths must be in $G - X$ also. Since x can lie only on the border of at most two cells having common vertices with C_x, we get that there is a path P' in $G - X$ whose endpoints are a_x and a_y (lying on C_x and C_y, resp.), and there exist no cell of B_x whose border contains both a_x and x.

Let us suppose that $B_x \cup B_y \cup P$ is a subgraph of $G - X$. Since all embeddings of a 3-connected planar graph are equivalent, we know that if we restrict an arbitrary planar embedding of $G - X$ to B_x, then all faces having x on their border correspond to a cell in B_x. Since x and y are connected through P and $V(P) \cap V(B_x) = \{x\}$, we get that y must be embedded in a region F corresponding to a cell C_F of B_x. But this implies that B_y must entirely be embedded also in F.

Since $V(P' - a_x - a_y) \cap V(B_x) = \emptyset$ and P' connects $a_x \in V(B_x)$ and $a_y \in V(B_y)$ we have that a_y must lie on the border of F. But then C_F is a cell of B_x containing both a_x and x on its border, which yields the contradiction. $\square$

Using this lemma we can identify certain vertices that have to be deleted. Let x be a *well-attached* vertex in G if there exist paths $P_1, P_2, \ldots, P_{k+2}$ and disjoint blocks $B_1, B_2, \ldots, B_{k+2}$ such that P_i connects x with an inner vertex of B_i $(1 \leq i \leq k + 2)$, the inner vertices of P_i are not in R, and if $i \neq j$ then the only common vertex of P_i and P_j is x.

Lemma 4. *Let $X \in \mathrm{PlanarDel}(G, k)$. If x is well-attached then $x \in X$.*

Proof. If $x \notin X$, then after deleting X from G (which means deleting at most k vertices) there would exist indices i and j such that no vertex from P_i, P_j, B_i, and B_j was deleted. But then the disjoint blocks B_i and B_j were connected by the path $P_i - x - P_j$, and by the previous lemma, this is a contradiction. $\square$

We can decide whether a vertex v is well-attached in time $O(f'(k)e)$ using standard flow techniques, where $e = |E(G)|$. This can be done by simply testing for each possible set of $k+2$ disjoint blocks whether there exist the required disjoint paths that lead from x to these blocks. Since the number of blocks in R depends only on k, and we can find p disjoint paths starting from a given vertex of a graph G in time $O(p|E(G)|)$, we can observe that this can be done indeed in time $O(f'(k)e)$.

Finding flat zones. Now we show that if there are no well-attached vertices in the graph G, then a flat zone exists in our grid.

Lemma 5. *Let $X \in \mathrm{PlanarDel}(G, k)$, and let G not include any well-attached vertices. If K is a grid-component, then there cannot exist $(k+1)^2$ disjoint blocks such that K is attached to an inner vertex of each block.*

Proof. Let us assume for contradiction that there exist $(k+1)^2$ such blocks. Since $|X| \leq k$, at least $(k+1)^2 - k$ of these blocks do not contain any vertex of X. So let $x_1, x_2, \ldots x_{(k+1)^2 - k}$ be adjacent to K and let $B_1, B_2, \ldots, B_{(k+1)^2 - k}$ be disjoint blocks of $G - X$ such that x_i is an inner vertex of B_i.

Since $G - X$ is planar, it follows from Lemma 3 that a component of $K - X$ cannot be adjacent to different vertices from $\{x_i | 1 \leq i \leq (k+1)^2 - k\}$. So let K_i be the connected component of $K - X$ that is attached to x_i in $G - X$. K is connected in G, hence for every K_i there is a vertex of $T = K \cap X$ that is adjacent to it in G. Since there are no well-attached vertices in G, every vertex of T can be adjacent to at most $k + 1$ of these subgraphs. But then $|T| \geq ((k+1)^2 - k)/(k+1) > k$ which is a contradiction since $T \subseteq X$.   □

Let us now fix the constant $d = (k+1)((k+1)^2 - 1)$.

Lemma 6. *Let $X \in \mathrm{PlanarDel}(G, k)$, let G not include any well-attached vertices, and let x be a vertex of the grid R. Then there cannot exist $B_1, B_2, \ldots, B_{d+1}$ disjoint blocks such that for all i $(1 \leq i \leq d+1)$ an inner vertex of B_i and x are both attached to some grid-component K_i.*

Proof. As a consequence of Lemma 5, each of the grid-components K_i can be attached to at most $(k+1)^2 - 1$ disjoint blocks. But since x is not a well-attached vertex, there can be only at most $k+1$ different grid-components among the grid-components K_i, $1 \leq i \leq d+1$. So the total number of disjoint blocks that are attached to x through a grid-component is at most $(k+1)((k+1)^2 - 1) = d$.   □

Lemma 7. *Let $X \in \mathrm{PlanarDel}(G, k)$, and let G not include any well-attached vertices. Then there exists a flat zone Z in G.*

Proof. Let Z be an open zone which has a vertex z in its core that is attached to a vertex v of another zone $(v \notin V(Z))$ through a grid-component K. By the choice of the size of the zones we have disjoint blocks B_z and B_v containing z and v respectively as inner points. We can also assume that B_z is a subgraph of Z which does not intersect the outer circle of Z.

By Lemma 3 we know that B_z, B_v or K contains a vertex from X. Let $\mathcal{Z}_1$ denote the set of zones with a core vertex in X, let $\mathcal{Z}_2$ denote the set of open zones with a core vertex to which a grid-component, having a common vertex with X, is attached, and finally let $\mathcal{Z}_3$ be the set of the remaining open zones. Since $|X| \leq k$ and a grid-component can be attached to inner vertices of at most $(k+1)^2$ disjoint blocks by Lemma 5, we have that $|\mathcal{Z}_1| \leq k$ and $|\mathcal{Z}_2| \leq k(k+1)^2$.

Let us count the number of zones in $\mathcal{Z}_3$. To each zone Z in $\mathcal{Z}_3$ we assign a vertex $u(Z)$ of the grid not in Z, which is connected to the core of Z by a grid-component. First let us bound the number of zones in $\mathcal{Z}_3$ to which we assigned a vertex in X. Lemma 6 implies that $v \in X$ can be connected this way to at most d zones, so we can have only at most kd such zones.

Now let $U = \{v \mid v = u(Z), Z \in \mathcal{Z}_3\}$. Let a and b be different members of U, and let a be connected through the grid-component K_a with the core vertex z_a of $Z_a \in \mathcal{Z}_3$. Let B_a denote a block which only contains vertices that are inner vertices of Z_a, and contains z_a as inner vertex. Such a block can be given due to the size of a zone and its core. Let us define K_b, z_b, Z_b, and B_b similarly. Note that $V(Z_a) \cap X = V(Z_b) \cap X = \emptyset$.

Now let us assume that a and b are in the same component of $R - X$. Let P be a path connecting them in $R - X$. If P has common vertices with B_a (or B_b) then we modify P the following way. If the first and last vertices reached by P in Z_a (or Z_b, resp.) are w and w', then we swap the $w - w'$ section of P using the outer circle of Z_a (or Z_b, resp.). This way we can fix a path in $R - X$ that connects a and b, and does not include any vertex from B_a and B_b. But this path together with K_a and K_b would yield a path in $G - X$ that connects two inner vertices of B_a and B_b, contradicting Lemma 3.

Therefore, each vertex of U lies in a different component of $R - X$. But we can only delete at most k vertices, and each vertex in a hexagonal grid has at most 3 neighbors, thus we can conclude that $|U| \leq 3k$. As for different zones Z_1 and Z_2 we cannot have $u(Z_1) = u(Z_2)$ (which is also a consequence of Lemma 3) we have that $|\mathcal{Z}_3| \leq 3k$. So if we choose the number of zones in R to be $z = 7k + k(k+1)^2 + kd + 1$ we have that there are at least $3k + 1$ zones in R which are not contained in $\mathcal{Z}_1 \cup \mathcal{Z}_2 \cup \mathcal{Z}_3$, indicating that they are closed. Since a vertex can be contained by at most 3 zones, $|X| \leq k$ implies that there exist a closed zone Z^*, which does not contain any vertex from X, and all grid-components attached to Z^* are also disjoint from X. This immediately implies that $T(Z^*)$ is a subgraph of $G - X$, and thus $T(Z^*)$ is planar.   □

Algorithm for Phase I. The exact steps of Phase I of the algorithm $\mathcal{A}$ are shown in Fig. 2. It starts with running algorithm $\mathcal{B}$ on the graph G and integers $w(r)$ and r. If $\mathcal{B}$ returns a hexagonal grid as a topological minor, then the algorithm proceeds with the next step. If $\mathcal{B}$ returns a tree decomposition T of width $w(r)$, then Phase I returns the triple (G, W, T). Otherwise G does not have H_r as minor and its treewidth is larger than $w(r)$, so by Lemma 1 we can conclude that $G \notin$ Planar $+ kv$.

In the next step the algorithm tries to find a flat zone Z. If such a zone is found, then the algorithm executes a deletion, whose correctness is implied

---

Phase I of algorithm $\mathcal{A}$:

Input: $G = (V, E)$.

Let $W = \emptyset$.

1. Run algorithm $\mathcal{B}$ on G, $w(r)$, and r.
   If it returns a subgraph H'_r topologically isomorphic to H_r then go to Step 2. If it returns a tree decomposition $\mathcal{T}$ of G, then output $(G, W, \mathcal{T})$. Otherwise output("No solution.").

2. For all zones Z do:
   If Z is flat then $G := G - T(R_0)$, and go to Step 1.

3. Let $U = \emptyset$. For all $x \in V$: if x is well-attached then $U := U \cup \{x\}$.
   If $|U| = \emptyset$ or $|W| + |U| > k$ then output("No solution.").
   Otherwise $W := W \cup U$, $G := G - U$ and go to Step 1.

Fig. 2. Phase I of algorithm $\mathcal{A}$

by Lemma 2. Note that after altering the graph, the algorithm must find the hexagonal grid again and thus has to run $\mathcal{B}$ several times.

If no flat zone was found in Step 2, the algorithm removes well-attached vertices from the graph in Step 3. The vertices already removed this way are stored in W, and U is the set of vertices to be removed in the actual step. By Lemma 4, if $X \in \text{PlanarDel}(G, k)$ then $W \cup U \subseteq X$, so $|W| + |U| > k$ means that there is no solution. By Lemma 7 the case $U = \emptyset$ means also that there is no solution for the problem instance. In these cases the algorithm stops with the output "No solution." Otherwise it proceeds with updating the variables W and G, and continues with Step 1.

The output of the algorithm can be of two types: it either refuses the instance (outputting "No solution.") or it returns an instance for Phase II. For the above mentioned purposes the new instance is equivalent with the original problem instance in the following sense:

Theorem 1. *Let* $(G', W, \mathcal{T})$ *be the triple returned by* $\mathcal{A}$ *at the end of Phase I. Then for all* $X \subseteq V(G)$ *it is true that* $X \in \text{PlanarDel}(G, k)$ *if and only if* $W \subseteq X$ *and* $(X \setminus W) \in \text{PlanarDel}(G', k - |W|)$.

Now let us examine the running time of this phase. The first step can be done in time $O(f''(k)n)$ according to [17,2,16] where $n = |V(G)|$. Since the algorithm only runs algorithm $\mathcal{B}$ again after reducing the number of the vertices in G, we have that $\mathcal{B}$ runs at most n times. This takes $O(f''(k)n^2)$ time. The second step requires only linear time (a breadth first search and a planarity test). Deciding whether a vertex is well-attached can be done in time $O(f'(k)e)$ (where $e = |E(G)|$), so we need $O(f'(k)ne)$ time to check every vertex at a given iteration in Step 3. Note that the third step is executed at most $k + 1$ times, since at each iteration $|W|$ increases. Hence, this phase of algorithm $\mathcal{A}$ uses total time $O(f''(k)n^2 + f'(k)kne) = O(f(k)n^2)$, as the number of edges is $O(kn)$.

5   Phase II of Algorithm $\mathcal{A}$

At the end of Phase I of algorithm $\mathcal{A}$ we either conclude that there is no solution, or we have a triple $(G', W, \mathcal{T})$ for which Theorem 1 holds. Here $\mathcal{T}$ is a tree decomposition for G' of width at most $w(r)$. This bound only depends on r which is a function of k. From the choice of the constants $r, q, z,$ and d we can easily derive that $\mathrm{tw}(G') \leq w(r) \leq 100(k+2)^{7/2}$.

In order to solve our problem we only have to find out if there is a set $Y \in \mathrm{PlanarDel}(G', k')$ where $k' = k - |W|$. For such a set, $Y \cup W$ would yield a solution for the original PLANAR + k VERTEX problem.

A theorem by Courcelle states that every graph property defined by a formula in monadic second-order logic (MSO) can be evaluated in linear time if the input graph has bounded treewidth. Here we consider graphs as relational structures of vocabulary $\{V, E, I\}$, where V and E denote unary relations interpreted as the vertex set and the edge set of the graph, and I is a binary relation interpreted as the incidence relation. We will denote by U^G the universe of the graph G, i.e., $U^G = V(G) \cup E(G)$. Variables in monadic second-order logic can be element or set variables. For a survey on MSO logic on graphs see [5].

Following Grohe [11], we use a strengthened version of Courcelle's Theorem:

Theorem 2. ([9]) *Let $\varphi(x_1, \ldots, x_i, X_1, \ldots, X_j, y_1, \ldots, y_p, Y_1, \ldots, Y_q)$ denote an MSO-formula and let $w \geq 1$. Then there is a linear-time algorithm that, given a graph G with $\mathrm{tw}(G) \leq w$ and $b_1, \ldots, b_p \in U^G, B_1, \ldots, B_q \subseteq U^G$, decides whether there exist $a_1, \ldots, a_i \in U^G, A_1, \ldots, A_j \subseteq U^G$ such that*
$$G \vDash \varphi(a_1, \ldots, a_i, A_1, \ldots, A_j, b_1, \ldots, b_p, B_1, \ldots, B_q),$$
and, if this is the case, computes such elements $a_1, \ldots, a_i$ and sets $A_1, \ldots, A_j$.

It is well-known that there is an MSO-formula $\varphi_{\mathrm{planar}}$ which describes the planarity of graphs, i.e., for every graph G the statement $G \vDash \varphi_{\mathrm{planar}}$ holds if and only if G is planar. This can be easily seen thanks to the simple characterization of planar graphs by Kuratowski's Theorem: a graph is planar if and only if it does not contain any subgraph topologically isomorphic to K_5 or $K_{3,3}$. The existence of these subgraphs can be formulated using vertex sets as variables.

It is easy to modify $\varphi_{\mathrm{planar}}$ so that we obtain a formula $\varphi^*(x_1, \ldots, x_{k'})$ that expresses the following: if we delete the vertices $x_1, \ldots, x_{k'}$ from the graph, then the resulting graph is planar. All we have to ensure is that the subgraphs that we obstruct in $\varphi_{\mathrm{planar}}$ (i.e., the subdivisions of the graphs K_5 and $K_{3,3}$) are disjoint from the vertices $x_1, \ldots, x_{k'}$. So we can state the following:

Theorem 3. *There exists an MSO-formula $\varphi^*(x_1, \ldots, x_{k'})$ for which the statement $G \vDash \varphi^*(v_1, \ldots, v_{k'})$ holds if and only if $G - \{v_1, \ldots, v_{k'}\}$ is planar.*

Now let us apply Theorem 2. Let $\mathcal{C}$ be the algorithm which, given a graph G of bounded treewidth, decides whether there exist $v_1, \ldots, v_{k'} \in U^G$ such that $G \vDash \varphi*(v_1, \ldots, v_{k'})$ is true, and if possible, also produces such variables. By Theorem 3, running $\mathcal{C}$ on G' either returns a set of vertices $U \in \mathrm{PlanarDel}(G', k')$, or reports that this is not possible. Hence, we can finish algorithm $\mathcal{A}$ in the following way: if $\mathcal{C}$ returns U then output($U \cup W$), otherwise output("No solution").

The running time of Phase II is $O(g(k)n)$ for some function g.

Remark 1. Phase II of the algorithm can also be done by applying dynamic programming, using the tree decomposition $\mathcal{T}$ returned by $\mathcal{B}$. This also yields a linear-time algorithm, with a double exponential dependence on tw(G'). Since the proof is quite technical and detailed, we omit it.

References

1. Baker, B.S.: Approximation algorithms for NP-complete problems on planar graphs. J. ACM 41, 153–180 (1994)
2. Bodlaender, H.L.: A linear-time algorithm for finding tree-decompositions of small treewidth. SIAM J. Comput. 25, 1305–1317 (1996)
3. Cai, L.: Fixed-parameter tractability of graph modification problems for hereditary properties. Inform. Process. Lett. 58, 171–176 (1996)
4. Courcelle, B.: Graph rewriting: an algebraic and logic approach. In: Handbook of Theoretical Computer Science, vol. 2, pp. 194–242. Elsevier, Amsterdam (1990)
5. Courcelle, B.: The expression of graph properties and graph transformations in monadic second-order logic. In: Handbook of graph grammars and computing by graph transformations, ch. 5, vol. 1, pp. 313–400. World Scientific, New-Jersey (1997)
6. Diestel, R.: Graph Theory. Springer, Berlin (2000)
7. Downey, R.G., Fellows, M.R.: Parameterized Complexity. Springer, New York (1999)
8. Eppstein, D.: Subgraph isomorphism in planar graphs and related problems. J. Graph Algorithms Appl. 3, 1–27 (1999)
9. Flum, J., Frick, M., Grohe, M.: Query evaluation via tree-decompositions. J. ACM 49, 716–752 (2002)
10. Garey, M.R., Johnson, D.S.: Crossing Number is NP-Complete. SIAM J. Algebraic Discrete Methods 4, 312–316 (1983)
11. Grohe, M.: Computing crossing number in quadratic time. J. Comput. System Sci. 68, 285–302 (2004)
12. Hadlock, F.: Finding a maximum cut of a planar graph in polynomial time. SIAM J. Comput. 4, 221–225 (1975)
13. Hopcroft, J.E., Tarjan, R.E.: Efficient planarity testing. J. ACM 21, 549–568 (1974)
14. Lewis, J.M., Yannakakis, M.: The vertex-deletion problem for hereditary properties is NP-complete. J. Comput. System Sci. 20, 219–230 (1980)
15. Lipton, R.J., Tarjan, R.E.: Applications of a planar separator theorem. SIAM J. Comput. 9, 615–627 (1980)
16. Perkovic, L., Reed, B.: An improved algorithm for finding tree decompositions of small width. Internat. J. Found. Comput. Sci. 11, 365–371 (2000)
17. Robertson, N., Seymour, P.D., Thomas, R.: Quickly excluding a planar graph. J. Combin. Theory Ser. B 62, 323–348 (1994)
18. Robertson, N., Seymour, P.D.: Graph minors. J. Combin. Theory Ser. B 92, 325–357 (2004)
19. Robertson, N., Seymour, P.D.: Graph minors. XIII. The disjoint paths problem. J. Combin. Theory Ser. B 63, 65–110 (1995)
20. Thomassen, C.: The graph genus problem is NP-complete. J. Algorithms 10, 568–576 (1989)

Mixed Search Number and Linear-Width of Interval and Split Graphs

Fedor V. Fomin, Pinar Heggernes, and Rodica Mihai

Department of Informatics, University of Bergen, N-5020 Bergen, Norway
{fedor.fomin,pinar.heggernes,rodica.mihai}@ii.uib.no

Abstract. We show that the mixed search number and the linear-width of interval graphs and of split graphs can be computed in linear time and in polynomial time, respectively.

1 Introduction

In the graph searching problem, a team of searchers (pursuers) is trying to catch a fugitive moving along the edges of a graph. The problem is to find the minimum number of searchers that can guarantee the capture of the fugitive in the worst case scenario for the searchers (assuming that fugitive is very fast, invisible, and knows the strategy of the searchers). The study of this problem started in 1970s when it was independently introduced by Parsons [19] and Petrov [23], and since that time it has been studied extensively [2,3,13,17,18,22]. It fits into the broader class of pursuit-evasion/search/rendezvous problems on which hundreds of papers have been written (see e.g., the book [1]).

The classical Parson-Petrov formulation of the problem is referred to as edge searching, and there are two modifications of it giving two other models. The first model, node searching, was introduced by Kirousis and Papadimitriou [13], and the second model, mixed searching, by Bienstock and Seymour [3]. The difference between the models is in the way the searchers are allowed to catch the fugitive or clear the edges of the graph. (We give formal definitions of the problems in the next section.) The minimum number of the searchers sufficient to perform searching and ensure capture for each of the models are respectively the edge, node, and mixed search numbers, and computations of these are all NP-hard [3,18,13].

The node search number of a graph is known to be equal to its pathwidth plus one. Similarly, the mixed search number of a graph is equal to its proper pathwidth [24]. Also for every graph G of minimum vertex degree at least 2, the proper pathwidth of G is equal to the linear-width of G. While pathwidth can be seen as a "linear" variant of treewidth, linear-width is a "linear" variant of branchwidth. Whereas the computation of treewidth and pathwidth of interval and split graphs is almost straightforward, the polynomial time algorithms [15,20] computing the branchwidth of an interval graph are not trivial and computing the branchwidth of a split graph is NP-hard [15]. Such a difference between computational complexities of computing treewidth and branchwidth of split graphs was one of the motivations for our study.

A. Brandstädt, D. Kratsch, and H. Müller (Eds.): WG 2007, LNCS 4769, pp. 304–315, 2007.
© Springer-Verlag Berlin Heidelberg 2007

Graph searching problems can be seen as vertex and edge ordering problems and computational complexity of different ordering problems on interval and split graphs is a well studied area (see e.g., the survey [6]). For example, the bandwidth minimization problem is solved in polynomial time on interval graphs [14] and is NP-hard on split graphs [16]. Profile (or SumCut) problem is trivially solvable in polynomial time on interval graphs and is NP-hard on split graphs [21], while Optimal Linear Arrangement is NP-hard on interval graphs [5].

It is easy to show that the node search number of an interval or split graph can be computed in linear time. Independently, Peng et al. [22] and Golovach et al. [10] show that similar result holds for edge search number. In this paper we show that the mixed search number (the proper pathwidth and the linear-width) of interval and split graphs can be computed in polynomial time. In fact, for interval graphs we obtain a linear time algorithm. For interval graphs our algorithm resembles the algorithm for the edge search number [22,10], but for split graphs there is a substantial difference between edge search and mixed search numbers, and the situation with mixed search is much more involved.

Due to space restrictions the proofs of some of our results are omitted.

2   Background, Definitions, and Notations

We work with simple and undirected graphs $G = (V, E)$, with vertex set $V(G) = V$ and edge set $E(G) = E$, and we let $n = |V|$, $m = |E|$. The set of *neighbors* of a vertex x is denoted by $N(x) = \{y \mid xy \in E\}$. A vertex set C is a *clique* if every two vertices in C are adjacent, and a *maximal clique* if no superset of C is a clique. The maximum size of a clique in G is denoted by $\omega(G)$. A vertex set I is an *independent set* if no two vertices of I are adjacent. The *degree* of a vertex v is $d(v) = |N(v)|$. A vertex is *isolated* if it has degree 0.

A *path* is a sequence $v_1, v_2, ..., v_p$ of distinct vertices of G, where $v_i v_{i+1} \in E$ for $1 \leq i < p$, in which case we say that this is a path *between* v_1 and v_p. A path $v_1, v_2, ..., v_p$ is called a *cycle* if $v_1 v_p \in E$. A *chord* of a cycle (path) is an edge connecting two non-consecutive vertices of the cycle (path).

A graph is *chordal* if every cycle of length at least 4 has a chord. A graph is an *interval graph* if intervals of the real line can be associated to its vertices such that two vertices are adjacent if and only if their corresponding intervals overlap. Interval graphs are chordal. A *caterpillar* is a graph that contains a chordless path such that if the vertices on this path are deleted only isolated vertices remain. Caterpillars are interval graphs and trees. A graph is a *split graph* if its vertices can be partitioned into a clique C and an independent set I, in which case (C, I) is called a *split partition* of G. A split partition of a split graph is not unique, but it is always possible to choose a partition such that C is a clique of maximum size. In this paper we will always assume that C is a clique of maximum size. For a vertex $u \in C$, we denote by $N_I(u)$ the neighbors of u in I. Split graphs are also chordal. Let us remark that split and interval graphs can be recognized in linear time. A clique of maximum size in these graphs can be found in linear time as well. (See e.g., [11]).

The parameters that we study in this paper, mixed search number and linear-width, are both closely related to a parameter called pathwidth, defined through path decompositions.

A *path-decomposition* of a graph $G = (V, E)$ is a linearly ordered sequence of subsets of V, called *bags*, such that the following three conditions are satisfied: 1. Every vertex $x \in V$ appears in some bag. 2. For every edge $xy \in E$ there is a bag containing both x and y. 3. For every vertex $x \in V$, the bags containing x appear consecutively. The *width* of a decomposition is the size of the largest bag minus one, and the *pathwidth* of a graph G, $pw(G)$, is the minimum width over all possible path decompositions. A path decomposition of width $pw(G)$ is called an *optimal* path decomposition of G.

By a classical result of Gillmore and Hoffman [9], a graph G is an interval graph if and only if it has an optimal path decomposition where every bag is a maximal clique of G. Such an optimal path decomposition is often called a *clique-path*. It is well known that the pathwidth of an interval graph is one less than the size of its largest clique. Clique-paths of interval graphs can be computed in linear time [4]. Consequently, the pathwidth of interval graphs can also be computed in linear time.

The *mixed search game* can be formally defined as follows. Let $G = (V, E)$ be a graph to be searched. A *search program* consists of a sequence of discrete steps which involves searchers. Initially there is no searcher on the graph. Every step is one of the following three types

- Some searchers are placed on some vertices of G (there can be several searchers located in one vertex);
- Some searchers are removed from G;
- A searcher slides from a vertex u to a vertex v along edge uv.

At every step of the search program the edge set of G is partitioned into two sets: *cleared* and *contaminated* edges. Intuitively, the agile and omniscient fugitive with unbounded speed who is invisible for the searchers, is located somewhere on a contaminated territory, and cannot be on cleared edges. Initially all edges of G are contaminated, i.e., the fugitive can be anywhere. A contaminated edge uv becomes cleared at some step of the search program either if both its endpoints contain searchers, or if at this step a searcher located in u slides to v along uv.

A cleared edge e is (re)contaminated at some step if at this step there exists a path P containing e and a contaminated edge and no internal vertex of P contains a searcher. For example, if a vertex u is incident to a contaminated edge e, there is only one searcher at u and this searcher slides from u to v along edge $uv \neq e$, then after this step the edge uv, which is cleared by sliding, is immediately recontaminated.

A search program is *winning* if after its termination all edges are cleared. The *mixed search number* of a graph G, denoted by $ms(G)$, is the minimum number of searchers required for a winning program of mixed searching on G.

The differences between mixed, edge, and node searching are in the way the edges can be cleared. In node searching an edge is cleared only if both its end-points are occupied (no clearing by sliding). In edge searching an edge can be cleared only by sliding. So mixed searching can be seen as a combination of node and edge searching. The *edge* and *node search numbers* of a graph G are defined similarly to the mixed search number, and are denoted by $es(G)$ and $ns(G)$, respectively.

A search program is called *monotone* if at any step of this program no recontamination occurs. For all three versions of graph searching, recontamination does not help to search the graph with fewer searchers [3,17], i.e., on any graph with {edge, mixed, node} search number k there exists a winning monotone {edge, mixed, node} search program using k searchers. Thus in this paper we consider only monotone search programs.

The linear-width was introduced by Thomas [26]. The *linear-width* of an arbitrary graph G, $lw(G)$, is defined to be the smallest integer $k \geq 0$ such that the edges of G can be ordered $e_1, ..., e_m$ in such a way that for every $i = 1, 2, ..., m-1$, there are at most k vertices incident both to an edge in $\{e_1, ..., e_i\}$ and to an edge in $\{e_{i+1}, ..., e_m\}$.

Takahashi et al. [24] proved that the mixed search number of any graph is equal to its proper pathwidth. Fomin [7] showed that proper pathwidth is also equivalent to another graph parameter, namely split bandwidth. Thus our algorithms for computing mixed search numbers of interval and split graphs can be applied to these parameters as well.

The next proposition follows directly from the results of Bienstock and Seymour [3], Fomin and Thilikos [8], and Takahashi, Ueno, and Kajitani [24].

Proposition 1. *For any graph G, $pw(G) \leq lw(G) \leq ms(G) \leq pw(G) + 1$.*

Furthermore, if G has no vertices of degree 1 then $lw(G) = ms(G)$ [3]. In fact, Thilikos showed the following stronger result [25].

Proposition 2 ([25]). *Let G be any graph, and let G' be the graph obtained by removing all vertices of degree 1 from G. Then $lw(G) = ms(G')$.*

3   Interval Graphs

By the characterization of interval graphs through clique paths, $pw(G) = \omega(G) - 1$ for every interval graph G. Hence, by Proposition 1 we know that for an interval graph G, $ms(G)$ is either $\omega(G) - 1$ or $\omega(G)$. We characterize interval graphs G with $ms(G) = \omega(G)$.

Observation 1. *If G is a complete graph then $ms(G) = n - 1$.*

Lemma 1. *Let G be a graph consisting of three cliques of size $n-2$ that intersect at the same $n - 3$ vertices. Then $ms(G) = \omega(G) = n - 2$.*

Proof. Clearly G is an interval graph and $\omega(G) = n - 2$. By Proposition 1, $n - 2$ searchers are enough to clear G. Let us show that $n-3$ searchers are not enough. Let C be the set of $n - 3$ vertices in the intersection of the three cliques, and let x, y, z be the remaining vertices. We need $n - 3$ searchers to clear $C \cup \{x\}$ by

Observation 1. If we start by placing all $n - 3$ searchers on vertices of C, then we will need to slide one of them to x, and this will allow recontamination of a vertex of C from y or z. Since we know that there is always an optimal program without recontamination, we can discard this approach. Let us place all $n - 3$ searchers on x and $n - 4$ vertices of C. Then slide the searcher on x to the single unguarded vertex of C. Now all edges between pairs of vertices in $C \cup \{x\}$ are cleared. To clear y or x, we need to slide a searcher from a vertex of C to one of these vertices, say y. But then this vertex of C will become recontaminated from x. Thus it is not possible to continue the search without recontamination with only $n - 3$ searchers, and hence $ms(G) = n - 2$.

For the following results, we need to give more details about clique-paths. An interval graph has at most n maximal cliques. We will let $(B_1, S_1, B_2, S_2, ..., S_{c-1}, B_c)$ denote a clique-path of G, where B_i is a bag of the clique-path and a maximal clique of G for each $1 \leq i \leq c$, and $S_i = B_i \cap B_{i+1}$ represents the edge between B_i and B_{i+1} for $1 \leq i < c$.

Lemma 2. *Let G be an interval graph with $\omega(G) = k + 1$. Then G contains three maximal cliques of size $k + 1$ that intersect at the same k vertices if and only if there are two consecutive edges S_{i-1} and S_i of cardinality k satisfying $S_{i-1} = S_i$ in every clique-path of G.*

Proof. Assume that G is an interval graph with $\omega(G) = k + 1$ and a clique-path of G contains two consecutive edges S_i and S_{i+1} of cardinality k satisfying $S_i = S_{i+1}$. This means that the maximal cliques of G appearing as bags B_{i-1}, B_i, B_{i+1} share the same k vertices. Since each such maximal clique is distinct, each has a vertex that is not in this intersection, and since $\omega(G) = k + 1$, each has exactly $k + 1$ vertices.

For the other direction, assume that G is an interval graph with $\omega(G) = k + 1$ and that G has three maximal cliques of size $k + 1$ that intersect at the same k vertices. Pick any clique-path of G, and let bags B_i, B_j, B_ℓ be the representatives of these three maximal cliques. They do not necessarily appear consecutively in the chosen clique-path. Let B_i be the one furthest to the left, and let B_j be the one furthest to the right. By the definition of clique-paths, the intersection $B_i \cap B_j$ must be a subset of every edge on the path between B_i and B_j. Since $\omega(G) = k + 1$ and $|B_i \cap B_j| = k$, this means that every edge on the path between B_i and B_j has cardinality exactly k and thus must be equal to $B_i \cap B_k$. Since B_ℓ also appears on the path between B_i and B_j, there are at least two such consecutive edges in the clique-path, and the proof is complete.

Theorem 2. *Let G be an interval graph. Then $ms(G) = \omega(G)$ if and only if G has three maximal cliques of size $\omega(G)$ that intersect at the same $\omega(G) - 1$ vertices. Otherwise, $ms(G) = \omega(G) - 1$.*

Proof. If G has three maximal cliques of size $\omega(G)$ that intersect at the same $\omega(G) - 1$ vertices, then the subgraph of G induced by the vertices of these three maximal cliques has mixed search number $\omega(G)$ by Lemma 1. Consequently,

$ms(G)$ cannot be smaller than $\omega(G)$. By Proposition 1, $ms(G) \leq pw(G) + 1 = \omega(G)$. Hence we can conclude that $ms(G) = \omega(G)$ in this case.

Assume now that there are no three maximal cliques of size $\omega(G)$ in G that intersect at the same $\omega(G) - 1$ vertices. We give a program to search the graph using $\omega(G) - 1$ searchers. Since $\omega(G) - 1$ searchers are the fewest possible by Proposition 1, the result will follow. Let $P = (B_1, S_1, B_2, S_2, ..., B_c)$ be any clique-path of G. By Lemma 2, we know that there is no index i with $2 \leq i \leq c - 1$ satisfying $S_{i-1} = S_i$ and $|S_{i-1}| = |S_i| = \omega(G) - 1$. We start searching the graph with bag B_1. Since $B_1 \not\subseteq B_2$, there is a vertex $x \in B_1 \setminus B_2$, and since $B_1 \cap B_2 = S_1 \neq \emptyset$, there is a vertex $y \in S_1$ with $y \neq x$. Furthermore $|B_1| \leq \omega(G)$. If $|B_1| \leq \omega(G) - 1$, then simply place searchers on all vertices of B_1. If $|B_1| = \omega(G)$, then place searchers on all vertices of $B_1 \setminus \{y\}$, and then slide the searcher on x to y. In both cases, all vertices of B_1 and all edges between them are cleared, since x has no neighbors in any other bag by the definition of a clique-path. Actually, by the same argument, no vertex of $B_1 \setminus B_2$ has any neighbors in any other bag than in B_1, hence searchers placed on these vertices can now be safely removed as long as we keep searchers on all vertices of S_1.

Observe that vertices of $(B_1 \cup ... \cup B_i) \setminus S_i$ have no neighbors belonging to $(B_{i+1} \cup ... \cup B_c) \setminus S_i$. Hence we can assume by induction that we have already cleared the subgraph induced by the vertices of $B_1, ..., B_i$ that do not belong to S_i, and we have searchers placed on all vertices of $S_i = B_i \cap B_{i+1}$. We will now show how to proceed so that the subgraph induced by the vertices of $B_1, ..., B_i, B_{i+1}$ is cleared and searchers are kept on all vertices of S_{i+1}. If $|B_{i+1}| \leq \omega(G) - 1$ then $|S_i| \leq \omega(G) - 2$ and there are available searchers not guarding the vertices of S_i such that we can place them on vertices of B_{i+1} and all vertices of B_{i+1} will be occupied by searchers. Consequently, all vertices of S_{i+1} are also occupied by searchers and the proof for this case is complete. Assume for the remaining part that $|B_{i+1}| = \omega(G)$. Thus we have enough available searchers to occupy all vertices of B_{i+1} except one, without removing any searcher from S_i. We distinguish between two cases: $S_i = S_{i+1}$ and $S_i \neq S_{i+1}$. If $S_i = S_{i+1}$, then we know by our assumption and Proposition 1 that $|S_i| = |S_{i+1}| \leq \omega(G) - 2$. Hence there are at least two vertices $x, y \in B_{i+1} \setminus S_{i+1}$. We place the available searchers on all unguarded vertices of B_{i+1} except y. Then we can safely slide the searcher on x to y, which clears whole B_{i+1} while guarding all vertices of S_{i+1} by keeping searchers on them. What remains is the case when $S_i \neq S_{i+1}$. Assume first that there is a vertex $x \in S_i \setminus S_{i+1}$. We place searchers on all unguarded vertices of B_{i+1} except one in an arbitrary way. Then we can safely slide the searcher on x to the single vertex of B_i that is not occupied by a searcher, since x has no neighbors in the bags appearing on the other side of S_{i+1}. This will again clear whole B_{i+1} while keeping searchers on all vertices of S_{i+1}. If there is no vertex $x \in S_i \setminus S_{i+1}$ then $S_i \subset S_{i+1}$, which means that $|S_i| \leq \omega(G) - 2$, and there is at least one vertex x in B_{i+1} that does not belong to any other bag since $S_{i+1} \subset B_{i+1}$. Hence we have at least one available searcher that we can place on x without removing any searchers from S_i. We place the available searchers on all vertices of B_{i+1} except one vertex different from x. Now, we can safely slide the searcher on x to the single unguarded vertex of B_{i+1}, which clears whole B_{i+1}, and we have searchers on all vertices of S_{i+1}, since $x \notin S_{i+1}$.

We have thus arrived at the main result of this section.

Theorem 3. *The mixed search number of an interval graph can be computed in linear time.*

Proof. Let G be an interval graph. A clique-path of G can be computed in linear time, and it has at most n maximal cliques [4]. This means that the sum of the sizes of all bags and all edges of the clique-path is $O(n + m)$. Given the clique-path, to find $ms(G)$, by Theorem 2 we need to check for every triple of consecutive maximal cliques B_{i-1}, B_i, B_{i+1} having each $\omega(G)$ vertices whether $B_{i-1} \cap B_i = B_i \cap B_{i+1}$.

We argue that this can be done in overall linear time. First sort all vertices in all bags according to the same order. Since the sum of the sizes of all bags is $O(n + m)$ and the largest value is n, this can be done in $O(n + m)$ time. Then comparing three bags can be done in $O(\omega(G))$ time. We do this only when all three bags have $\omega(G)$ vertices. There are $\omega(G) - 1$ edges in B_{i-1} that do not appear in B_i or B_{i+1} since B_{i-1} has one vertex not appearing in B_i or B_{i+1}. Hence we can let these edges of B_{i-1} "pay for" the comparison between B_{i-1}, B_i, B_{i+1}. Thus each edge of G will pay for at most 3 of comparisons, and therefore we can bound the total time of all comparisons by $O(m)$. $\square$

Corollary 4. *The linear-width of an interval graph can be computed in linear time.*

As an additional structural result, we characterize interval graphs G with $lw(G) \neq ms(G)$.

Theorem 5. *Let G be an interval graph with at least two edges. Then $lw(G) = ms(G) - 1$ if and only if G is a caterpillar with maximum degree at least 3. Otherwise $lw(G) = ms(G)$.*

Proof. By Proposition 1, $lw(G)$ is either equal to $ms(G)$ or to $ms(G) - 1$.

Let G be a caterpillar with maximum degree at least 3. Then it has three maximal cliques of size 2 that intersect at the same 1 vertex. By Theorem 2, $ms(G) = \omega(G) = 2$. Since G is a caterpillar, $lw(G) = pw(G) = 1$. Therefore $lw(G) = ms(G) - 1$.

For the other direction, let G be an interval graph with $lw(G) \neq ms(G)$. By Proposition 1, $ms(G) = pw(G) + 1 = \omega(G)$ and $lw(G) = pw(G) = \omega(G) - 1$. Since $ms(G) = \omega(G)$ by Theorem 2, G has three cliques of size $\omega(G)$ that intersect at the same $\omega(G) - 1$ vertices. If $\omega(G) = 2$ then G is a caterpillar with maximum degree at least three, and the proof is complete. Assume for contradiction that $\omega(G) > 2$. Then the three cliques mentioned above are of size at least 3. The subgraph of G induced by the union of the vertices of these three cliques has mixed search number $\omega(G)$ by Lemma 1. Since this subgraph has no vertices of degree 1, its linear-width is also $\omega(G)$. Consequently, $lw(G)$ and $ms(G)$ cannot be less than $\omega(G)$, and we conclude that they are equal, which gives a contradiction. Hence $\omega(G)$ cannot be more than 2 if $lw(G) \leq ms(G)$.

4   Split Graphs

It is easy to show that the pathwidth of a split graph G is either $\omega(G) - 1$ or $\omega(G)$ [12]. Hence it follows from this and Proposition 1 that $ms(G)$ is equal to one of the following: $\omega(G) - 1$, $\omega(G)$, or $\omega(G) + 1$. We characterize each of these three cases.

Theorem 6. *Let $G = (V, E)$ be a split graph with a split partition (C, I) of its vertices where C is a clique of maximum size. Then $ms(G) = \omega(G) - 1 = |C| - 1$ if and only if one of the following three conditions holds.*

1. *There are 2 vertices $u, v \in C$ such that $N_I(u) \cap N_I(v) = \emptyset$ and $|N_I(v)| \leq 1$ and one of the following is true:*
   - *(a) $|N_I(u)| \leq 1$, or*
   - *(b) there is an additional vertex $x \in C$ such that $|N_I(u) \cap N_I(x)| \leq 1$.*
2. *There are 3 vertices $x, u, v \in C$ such that $N_I(u) \cap N_I(x) = \emptyset$ and $|N_I(v) \cap N_I(x)| \leq 1$ and one of the following is true:*
   - *(a) $|N_I(u)| \leq 1$, or*
   - *(b) $|N_I(v)| \leq 2$ and $|N_I(u) \cap N_I(v)| \leq 1$.*
3. *There are 4 vertices $x, y, u, v \in C$ such that $(N_I(x) \cup N_I(y)) \cap (N_I(u) \cup N_I(v))$ contains at most 2 vertices u_1, v_1, and $u_1 x \notin E$ and $v_1 u \notin E$.*

Proof. Assume that $ms(G) = |C| - 1$. Hence there is a monotone mixed search program that is able to clear G with $|C| - 1$ searchers. At some point of this search, all $|C| - 1$ searchers must occupy $|C| - 1$ vertices of C, since only from such a situation, by sliding one of the searchers to the single vertex of C without a searcher, we can clear any clique C and the edges with both endpoints in C. Let us consider the first time when $|C| - 1$ vertices of C are occupied by searchers during this search program. Let v be the vertex that got occupied by a searcher at this point, and let u be the vertex of C without a searcher. Without loss of generality, we can assume that all vertices of I that are not adjacent to u or v are cleared, as well as all edges between such vertices and C, since we can always do this before placing the last searcher on v. Furthermore, no neighbors of u in I are cleared (otherwise recontamination would be allowed), and at most one neighbor of v in I is cleared (if we placed the last searcher on a neighbor and slided to v). Let v_1 this possible neighbor of v; consequently v_1 is not adjacent to u. The next step must be to slide a searcher from one of the cleared vertices of C to u, because all vertices of C are pairwise adjacent and u is not cleared, so we would get recontamination otherwise. Let *step i* be this next step. Hence step i is either to slide from v to u, or to slide from another vertex $x \neq v$ of C to u. We will show that in either case, the possible ways of completing this search successfully without any more searchers all lead to the conditions of the theorem, proving the *only if* direction. For each condition, we will also explain how to complete the search with $|C| - 1$ searchers, hence the *if* direction will be proved at the same time.

Assume that step i of the search is to slide from v to u. This can only be done if v_1 is cleared (or does not exist), hence a searcher must have slided from v_1 to v before this step. Then we know that $|N_I(v)| \leq 1$ and that u and v have no common neighbors. After step i, v is the only vertex of C without a searcher.

If step $i + 1$ is to slide from u to a neighbor u_1 of u in I, then u cannot have more neighbors (otherwise recontamination), and we have Condition 1 (a). The search is complete.

If step $i + 1$ is to slide (or move) from another vertex $x \neq u$ to a neighbor u_1 of u in I, then x and u can have at most this single neighbor u_1 in common in I. Also x and v can have at most v_1 as a common neighbor in I, but this is already covered since $|N_I(v)| \leq 1$. Hence we have Condition 1 (b). Now any other neighbor that u might have in I can be cleared with the searcher on u_1, to complete the search.

Assume that step i of the search is to slide from a vertex $x \neq v$ of C to u. Then we know that $N_I(v) \cap N_I(x) \subseteq \{v_1\}$ and $N_I(u) \cap N_I(x) = \emptyset$. Hence, if x is adjacent to v_1, we must have slided a searcher from v_1 to v to place this searcher on v, and v_1 is cleared along with all edges between v_1 and C. After step i, x is the only vertex of C without a searcher.

If step $i + 1$ is to slide from u to a neighbor $u_1 \in I$ of u, then $|N_I(u)| \leq 1$. Hence we have Condition 2 (a). Now we can use the searcher on u_1 to clear all neighbors of v and all edges between these and C, one by one.

If step $i + 1$ is to move from u to a neighbor of v in I which is not adjacent to u, then $N_I(u) = \emptyset$ and we can use the searcher on u to clear all neighbors of v in I one by one. This case is covered by the previous case, since u could have a neighbor in I which we could have cleared before moving on to the neighbors of v. Hence we have again Condition 2 (a).

If step $i + 1$ is to slide from v to a neighbor of v in I then $|N_I(v)| \leq 2$. Furthermore, if v has two neighbors in I, then at most one of these can be adjacent to u, because to clear both neighbors of v we must have slided to v from v_1, and v_1 is not adjacent to u as argued above. Hence we have Condition 2 (b). Now we can use the searcher on the second neighbor of v to search all remaining neighbors of u in I one by one.

If step $i + 1$ is to move from v to a neighbor $u_1 \in I$ of u with $u_1 \notin N_I(v)$, then $N_I(v)$ does not contain any other vertices than v_1. Now we can use the searcher on u_1 to clear all neighbors of u. Note that this situation is just a special case of the previous situation, hence it is covered by Condition 2 (b).

If step $i + 1$ is to move from a vertex $y \neq u, x$ of C to a neighbor $u_1 \in I$ of u, then $N_I(y) \cap N_I(u) \subseteq \{u_1\}$ and $N_I(y) \cap N_I(v) \subseteq \{v_1, u_1\}$. In this case, we must have slided from v_1 to v before placement on v, and after the slide (or move) from y to u_1, u_1 will be cleared and it can be adjacent to both u and v. After step $i + 1$, the searcher on u_1 can be used to clear all remaining uncleared vertices of $N_I(u) \cup N_I(v)$. We should also remember that $N_I(v) \cap N_I(x) \subseteq \{v_1\}$ and $v_1 u \notin E$. Hence we have Condition 3.

If step $i + 1$ is to move from a vertex $y \neq u, x$ of C to a neighbor of v in I then have the exact same situation as the previous situation (because u_1 can be adjacent to v and hence serve as the assumed neighbor of v), and hence again Condition 3. The search can be completed in the same way.

Lemma 3. *Let G be a split graph with a split partition (C, I) of its vertices where C is a clique of maximum size. If there are two vertices $u, v \in C$ such that $|N_I(u) \cap N_I(v)| \leq 2$ then $ms(G) \leq \omega(G) = |C|$.*

Theorem 7. *Let G be a split graph with a split partition (C, I) of its vertices where C is a clique of maximum size. Then $ms(G) = \omega(G) + 1 = |C| + 1$ if and only if $|N_I(u) \cap N_I(v)| \geq 3$ for every pair of vertices $u, v \in C$.*

Proof. We know that $|C| + 1$ searchers are always enough to clear any split graph, hence if $ms(G) > |C|$ then $ms(G) = |C| + 1$. Consequently Lemma 3 readily states that if $ms(G) = |C| + 1$ then every pair of vertices in C must have at least three common neighbors in I. For the opposite direction we need to show that $|C|$ searchers cannot clear the graph under this condition. Assume on the contrary that every two vertices in C have at least three common neighbors in I, and that there is a monotone mixed search program that is able to clear the whole graph with $|C|$ searchers. As argued previously, at some point of the search searchers must be placed on $|C| - 1$ vertices of C. Consider the first point in time when this happens, and let u be the vertex of C without a searcher. Without loss of generality, we can use the last searcher to clear all vertices of I that are not neighbors of u, and assume that all edges between pairs of vertices in $(C \setminus \{u\}) \cup (I \setminus N(u))$ are cleared. We need to clear the edges between u and its neighbors, and the edges between $N_I(u)$ and $C \setminus \{u\}$. We know that there can be at most one vertex in $N_I(u)$ that can have a searcher on it, and no other vertices in $N_I(u)$ are cleared, because otherwise they would be subject to recontamination from u. We know that every vertex in $C \setminus \{u\}$ has at least 3 neighbors in $N_I(u)$. If we slide a searcher from a vertex of C to another vertex, recontamination will occur. Assuming without loss of generality that the last searcher is placed on a vertex u_1 of $N_I(u)$, the next step must be to slide the last searcher from u_1 to u. This will clear all edges between u_1 and C, and all edges with both endpoints in C. After this u has still uncleared neighbors, and for each $v \in C \setminus \{u\}$, u and v have at least two common uncleared neighbors v_1 and v_2. If we slide a searcher from v to v_1, v will get recontaminated from v_2, and if we slide a searcher from u to v_1, u will become recontaminated from v_2. Hence we cannot complete the search with only $|C|$ searchers, which gives the desired contradiction.

Corollary 8. *For a split graph G, $ms(G) = \omega(G)$ if and only if neither any of the three conditions of Theorem 6 nor the condition of Theorem 7 are satisfied.*

We are ready to state the main result of this section, which follows almost immediately from the above results.

Theorem 9. *The mixed search number of a split graph can be computed in polynomial time.*

Corollary 10. *The linear-width of a split graph can be computed in polynomial time.*

Finally, like Theorem 5 on interval graphs, we conclude with a structural result on the linear-width of split graphs.

Lemma 4. *Let G be a split graph.*

1. *If $lw(G) \neq ms(G)$ then $lw(G) = \omega(G) - 1$ and $ms(G) = \omega(G)$.*
2. *If G' is the graph obtained by removing all vertices of degree 1 from G, and $\omega(G) \geq 3$, then $lw(G) = lw(G')$.*

Proof. By Proposition 1, the condition $lw(G) \neq ms(G)$ yields one of the following

(a) $lw(G) = \omega(G) - 1$ and $ms(G) = \omega(G)$;
(b) $lw(G) = \omega(G)$ and $ms(G) = \omega(G) + 1$.

Let G' be a graph obtained from G by removing all vertices of degree one.

1. For sake of contradiction, let us assume that the first statement of Lemma does not hold. Then G satisfies (b), i.e. $ms(G) = \omega(G) + 1$. By Theorem 7, $ms(G') = ms(G) = \omega(G) + 1$. Since G' has no vertices of degree 1, we have that $ms(G') = lw(G)$. Therefore,

$$ms(G) \geq lw(G) \geq lw(G') = ms(G') = ms(G) = \omega(G) + 1.$$

This is a contradiction, because $lw(G) \neq ms(G)$. Therefore, G does not satisfy (b), and thus should satisfy (a).

2. Because $\omega(G) \geq 3$, we have that G' has no vertices of degree 1. Consequently, by Proposition 2, $lw(G) = ms(G') = lw(G')$.

References

1. Alpern, S., Gal, S.: The theory of search games and rendezvous. International Series in Operations Research & Management Science, vol. 55. Kluwer Academic Publishers, Boston, MA (2003)
2. Bienstock, D.: Graph searching, path-width, tree-width and related problems (a survey). In: Reliability of computer and communication networks. DIMACS Ser. Discrete Math. Theoret. Comput. Sci., vol. 5, pp. 33–49. Amer. Math. Soc., Providence, RI (1991)
3. Bienstock, D., Seymour, P.: Monotonicity in graph searching. J. Algorithms 12, 239–245 (1991)
4. Booth, K.S., Lueker, G.S.: Testing for the consecutive ones property, interval graphs, and graph planarity using pq-tree algorithms. J. Comp. Syst. Sc. 13, 335–379 (1976)
5. Cohen, J., Fomin, F.V., Heggernes, P., Kratsch, D., Kucherov, G.: Optimal linear arrangement of interval graphs. In: Královič, R., Urzyczyn, P. (eds.) MFCS 2006. LNCS, vol. 4162, pp. 267–279. Springer, Heidelberg (2006)

6. Díaz, J., Petit, J., Serna, M.: A survey of graph layout problems. ACM Computing Surveys 34, 313–356 (2002)
7. Fomin, F.V.: A generalization of the graph bandwidth. Vestnik St. Petersburg Univ. Math. 34(2001), 15–19 (2002)
8. Fomin, F.V., Thilikos, D.M.: A 3-approximation for the pathwidth of halin graphs. J. Discrete Algorithms 4, 499–510 (2006)
9. Gilmore, P.C., Hoffman, A.J.: A characterization of comparability graphs and of interval graphs. Canad. J. Math. 16, 539–548 (1964)
10. Golovach, P.A., Petrov, N.N.: Some generalizations of the problem on the search number of a graph. Vestn. St. Petersbg. Univ., Math. 28(3), 18–22 (1995), Tanslation from Vestn. St-Peterbg. Univ., Ser. I, Mat. Mekh. Astron. 3, 21–27 (1995)
11. Golumbic, M.C.: Algorithmic Graph Theory and Perfect Graphs, 2nd edn. Annals of Discrete Mathematics, vol. 57. Elsevier, Amsterdam (2004)
12. Gustedt, J.: On the pathwidth of chordal graphs. Disc. Appl. Math. 45, 233–248 (1993)
13. Kirousis, L.M., Papadimitriou, C.H.: Interval graphs and searching. Discrete Math. 55, 181–184 (1985)
14. Kleitman, D.J., Vohra, R.V.: Computing the bandwidth of interval graphs. SIAM J. Disc. Math. 3, 373–375 (1990)
15. Kloks, T., Kratochvíl, J., Müller, H.: Computing the branchwidth of interval graphs. Disc. Appl. Math. 145, 266–275 (2005)
16. Kloks, T., Kratsch, D., Borgne, Y.L., Müller, H.: Bandwidth of split and circular permutation graphs. In: Brandes, U., Wagner, D. (eds.) WG 2000. LNCS, vol. 1928, pp. 243–254. Springer, Heidelberg (2000)
17. LaPaugh, A.S.: Recontamination does not help to search a graph. J. Assoc. Comput. Mach. 40, 224–245 (1993)
18. Megiddo, N., Hakimi, S.L., Garey, M.R., Johnson, D.S., Papadimitriou, C.H.: The complexity of searching a graph. J. Assoc. Comput. Mach. 35, 18–44 (1988)
19. Parsons, T.D.: Pursuit-evasion in a graph. In: Theory and applications of graphs. LNCM, vol. 642, pp. 426–441. Springer, Heidelberg (1978)
20. Paul, C., Telle, J.A.: New tools and simpler algorithms for branchwidth. In: Brodal, G.S., Leonardi, S. (eds.) ESA 2005. LNCS, vol. 3669, pp. 379–390. Springer, Heidelberg (2005)
21. Peng, S.-L., Chen, C.-K.: On the interval completion of chordal graphs. Discrete Appl. Math. 154, 1003–1010 (2006)
22. Peng, S.-L., Ko, M.-T., Ho, C.-W., Hsu, T.-s., Tang, C.Y.: Graph searching on some subclasses of chordal graphs. Algorithmica 27, 395–426 (2000)
23. Petrov, N.N.: A problem of pursuit in the absence of information on the pursued. Differentsial'nye Uravneniya 18, 1345–1352, 1468 (1982)
24. Takahashi, A., Ueno, S., Kajitani, Y.: Mixed searching and proper-path-width. Theoret. Comput. Sci. 137, 253–268 (1995)
25. Thilikos, D.M.: Algorithms and obstructions for linear-width and related search parameters. Discrete Appl. Math. 105, 239–271 (2000)
26. Thomas, R.: Tree-decompositions of graphs (lecture notes). School of Mathematics. Georgia Institute of Technology, Atlanta, Georgia 30332, USA (1996)

Lower Bounds for Three Algorithms for the Transversal Hypergraph Generation

Matthias Hagen*

University of Kassel, Research Group Programming Languages / Methodologies,
Wilhelmshöher Allee 73, D–34121 Kassel, Germany
hagen@uni-kassel.de

Abstract. The computation of all minimal transversals of a given hypergraph in output-polynomial time is a long standing open question known as the transversal hypergraph generation. One of the first attempts on this problem—the sequential method [Ber89]—is not output-polynomial as was shown by Takata [Tak02]. Recently, three new algorithms improving the sequential method were published and experimentally shown to perform very well in practice [BMR03, DL05, KS05]. Nevertheless, a theoretical worst-case analysis has been pending. We close this gap by proving lower bounds for all three algorithms. Thereby, we show that none of them is output-polynomial.

1 Introduction

The transversal hypergraph generation is the problem to compute, for a given hypergraph $\mathcal{H} \subseteq 2^V$ with vertex set V, the transversal hypergraph $Tr(\mathcal{H})$ that consists of all minimal subsets of V having a non-empty intersection with each hyperedge of $\mathcal{H}$. This problem has many applications in such different fields like artificial intelligence and logic [EG95, EG02], computational biology [Dam06], database theory [MR92], data mining and machine learning [GKMT97], mobile communication systems [SS98], and distributed computing [GB85].

Due to the importance of the transversal hypergraph generation there have been various approaches to solve it. But since the size of $Tr(\mathcal{H})$ may be exponential in the size of $\mathcal{H}$, we cannot find an algorithm that runs in time polynomial in the size of the input $\mathcal{H}$. Therefore, another notion of fast solvability has to be used. An algorithm is said to be output-polynomial if its running time is bounded polynomially in the size of the input and output [JPY88]. Finding an output-polynomial algorithm for the transversal hypergraph generation is a long standing open problem [Pap97].

One of the earliest approaches is the sequential method [Ber89]. It computes the transversal hypergraph by iteratively combining transversals of specific subhypergraphs of the input in a brute-force manner. The worst-case analysis of the

---

* Partially supported by a Landesgraduiertenstipendium Thüringen and the Deutsche Forschungsgemeinschaft (DFG) through project OPAL (optimal algorithms for hard problems in computational biology), NI-369/2.

A. Brandstädt, D. Kratsch, and H. Müller (Eds.): WG 2007, LNCS 4769, pp. 316–327, 2007.

sequential method took many years until Takata showed that it is not output-polynomial [Tak02]. So far, this is the only proven nontrivial lower bound for any algorithm for the transversal hypergraph generation.

In recent years, several improvements of the sequential method have been published. We focus on the DL-algorithm of Dong and Li [DL05], the BMR-algorithm of Bailey, Manoukian, and Ramamohanarao [BMR03], and the KS-algorithm of Kavvadias and Stavropoulos [KS05]. All three algorithms have been empirically tested on practical instances. Especially the BMR-algorithm performs very well on instances from the data mining field. But while the practical performance of the algorithms has been examined, a theoretical worst-case analysis of their running times has been pending. We close this gap by giving nontrivial lower bounds for all three algorithms. Furthermore, the bounds show that none of the three algorithms is output-polynomial.

The paper is organized as follows. Section 2 contains some basic definitions, a brief recapitulation of the sequential method and its analysis by Takata. In Section 3 we show the DL- and the BMR-algorithm not to be output-polynomial. Section 4 contains the analysis of the KS-algorithm. Some concluding remarks follow in Section 5.

2   Basic Definitions and the Sequential Method

A *hypergraph* $\mathcal{H} = (V, E)$ consists of a set V of vertices and a finite family E of subsets of V—the edges. If there is no danger of ambiguity, we also use the edge set to refer to $\mathcal{H}$. The *size* of $\mathcal{H}$ is the number of occurrences of vertices in the edges. A *transversal* of $\mathcal{H}$ is a set $t \subseteq V$ that has a non-empty intersection with each edge of $\mathcal{H}$. A transversal t is *minimal* if no proper subset of t is a transversal. The set of all minimal transversals of $\mathcal{H}$ forms the *transversal hypergraph* $Tr(\mathcal{H})$. A hypergraph $\mathcal{H}$ is *simple* if it does not contain two hyperedges e, f with $e \subseteq f$. By $\min(\mathcal{H})$ we denote the simple hypergraph consisting of the minimal hyperedges of $\mathcal{H}$ with respect to set inclusion. Since $\min(\mathcal{H})$ can be easily computed in polynomial time and $Tr(\mathcal{H}) = Tr(\min(\mathcal{H}))$ holds for every hypergraph $\mathcal{H}$, we concentrate on the transversal hypergraph generation for simple hypergraphs. But even for simple hypergraphs the size of the transversal hypergraph may be exponential. Hence, there cannot be an algorithm computing the transversal hypergraph in polynomial time. A suitable notion of fast solvability for such kind of problems is that of output-polynomial time [JPY88]. An algorithm is said to be *output-polynomial* if its running time is bounded polynomially in the sum of the sizes of the input and output.

Given simple hypergraphs $\mathcal{H} = \{e_1, e_2, \ldots, e_m\}$ and $\mathcal{H}' = \{e_1', e_2', \ldots, e_{m'}'\}$ there are two different "unions", namely

$$\mathcal{H} \cup \mathcal{H}' = \{e_1, e_2, \ldots, e_m, e_1', e_2', \ldots, e_{m'}'\} \text{ and}$$
$$\mathcal{H} \vee \mathcal{H}' = \{e_i \cup e_j' : i = 1, 2, \ldots, m, \ j = 1, 2, \ldots, m'\}.$$

Proposition 2.1 ([Ber89]). *Let $\mathcal{H}$ and $\mathcal{H}'$ be two simple hypergraphs. Then $Tr(\mathcal{H} \cup \mathcal{H}') = \min(Tr(\mathcal{H}) \vee Tr(\mathcal{H}'))$.*

Algorithm 1. The Sequential Method

1: $Tr(\mathcal{H}_1) \leftarrow \{\{v\} : v \in e_1\}$
2: **for** $i \leftarrow 2, \ldots, m$ **do**
3:     $Tr(\mathcal{H}_i) \leftarrow \min(Tr(\mathcal{H}_{i-1}) \vee \{\{v\} : v \in e_i\})$
4: **end for**
5: **output** $Tr(\mathcal{H}_m)$

The sequential method [Ber89] uses Proposition 2.1 to generate the transversal hypergraph as follows. For a hypergraph $\mathcal{H} = \{e_1, e_2, \ldots, e_m\}$ let $\mathcal{H}_i = \{e_1, e_2, \ldots, e_i\}$, $i = 1, 2, \ldots, m$. We then have

$$Tr(\mathcal{H}_i) = \min(Tr(\mathcal{H}_{i-1}) \vee Tr(\{e_i\})) = \min(Tr(\mathcal{H}_{i-1}) \vee \{\{v\} : v \in e_i\})$$

and $Tr(\mathcal{H}) = Tr(\mathcal{H}_m)$. This implies a straightforward iterative computation process—the sequential method. A pseudocode listing is given in Algorithm 1.

Despite the simplicity of the sequential method it took a couple of years until Takata [Tak02] presented a nontrivial lower bound using the following inductively defined family of hypergraphs.

$$\mathcal{G}_0 = \{\{v_1\}\} \qquad \text{and}$$
$$\mathcal{G}_i = (\mathcal{A} \cup \mathcal{B}) \vee (\mathcal{C} \cup \mathcal{D}), \text{ where } \mathcal{A}, \mathcal{B}, \mathcal{C}, \mathcal{D} \text{ are vertex-disjoint copies of } \mathcal{G}_{i-1}.$$

Takata showed the sequential method not to be output-polynomial based on the following observations.

Lemma 2.2 ([Tak02]). We have $|V_{\mathcal{G}_i}| = 4^i$, $|\mathcal{G}_i| = 2^{2(2^i-1)}$, $|Tr(\mathcal{G}_i)| = 2^{2^i-1}$. For $i \geq 2$ and any $e \in \mathcal{G}_i$, it holds that $|Tr(\mathcal{G}_i \setminus \{e\}) \setminus Tr(\mathcal{G}_i)| \geq 2^{(i-2)2^i+2}$.

From Lemma 2.2 it follows that, independent of the edge ordering, the penultimate (intermediate) result computed by the sequential method on input $\mathcal{G}_i$ is superpolynomial in the size of the input and output (cf. the original paper [Tak02] for more details).

3   The Algorithms of Dong and Li, and Bailey, Manoukian and Ramamohanarao

The border-differential algorithm of Dong and Li [DL05] comes from the data mining field and is intended for mining emerging patterns. The analogy to the generation of hypergraph transversals was already pointed out by Bailey, Manoukian, and Ramamohanarao [BMR03]. A pseudocode listing of the DL-algorithm is given in Algorithm 2.

The algorithm was experimentally evaluated on many practical data mining cases [DL05] whereas a theoretical analysis of the running time was left open. For this purpose the conversion of the algorithm to the hypergraph setting is very fruitful. The only observable difference between the sequential method and

Algorithm 2. The DL-Algorithm

1: $Tr(\mathcal{H}_1) \leftarrow \{\{v\} : v \in e_1\}$
2: **for** $i \leftarrow 2, \ldots, m$ **do**
3:     $Tr_{guaranteed} \leftarrow \{t \in Tr(\mathcal{H}_{i-1}) : t \cap e_i \neq \emptyset\}$
4:     $e_i^{covered} \leftarrow \{v \in e_i : \{v\} \in Tr_{guaranteed}\}$
5:     $Tr(\mathcal{H}_{i-1})' \leftarrow Tr(\mathcal{H}_{i-1}) \setminus Tr_{guaranteed}$
6:     $e_i' \leftarrow e_i \setminus e_i^{covered}$
7:     **for all** $t' \in Tr(\mathcal{H}_{i-1})'$ in increasing cardinality order **do**
8:         **for all** $v \in e_i'$ **do**
9:             **if** $t' \cup \{v\}$ is not superset of any $t \in Tr_{guaranteed}$ **then**
10:                $Tr_{guaranteed} \leftarrow Tr_{guaranteed} \cup \{t' \cup \{v\}\}$
11:             **end if**
12:         **end for**
13:     **end for**
14:     $Tr(\mathcal{H}_i) \leftarrow Tr_{guaranteed}$
15: **end for**
16: **output** $Tr(\mathcal{H}_m)$

---

the DL-algorithm is that the DL-algorithm takes special care on how to perform the minimization of $Tr(\mathcal{H}_{i-1}) \vee \{\{v\} : v \in e_i\}$. But as Takata's analysis showed, the minimization is not the bottleneck of the sequential method. Thus, we can extend Takata's analysis of the sequential method in a straightforward way to the DL-algorithm and get the same lower bound.

Theorem 3.1. *The DL-algorithm is not output-polynomial. Its running time is at least $n^{\Omega(\log \log n)}$, where n denotes the size of the input and output.*

Nevertheless, for hypergraphs with only a few edges of small size the DL-algorithm has been shown experimentally to perform well [DL05]. This property is exploited by the BMR-algorithm [BMR03] (cf. Algorithm 3 for the listing) as it uses the DL-algorithm as a subroutine that computes all minimal transversals for small hypergraphs (line 16 of the listing). The BMR-algorithm on input $\mathcal{H}$ is invoked by the top-level call with the set E of edges of $\mathcal{H}$ and an empty set V_{part}. The global variable Tr is initially empty.

A bottleneck for the running time of the BMR-algorithm is that possibly many of the recursively computed transversals—the set Tr' in the listing—actually are not minimal for the input hypergraph $\mathcal{H}$. We concentrate on this issue and construct a family $\mathcal{G}_i'$ of hypergraphs for which the BMR-algorithm computes too many such non-minimal transversals to run in output-polynomial time. Let $\mathcal{G}'(i) = \{e_i, f_i\}$, where $e_i = \{v_{i^2-i+1}, \ldots, v_{i^2}\}$ and $f_i = \{v_{i^2+1}, \ldots, v_{i^2+i}\}$. We inductively define

$$\mathcal{G}_1' = \{\{v_1\}, \{v_2\}\}, \text{ and}$$
$$\mathcal{G}_i' = (\mathcal{G}_{i-1}' \cup \{\{w_i\}\}) \vee \mathcal{G}'(i), \text{ for } i \geq 2.$$

Note that $\mathcal{G}_{i-1}'$, $\{\{w_i\}\}$ and $\mathcal{G}'(i)$ are pairwise vertex-disjoint simple hypergraphs for $i \geq 2$. To calculate the size of $\mathcal{G}_i'$ and of $Tr(\mathcal{G}_i')$ we have to solve the recurrences

---

Algorithm 3. The BMR-Algorithm

Input: a simple hypergraph, given by the set E of its hyperedges, and a set V_{part} of partitioning vertices

1: $V \leftarrow$ set of all vertices in E
2: order vertices in increasing frequency $\Rightarrow [v_1, \dots, v_k]$
3: **for** $i \leftarrow 1, \dots, k$ **do**
4:      $E_{part} \leftarrow \emptyset$
5:      $V \leftarrow V \setminus \{v_i\}$
6:      **for all** $e \in E$ **do**
7:          **if** $v_i \notin e$ **then**
8:              $E_{part} \leftarrow \min(E_{part} \cup (e \setminus V))$
9:          **end if**
10:     **end for**
11:     $V_{part} \leftarrow V_{part} \cup \{v_i\}$
12:     $a \leftarrow$ average edge cardinality of E_{part} multiplied by $|E_{part}|$
13:     **if** $|E_{part}| \geq 2$ and $a \geq 50$ **then**
14:         recursively call the BMR-algorithm on input E_{part}, V_{part}
15:     **else**
16:         compute $Tr(E_{part})$ via the DL-algorithm
17:         $Tr' \leftarrow Tr(E_{part}) \vee \{V_{part}\}$
18:         $Tr \leftarrow \min(Tr \cup Tr')$
19:     **end if**
20:     $V_{part} \leftarrow V_{part} \setminus \{v_i\}$
21: **end for**
22: **return** Tr

$|\mathcal{G}'_i| = 2 \cdot |\mathcal{G}'_{i-1}| + 2$ and $|Tr(\mathcal{G}'_i)| = |Tr(\mathcal{G}'_{i-1})| + i^2$. With the initial conditions $|\mathcal{G}'_1| = 2$ and $|Tr(\mathcal{G}'_1)| = 1$ we obtain

$$|\mathcal{G}'_i| = 2^{i+1} - 2 \quad \text{and} \quad |Tr(\mathcal{G}'_i)| = \frac{2i^3 + 3i^2 + i}{6}$$

by iteration. As for the number $|V_{\mathcal{G}'_i}|$ of vertices of $\mathcal{G}'_i$, we have $|V_{\mathcal{G}'_i}| = i^2 + 2i - 1$.

The BMR-algorithm iteratively partitions the input hypergraph to obtain smaller hypergraphs where the transversal generation is feasible. The partitioning depends on the vertex frequencies. Hence, we first have to analyze the frequencies of the vertices in $\mathcal{G}'_i$.

Lemma 3.2. *For $i \geq 2$ let $\#_v(i,j)$ and $\#_w(i,j)$ respectively denote the number of occurrences of vertex v_j and w_j in $\mathcal{G}'_i$. Then*

$$\#_w(i,j) = 0 \qquad \text{for } j \geq i,$$
$$\#_w(i,j) > \#_w(i,j+1) \quad \text{for } 2 \leq j < i,$$
$$\#_w(i,2) = \#_v(i,1) = \#_v(i,2),$$
$$\#_v(i,j) = 0 \qquad \text{for } j \geq i^2 + i,$$
$$\#_v(i,j) = \#_v(i,k) \qquad \text{for } l^2 - l + 1 \leq j \leq k \leq l^2 + l, \text{ with } 1 \leq l \leq i,$$
$$\#_v(i,j) < \#_v(i,k) \qquad \text{for } 1 \leq j < l^2 - l + 1 \leq k \leq l^2 + l, \text{ with } 2 \leq l \leq i.$$

All of the above (in)equalities follow directly from the definition of $\mathcal{G}'_i$ or can be easily proven by induction. From Lemma 3.2 it follows that the vertices from $\mathcal{G}'(i)$ are the last vertices in the vertex ordering computed by the BMR-algorithm on input $\mathcal{G}'_i$. This is crucial for the next step of our analysis in which we examine the recursive calls produced by the BMR-algorithm on input $\mathcal{G}'_i$.

Lemma 3.3. *For $i \geq 4$, the BMR-algorithm on input $\mathcal{G}'_i$ recursively calls the BMR-algorithm at least $2i$ times with a modified $\mathcal{G}'_{i-1} \cup \{\{w_i\}\}$ as input. Here, modified means that all edges of $\mathcal{G}'_{i-1} \cup \{\{w_i\}\}$ may additionally include at most half of the vertices of $\mathcal{G}'(i)$.*

Proof. We only examine the last $2i$ vertices processed by the BMR-algorithm. From Lemma 3.2 we know that these are exactly the vertices from $\mathcal{G}'(i)$— contained in the edges e_i and f_i. Let $v'_1, v'_2, \ldots, v'_{2i}$ be any ordering of these vertices. We consider the BMR-algorithm on that ordering.

Let the j-th vertex v'_j, $1 \leq j \leq 2i$, from the above ordering be the current partitioning vertex (line 3 of the BMR-algorithm). After partitioning (lines 5 to 10), the remaining hypergraph has the form

$$(\mathcal{G}'_{i-1} \cup \{\{w_i\}\}) \vee (\{v'_1, \ldots, v'_{j-1}\} \cap x_i),$$

where $x_i = f_i$ if $v'_j \in e_i$, and $x_i = e_i$ if $v'_j \in f_i$. Hence, the remaining hypergraph always is a $\mathcal{G}'_{i-1} \cup \{\{w_i\}\}$ with at most half of the vertices from $\mathcal{G}'(i)$ in every edge.

Altogether, for each of the last $2i$ vertices the minimal transversals of a modified $\mathcal{G}'_{i-1} \cup \{\{w_i\}\}$ have to be computed. Note that a modified $\mathcal{G}'_3$ has 15 edges of average size at least 5.4 and thus $a \geq 81$ (line 12). Hence, for $i \geq 4$ the last $2i$ vertices invoke recursive calls of the BMR-algorithm with a modified $\mathcal{G}'_{i-1} \cup \{\{w_i\}\}$ as input.   □

With Lemma 3.3 at hand we can analyze the number of non-minimal transversals computed by the BMR-algorithm.

Lemma 3.4. *Let $i \geq 4$. For the number $\eta(i)$ of non-minimal transversals computed by the BMR-algorithm on input $\mathcal{G}'_i$ we have $\eta(i) \geq 2^{i-1} \cdot i!$.*

Proof. From Lemma 3.3 it follows that there are $2i$ recursive calls with a modified $\mathcal{G}'_{i-1} \cup \{\{w_i\}\}$ as input. Such a recursive call produces at least all of the minimal and non-minimal transversals of $\mathcal{G}'_{i-1} \cup \{\{w_i\}\}$ augmented by the current partitioning vertex as transversals for $\mathcal{G}'_i$. But since at least the partitioning vertex is dispensable in these transversals, none of them is minimal for $\mathcal{G}'_i$. There are at least $\eta(i-1) + |Tr(\mathcal{G}'_{i-1})|$ such non-minimal transversals per recursive call. Hence, we have to solve the recurrence

$$\eta(i) \geq 2i \cdot (\eta(i-1) + |Tr(\mathcal{G}'_{i-1})|)$$
$$\geq 2i \cdot \eta(i-1).$$

A straightforward computation yields $\eta(3) = 34$. Hence, $\eta(3) \geq 2^2 \cdot 3!$ and we get $\eta(i) \geq 2^{i-1} \cdot i!$ by iteration.   □

Putting all pieces together we are able to give a superpolynomial lower bound on the running time of the BMR-algorithm.

Theorem 3.5. *The BMR-algorithm is not output-polynomial. Its running time is at least $n^{\Omega(\log\log n)}$, where n denotes the size of the input and output.*

Proof. We consider the BMR-algorithm on input $\mathcal{G}'_i$. By $m_i = |V_{\mathcal{G}'_i}| \cdot (|\mathcal{G}'_i| + |Tr(\mathcal{G}'_i)|)$ we denote an upper bound on the size of the input and output. For $i \geq 22$ we have

$$m_i = (i^2 + 2i - 1) \cdot \left(2^{i+1} - 2 + \frac{2i^3 + 3i^2 + i}{6}\right) \leq 2^{3i}.$$

The running time of the BMR-algorithm on input $\mathcal{G}'_i$ is at least $\eta(i)$, the number of non-minimal transversals generated. Thus, to analyze the running time we will show that $\eta(i)$ is superpolynomial in m_i. It suffices to show that

$$2^{i-1} \cdot i! > (2^{3i})^c, \quad \text{for any constant } c.$$

This is equivalent to $i - 1 + \log(i!) > c \cdot 3i$, for any constant c. Using Stirling's formula we have $\log(i!) \geq i \cdot \log i - i$ and thus it suffices to show $i - 1 + i \cdot \log i - i > c \cdot 3i$, for any constant c. This is equivalent to

$$\frac{\log i}{3} - \frac{1}{3i} > c, \quad \text{for any constant } c.$$

Since the last equation obviously holds for sufficiently large i, we have proven that $\eta(i)$ is superpolynomial in m_i, namely $\eta(i) = m_i^{\Omega(\log\log m_i)}$. $\square$

4   The Algorithm of Kavvadias and Stavropoulos

A first drawback of the sequential method or the BMR-algorithm that Kavvadias and Stavropoulos [KS05] observe is the memory requirement. Since newly computed transversals have to be checked for minimality against the previously computed minimal transversals, all the previously generated minimal transversals have to be stored. The KS-algorithm tries to overcome this potentially exponential memory requirement by two techniques. The first is to combine vertices that belong exactly to the same hyperedges.

Definition 4.1 (generalized vertex, [KS05]). *Let $\mathcal{H}$ be a hypergraph with vertex set V. The set $X \subseteq V$ is a generalized vertex of $\mathcal{H}$ if all vertices in X belong to exactly the same hyperedges of $\mathcal{H}$.*

While adding edge e_i, and hence generating the minimal generalized transversals of $\mathcal{H}_i$ out of the minimal generalized transversals of $\mathcal{H}_{i-1}$, the generalized vertices have to be updated according to e_i. Kavvadias and Stavropoulos characterize the following three types of generalized vertices X of a minimal generalized transversal t of $\mathcal{H}_{i-1}$.

- type α: $X \cap e_i = \emptyset$. Hence, X is a generalized vertex of $\mathcal{H}_i$.
- type β: $X \subset e_i$. Hence, X is a generalized vertex of $\mathcal{H}_i$.
- type γ: $X \cap e_i \neq \emptyset$ and $X \not\subset e_i$. Here, X is divided into $X_1 = X \setminus (X \cap e_i)$ and $X_2 = X \cap e_i$. Both X_1 and X_2 are generalized vertices of $\mathcal{H}_i$.

Let $\kappa_\alpha(t, i)$, $\kappa_\beta(t, i)$, and $\kappa_\gamma(t, i)$ denote the number of generalized vertices of type α, β, and γ in t according to e_i. When edge e_i is added, the minimal generalized transversal t of $\mathcal{H}_{i-1}$ has to be split into $2^{\kappa_\gamma(t,i)}$ generalized transversals of $\mathcal{H}_{i-1}$—the so-called *offsprings* of t—since all combinations of newly generalized vertices have to be generated. If $\kappa_\beta(t, i) \neq 0$, all these newly generated offsprings are also minimal transversals of $\mathcal{H}_i$. But if $\kappa_\beta(t, i) = 0$, there is a special offspring t_0 of t that contains all the X_1-parts of the γ-type generalized nodes of t. Hence, $t_0 \cap e_i = \emptyset$ and t_0 has to be augmented by a vertex from e_i to be a transversal of $\mathcal{H}_i$. All the other offsprings of t already are minimal transversals of $\mathcal{H}_i$ since they contain at least one X_2-part of a generalized vertex from t.

The second technique to overcome the potentially exponential memory requirement is based on the observation that the sequential method is a form of breadth-first search through a "tree" of minimal transversals. At the ith-level of the "tree" the nodes are the minimal transversals of the partial hypergraph $\mathcal{H}_i$. The descendants of a minimal transversal t at level i are the minimal transversals of $\mathcal{H}_{i+1}$ that include t. Note that, since a node at level $i + 1$ may have several ancestors at level i, the structure is not really a tree but very tree-like. The bottom level consists of the minimal transversals of $\mathcal{H}$. When cycling through this "tree" breadth-first, one has to wait very long for the first minimal transversal to be output and some nodes are visited several times because they have more than one ancestor. To overcome the long time that may pass till the first minimal transversal is output, the KS-algorithm uses a depth-first strategy. And to really cycle through a tree and not a tree-like structure with some cycles, Kavvadias and Stavropoulos introduce the notion of so-called appropriate vertices.

Definition 4.2 (appropriate vertex, [KS05]). *Let $\mathcal{H} = \{e_1, \ldots, e_m\}$ be a hypergraph with vertex set V and let t be a minimal transversal of the partial hypergraph $\mathcal{H}_i$ of $\mathcal{H}$. A generalized vertex $v \in V \setminus t$ is an appropriate vertex for t if no other vertex in $t \cup \{v\}$ except v can be removed and the remaining set still be a transversal of $\mathcal{H}_i$. The set $appr(t, e)$ contains all appropriate vertices for t in edge e.*

Note that the special offspring t_0 of a minimal generalized transversal t of $\mathcal{H}_{i-1}$ has to be augmented by a vertex from $appr(t, e_i)$ only. All the other vertices from e_i can be skipped. Expanding only with appropriate vertices ensures that no non-minimal transversals are generated and avoids regenerations. Another advantage is that the previously described transversal "tree" structure becomes a real tree (cf. the original paper [KS05] for more details).

All the described techniques—generalized vertices, depth-first strategy, appropriate vertices—together with the main idea of the sequential method—processing the edges one after the other—are used in the KS-algorithm (cf. Algorithm 4 for the listing).

Algorithm 4. The KS-Algorithm

1: express e_1 as a set of one generalized vertex
2: compute the transversal $t = Tr(e_1)$
3: ADDNEXTHYPEREDGE(t, e_2)

4: **procedure** ADDNEXTHYPEREDGE(t, e_i)
5:     update the set of generalized vertices
6:     express t and e_i as sets of generalized vertices of level i
7:     $l \leftarrow 1$
8:     **while** GENERATENEXTTRANSVERSAL(t, l) **do**
9:         **if** e_i is the last hyperedge **then**
10:            **output** t' without using generalized vertices
11:        **else**
12:            ADDNEXTHYPEREDGE(t', e_{i+1})
13:            $l \leftarrow l + 1$
14:        **end if**
15:    **end while**
16: **end procedure**

17: **function** GENERATENEXTTRANSVERSAL(t, l)
18:    **if** $\kappa_\beta(t, i) \neq 0$ **then**
19:        **if** $l \leq 2^{\kappa_\gamma(t,i)}$ **then**
20:            $t' \leftarrow$ the l-th offspring of t
21:            **return true**
22:        **else**
23:            **return false**
24:        **end if**
25:    **else if** $\kappa_\beta(t, i) = 0$ **then**
26:        **if** $l \leq 2^{\kappa_\gamma(t,i)} - 1$ **then**
27:            $t' \leftarrow$ the l-th offspring of t except t_0
28:            **return true**
29:        **else if** $2^{\kappa_\gamma(t,i)} \leq l \leq 2^{\kappa_\gamma(t,i)} - 1 + |appr(t, e_i)|$ **then**
30:            $t' = t_0$ augmented by the $(l - 2^{\kappa_\gamma(t,i)} + 1)$-th vertex of $appr(t, e_i)$
31:            **return true**
32:        **end if**
33:    **else**
34:        **return false**
35:    **end if**
36: **end function**

As for the running time, the KS-algorithm is experimentally shown [KS05] to be competitive to the sequential method, the BMR-algorithm, and an implementation of Algorithm A of Fredman and Khachiyan [BEGK03, FK96]. We will show that the KS-algorithm is not output-polynomial.

First, we note that there are situations in which the KS-algorithm cannot find an appropriate vertex. Consider for example the hypergraph $\mathcal{H} = \{\{v_1, v_5\}, \{v_2, v_5\}, \{v_3, v_6\}, \{v_4, v_6\}, \{v_5, v_6\}\}$. Having processed all but the last edge, there are no generalized vertices left. We concentrate on the path down the transversal tree that corresponds to choosing v_1, v_2, v_3, and v_4. The intermediate transversal is $t = \{v_1, v_2, v_3, v_4\}$. The only edge left is $\{v_5, v_6\}$. But the KS-algorithm cannot find an appropriate vertex in this edge for t. Hence, there are dead ends in the tree, namely leaves that do not contain a minimal transversal of the input $\mathcal{H}$. The next step is to find hypergraphs with too many such dead ends.

Lemma 4.3. *For $i \geq 3$, the number of dead ends the KS-algorithm has to visit for any of Takata's hypergraphs $\mathcal{G}_i$ as input is at least $2^{(i-2)2^i+1}$, independent of the edge ordering.*

Proof. Consider the hypergraph family $\mathcal{G}_i$ of Takata defined in Section 2. From Lemma 2.2 it follows that, whatever ordering of the edges is chosen, there are at least $2^{(i-2)2^i+2}$ nodes in the penultimate level of the transversal tree described above Definition 4.2. The bottom level of the tree obviously contains $|Tr(\mathcal{G}_i)|$ many nodes—one for each minimal transversal. Since $|Tr(\mathcal{G}_i)| = 2^{2^i-1}$ (cf. Lemma 2.2), there is a decrease in the number of nodes from the penultimate level to the bottom level for $i \geq 3$. This decrease can only be caused by dead ends in the penultimate level. Hence, for $i \geq 3$ there are at least $2^{(i-2)2^i+2} - 2^{2^i-1} \geq 2^{(i-2)2^i+1}$ many dead ends in the penultimate level. $\square$

Using Lemma 4.3 we can show that the KS-algorithm is not output-polynomial.

Theorem 4.4. *The KS-algorithm is not output-polynomial. Its running time is at least $n^{\Omega(\log \log n)}$, where n denotes the size of the input and output.*

Proof. We consider the KS-algorithm on input $\mathcal{G}_i$. By $m_i = |V_{\mathcal{G}_i}| \cdot (|\mathcal{G}_i| + |Tr(\mathcal{G}_i)|)$ we denote an upper bound on the size of $\mathcal{G}_i$ and $Tr(\mathcal{G}_i)$. From Lemma 2.2 we have $m_i = 4^i \cdot (2^{2(2^i-1)} + 2^{2^i-1})$, which results in $m_i \leq 2^{2^{i+2}}$.

Let $\hat{\eta}(i)$ denote the number of dead end situations visited by the KS-algorithm on input $\mathcal{G}_i$. The time, the KS-algorithm needs to compute $Tr(\mathcal{G}_i)$, is at least the number of dead end situations visited. Since the KS-algorithm visits the transversal tree depth-first, it visits all the dead end situations in the penultimate level of the tree. With Lemma 4.3 we have $\hat{\eta}(i) \geq 2^{(i-2)2^i+1}$ for $i \geq 3$. Thus, to analyze the running time we will show that $\hat{\eta}(i)$ is superpolynomial in m_i. It suffices to show that $2^{(i-2)2^i} > (2^{2^{i+2}})^c$, for any constant c. This is equivalent to $i - 2 > 4c$, for any constant c. Since this obviously holds for large enough i, we have proven that $\hat{\eta}(i)$ is superpolynomial in m_i, namely $\hat{\eta}(i) = m_i^{\Omega(\log \log m_i)}$. $\square$

5 Concluding Remarks

We have proven superpolynomial lower bounds for the DL-, the BMR-, and the KS-algorithm in terms of the size of the input and output. Thus, like the underlying sequential method, these three algorithms are not output-polynomial.

We are not aware of any other nontrivial lower bounds for algorithms generating the transversal hypergraph although we suppose that none of the known algorithms is output-polynomial. Extending the existing lower bounds to other algorithms seems to be not that straightforward.

Consider for instance the multiplication method suggested by Takata [Tak02]. Very recently Elbassioni proved a quasi-polynomial upper bound on the running time [Elb06]. But giving a superpolynomial lower bound for the multiplication method requires the construction of new hypergraphs. Takata's hypergraphs $\mathcal{G}_i$ and our hypergraphs $\mathcal{G}'_i$ are solved too fast by the multiplication method.

There are also no nontrivial lower bounds known for Algorithms A and B of Fredman and Khachiyan [FK96]. Though Gurvich and Khachiyan [GK97] note that it should be possible to give a superpolynomial lower bound for Algorithm A using hypergraphs very similar to the $\mathcal{G}_i$, the proof is still open. Giving a lower bound for Algorithm B—considered to be the fastest known transversal hypergraph algorithm—seems to be even more involved.

Acknowledgments. I thank Martin Mundhenk and the anonymous referees for their valuable comments and suggestions.

References

[BEGK03] Boros, E., Elbassioni, K.M., Gurvich, V., Khachiyan, L.: Extending the Balas-Yu bounds on the number of maximal independent sets in graphs to hypergraphs and lattices. Mathematical Programming 98(1-3), 355–368 (2003)

[Ber89]   Berge, C.: Hypergraphs. North-Holland Mathematical Library, vol. 45. North-Holland, Amsterdam (1989)

[BMR03]   Bailey, J., Manoukian, T., Ramamohanarao, K.: A fast algorithm for computing hypergraph transversals and its application in mining emerging patterns. In: ICDM 2003. Proceedings of the 3rd IEEE International Conference on Data Mining, Melbourne, Florida, USA, 19-22 December 2003, pp. 485–488. IEEE Computer Society, Los Alamitos (2003)

[Dam06]   Damaschke, P.: Parameterized enumeration, transversals, and imperfect phylogeny reconstruction. Theoretical Computer Science 351(3), 337–350 (2006)

[DL05]   Dong, G., Li, J.: Mining border descriptions of emerging patterns from dataset pairs. Knowledge and Information Systems 8(2), 178–202 (2005)

[EG95]   Eiter, T., Gottlob, G.: Identifying the minimal transversals of a hypergraph and related problems. SIAM Journal on Computing 24(6), 1278–1304 (1995)

[EG02]   Eiter, T., Gottlob, G.: Hypergraph transversal computation and related problems in logic and AI. In: Flesca, S., Greco, S., Leone, N., Ianni, G. (eds.) JELIA 2002. LNCS (LNAI), vol. 2424, Springer, Heidelberg (2002)

[Elb06]   Elbassioni, K.M.: On the complexity of the multiplication method for monotone CNF/DNF dualization. In: Azar, Y., Erlebach, T. (eds.) ESA 2006. LNCS, vol. 4168, Springer, Heidelberg (2006)

[FK96]   Fredman, M.L., Khachiyan, L.: On the complexity of dualization of monotone disjunctive normal forms. Journal of Algorithms 21(3), 618–628 (1996)

[GB85]   Garcia-Molina, H., Barbará, D.: How to assign votes in a distributed sys-
            tem. Journal of the ACM 32(4), 841–860 (1985)
[GK97]   Gurvich, V., Khachiyan, L.: On the frequency of the most frequently oc-
            curring variable in dual monotone DNFs. Discrete Mathematics 169(1-3),
            245–248 (1997)
[GKMT97]   Gunopulos, D., Khardon, R., Mannila, H., Toivonen, H.: Data mining,
            hypergraph transversals, and machine learning. In: Proceedings of the
            Sixteenth ACM SIGACT-SIGMOD-SIGART Symposium on Principles
            of Database Systems, Tucson, Arizona, May 12-14, 1997, pp. 209–216.
            ACM Press, New York (1997)
[JPY88]   Johnson, D.S., Papadimitriou, C.H., Yannakakis, M.: On generating all
            maximal independent sets. Information Processing Letters 27(3), 119–123
            (1988)
[KS05]   Kavvadias, D.J., Stavropoulos, E.C.: An efficient algorithm for the
            transversal hypergraph generation. Journal of Graph Algorithms and Ap-
            plications 9(2), 239–264 (2005)
[MR92]   Mannila, H., Räihä, K.-J.: On the complexity of inferring functional de-
            pendencies. Discrete Applied Mathematics 40(2), 237–243 (1992)
[Pap97]   Papadimitriou, C.H.: NP-completeness: A retrospective. In: Degano,
            P., Gorrieri, R., Marchetti-Spaccamela, A. (eds.) ICALP 1997. LNCS,
            vol. 1256, Springer, Heidelberg (1997)
[SS98]   Sarkar, S., Sivarajan, K.N.: Hypergraph models for cellular mobile com-
            munication systems. IEEE Transactions on Vehicular Technology 47(2),
            460–471 (1998)
[Tak02]   Takata, K.: On the sequential method for listing minimal hitting sets.
            In: Proceedings Workshop on Discrete Mathematics and Data Mining.
            2nd SIAM International Conference on Data Mining, Arlington, Virginia,
            USA, April 11-13, 2002, pp. 109–120 (2002)

The Complexity of Bottleneck Labeled
Graph Problems*

Refael Hassin[1], Jérôme Monnot[2], and Danny Segev[1]

[1] School of Mathematical Sciences, Tel-Aviv University, Tel-Aviv 69978, Israel
{hassin,segevd}@post.tau.ac.il
[2] CNRS LAMSADE, Université Paris-Dauphine, Place du Maréchal de Lattre de
Tassigny, 75775 Paris Cedex 16, France
monnot@lamsade.dauphine.fr

Abstract. We present hardness results, approximation heuristics, and
exact algorithms for bottleneck labeled optimization problems arising in
the context of graph theory. This long-established model partitions the
set of edges into classes, each of which is identified by a unique color.
The generic objective is to construct a subgraph of prescribed structure
(such as that of being an s-t path, a spanning tree, or a perfect matching)
while trying to avoid over-picking or under-picking edges from any given
color.

1 Introduction

Let $G = (V, E)$ be a directed or undirected graph, with a weight function $w :
E \rightarrow \mathbb{R}_+$ and a labeling function $\mathcal{L} : E \rightarrow \{c_1, \ldots, c_q\}$. We interchangeably refer
to the elements of $\mathcal{L}(E)$ as labels or colors. In addition, for $E' \subseteq E$ and $1 \leq i \leq q$,
we use $\mathcal{L}_i(E') = \{e \in E' : \mathcal{L}(e) = c_i\}$ to denote the collection of c_i-colored edges
in E'. With this notation in mind, the c_i-*color weight* of an edge set $E' \subseteq E$ is
defined as $\sum_{e \in \mathcal{L}_i(E')} w(e)$, i.e., the total weight of all c_i-colored edges in E'.

Now let $\mathcal{P}$ be a given graph property defined on subsets of E, such as that of
inducing a spanning tree, an s-t path, an s-t cut, or a perfect matching. The *min-
max weighted labeled* $\mathcal{P}$ problem (henceforth, WL-min-max $\mathcal{P}$) asks to compute
an edge set $E' \subseteq E$ satisfying $\mathcal{P}$ that minimizes $\max_i \sum_{e \in \mathcal{L}_i(E')} w(e)$, the maximum color weight of E'. Similarly, in *max-min weighted labeled* $\mathcal{P}$ (WL-max-min
$\mathcal{P}$), the minimum color weight should be maximized. We refer to both versions
as *weighted labeled bottleneck* $\mathcal{P}$ problems. Furthermore, for ease of presentation,
we denote by UL-min-max $\mathcal{P}$ the unweighted special case of WL-min-max $\mathcal{P}$,
that asks to minimize the maximum color frequency. Analogous notation will
also be used for the corresponding max-min variant.

The complexity of WL-min-max $\mathcal{P}$ has been investigated for several graph
properties by Richey and Punnen [23], Punnen [21,22], and Averbakh and

---

* Due to space limitations, some proofs were omitted from this extended abstract.
We refer the reader to the full version of this paper (currently available online at
http://www.lamsade.dauphine.fr/~monnot), in which all missing details are provided.

A. Brandstädt, D. Kratsch, and H. Müller (Eds.): WG 2007, LNCS 4769, pp. 328–340, 2007.
© Springer-Verlag Berlin Heidelberg 2007

Berman [5], in the context of "optimization problems under categorization". As indicated in [23,5], WL-min-max $\mathcal{P}$ contains both min-max weighted $\mathcal{P}$ and min-sum weighted $\mathcal{P}$ as special cases. One simply has to assign a distinct label to each edge in the former variant, and a single label for all edges in the latter variant. Similar arguments lead to an analogous result, stating that max-sum weighted $\mathcal{P}$ can be formulated in terms of WL-max-min $\mathcal{P}$. Consequently, whenever min-sum weighted (respectively, max-sum weighted) $\mathcal{P}$ is NP-hard, so is WL-min-max (respectively, max-min) $\mathcal{P}$.

1.1 Our Results

We now provide, for each problem considered in this paper, a brief description of our main findings, accompanied by a concise summary of previous work.

Labeled bottleneck s-t path. Previous work:

1. Averbakh and Berman [5] showed that WL-min-max s-t path is weakly NP-hard, even in bicolored graphs. Moreover, they proved that UL-min-max s-t path is NP-hard for an arbitrary number of colors. These results apply to both directed and undirected graphs.
2. In [12] (problem [GT54], p. 203), it was mentioned that the *pair-choice vertex* problem is NP-hard. Here, we are given a directed graph $G = (V, E)$, two specified nodes $\{s, t\} \subseteq V$, and a collection of pairwise-disjoint pairs of arcs. The objective is to determine whether there exists an s-t path traversing at most one arc from any given pair. Since UL-min-max directed s-t path can be viewed as a special case of this problem (pairs correspond to colors), the former cannot be approximated within a factor of $2 - \epsilon$ for any fixed $\epsilon > 0$, unless P=NP.
3. It is not difficult to verify that UL-max-min s-t path generalizes the *longest path* problem, even in monochromatic graphs. Therefore, the results of Karger, Motwani and Ramkumar [15] imply that UL-max-min s-t path cannot be approximated within a factor of $2^{O(\log^{1-\epsilon} n)}$ for any fixed $\epsilon > 0$, unless $\text{NP} \subseteq \text{DTIME}(2^{O(\log^{1/\epsilon} n)})$.

New results:

1. UL-max-min s-t path is not approximable at all, unless P=NP (Theorem 6).
2. For a fixed number of colors, there is a fully polynomial-time approximation scheme for WL-min-max s-t path (Corollary 5).
3. For an arbitrary number of colors, there is an efficient algorithm that constructs a feasible solution to UL-min-max s-t path in undirected graphs using $O(\sqrt{n\text{OPT}})$ edges from any given color (Section 4.2). Here, $n = |V|$ and OPT denotes the objective value of an optimal solution. For directed graphs, the path we construct traverses $O(\sqrt{m\text{OPT}})$ edges from any color, where $m = |E|$ (Section 4.3).

Labeled bottleneck spanning tree. Previous work: Richey and Punnen [23] showed that WL-min-max spanning tree is weakly NP-hard, even in bicolored graphs. We are not aware of previous work regarding the max-min version of this problem.

New results:

1. WL-min-max spanning tree is strongly NP-hard (Theorem 9); it can be approximated within a factor of $O(\log n)$ (Section 5.3).
2. UL-min-max spanning tree can be solved in polynomial time (Theorem 11).
3. UL-max-min spanning tree can be solved in polynomial time (Theorem 10). WL-max-min spanning tree is strongly NP-hard (Theorem 9), and it is also weakly NP-hard in planar bicolored graphs (Theorem 1).
4. For a fixed number of colors, there is a fully polynomial-time approximation scheme for both versions of weighted labeled bottleneck spanning tree (Corollary 5).

Labeled bottleneck perfect matching. Previous work:

1. Richey and Punnen [23] showed that WL-min-max perfect matching is weakly NP-hard, even in bicolored graphs. A stronger result has recently been obtained by Punnen [22], who proved that even the simpler WL-min-max assignment problem is strongly NP-hard.
2. Itai, Rodeh, and Tanimoto [14] proved that the following problem is NP-complete: Given a bipartite graph and a collection of pairs of edges, decide whether there exists a perfect matching that picks at most one edge from any given pair. This problem remains NP-complete for a collection of disjoint pairs [12] (problem [GT59], p. 203). Since UL-min-max perfect matching can be viewed as a special case of this problem, the former cannot be approximated within a factor of $2 - \epsilon$ for any fixed $\epsilon > 0$, unless P=NP.
3. Karzanov [16], and Yi, Murty and Spera [26] proved that, given a complete bipartite graph $K_{n,n}$ with edges colored either red or blue, the problem of finding a perfect matching consisting of exactly r red edges and $n - r$ blue edges is polynomial-time solvable[1]. Therefore, UL-min-max and UL-max-min perfect matching in complete bipartite bicolored graphs can be solved to optimality in polynomial time.
4. To our knowledge, WL-max-min perfect matching has not been studied in the literature.

New results:

1. WL-max-min perfect matching is weakly NP-hard in bicolored planar graphs (Theorem 1). UL-max-min perfect matching is not approximable at all in general graphs, unless P=NP.
2. There is an approximation-preserving reduction from UL-min-max directed s-t path to UL-min-max perfect matching.
3. For a fixed number of colors, there is a fully polynomial-time approximation scheme for both versions of weighted labeled bottleneck perfect matching (Corollary 5).

---

[1] On the other hand, the complexity of this problem in general bipartite graphs is still open.

Due to space limitations, these results appear in the full version of this paper.

Labeled bottleneck s-t cut. Previous work: To our knowledge, both versions of this problem have not been studied yet.

New results:

1. UL-min-max s-t cut is NP-hard in bicolored graphs (Theorem 3). When the underlying graph is planar, UL-min-max s-t cut cannot be approximated within a factor of $2 - \epsilon$ for any fixed $\epsilon > 0$, unless P=NP, and the weighted version of this problem is weakly NP-hard when the graph is bicolored as well (Theorem 1).
2. WL-max-min s-t cut is weakly NP-hard in planar bicolored graphs (Theorem 1). For an arbitrary number of colors, this problem is not approximable at all in planar multigraphs, unless P=NP.

Due to space limitations, these results appear in the full version of this paper.

1.2   Related Work

In this section, we provide a brief survey of several frameworks to which our contributions are related. Since some of the settings under consideration have received a great deal of attention in recent years, it is beyond the scope of this writing to present an exhaustive overview. We refer the reader to the undermentioned papers and to the references therein for a more comprehensive review of the literature.

Multiobjective combinatorial optimization [11,24,25]. The basic ingredients of a multiobjective optimization problem are typically: A set of instances $\mathcal{I}$; a set of feasible solutions $\mathcal{F}(x)$ associated with every instance $x \in \mathcal{I}$; and a collection of cost functions $w_1(x, y), \ldots, w_k(x, y)$ associated with every instance $x \in \mathcal{I}$ and feasible solution $y \in \mathcal{F}(x)$. Given an instance $x \in \mathcal{I}$, the goal is to solve $\min_{y \in \mathcal{F}(x)} \{w_1(x, y), \ldots, w_k(x, y)\}$, where the exact meaning of "min" depends on the particular setting in question. For example, it may stand for Pareto optimality (see Section 3), for aiming to minimize the worst cost function, or for lexicographically minimizing the vector of cost functions. It is not difficult to verify that WL-min-max $\mathcal{P}$ is actually a multiobjective optimization problem in disguise: The set of feasible solutions consists of all edge sets that satisfy $\mathcal{P}$; for every color c_i there is a corresponding cost function w_i which is exactly the c_i-color weight; and the goal is to minimize the maximum cost function. Minor adjustments allow us to treat WL-max-min $\mathcal{P}$ in a similar way.

Robust discrete optimization [17,6]. Very informally, robust optimization deals with decision making in environments of considerable data uncertainty, trying to come up with solutions that hedge against the worst contingency that may arise. Several alternative approaches for coping with uncertainty have been explored and exploited; however, the *scenario-based* framework of Kouvelis and Yu [17] seems most relevant to our paper. In this context, future developments are described by a finite number of scenarios, each of which corresponds to a possible realization of the unknown model parameters. The objective is to optimize against the worst possible scenario by using a min-max objective. Once again, we note that WL-min-max $\mathcal{P}$ can be easily cast as a scenario-based robust optimization problem: For every color c_i there is an analogous scenario s_i, in which the weight $w_{s_i}(e)$ of an edge $e \in E$ is set to $w(e)$ if its color is c_i, and to 0 otherwise. In addition, the cost of an edge set $E' \subseteq E$ in scenario s_i is given by $\sum_{e \in E'} w_{s_i}(e)$, which is exactly the c_i-color weight of this set.

The min-sum-max setting. A complementary line of work [23,5,22] on edge-colored graphs attempts to minimize the sum of the maximal edge weight picked from every given color. In particular, when all edges are associated with unit weights, a problem of this nature reduces to that of constructing subgraphs satisfying a required property while minimizing the number of colors used. Some properties that have recently been studied in this context include spanning trees [10,7,9,13], s-t paths [8,13], and perfect matchings [19].

2   Fixed Number of Colors: Hardness Results

2.1   Weak NP-Hardness in Bicolored Graphs

In what follows, we prove that several weighted labeled bottleneck problems are NP-hard, even in planar bicolored graphs. As noted in Section 1.1, WL-min-max $\mathcal{P}$ is known to be NP-hard in bicolored graphs for $\mathcal{P} \in \{$spanning tree, s-t path, perfect matching$\}$ [23,5].

Theorem 1. *WL-min-max $\mathcal{P}$ and WL-max-min $\mathcal{P}$ are NP-hard, even in planar bicolored graphs, for $\mathcal{P} \in \{s$-t path, s-t cut, perfect matching, spanning tree$\}$.*

2.2   Strong NP-Hardness for s-t Cuts

Aissi, Bazgan and Vanderpooten [4] proved that min-max robust $\mathcal{P}$ with a fixed number of scenarios admits pseudo-polynomial algorithms for s-t paths and spanning trees in general graphs and for perfect matchings in planar graphs. Since WL-min-max $\mathcal{P}$ can be viewed as a special case of these settings (see Section 1.2), it follows that the corresponding min-max labeled problems have pseudo-polynomial algorithms for a fixed number of colors.

In contrast, we proceed by proving that WL-min-max s-t cut is strongly NP-hard in bicolored graphs. A similar result was established for bi-criteria s-t cut [20, Thm. 6], and more recently for min-max robust s-t cut with two scenarios [3, Cor. 1]. Unfortunately, in their reductions the resulting instances do not correspond to WL-min-max s-t cut instances, and it appears as if we cannot conclude the desired result for WL-min-max s-t cut in an obvious way. However, we can slightly modify the construction of Papadimitriou and Yannakakis [20].

Theorem 2. *WL-min-max s-t cut is strongly NP-hard in bicolored graphs.*

Proof. We propose a reduction from the *bisection width* problem. Given a connected graph $G = (V, E)$ on $2n$ vertices, a bisection is a cut (V_1, V_2) in G with $|V_1| = |V_2| = n$. The decision version of bisection width asks to determine, for a given integer k, whether there exists a bisection with at most k edges. This problem is known to be NP-complete [12] (problem [ND17], p. 210).

Given an instance of bisection width, as described above, we construct an instance $I = (G', w, \mathcal{L})$ of WL-min-max s-t cut, with $G' = (V', E')$ and $\mathcal{L}(E') = \{c_1, c_2\}$, as follows:

– G' has two additional vertices, s and t, each of which is connected to every vertex of G.
– $\mathcal{L}(s, v) = c_1$ for every $v \in V$; all other edges have color c_2.
– $w(s, v) = k(n + 1)$ and $w(t, v) = kn$ for every $v \in V$; $w(e) = n$ for every original edge $e \in E$.

We now argue that G has a bisection of size at most k if and only if I has an s-t cut whose min-max value is at most $kn(n + 1)$. If (V_1, V_2) is a bisection with at most k edges, then $(\{s\} \cup V_1, \{t\} \cup V_2)$ is an s-t cut in G' that picks c_1-colored edges of total weight $\sum_{e \in (\{s\}, V_2)} w(e) = kn(n + 1)$ and c_2-colored edges of total weight $\sum_{e \in (\{t\}, V_1)} w(e) + \sum_{e \in (V_1, V_2)} w(e) = kn^2 + kn = kn(n+1)$. Conversely, let $(\{s\} \cup V_1, \{t\} \cup V_2)$ be an s-t cut in G' with min-max value of at most $kn(n + 1)$. Since each c_1-colored edge in this cut has a weight of $k(n + 1)$, it follows that $|V_2| \leq n$. In addition, the c_2-colored edges in this cut have a total weight of $n|E''| + kn|V_1|$, where $E'' = (V_1, V_2)$, and we conclude that $|V_1| \leq n+1$. Now, if $|V_1| = n+1$ the inequality $n|E''| + kn|V_1| \leq kn(n+1)$ implies $E'' = \emptyset$, so G is clearly disconnected (contradicting our initial assumption); thus $|V_1| \leq n$. Finally, since $|V_1| \leq n$ and $|V_2| \leq n$, we have $|V_1| = |V_2| = n$, and therefore (V_1, V_2) is a bisection with at most k edges.   □

Theorem 3. *UL-min-max s-t cut is NP-hard in bicolored graphs.*

Proof. To prove the theorem, we show that a ρ-approximation for UL-min-max s-t cut can be converted in polynomial time into a ρ-approximation for WL-min-max s-t cut when the edge weights are integers upper bounded by a polynomial in n. The theorem follows from the combination of this result and Theorem 2.

Let $I = (G, w, \mathcal{L})$ be an instance of WL-min-max s-t cut, where $G = (V, E)$ has n vertices and $\max_{e \in E} w(e) = O(n^{O(1)})$. We replace each edge $e = (u, v) \in E$ by a collection $H(e)$ of $w(e)$ edge-disjoint paths of length two (connecting u and v), each edge of which is colored by $\mathcal{L}(e)$. The vertices u and v will be called *extreme vertices* of $H(e)$, whereas other vertices of $H(e)$ will be called *inner vertices*. We refer to the resulting UL-min-max s-t cut instance as $I' = (G', \mathcal{L}')$.

Consider an s-t cut (S', T') in G', with $s \in S'$ and $t \in T'$. We iteratively apply the following procedure for each original edge $e \in E$: If the extreme vertices of $H(e)$ appear in the same set of the partition, assign all inner vertices of $H(e)$ to that set. These changes can only decrease the total weight of $\mathcal{L}(e)$-colored edges in the current s-t cut, and therefore also its min-max value. From the resulting s-t cut (S', T'), we can find an s-t cut in G of identical min-max value by considering $(S' \cap V, T' \cap V)$, the restriction of this cut to G.   □

3   Fixed Number of Colors: An FPTAS

In what follows, we present a fully polynomial-time approximation scheme for weighted labeled bottleneck s-t path, spanning tree, and perfect matching, for a fixed number of colors.

Approximate Pareto curves. Let $\mathcal{P}$ be a property described in Section 1, and consider the multiobjective version of $\mathcal{P}$ (henceforth, Multi$_k\mathcal{P}$). An instance I of this problem consists of a graph $G = (V, E)$, and a weight vector $w(e) = (w_1(e), \dots, w_k(e))$ for each edge $e \in E$. An edge set $E' \subseteq E$ forms a feasible solution to Multi$_k\mathcal{P}$ if it satisfies $\mathcal{P}$, and the objective value of E' is given by the vector $(\sum_{e \in E'} w_1(e), \dots, \sum_{e \in E'} w_k(e))$. In a minimization problem, we say that a solution E' is *dominated* by E'' if $\sum_{e \in E''} w_i(e) \leq \sum_{e \in E'} w_i(e)$ for every $1 \leq i \leq k$, and the inequality is strict for at least one index; the inequalities are reversed for a maximization problem. The goal is to compute the *Pareto curve* $\mathcal{C}(I)$, which is the set of all undominated solutions to I. Finally, an ϵ-*approximate Pareto curve* for the minimization (respectively, maximization) version of Multi$_k\mathcal{P}$ is a set $\mathcal{C}_\epsilon(I)$ of solutions such that

1. $|\mathcal{C}_\epsilon(I)|$ is polynomially bounded in terms of the input size and $1/\epsilon$.
2. For every $E^* \in \mathcal{C}(I)$, there exists $E' \in \mathcal{C}_\epsilon(I)$ with $\sum_{e \in E'} w_i(e) \leq (1 + \epsilon) \sum_{e \in E^*} w_i(e)$ for every $1 \leq i \leq k$ (respectively, $\sum_{e \in E'} w_i(e) \geq (1 - \epsilon) \sum_{e \in E^*} w_i(e)$).

When k is fixed, Papadimitriou and Yannakakis [20, Cor. 5] proposed an FP-TAS for constructing ϵ-approximate Pareto curves of multiobjective s-t walk, spanning tree, and perfect matching.

The approximation scheme. We now relate the approximability of several weighted labeled bottleneck problems to that of their multiobjective counterparts. This approach has already been suggested in the context of robust optimization [17,2], implying that results similar to those described in the next theorem can be immediately derived for the min-max variants.

Theorem 4. *For a fixed number of colors, the efficient construction of an ϵ-approximate Pareto curve for the maximization version of Multi$_k\mathcal{P}$ implies a $(1 - \epsilon)$-approximation to WL-max-min $\mathcal{P}$. A similar result for the minimization version leads to a $(1 + \epsilon)$-approximation to WL-min-max $\mathcal{P}$.*

By combining Theorem 4 and the results of Papadimitriou and Yannakakis [20] mentioned earlier, Corollary 5 follows. However, an important remark is in place. Even though the algorithm in [20] constructs an ϵ-approximate Pareto curve of multiobjective s-t walk, note that any such walk can be converted (by eliminating cycles) to an s-t path of no greater min-max objective value. An analogous claim regarding the max-min version is incorrect.

Corollary 5. *For a fixed number of colors, weighted labeled bottleneck spanning tree and perfect matching admit a fully polynomial-time approximation scheme. A similar result also holds for WL-min-max s-t path.*

4 Arbitrary Number of Colors: *s-t* Paths

For a fixed number of colors, UL-min-max *s-t* path is polynomial time solvable. This claim follows from the observation that we can decide whether there exists a walk connecting *s* and *t* whose objective value is exactly $p \in \{1, \ldots, n-1\}$ by means of dynamic programming. In contrast, we proceed by showing that both versions of the problem under consideration become NP-hard for an arbitrary number of colors. We complement these results by devising efficient approximation algorithms.

4.1 Hardness Results

We now derive new inapproximability bounds for both versions of labeled bottleneck *s-t* path, in undirected as well as directed graphs. To our knowledge, these results do not follow from existing work.

Theorem 6. *UL-min-max s-t path is not $(2 - \epsilon)$-approximable for any fixed $\epsilon > 0$, and UL-max-min s-t path is not approximable at all, unless P=NP. Similar results hold for directed graphs.*

4.2 UL-Min-Max *s-t* Path: Approximating the Undirected Case

In what follows, we show how to efficiently construct an undirected *s-t* path using $O(\sqrt{n\text{OPT}})$ edges from any given color, where $n = |V|$ and OPT denotes the cost of an optimal solution. An essential building block of our algorithm is a constant-factor approximation for *multi-budget maximum coverage*. An instance of this problem consists of a ground set U and a collection of subsets $\mathcal{S} \subseteq 2^U$, which is partitioned into $\mathcal{S}_1, \ldots, \mathcal{S}_r$. Given an integral budget b_t for each part $\mathcal{S}_t$, the objective is to find a subcollection $\mathcal{S}' \subseteq \mathcal{S}$ such that $\mathcal{S}'$ picks at most b_t sets from each $\mathcal{S}_t$ and such that the number of elements covered by $\mathcal{S}'$ is maximized. For these particular settings, a performance guarantee of $1 - 1/e$ can be achieved by adopting the maximum coverage heuristic of Ageev and Sviridenko [1, Rem. 2].

The algorithm. For simplicity of presentation, it would be convenient to assume that OPT is known in advance. Clearly, this assumption can be enforced by testing $1, \ldots, n-1$ as candidate values, and returning the best solution found. We also make use of $\Delta = \Delta(n, \text{OPT})$ as a parameter whose value will be determined later.

1. $F \leftarrow \emptyset$, $H \leftarrow G$.
2. While $\text{dist}_H(s, t) > \Delta$
   (a) Create a multi-budget maximum coverage instance by: The ground set is $V(H)$; for each edge $e \in E(H)$ there is a corresponding subset V_e, consisting of the endpoints of e; these subsets are partitioned into $\{\mathcal{S}_1, \ldots, \mathcal{S}_q\}$, where $\mathcal{S}_i = \{V_e : \mathcal{L}(e) = c_i\}$; each $\mathcal{S}_i$ has a budget of OPT.

(b) Approximate the instance defined above, and let F^+ be the collection of edges $e \in E(H)$ for which V_e is picked by the resulting solution.

(c) $F \leftarrow F \cup F^+$, $H \leftarrow$ the contraction of F^+ in H.

3. Let P be a shortest s-t path in H. Return $F \cup P$.

Theorem 7. *By setting* $\Delta = \sqrt{n\text{OPT}}$, *the subgraph induced by* $F \cup P$ *picks* $O(\sqrt{n\text{OPT}})$ *edges from any given color.*

Proof. We begin by showing that, for any value of Δ, step 2 terminates within no more than $4n/\Delta$ iterations. For this purpose, it is sufficient to prove that the number of vertices in H decreases by at least $\Delta/4$ whenever an edge set is contracted. Let $E^* \subseteq E$ be an optimal solution, with $\max_i |\mathcal{L}_i(E^*)| = \text{OPT}$, and consider a single iteration. Since the edges $E^* \cap E(H)$ form a subgraph of H containing an s-t path, it follows that $\{V_e : e \in E^* \cap E(H)\}$ is a feasible solution to the multi-budget maximum coverage instance defined in step 2a. Moreover, as the s-t distance in H is at least Δ, the latter solution satisfies $|\bigcup_{e \in E^* \cap E(H)} V_e| \geq \Delta$. Consequently, for the current F^+ we must have $|\bigcup_{e \in F^+} V_e| \geq (1 - 1/e)\Delta$, implying that the contraction of F^+ decreases the number of vertices by at least $(1 - 1/e)\Delta/2 > \Delta/4$.

Now, starting with an empty set of edges, in each iteration of step 2 we augment F with an edge set F^+ that contains at most OPT edges from each color. Therefore, by setting $\Delta = \sqrt{n\text{OPT}}$, the maximum number of edges we pick from any given color is at most $(4n/\Delta)\text{OPT}+|P| \leq (4n/\Delta)\text{OPT}+\Delta = 5\sqrt{n\text{OPT}}$. □

4.3   UL-Min-Max s-t Path: Approximating the Directed Case

In the following, we demonstrate that ideas similar to those presented in Section 4.2 can be employed to construct a directed s-t path using $O(\sqrt{m\text{OPT}})$ arcs from any given color. Here, $m = |E|$ and OPT denotes the cost of an optimal solution.

The algorithm. Once again, we assume that OPT is known in advance, and let $\Delta = \Delta(m, \text{OPT})$ be a parameter whose value will be determined later.

1. $F \leftarrow \emptyset$, $\chi_{E \setminus F} \leftarrow$ characteristic function of $E \setminus F$.

2. While $\text{dist}_{\chi_{E \setminus F}}(s, t) > \Delta$

(a) Create a multi-budget maximum coverage instance by: The ground set is V; for each arc $e = (u, v) \in E \setminus F$ there is a corresponding singleton $V_e = \{v\}$; these subsets are partitioned into $\{S_1, \ldots, S_q\}$, where $S_i = \{V_e : \mathcal{L}(e) = c_i\}$; each S_i has a budget of OPT.

(b) Approximate the instance defined above, and let F^+ be the collection of arcs $e \in E \setminus F$ for which V_e is picked by the resulting solution.

(c) $F \leftarrow F \cup F^+$.

3. Let P be a shortest s-t path (with respect to $\chi_{E \setminus F}$). Return P.

Theorem 8. *By setting* $\Delta = \sqrt{m\text{OPT}}$, *the path* P *traverses* $O(\sqrt{m\text{OPT}})$ *arcs from any given color.*

Proof. We first demonstrate that step 2 consists of at most $2m/\Delta$ iterations, by showing that we always have $|F^+| \geq \Delta/2$. Let P^* be an optimal solution, with $\max_i |\mathcal{L}_i(P^*)| = \text{OPT}$. In each iteration, $\{V_e : e \in P^* \setminus F\}$ is a feasible solution to the multi-budget maximum coverage instance defined in step 2a. Moreover, as $\text{dist}_{\chi_{E \setminus F}}(s, t) > \Delta$, the latter solution satisfies $|\bigcup_{e \in P^* \setminus F} V_e| \geq \Delta$. Consequently, we must have $|F^+| \geq |\bigcup_{e \in F^+} V_e| \geq (1 - 1/e)\Delta > \Delta/2$.

Now, starting with an empty set of arcs, in each iteration of step 2 we augment F with an arc set F^+ that contains at most OPT arcs from each color. Therefore, by setting $\Delta = \sqrt{m\text{OPT}}$, the maximum number of edges P traverses from any given color is at most $|F| + \Delta \leq (2m/\Delta)\text{OPT} + \Delta \leq 3\sqrt{m\text{OPT}}$.   □

5   Arbitrary Number of Colors: Spanning Trees

In Corollary 5 we have shown that, for a fixed number of colors, both versions of weighted labeled spanning tree admit an FPTAS. In this section, we provide hardness results, exact algorithms, and approximation algorithms for the general case of an arbitrary number of colors.

5.1   Hardness Results

As indicated in Section 1.1, WL-min-max spanning tree is known to be weakly NP-hard [23]. Here, we show that both weighted labeled bottleneck spanning tree problems are in fact strongly NP-hard.

Theorem 9. *Both weighted labeled bottleneck spanning tree problems are strongly NP-hard.*

5.2   Exact Algorithms

Broersma and Li [7] devised a polynomial-time algorithm based on matroid intersection for computing a spanning tree using a maximum number of colors. Here, we prove that both unweighted labeled bottleneck spanning tree problems can also be solved in polynomial time by utilizing matroid intersection. It is interesting to observe that this result is in contrast to the weighted case, which was shown to be strongly NP-hard in Theorem 9.

Theorem 10. *UL-max-min spanning tree can be solved to optimality in polynomial time.*

Proof. Given an instance $(G, \mathcal{L})$ of UL-max-min spanning tree, with $G = (V, E)$, we may assume without loss of generality that OPT is known in advance, since we can test $0, \ldots, n - 1$ as candidate values for this parameter, and return the best solution found. Now, since the optimal tree picks at least OPT edges from every color in $\mathcal{L}(E) = \{c_1, \ldots, c_q\}$, it follows that there exists a forest picking exactly OPT edges from any given color. Moreover, such a forest can be efficiently constructed by computing a maximum cardinality intersection[2] of the matroids M_1 and M_2, where:

- $M_1 = (E, \mathcal{I}_1)$ is the graphic matroid, that is, $\mathcal{I}_1 = \{F \subseteq E : F \text{ is a forest}\}$.
- $M_2 = (E, \mathcal{I}_2)$ is a partition matroid, with $\mathcal{I}_2 = \{F \subseteq E : |\mathcal{L}_i(F)| \leq \text{OPT for every } 1 \leq i \leq q\}$.

We complete the resulting forest into a spanning tree in an arbitrary way, noting that this augmentation leaves the objective value unchanged.   □

Theorem 11. *UL-min-max spanning tree can be solved to optimality in polynomial time.*

Proof. The algorithm for this version is nearly identical to the one given for UL-max-min spanning tree; however, an important remark is in place. After we "guess" OPT and compute a maximum cardinality intersection $F \subseteq E$ of M_1 and M_2, there is no need to complete the subgraph induced by F into a spanning tree, implying that its objective value remains unchanged. This claim follows from observing that $|F| = |V| - 1$, since the edge set of the optimal spanning tree forms a feasible solution to the matroid intersection problem we solve.   □

5.3   WL-Min-Max Spanning Tree: A Logarithmic Approximation

In what follows, we show that a matroid intersection algorithm is not only a useful tool for solving the unweighted version to optimality; rather, it can also be applied to approximate the weighted min-max version.

The algorithm. For ease of exposition, we assume without loss of generality that an estimator of the optimum $W \in [\text{OPT}, 2 \cdot \text{OPT}]$ is known in advance. Otherwise, for every $0 \leq k \leq \lceil \log(n w_{max}/w_{min}) \rceil$, we can test $2^k w_{min}$ as a candidate value and return the best solution found, where w_{min} and w_{max} denote the minimum and maximum non-zero edge weights, respectively.

1. Delete all edges of weight greater than W, and define a partition of the undeleted edges as follows:
   (a) For every $1 \leq i \leq q$ and $0 \leq k \leq \lfloor \log n \rfloor$, let $\mathcal{E}_{i,k}$ be the set of edges e with $\mathcal{L}(e) = c_i$ and $w(e) \in (W/2^{k+1}, W/2^k]$.
   (b) In addition, let $\mathcal{E}_{\text{free}}$ be the set of remaining edges (of weight at most W/n).
2. By applying a matroid intersection algorithm, find a spanning tree T that picks at most 2^{k+1} edges from each $\mathcal{E}_{i,k}$ and any number of edges from $\mathcal{E}_{\text{free}}$. Return T.

Note that the suggested algorithm is well-defined. To establish this claim, it is sufficient to show that a spanning tree satisfying the constraints of step 2 indeed exists. It is easy to verify that all edges of the optimal tree T^* survive step 1 and that $|T^* \cap \mathcal{E}_{i,k}| \leq 2^{k+1}$, or otherwise there is a color c_i from which T^* picks edges of total weight strictly greater than $W \geq \text{OPT}$.

---

[2] See, for example, [18, Chap. 8].

Theorem 12. *The edges picked by T from any given color have an overall weight of $O(\log n) \cdot \mathrm{OPT}$.*

Proof. Consider some color c_i. Then,

$$
\sum_{e \in \mathcal{L}_i(T)} w(e) = \sum_{k=0}^{\lfloor \log n \rfloor} \sum_{e \in T \cap \mathcal{E}_{i,k}} w(e) + \sum_{e \in \mathcal{L}_i(T \cap \mathcal{E}_{\text{free}})} w(e)
$$

$$
\leq \sum_{k=0}^{\lfloor \log n \rfloor} \left(|T \cap \mathcal{E}_{i,k}| \cdot \max_{e \in T \cap \mathcal{E}_{i,k}} w(e) \right) + |T \cap \mathcal{E}_{\text{free}}| \cdot \max_{e \in T \cap \mathcal{E}_{\text{free}}} w(e)
$$

$$
\leq \sum_{k=0}^{\lfloor \log n \rfloor} 2^{k+1} \frac{W}{2^k} + (n-1) \frac{W}{n} \leq (2\lfloor \log n \rfloor + 3)W \leq (4 \lfloor \log n \rfloor + 6)\mathrm{OPT} .
$$

The second inequality holds since $|T \cap \mathcal{E}_{i,k}| \leq 2^{k+1}$ for every $0 \leq k \leq \lfloor \log n \rfloor$, and since $|T \cap \mathcal{E}_{\text{free}}| \leq n - 1$. The last inequality follows from the assumption $W \leq 2 \cdot \mathrm{OPT}$. $\qquad\square$

References

1. Ageev, A.A., Sviridenko, M.: Pipage rounding: A new method of constructing algorithms with proven performance guarantee. Journal of Combinatorial Optimization 8(3), 307–328 (2004)
2. Aissi, H., Bazgan, C., Vanderpooten, D.: Approximation complexity of min-max (regret) versions of shortest path, spanning tree, and knapsack. In: Brodal, G.S., Leonardi, S. (eds.) ESA 2005. LNCS, vol. 3669, pp. 862–873. Springer, Heidelberg (2005)
3. Aissi, H., Bazgan, C., Vanderpooten, D.: Complexity of the min-max (regret) versions of cut problems. In: Deng, X., Du, D.-Z. (eds.) ISAAC 2005. LNCS, vol. 3827, pp. 789–798. Springer, Heidelberg (2005)
4. Aissi, H., Bazgan, C., Vanderpooten, D.: Pseudo-polynomial algorithms for min-max and min-max regret problems. In: 5th ISORA, pp. 171–178 (2005)
5. Averbakh, I., Berman, O.: Categorized bottleneck-minisum path problems on networks. Operations Research Letters 16(5), 291–297 (1994)
6. Bertsimas, D., Sim, M.: Robust discrete optimization and network flows. Mathematical Programming Series B 98(1), 49–71 (2003)
7. Broersma, H., Li, X.: Spanning trees with many or few colors in edge-colored graphs. Discussiones Mathematicae Graph Theory 17(2), 259–269 (1997)
8. Broersma, H., Li, X., Woeginger, G., Zhang, S.: Paths and cycles in colored graphs. Australasian Journal on Combinatorics 31, 299–311 (2005)
9. Brüggemann, T., Monnot, J., Woeginger, G.: Local search for the minimum label spanning tree problem with bounded color classes. Operations Research Letters 31(3), 195–201 (2003)
10. Chang, R.-S., Leu, S.-J.: The minimum labeling spanning trees. Information Processing Letters 63(5), 277–282 (1997)
11. Ehrgott, M., Gandibleux, X. (eds.): Multiple Criteria Optimization: State of the Art Annotated Bibliographic Survey. International Series in Operations Research and Management Science, vol. 52. Kluwer Academic Publishers, Dordrecht (2002)

12. Garey, M.R., Johnson, D.S.: Computers and Intractability: A Guide to the Theory of NP-Completeness. W. H. Freeman and Company, New York (1979)
13. Hassin, R., Monnot, J., Segev, D.: Approximation algorithms and hardness results for labeled connectivity problems. In: Královič, R., Urzyczyn, P. (eds.) MFCS 2006. LNCS, vol. 4162, pp. 480–491. Springer, Heidelberg (2006)
14. Itai, A., Rodeh, M., Tanimoto, S.L.: Some matching problems for bipartite graphs. Journal of the ACM 25(4), 517–525 (1978)
15. Karger, D.R., Motwani, R., Ramkumar, G.D.S.: On approximating the longest path in a graph. Algorithmica 18(1), 82–98 (1997)
16. Karzanov, A.V.: Maximum matchings of given weight in complete and complete bipartite graphs. Kibernetika 1, 7–11 (1987), English translation in CYBNAW, 23, 8–13
17. Kouvelis, P., Yu, G.: Robust Discrete Optimization and its Applications. Kluwer Academic Publishers, Dordrecht (1997)
18. Lawler, E.L.: Combinatorial Optimization: Networks and Matroids. Holt, Rinehart and Winston, New York (1976)
19. Monnot, J.: The labeled perfect matching in bipartite graphs. Information Processing Letters 96(3), 81–88 (2005)
20. Papadimitriou, C.H., Yannakakis, M.: On the approximability of trade-offs and optimal access of web sources. In: 41st FOCS, pp. 86–92 (2000)
21. Punnen, A.P.: Traveling salesman problem under categorization. Operations Research Letters 12(2), 89–95 (1992)
22. Punnen, A.P.: On bottleneck assignment problems under categorization. Computers and Operations Research 31(1), 151–154 (2004)
23. Richey, M.B., Punnen, A.P.: Minimum perfect bipartite matchings and spanning trees under categorization. Discrete Applied Mathematics 39(2), 147–153 (1992)
24. Ulungu, E.L., Teghem, J.: Multi-objective combinatorial optimization problems: A survey. Journal of Multi-Criteria Decision Analysis 3, 83–104 (1994)
25. White, D.J.: A bibliography on the applications of mathematical programming multiple-objective methods. Journal of the Operational Research Society 41(8), 669–691 (1990)
26. Yi, T., Murty, K.G., Spera, C.: Matchings in colored bipartite networks. Discrete Applied Mathematics 121(1-3), 261–277 (2002)

Author Index

Lecture Notes in Computer Science

Sublibrary 1: Theoretical Computer Science and General Issues

For information about Vols. 1– 4527
please contact your bookseller or Springer

Vol. 4688: K. Li, M. Fei, G.W. Irwin, S. Ma (Eds.), Bio-Inspired Computational Intelligence and Applications. XIX, 805 pages. 2007.

Vol. 4684: L. Kang, Y. Liu, S. Zeng (Eds.), Evolvable Systems: From Biology to Hardware. XIV, 446 pages. 2007.

Vol. 4683: L. Kang, Y. Liu, S. Zeng (Eds.), Advances in Computation and Intelligence. XVII, 663 pages. 2007.

Vol. 4681: D.-S. Huang, L. Heutte, M. Loog (Eds.), Advanced Intelligent Computing Theories and Applications. XXVI, 1379 pages. 2007.

Vol. 4672: K. Li, C. Jesshope, H. Jin, J.-L. Gaudiot (Eds.), Network and Parallel Computing. XVIII, 558 pages. 2007.

Vol. 4671: V.E. Malyshkin (Ed.), Parallel Computing Technologies. XIV, 635 pages. 2007.

Vol. 4669: J.M. de Sá, L.A. Alexandre, W. Duch, D. Mandic (Eds.), Artificial Neural Networks – ICANN 2007, Part II. XXXI, 990 pages. 2007.

Vol. 4668: J.M. de Sá, L.A. Alexandre, W. Duch, D. Mandic (Eds.), Artificial Neural Networks – ICANN 2007, Part I. XXXI, 978 pages. 2007.

Vol. 4666: M.E. Davies, C.J. James, S.A. Abdallah, M.D. Plumbley (Eds.), Independent Component Analysis and Blind Signal Separation. XIX, 847 pages. 2007.

Vol. 4665: J. Hromkovič, R. Královič, M. Nunkesser, P. Widmayer (Eds.), Stochastic Algorithms: Foundations and Applications. X, 167 pages. 2007.

Vol. 4664: J. Durand-Lose, M. Margenstern (Eds.), Machines, Computations, and Universality. X, 325 pages. 2007.

Vol. 4661: U. Montanari, D. Sannella, R. Bruni (Eds.), Trustworthy Global Computing. X, 339 pages. 2007.

Vol. 4649: V. Diekert, M.V. Volkov, A. Voronkov (Eds.), Computer Science – Theory and Applications. XIII, 420 pages. 2007.

Vol. 4647: R. Martin, M.A. Sabin, J.R. Winkler (Eds.), Mathematics of Surfaces XII. IX, 509 pages. 2007.

Vol. 4646: J. Duparc, T.A. Henzinger (Eds.), Computer Science Logic. XIV, 600 pages. 2007.

Vol. 4644: N. Azémard, L. Svensson (Eds.), Integrated Circuit and System Design. XIV, 583 pages. 2007.

Vol. 4641: A.-M. Kermarrec, L. Bougé, T. Priol (Eds.), Euro-Par 2007 Parallel Processing. XXVII, 974 pages. 2007.

Vol. 4639: E. Csuhaj-Varjú, Z. Ésik (Eds.), Fundamentals of Computation Theory. XIV, 508 pages. 2007.

Vol. 4638: T. Stützle, M. Birattari, H. H. Hoos (Eds.), Engineering Stochastic Local Search Algorithms. X, 223 pages. 2007.

Vol. 4630: H.J. van den Herik, P. Ciancarini, H.H.L.M.(J.) Donkers (Eds.), Computers and Games. XII, 283 pages. 2007.

Vol. 4628: L.N. de Castro, F.J. Von Zuben, H. Knidel (Eds.), Artificial Immune Systems. XII, 438 pages. 2007.

Vol. 4627: M. Charikar, K. Jansen, O. Reingold, J.D.P. Rolim (Eds.), Approximation, Randomization, and Combinatorial Optimization. XII, 626 pages. 2007.

Vol. 4624: T. Mossakowski, U. Montanari, M. Haveraaen (Eds.), Algebra and Coalgebra in Computer Science. XI, 463 pages. 2007.

Vol. 4623: M. Collard (Ed.), Ontologies-Based Databases and Information Systems. X, 153 pages. 2007.

Vol. 4621: D. Wagner, R. Wattenhofer (Eds.), Algorithms for Sensor and Ad Hoc Networks. XIII, 415 pages. 2007.

Vol. 4619: F. Dehne, J.-R. Sack, N. Zeh (Eds.), Algorithms and Data Structures. XVI, 662 pages. 2007.

Vol. 4618: S.G. Akl, C.S. Calude, M.J. Dinneen, G. Rozenberg, H.T. Wareham (Eds.), Unconventional Computation. X, 243 pages. 2007.

Vol. 4616: A.W.M. Dress, Y. Xu, B. Zhu (Eds.), Combinatorial Optimization and Applications. XI, 390 pages. 2007.

Vol. 4614: B. Chen, M. Paterson, G. Zhang (Eds.), Combinatorics, Algorithms, Probabilistic and Experimental Methodologies. XII, 530 pages. 2007.

Vol. 4613: F.P. Preparata, Q. Fang (Eds.), Frontiers in Algorithmics. XI, 348 pages. 2007.

Vol. 4600: H. Comon-Lundh, C. Kirchner, H. Kirchner (Eds.), Rewriting, Computation and Proof. XVI, 273 pages. 2007.

Vol. 4599: S. Vassiliadis, M. Bereković, T.D. Hämäläinen (Eds.), Embedded Computer Systems: Architectures, Modeling, and Simulation. XVIII, 466 pages. 2007.

Vol. 4598: G. Lin (Ed.), Computing and Combinatorics. XII, 570 pages. 2007.

Vol. 4596: L. Arge, C. Cachin, T. Jurdziński, A. Tarlecki (Eds.), Automata, Languages and Programming. XVII, 953 pages. 2007.

Vol. 4595: D. Bošnački, S. Edelkamp (Eds.), Model Checking Software. X, 285 pages. 2007.

Vol. 4590: W. Damm, H. Hermanns (Eds.), Computer Aided Verification. XV, 562 pages. 2007.

Vol. 4588: T. Harju, J. Karhumäki, A. Lepistö (Eds.), Developments in Language Theory. XI, 423 pages. 2007.

Vol. 4583: S.R. Della Rocca (Ed.), Typed Lambda Calculi and Applications. X, 397 pages. 2007.

Vol. 4580: B. Ma, K. Zhang (Eds.), Combinatorial Pattern Matching. XII, 366 pages. 2007.

Vol. 4576: D. Leivant, R. de Queiroz (Eds.), Logic, Language, Information and Computation. X, 363 pages. 2007.

Vol. 4547: C. Carlet, B. Sunar (Eds.), Arithmetic of Finite Fields. XI, 355 pages. 2007.

Vol. 4546: J. Kleijn, A. Yakovlev (Eds.), Petri Nets and Other Models of Concurrency – ICATPN 2007. XI, 515 pages. 2007.

Vol. 4545: H. Anai, K. Horimoto, T. Kutsia (Eds.), Algebraic Biology. XIII, 379 pages. 2007.

Vol. 4533: F. Baader (Ed.), Term Rewriting and Applications. XII, 419 pages. 2007.

Vol. 4528: J. Mira, J.R. Álvarez (Eds.), Nature Inspired Problem-Solving Methods in Knowledge Engineering, Part II. XXII, 650 pages. 2007.